BIOTECHNOLOGY
NEW CONCEPTS AND INNOVATIONS

Kruger Brentt
Publishers

BIOTECHNOLOGY
NEW CONCEPTS AND INNOVATIONS

Luis A. Felix
Mark Moldwin

Kruger Brentt
Publishers
2024

Kruger Brentt Publishers UK. LTD.
Company Number 9728962

Regd. Office: 68 St Margarets Road, Edgware, Middlesex HA8 9UU

ISBN: 978-1-78715-258-8

Disclaimer:

Every possible effort has been made to ensure that the information contained in this book is accurate at the time of going to press, and the publisher and author cannot accept responsibility for any errors or omissions, however caused. No responsibility for loss or damage occasioned to any person acting, or refraining from action, as a result of the material in this publication can be accepted by the editor, the publisher or the author. The Publisher is not associated with any product or vendor mentioned in the book. The contents of this work are intended to further general scientific research, understanding and discussion only. Readers should consult with a specialist where appropriate.

Every effort has been made to trace the owners of copyright material used in this book, if any. The author and the publisher will be grateful for any omission brought to their notice for acknowledgement in the future editions of the book.

For information on all our publications visit our website at http://krugerbrentt.com/

PREFACE

Innovations in Biotechnology unfold as a compelling narrative that navigates the frontiers of scientific progress and transformative discoveries within the dynamic realm of biotechnology. This comprehensive anthology traverses the expansive landscape of plant, medical, microbial, animal, and general biotechnology, providing a panoramic view of the groundbreaking advancements that define the current era. As we embark on this intellectual journey, each chapter unveils a tapestry of innovation woven by the collaborative efforts of experts across diverse disciplines. From enhancing garlic propagation through biotechnological approaches to deciphering the intricacies of gene expression microarrays in microgravity research, the anthology reflects the kaleidoscopic diversity of biotechnological applications. It is a testament to the relentless pursuit of knowledge, where cutting-edge science converges with practical applications, offering a glimpse into the future of biotechnology and its profound impact on the way we understand and manipulate the living world. Biotechnology stands at the forefront of scientific and technological advancements, weaving a tapestry of innovation that spans across diverse realms of life sciences. This comprehensive compilation, "Innovations in Biotechnology," delves into the multifaceted landscape of biotechnological breakthroughs, offering a panoramic view of the transformative strides made in plant, medical, microbial, animal, and general biotechnology.

In the first section, "Plant Biotechnology: An Introduction," we embark on a journey through the world of plants, exploring biotechnological approaches that enhance garlic propagation and quality. The pivotal role of plant beneficial microbes in shaping plant biotechnology is also illuminated, underscoring the intricate relationships between plants and microorganisms. Moving into the realm of Medical Biotechnology, we delve into cutting-edge techniques such as the assessment of circular RNA expression using the permutated intron-exon technique and the innovative use of anti-restriction and pentapeptide repeat proteins to mimic DNA. The application of biotechnology in orthopedics and sports medicine through Platelet-Rich Plasma (PRP) takes center stage, showcasing its clinical uses and potential. Microbial Biotechnology unfolds in the third section, elucidating microbial expression platforms from both market and economic perspectives. The utilization of external catalase gene expression to amplify metabolic performance and biomaterial production in host microbes is explored, shedding light on the economic implications of such interventions. In the domain of Animal Biotechnology, we explore the intriguing role of acupuncture in managing common obesity and delve into the intricate world of spermatogonial stem cells and their pivotal role in animal transgenesis. The application of gene expression microarrays for microgravity research and the intersection of biotechnology patents in securing human health provide a holistic understanding of the field. The final section, "General Biotechnology," encompasses a range of topics, from the development of biotechnology virtual labs for on-demand laboratory access in classroom education to the exploration of attitudes towards biotechnology at the intersection of gender, knowledge, and scientific expertise. Structural bioinformatics in protein engineering and the use of high-performance ion exchange chromatography for monoclonal antibody development and analysis round out the comprehensive exploration of general biotechnological applications.

The book provides a thorough exploration of various branches of biotechnology, including Plant Biotechnology, Medical Biotechnology, Microbial Biotechnology, Animal Biotechnology, and General Biotechnology. The book addresses ethical considerations in biotechnology, especially in chapters related to human health, gender perspectives, and patenting. Emphasis is placed on translating research into practical applications, demonstrating the real-world impact of biotechnological innovations. Readers gain insights into the integration of cutting-edge technologies, such as circular RNA expression techniques, anti-restriction and pentapeptide repeat proteins, and high-performance ion exchange chromatography. The book serves as a valuable resource for researchers and professionals in the field of biotechnology, offering them in-depth knowledge and updates on the latest advancements. Graduate and postgraduate students studying biotechnology or related disciplines will find the book beneficial for gaining a deeper understanding of specialized areas within the field.

We are grateful to all those persons as well as various books, manuals, periodicals, newsletters, magazines, journals etc. that helped in the preparation of this book. In spite of the best efforts, it is possible that some errors may have occurred into the compilation and editing of the book. Further queries, constructive suggestions and criticisms for the further improvement of the book are always welcome and shall be thankfully acknowledged.

Luis A. Felix | Mark Moldwin

CONTENTS

PART-I
PLANT BIOTECHNOLOGY

CHAPTER-1

PLANT BIOTECHNOLOGY: AN INTRODUCTION

INTRODUCTION: AN OVERVIEW

Plant biotechnology, a groundbreaking field at the intersection of biology and technology, holds the promise of revolutionizing agriculture, environmental conservation, and even medicine. It leverages the power of genetic manipulation to enhance plant traits, boost crop yields, and develop novel plant-based products for diverse applications. In this introductory exploration of plant biotechnology, we will delve into the fundamental principles, methodologies, and potential applications that are reshaping the way we interact with the plant kingdom. From genetically modified crops that resist pests and adapt to changing climate conditions to the development of medicinal plants for the pharmaceutical industry, plant biotechnology offers a spectrum of opportunities for innovation and sustainable solutions in a rapidly changing world. Join us on a journey into this captivating realm where science, technology, and the natural world converge to address some of our most pressing challenges and improve the quality of life for both humanity and the planet.

Actinidia is a genus of 55 species and about 76 taxa native to central China and with a wide geographic distribution throughout China and South Eastern Asia (X. Li et al., 2009). Palaeobiological studies estimate *Actinidia* to be at least 20–26 million years old (Qian & Yu, 1991). *Actinidia* species are vigorous and long-lived perennial vines, producing oblong or spherical berries that vary considerably in shape and colour (Fig. 1). *Actinidia* are normally dioecious, but occasional plants have perfect flowers (A. R. Ferguson, 1984). The basic chromosome number in *Actinidia* is X=29, with a diploid number of 58. During evolution a chromosome may have duplicated (McNeilage & Considine, 1989), followed by an aneuploid event, such as breakage of a centromere, to give an additional chromosome (He et al., 2005). The genus has a reticulate polyploidy structure, with diploids, tetraploids, hexaploids and octaploids occurring in diminishing frequency (A. R. Ferguson et al., 1997). The genus has unusual inter- and intra-taxal variation in ploidy (A. R. Ferguson & Huang, 2007; A. R. Ferguson et al., 1997), with, for example, *A. chinensis* found as both diploid and tetraploid and *A. arguta* as usually tetraploid, but also found as diploid, hexaploid or octaploid. In this chapter, we will describe advances in *Actinidia* plant tissue culture and molecular biology and the present and future applications of these biotechnology techniques in kiwifruit breeding and germplasm improvement.

SIGNIFICANCE OF KIWIFRUIT ON A WORLDWIDE SCALE

Actinidia species were introduced to Europe, the U.S.A., and New Zealand in the late 19th and early 20th century (A.R. Ferguson & Bollard, 1990). New Zealand was largely responsible for the initial development and commercial growing of kiwifruit, with the first commercial orchards established in the 1930s. Domestication and breeding of firstly *Actinidia deliciosa,* and more recently, *A. chinensis,* from wild germplasm has resulted in varieties now cultivated commercially in a number of continents. The inherent qualities of novel appearance, attractive flesh colour, texture and flavour, high vitamin C content and favourable handling and storage characteristics make kiwifruit a widely acceptable and popular fruit crop for producers and consumers.

Commercial kiwifruit growing areas have expanded rapidly and consistently since the 1990s. By 2010, the global kiwifruit planting area had reached over 150,000 ha. China (70,000 ha), Italy (27,000 ha), New Zealand (14,000 ha) and Chile (14,000 ha) account for about 83%of world kiwifruit plantings, and global kiwifruit production represents about 0.22% of total production for major fruit crops, with the majority of kiwifruit consumed as fresh fruit. Science has made a significant contribution to the success of the New Zealand kiwifruit industry, particularly in developing excellent breeding programmes and technologies for optimal plant growth, orchard management, fruit handling and storage, and transport to the global market, to ensure high quality premium fruit reach the consumer.

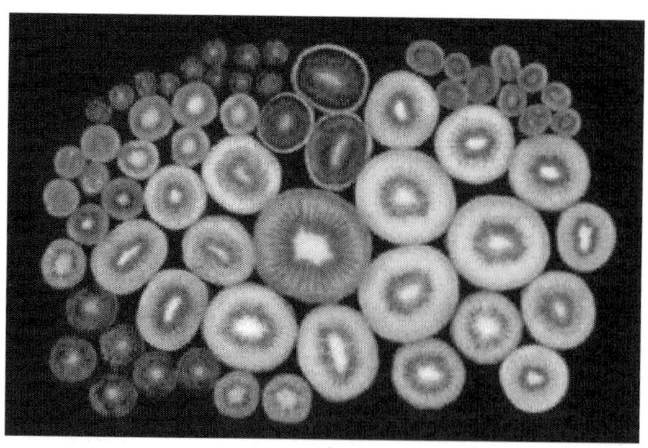

Fig. 1. Fruit of the *Actinidia* genus showing variation in flesh colour, size and shape

Kiwifruit have a reputation for being a highly nutritious food. A typical commercial *A. deliciosa* 'Hayward' kiwifruit contains about 85 mg/100 g fresh weight of vitamin C, which is 50% more than an orange, or 10 times that of an apple (A. R. Ferguson & Ferguson, 2003). The fruit of some *Actinidia* species, such as *A. latifolia, A. eriantha* and *A. kolomikta,* have in excess of 1000 mg of vitamin C per 100 g fresh weight (A. R. Ferguson, 1990; A. R. Ferguson& MacRae, 1992). Kiwifruit are also an excellent source of potassium, folate and vitamin E (Ferguson & Ferguson, 2003), and are high amongst fruit for their antioxidant capacity (H. Wang et al., 1996).

Breeding and Commercial Cultivars of Kiwifruit

The extensive *Actinidia* germplasm resources, with tremendous genetic and phenotypic diversity at both the inter- and intra-specific levels, offer kiwifruit breeders infinite opportunities for developing new products. Since its development in the 1920s, *A. deliciosa* 'Hayward' has continued to perform extraordinarily well on the global market in terms of production and sales; it remains the dominant commercial kiwifruit cultivar. Advances in *Actinidia* breeding have seen the appearance of a number of new commercial kiwifruit varieties. In 1999 an *A. chinensis* cultivar named 'Hort16A', developed in New Zealand by HortResearch (now Plant & Food Research), entered the international market, with fruit sold under the name of ZESPRI® GOLD Kiwifruit, reflecting the distinctive golden-yellow fruit flesh. 'Hort16A' fruit are sweet tasting and the vine is more subtropical than 'Hayward'. Subsequently, a range of new cultivars were commercialised in China and Japan, some of which have become significant internationally. Jintao®, a yellow-fleshed cultivar selected inWuhan, China (H.W. Huang et al., 2002b), is now widely planted in Italy (Ferguson & Huang, 2007) and more recently, the *A. chinensis* cultivar 'Hongyang' selected in China, and with a distinctive yellow-fleshed fruit with brilliant red around the central core, is widely cultivated for the export market, particularly Japan (M. Wang et al., 2003). Most cultivars to date have been selected from *A. chinensis* and *A. deliciosa;* however, *A. arguta* are now commercially cultivated in USA, Chile and New Zealand (Ferguson & Huang, 2007). The fruit of *A. arguta* are small, smooth-skinned, with a rich and sweet flavour, and can be eaten whole (Williams et al., 2003). Internationally, kiwifruit breeding programmes are directed primarily at producing varieties mainly from *A. deliciosa* and *A. chinensis,* with large fruit size, good flavour, novel flesh colour, variations in harvest period, improved yield and growth habit, hermaphroditism, tolerance to adverse conditions and resistance to disease (A. R. Ferguson et al., 1996). Although kiwifruit cultivars currently on the commercial market have been developed using traditional breeding techniques (MacRae, 2007), the expansion of genetic, physiological and biochemical knowledge and the application of biotechnology tools are being used increasingly to assist breeders in the development of novel cultivars.

TISSUE CULTURE AND CROP IMPROVEMENT

Although the genetic diversity of *Actinidia* provides tremendous potential for cultivar improvement, there are features (including the vigorous nature of climbing vines, the 3- to 5-year juvenile period, the dioecious nature and the reticulate polyploidy structure) that make *Actinidia* less amenable to achieving certain breeding goals, compared with many other agronomic crops. Plant tissue culture, the *in vitro* manipulation of plant cells, tissues and organs, is an important technique for plant biotechnology, and a number of plant tissue culture techniques have been employed to overcome some of the limitations that *Actinidia* presents to classical breeding.

Multiplications

Plant tissue culture for kiwifruit propagation was first reported by Harada (1975), followed by numerous reports using a range of explant types and genotypes (Gui, 1979; M. Kim et al., 2007; Kumar & Sharma, 2002; Q.L. Lin et al., 1994; Monette, 1986). Murashige & Skoog (MS) basal salts are the most widely used media for shoot regeneration and callus formation. However, other media have been used successfully, including Gamborg B5 medium (Barbieri & Morini, 1987) and N6 medium (Q.L. Lin et al., 1994).

Multiplication protocols essentially follow three steps: (1) surface sterilization of explants with 0.5–1.5% sodium hypochlorite; (2) shoot multiplication from explants (e.g. buds, nodal sections or young leaves) on MS medium, supplemented with 2–3% sucrose, 0.1–1.0 mg/l zeatin and 0.01–0.1 mg/l naphthalene acetic acid (NAA), solidified with 0.7% agar, at pH 5.8; and (3) rooting on half strength MS medium containing 0.5–1.0 mg/l indole-3-butyric acid (IBA). Generally, cultures are incubated at 24±2ºC under a 16 h photoperiod (20–30μmol/m^2/s of light intensity applied). Shoot proliferation rates vary depending upon species, cultivar, explant type, plant growth regulator combinations and culture conditions. Standardi & Catalano (1984) achieved a multiplication rate of 5.3 shoots per bud explant using a 30-day subculture period, and 90% of shoots rooted after three weeks, developinginto 150–200 mm high plantlets, with 6–10 leaves within 60 days. A multiplication rate of2.61 at seven weeks was achieved using 800 μm or 1200 μm transversal micro-cross section (MCS) of *A. deliciosa* 'Hayward' explants, cultured on ½ MS medium supplemented with 3% (w/v) sucrose, 4.5 x 10^{-3} μM 2,4-dichlorophenoxyacetic acid (2,4-D) and 4.6 x 10^{-1} μM zeatin in 0.8% agar (w/v), pH 5.8 (Kim et al., 2007).

Process of Protoplast culture and somatic hybridization

As dioecy and polyploidy of *Actinidia* can often restrict breeding possibilities, somatic hybridization provides an approach to combine different genetic backgrounds of the same gender or to overcome inter-specific incompatibility, to produce valuable material with desirable traits from two species. Somatic hybridization is generally achieved through protoplast fusion, and methods of protoplast isolation from callus, suspension cultures, leaf mesophyll and cotyledons of various *Actinidia* genotypes and species have been developed. Tsai (1988) isolated protoplasts from calli derived from *A. deliciosa* leaves and stems and used TCCM medium with 0.23 µM 2,4-D, 0.44 µM 6-benzylaminopurine (BAP), 2% coconut milk, 10 g/l sucrose, 1 g/l glucose, 0.3 M mannitol and 0.1 M sorbitol, for preconditioning. Enzymatic degradation of cell walls was achieved in 2% Cellulase Onozuka R-10, 0.5% Macerozyme R-10, 0.5 M mannitol and 3 mM MES. *A. eriantha* protoplasts were isolated from newly growing leaves of *in vitro* culture seedlings, by preconditioning in MS liquid (without NH4NO3), supplemented with 1.0 mg/l 2,4-D and 0.4 M glucose and isolated using 1% Cellulase R-10, 0.5% Macerozyme R-10, 0.05% Pectolyase Y-23 and 3 mM MES (Y.J. Zhang et al., 1998). Plating efficiency after 3 weeks of culture was 19.4%, and calli subsequently recovered and regenerated shoots when cultured on MS media containing 2.28µM zeatin and 0.57 µM indole-3-acetic acid (IAA).

Xiao & Han (1997) reported successful protoplast fusion of *A. chinensis* and *A. deliciosa*, demonstrating the potential of using this technique to aid breeding programmes. Isolated protoplasts from cotyledon-derived calli for *A. chinensis* ($2n = 2x = 58$) and *A. deliciosa* ($2n = 6x = 174$) were fused, using a PEG (polyethylene glycol) method and plantlets were regenerated from the fused calli. Xiao et al. (2004), in an attempt to introduce the chilling tolerance characteristics of *A. kolomikta* into *A. chinensis*, fused protoplasts isolated from cotyledon-derived calli of *A. chinensis* ($2n = 2x = 58$) and the mesophyll cells of *A. kolomikta* ($2n = 2x = 58$). A number of techniques were employed to confirm that the regenerated plantlets were an inter-specific somatic hybrid ($2n = 4x = 116$) and assessment of the chilling tolerance of *in vitro* leaves suggested that the somatic hybrid was more similar to *A. kolomikta*, with a higher capacity of cold resistance than *A. chinensis*.

Other Embryo culture techniques

Embryo culture techniques, for embryo rescue were developed to recover hybrids from inter-specific crosses in *Actinidia*. From an *A. chinensis* (2*x*) × *A. melanandra* (4*x*) cross, embryo rescue was used successfully to transfer hybrid embryos to *in vitro* culture at an early stage of their development (Mu et al., 1990). Nutrient and hormone requirements were dependent on the stage of embryo development and the endosperm, and nursing tissue was beneficial when globular embryos were cultured. Embryo size and their genetic background are major factors in determining the success of the procedure (Harvey et al., 1995; Kin et al.,1990). Hirsch et al. (2001) carried out inter-specific hybridizations of different *Actinidia* species and ploidy races, using embryo rescue to obtain hybrid plantlets of *A. kolomikta X A. chinensis, A. polygama X A. valvata, A. arguta X A. polygama* and *A. kolomikta X A. deliciosa*. When optimal media were used, the immature embryos that reached the torpedo stage could be rescued. A series of culture media were developed, which performed as the hybrid embryo's deficient endosperm to ensure embryo survival at the globular and heart stages. Ovule culture has been used also to obtain hybrid plantlets from the inter-specific cross of *A. chinensis X A. kolomikta* (X. Chen et al., 2006).

Endosperm culture is another approach to generating *Actinidia* inter-specific hybrids. Endosperms from F1 and F2 seeds from three inter-specific hybrids (*A. chinensis X A. melanandra; A. arguta X A. melanandra*; and an open pollinated *A. arguta X A. deliciosa*) were induced to form calli, from which plants were recovered by induction of organogenesis or embryogenesis. Media for callus induction and differentiation varied with genotype, and chromosome counts showed evidence of extensive mixoploidy in all hybrids (Mu et al., 1990).

Recently, *in vitro* chromosome doubling using colchicine treatment was reported (J. Wu et al., 2009; 2011). Petiole segments of five diploid *A. chinensis* genotypes, including 'Hort16A', were cultured on half-strength MT basal salt medium, supplemented with 3.0 mg/l BAP, 0.4 mg/l zeatin and 0.5 mg/l IBA for four weeks. Resulting microshoots were treated with 0.05– 0.1% colchicine, and over one-third of the regenerated shoots were confirmed as tetraploid by flow cytometry, with orchard-grown autotetraploid 'Hort16A' plants showing polyploid characteristics such as thicker leaves and flatter flowers, and some plants producing fruit almost double the weight of the original diploid 'Hort16A' fruit (J. Wu et al., 2009).

Cryopreservation is an excellent means of preserving germplasm for long-term storage, and various techniques and methods have been investigated for *Actinidia* germplasm (Bachiri et al., 2001; Hakozaki et al., 1996; Jian & Sun, 1989; Y. Wu et al., 2001; X. Xu et al., 2006; Zhai et al., 2003). Shoot tips from *in vitro* culture of a dwarf *A. chinensis* genotype were pre-cultured in MS medium containing 5% dimethyl sulfoxide (DMSO) and 5% sucrose for four days, followed by dehydration with PVP2 solution (30% glycerol, 15% DMSO, 15% PEG and 13.7% sucrose) for 40 min at 0ºC, and then transferred to liquid nitrogen for storage, with a survival rate of 56.7% upon defrosting shoots (X. Xu et al., 2006). Encapsulation-dehydration protocols used for the preservation of *in vitro* cultured hybrids of *A. arguta X A. deliciosa, A. chinensis* and *A. eriantha* gave even higher survival rates, of 85–95% (Bachiri et al., 2001; Y. Wu et al., 2001).

TRANSFORMATION SYSTEMS

Since the first report of a transgenic *Actinidia* plant two decades ago (Matsuta et al., 1990), six *Actinidia* species having been transformed, almost exclusively by *Agrobacterium*-mediated transformation. Initially, the development of *Actinidia* transformation focused on the integration into the plant genome of reporter and selectable marker genes (Fraser et al., 1995; Janssen & Gardner, 1993; Uematsu et al., 1991), but transformation of various heterologous genes has followed. These include: *A. rhizogenes rol* genes (Rugini et al., 1991); a soybean β-1,3 endoglucanase cDNA (Nakamura et al., 1999); a rice *OSH1* homeobox gene (Kusaba et al., 1999), and an *Arabidopsis* Na^+/H^+ antiporter gene (Tian et al., 2011), in attempts to improve kiwifruit disease resistance or drought tolerance; a synthetic gene encoding human epidermal growth factor (Kobayashi et al., 1996); and a grape stilbene synthase (Kobayashi et al., 2000), inattempts to accumulate bioactive compounds; citrus geranylgeranyl diphosphate synthase, phytoene desaturase, β-carotene desaturase, β-carotene hydroxylase and phytoene synthase, to modify the lutein or β-carotene content of kiwifruit (MiSun Kim et al., 2010) and the *A. tumefaciens* isopentyl transferase (*ipt*) gene, to alter vine architecture (Honda et al., 2011).

Agrobacterium-mediated transformation

Agrobacterium-mediated transformation of *Actinidia* is a component of the Plant & Food Research functional genomics platform and has been used to introduce over 100 *Actinidia* genes into various *Actinidia* species. In general, Plant & Food Research *Actinidia* transformation protocols are as follows: Orchard-grown winter mature and dormant canes are maintained at 4°C for 4–6 weeks. To initiate bud break, one-third of a 40 cm cane (with>3 nodes) is immersed in water, and maintained at room temperature under normal light conditions. After four weeks, newly initiated shoots are removed from the canes and shoot sections with a single node (1–2 cm) are soaked in 70% ethanol for 30 s, then surface sterilized with 25% (v/v) commercial bleach (5% active chlorine). After a sterile water rinse, the node sections are cultured on MS media, supplemented with 0.1 mg/l IBA at 24°C ± 2, 16 h photoperiod, with cool white fluorescent light (40 $\mu mol/m^2/s$). Young leaves harvested from *in vitro* grown shoots are cut into 2 x 5 mm leaf strips. *Agrobacterium tumefaciens* EHA105, harbouring a pART27-derived binary vector (Gleave, 1992), is cultured in 50 ml MGL medium (Tingay et al., 1997) containing 100 mg/l spectomycin dihydrochloride, for 16–20 h at 28°C, with shaking at 250 rpm. At an OD600 nm =1.0-1.5, the bacterial cells are pelleted by centrifugation (5000 g for 10 min) and re-suspended in 10 ml MS media, supplemented 100 μM acetosyringone. Leaf strips are immersed in the *A. tumefaciens* suspension culture for 10 min, blotted dry with sterile filter paper and transferred onto co- cultivation media (MS supplemented with 3.0 mg/l zeatin, 0.1 mg/l naphthaleneacetic acid (NAA) and 50 μM of AS). After two days of co-cultivation, the leaf strips are transferred to regeneration and selection medium (MS supplemented with 3.0 mg/l zeatin, 0.1 mg/l NAA, 150 mg/l kanamycin sulphate, 300 mg/l timentin, 30 g/l sucrose and 2.5 g/l Phytagel). The leaf strips produce calli along the cut edges at about four weeks and excised calli are transplanted onto fresh regeneration and selection media. Adventitious buds regenerated from the calli are excised individually and transferred to shoot elongation medium (MS supplemented with 0.1 mg/l IBA, 100 mg/l kanamycin sulphate and 300 mg/l timentin). When shoots reach 1–2 cm in height, they are transplanted onto rooting medium (½ MS basal salts and vitamins supplemented with 1.0 mg/l IBA, 150 mg/l timentin, 50 mg/l kanamycin sulphate, 20 g/l sucrose and 7 g/l agar). Rooted transgenic plants are potted in a½-litre pot and placed in a containment glasshouse facility. The utility of a plant transformation system is very much dependent upon its efficiency, and several factors that affect *Actinidia* transformation efficiency are discussed below.

Agrobacterium tumefaciens strains

- *Tumefaciens* strains are defined by their chromosomal background and resident Ti plasmid, and exhibit differences in their capacity to transfer T-DNA to various plant species (Godwin et al., 1991). *A. tumefaciens* LBA4404, A281, C58, EHA101 and EHA105 are the strains commonly used in *Actinidia* transformation. Fraser et al. (1995) reported no marked difference in efficiency of *A. chinensis* transformation between strains A281 (a virulent L,L-succinamopine strain) and C58 (a virulent strain carrying the nopaline Ti plasmid pTiC58), which both harbour the binary vector pKIWI105. However, Janssen and Gardner (1993) showed A281 produced slightly higher rates of gene transfer than C58 and EHA101 in *A. deliciosa* transformation, and noted that because of source material variability, strain comparisons need to be repeated several times. Strain A281 harbours a tumour-inducing plasmid pTiBo542 (Hood et al., 1986) and an extra copy of the transcription activator of virulence (*vir*) genes, which may account for the higher transformation efficiency.
- Comparison of *A. chinensis* callus formation using *A. tumefaciens* A281, GV3101, EHA105 and LBA4404, all harbouring the pART27-10 binary vector, revealed that 27% of leaf strips produced calli using A281, compared with 22.2%, 18.1% or 13.9% when using EHA105, LBA4404 and GV3101, respectively (T. Wang et al., 2007). Both A281 and its non- oncogenic derivative, EHA105, have the Ti-plasmid pTiBo542 in a C58 chromosomal background (Hood et al., 1993; 1986), and have been shown to be superior in gene transfer in other plant species, e.g. apple (Bondt et al., 1994). However, high rates

of callus formation do not necessarily mean high efficiency of transgenic plant production, and Wang et al. (2007) also found differences in shoot regeneration related to whether co- cultivation had been with strains harbouring an oncogenic Ti plasmid (A281) or a non- oncogenic Ti plasmid (EHA105). Transformants derived from the use of the disarmed strains EHA105, LBA4404 and GV3101 had callus and regeneration patterns similar to those of control explants, not co-cultivated with *A. tumefaciens*, whereas the use of A281 tended to result in large calli and take about two weeks longer to initiate adventitious buds. Less than 20% of the calli derived from A281 co-cultivation had subsequent shoot and root development, whereas over 70% of calli derived from EHA105, GV3101 and LBA4404 co-cultivation regenerated shoots and roots. Over-proliferation of calli derived from A281 co-cultivation was even more severe in *A. eriantha* and no regenerated shoots were obtained (T. Wang et al., 2006). It is likely that high callus formation and poor adventitious bud and root initiation from the A281 co-cultivated tissue is related to the co- integration into plant genome of the oncogenes.

Species

Most *Actinidia* transformation systems have been developed for *A. chinensis* and *A. deliciosa*, thoughtransformationof *A.arguta*, *A.eriantha*, *A.kolomikta* and *A.latifolia* has been reported. All *Actinidia* genotypes tested have been found to be responsive to a range of tissue culture conditions, and relatively amenable to regeneration protocols (Fraser et al., 1995). Compared with other woody species, e.g. apple (James et al., 1989), relatively high *A. deliciosa* transformation and regeneration rates have been achieved (Uematsu et al., 1991), and *A. chinensis* transformation efficiencies of up to 27.8% have been reported (T. Wang et al., 2007). However, *A. arguta* transformation was less successful when applying the transformation protocols developed for *A. chinensis* or *A. eriantha*, with co-cultivated explants suffering considerable browning and necrosis during callus induction and shoot regeneration stages. Minimizing the extent of explant browning and necrosis was achieved through reducing the basal salt concentration to ½ MS medium, combined with lower light intensity (3.4μmol/m^2/s) during the callus induction and regeneration stages. This resulted in adventitious shoot development and an efficient and reproducible *Agrobacterium*-mediated transformation system for *A. arguta* (Han et al., 2010).

From the production of over 1000 transgenic *Actinidia* plants at Plant & Food Research, the salient features in comparing the transformation of four species are three-fold. *A. arguta* displays a relatively, low transformation efficiency of 1–10% compared with the 5–20% for*A. deliciosa* and *A. eriantha* and 5–30% for *A. chinensis*;

the induction of *A. eriantha* callus is relatively high compared with other species; but the regeneration of *A. eriantha* kanamycin- resistant shoots takes much longer than with the other three species.

Co-cultivation conditions

Agrobacterium-mediated DNA delivery to plant cells is initiated through a series of chemical signals exchanged between the host and pathogen, which may activate *vir* genes to signal the bacterium to enter virulence mode. Phenolics, sugars, temperature and pH can affect *Agrobacterium* virulence and presumably its capacity to transform plant cells (Alt-Moerbe et al., 1988). However, the degree to which these factors influence transformation efficiency varies with species and reports. Acetosyringone (AS), one of the phenolic compounds released by wounded plant tissue, and a signal molecule to ensure effective *vir*-induction and T-DNA transfer (Stachel et al., 1985; 1986), has been widely used to increase transformation efficiency in various crops (James et al., 1993; H. Wu et al., 2003). Janssen and Gardner (1993) found the addition of 20 μM AS to the *A. tumefaciens* growth and co-cultivation medium increased DNA transfer approximately 2-fold in *A. deliciosa* leaf pieces, whereas highest levels of *A. latifolia* transformation were achieved using 200 μM in the co- cultivation medium (Gao et al., 2007). Wang et al. (2006; 2007) used 100 μM AS in bacterial cultures for co-cultivation to improve the efficiency of *A. chinensis* and *A. eriantha* transformation. The inclusion of a suspension cell feeder layer during co-cultivation, separated from the explants by a layer of filter paper, has been used to improve *Actinidia* transformation frequency (Janssen & Gardner, 1993). In addition, as mentioned earlier, light intensity plays a role in the efficiency of *A. arguta* transformation (Han et al., 2010).

Plant regeneration

Selecting plant cell types or explants that have the ability to differentiate into whole plants is an essential step for the successful production of transgenic plants. Fortunately, *A. deliciosa* and *A. chinensis* callus induction and adventitious bud initiation are relatively straightforward after establishment in tissue culture if appropriate explant material is used. Young leaves, petioles and stem segments have been used successfully for *Actinidia* transformation, and, as with most other crops, the younger the explants, the easier regeneration will be. However, *A. arguta* transformation is one exception to this, as necrosis or browning occurs after *A. tumefaciens* co-cultivation if the explants used are too young (Han et al., 2010).

To maintain *Actinidia* explants in active and amenable condition for co-cultivation with *A. tumefaciens*, it is essential to subculture *in vitro* shoots at 3- to 4-week intervals (Fraser et al., 1995; Wang et al., 2006). MS basal medium has been used successfully for callus induction as well as regeneration in *Actinidia* (Kumar and Sharma 2002). However, optimum application of auxins and cytokinins, and combinations thereof, vary depending on the condition of the explant material used. Fraser et al. (1995) found that for *A. chinensis* regeneration, thidiazuron (TDZ) and kinetin, (0.1 and 10 mg/l) were clearly inferior to other cytokinins.

Differences between NAA and IAA (indole-3-acetic acid) were insignificant. The most satisfactory combination of growth regulator additives was found to be 5 mg/l zeatin combined with 0.1 mg/l of NAA, or 1 mg/l zeatin and 0.5 mg/l BAP combined with 0.1 mg/l of NAA. Zeatin was clearly superior to BAP, when either was used as the sole cytokinin, but a combination of the two cytokinins gave the best overall result, in terms of the numbers of normal-looking shoots produced. Wang et al. (2006) made similar observations with *A. eriantha* where the highest shoot regeneration rates were obtained using medium containing a combination of 2 mg/l zeatin and 3 mg/l BAP. Uematsu et al. (1991) reported that the regeneration frequency varied with the basal medium used, and B5 basal medium containing zeatin was most suitable for obtaining transformed *A. deliciosa* shoots. Using *A. deliciosa* MCS explants for transformation, Kim et al. (2010) used half-strength MS medium containing 0.001 mg/l 2,4-D and 0.1 gm/l zeatin, for callus induction and shoot regeneration. Calli formed on the surface of MCS segments after two weeks of culturing on selection medium and shoots were regenerated after four weeks. The transformation efficiencies ranged from 2.9 to 22.1% depending on the gene being transformed into the cells. The high degree of callus formation and shoot regeneration of *Actinidia* material from tissue culture makes it possible to obtain transformed shoots at a reasonably high frequency, although it is desirable to minimize callus development and maximize shoot development, to minimize the occurrence of somaclonal variation during these processes.

Particle bombardment

As opposed to the biological *Agrobacterium*-mediated transformation process, particle bombardment is a purely physical method for DNA delivery, using DNA-coated microscopic metal particles accelerated towards a target tissue. Qiu et al. (2002) used particle bombardment of *A. deliciosa* suspension cells, with a CaMV 35S transcribed maize *DHN1* gene (induced in response to abiotic stress) fused to the green fluorescent protein (GFP) reporter gene. GFP expression was localized within the cell nucleus after 10 h and was visualized in the cytoplasm (mainly around the plasma membranes) in response to increased osmotic stress (Qiu et al., 2002).

Various Types of DNA transfer methods

Although *Agrobacterium*- and particle bombardment-mediated DNA transfer are the most commonly used systems of gene transfer to plants, a polyethylene glycol (PEG)-mediated approach was frequently used in the early 1980s to deliver DNA into protoplasts. Oliveira et al. (1991) used the chloramphenicol acetyl transferase (CAT) gene as a reporter to optimize the conditions for PEG-mediated transfection of *Actinidia* protoplasts, finding that the greatest CAT activity was obtained using 30% PEG 4000 and submitting protoplasts to a 5- min 45°C heat shock, prior to transfection. Using *in vitro* cultured *A. deliciosa* leaves, Raquel & Oliveira (1996) found protoplasts originating from the epidermis and leaf veins had cell division and regeneration ability, and displayed transient expression of a GUS gene introduced by PEG-mediated DNA transfer. Zhu et al. (2003) successfully transferred a GFP gene into *A. arguta* protoplasts by PEG-mediated transfer, with transient GFP expression detected in calli generated from the protoplasts. The physiological conditions of the protoplasts, the PEG concentration, and the time of heat stimulus are factors affecting the efficiency of DNA transfer using this approach. Because of the low yields of transformants and the inability of many species to be regenerated from protoplasts into viable plants, direct DNA uptake methods of transformation are much less frequently adopted than *Agrobacterium*-mediated transformation. However, the successful regeneration of whole plants from *A. chinensis, A. deliciosa* and *A. eriantha* protoplasts has been published (see earlier). Future development of new commercial cultivars produced directly or indirectly via genetic manipulation may see a resurgence in direct DNA uptake methods and protoplast regeneration, as these approaches may be more amenable to some genetic manipulation technologies, such as Zinc finger nuclease targeted site-directed mutagenesis.

ACTINIDIA MOLECULAR BIOLOGY

Initial molecular studies of *Actinidia* concentrated on fruit tissue, with an emphasis on genes involved in ethylene biosynthesis, cell wall modification, and carbohydrate metabolism (Atkinson & MacRae, 2007 and references therein). The cloning and/or expression of 1- aminocyclopropane-1 carboxylic acid (ACC) oxidase, *S*-adenosyl-*L*-methionine (SAM) synthase and ACC synthase identified some of the key genes involved in ethylene biosynthesis, a control point of fruit ripening. Molecular studies on genes encoding key enzymes in carbohydrate metabolism have included: polygalacturanase; xyloglucan endotransglycosylase/hydrolase; polygalacturonase inhibitor protein; sucrose phosphate synthase; and sucrose synthase. The most widely studied genes in these early forays into *Actinidia* molecular biology were those encoding the cysteine protease, actinidin, which can account for up to 50% of fruit soluble protein. Actinidin genes have been cloned, expressed in transgenic tobacco, the promoter sequenced, and studied in transgenic petunia.

Expressed sequence tag (EST) databases

A significant watershed in advancing *Actinidia* molecular biology was the generation of a database of 132,577 expressed sequence tags (EST), from a variety of *Actinidia* species, (Crowhurst et al., 2008). This provided a significant increase in the availability of *Actinidia* transcriptomic data, which prior to this publication were represented by 511 sequences in GenBank (dbEST Jan. 2008). This genetic resource, derived primarily from four species (*A. chinensis*, *A. deliciosa*, *A. arguta* and *A. eriantha*), included a range of tissues and developmental time points (Table 1). The average sequence length of these EST sequences was 503 bases. As expected, a high frequency of redundancy was observed within the *Actinidia* EST dataset and clustering at a 95% threshold, resulting in 23,788 sequences remaining as singletons and 18,070 tentative consensus (TC) sequences, a combined total of 41,858 non-redundant clusters (NRs). Analysis revealed that 28,345 NRs had sufficient homology to *Arabidopsis* sequences ($E>1.0e^{-10}$) to be assigned a functional classification. Many of the NRs with no *Arabidopsis* homolog did however, have homologs in other crops. Crowhurst et al. (2008) also reported more specific analysis of ESTs of key genes related to distinctive features of *Actinidia* including flavour and aroma, colour, health-beneficial compounds, allergens, and cell wall structure.

Codon usage analysis revealed that *Actinidia* shared many similarities with other dicotyledonous plants, and although codon usage was similar among three *Actinidia* species, it was not identical. A higher GC ratio was seen in coding than in non-coding regions, and this was more marked in *A. deliciosa* and *A. eriantha* than in *A. chinensis*. A modest degree ofCpG suppression was also evident in the three *Actinidia* species, with an XCG/XGG ratio of 0.68–0.71. Analysis of overlapping regions of 3,901 TCs identified 32,764 bi-allelic single nucleotide polymorphisms (SNPs), with one SNP every 417 bp, although some of the SNPs were probably the result of homeologous or paralogous sequences, rather than allelic variation. The allelic SNPs have potential for the development of molecular markers for use in genetic mapping, population genetics and linkage disequilibrium studies or for marker- assisted selection. The inter-specific SNPs, identified in orthologous loci from different *Actinidia* species represent species–species variation and have utility in kiwifruit breeding using crosses between different species. Further analysis revealed that over 30% of the *Actinidia* EST NRs had at least one SSR, with dinucleotide repeats, predominantly in the 5' untranslated region, being twice as frequent as trinucleotide repeats, which were more evenly distributed across the gene.

Table 1. Numbers of ESTs derived from various *Actinidia* species and tissues

Actinidia sp.	Tissue type							
	Bud	Fruit	Leaf	Petal	Root	Cell	Stem	Total
A. deliciosa	34,519	13,282		9,950				57,751
A. chinensis	15,689	8,453	17,325	1,061		4,851		47,379
A. eriantha		11,259		1,388				12,647
A. arguta		5,421		1,836				7,257
A. hemsleyana					5,101			5,101
A. polygama				1,348				1,348
A. setosa							1,020	1,020
A. indochinensis				74				74
Total	50,208	38,415	17,325	15,657	5,101	4,851	1,020	132,577

An *Actinidia* microarray platform

Characterizing a gene's temporal and spatial expression is critical to understanding its function. Early *Actinidia* molecular studies characterized the expression of a limited number of genes, identified as being differentially expressed during a particular developmental phase (Ledger & Gardner, 1994) or members of a particular gene family (Langenkamper et al., 1998). The *Actinidia* EST database provided a resource for more global gene expression analysis, through the development of a 17,472-feature oligonucleotide microarray of *Actinidia* genes. This microarray represented genes from a variety of species: *A. chinensis* (51%); *A. deliciosa* (38%); *A. eriantha* (6%); *A. arguta* (3%); and other *Actinidia* species (2%).

Walton et al. (2009) used the *Actinidia* microarray to examine gene expression in *A. deliciosa* meristems and buds in response to the dormancy-breaking hydrogen cyanamide (HC) chemical treatment, over two growing seasons. Although most of the genes that responded early (1–3 days) to HC treatment differed between seasons, there was a high degree of commonality between seasons of genes that showed the greatest change in expression six days post treatment, with 123 genes up-regulated and 35 genes down-regulated at day 6 in both seasons. Quantitative PCR (qPCR) of 35 selected genes validated the microarray data for 97% of up-regulated and 60% of down-regulated genes. Genes that changed in expression upon HC-treatment were classified into distinct profiles, including: i) genes that reached a peak in expression at 3 or 6 days post treatment, then returned to baseline levelsby day 15; ii) genes that reached a peak in expression at 3 or 6 days post treatment, followed by a second burst of transcription at 25–40 day post treatment, iii) genes that decreased in expression prior to meristematic activity or external bud growth. Putative function of these HC-responsive *Actinidia* genes, based on homology to other plant genes, indicated that many had been identified in other plant stress-related studies, including a number of genes that had shown similar responses in HC-treated grape, suggesting similar mechanisms in response to HC-treatment in these two crops.

Actinidia species are a climacteric fruit, showing a dramatic increase in ethylene production and a high respiration rate during fruit ripening. Generally, kiwifruit are harvested firm, and then enter a period of softening, which is followed by the onset of autocatalytic ethylene production, when fruit soften to "eating ripe" firmness and develop their characteristic flavours and aromas. The final step of the ethylene biosynthetic pathway is the conversion of 1-aminocyclopropane-1 carboxylic acid (ACC) to ethylene by ACC oxidase. Atkinson et al. (2011) examined gene expression changes during the ripening process, using an ACC oxidase-silenced transgenic *Actinidia* line, the fruit of which produce no detectable climacteric ethylene, but could be induced to undergo softening, aroma and flavour development through the application of exogenous ethylene. Using the *Actinidia* microarray, expression of 401 genes changed significantly within 168 h of ethylene treatment, with 25 genes showing a response at 4 h, 81 genes at 12 h, and 183 genes 24 h after application. These ethylene-responsive genes could be grouped into functional categories, including: metabolism; oxidative stress; photosynthesis; regulation; cell wall; hormone; starch; other; and unknown functions. The expression patterns indicated that the majority of photosynthesis- and starch-related genes were down-regulated by ethylene, whereas up- and down-regulation of genes in other functional groups were observed in response to ethylene. Validation by qPCR confirmed significant changes in gene expression of a number of genes involved in cell wall modification in response to ethylene, including a polygalacturonase, a pectin lyase, a pectin methylesterase and a xylan-degrading enzyme, as well as genes involved in fruit flavour, ethylene production and perception.

The microarray platform has provided a useful tool for genome-wide gene expression, as is evident from the studies above. However, microarrays have a limited dynamic range, lack the sensitivity required to detect subtle changes in expression, and are essentially a 'closed' platform, limited to examining the expression of only those genes represented on the array. Second-generation sequencing (2^{nd}GS) is becoming the methodology of choice for many genome-wide expression studies (L. Wang et al., 2010), as this is an 'open' platform, capable of detecting any of the genes that are expressed within a particular tissue, organ or cell type at the time of RNA sampling. Analysis of *Actinidia* transcription has been initiated using Illumina 2^{nd}GS, with mRNA-sequence data generated from a range of *A. chinensis* tissues and stages of fruit development (A.P. Gleave & Z. Luo, unpublished).

Functional genomics in *Actinidia* and heterologous hosts

Prior to the initiation of generating the *Actinidia* EST resource in 2000, reports of functional genomics through expression of *Actinidia* genes in either a heterologous or an *Actinidia* host were somewhat limited (Guo et al., 1999; Lay et al., 1996; Paul et al., 1995; Schroder et al., 1998; Z.C. Xu et al., 1998). The EST resource has facilitated a significant increase in *Actinidia* functional genomics, through expression of genes in various microbial and plant hosts.

Actinidia genes encoding: a pectin methylesterase inhibitor, with applications in fruit juice production (Hao et al., 2008); Bet v 1 and profilin-homologous allergens (Bublin et al., 2010; Oberhuber et al., 2008); an *L*-galactose-1-phosphate phosphatase and *l*-galactose guanyltransferase, (Laing et al., 2004; 2007) and *L*-galactose dehyrogenase (Shang et al., 2009), involved in vitamin C production; a lycopene beta-cyclase, involved in carotenoid production (Ampomah-Dwamena et al., 2009); three xyloglucan endotransglucosylase/ hydrolases involved in cell wall structure (Atkinson et al., 2009); two terpene synthases, involved in the production of floral sesquiterpenes (Nieuwenhuizen et al., 2009); and three glycosyltransferases of the anthocyanin pathway (Montefiori et al., 2011), have all been successfully expressed in *Escherichia coli,* with the recombinant proteins being used to study protein/ enzyme function. The yeast species, *Pichia pastoris* or *Saccharomyces cerevisiae*, have also been used to express recombinant *Actinidia* proteins, a pectin methylesterase inhibitor (Mei et al., 2007), and three alcohol acyltransferases, involved in the production of volatile esters (Gunther et al., 2011) and actinidin, which was found to have a negative effect on *S. cervesiae* growth (Yuwono, 2004).

In planta functional genomics of *Actinidia* genes has been used to study genes involved in a variety of processes. Paul et al. (1995) expressed *A. deliciosa* preproactinidin in transgenic *Nicotiana tabacum*, showing that the protein was correctly processed and detrimental to plant growth when it accumulated to high levels. Yin et al. (2010) showed that expression of the *A. deliciosa ETHYLENE INSENSITIVE3*-like *EIL2* and *EIL3* transcription factor cDNAs in *Arabidopsis thaliana* stimulated ethylene production, and up-regulation of host ACC synthase and ACC oxidase gene family members, as well as a number of xyloglucan endotransglycoylase (*XET*) genes. Yin et al. (2010) also used the *N. benthamiana* transient expression system, described by Hellens et al. (2005), to demonstrate transactivation of *A. deliciosa* ripening-related *ACO1* and *XET5* promoters by *EIL2* and *EIL3*, confirming their role in the signal transduction pathway connecting ethylene signalling and ripening processes.

To understand the role of *Actinidia* lipoxygenase (*LOX*) genes, which in other plants are involved in a range of processes, including senescence and fruit ripening, B. Zhang et al. (2006) used transient expression of *A. deliciosa LOX1 and LOX2* genes in *N. benthamiana*. qPCR had shown that *LOX1* increased in expression in ethylene-treated fruit, in contrast to *LOX2 expression*, which was repressed by ethylene. The transient expression studies revealed that *LOX1* significantly accelerated chlorophyll degradation and chlorophyll fluorescence, whereas *LOX2* had no apparent effect on senescence.

Varkonyi-Gasic et al. (2011) expressed cDNAs of nine *Actinidia* MADS-box genes in *A. thaliana*, to determine their role in floral meristem and floral organ fate. Resulting transgenic plants showed a variety of phenotypes. *FUL-like* expression promoted floral transition in both long day (LD) and short day (SD) conditions, with a terminal flower phenotype evident in plants showing high levels of transgene expression. Expression of *FUL* promoted flowering, but less efficiently than *FUL-like*, and the floral phenotype was as wild-type. *SEP4* expression also promoted floral transition, with many plants showing small and curled leaves, and a reversion to vegetative growth and aerial rosettes during SD conditions. *SEP3* expression had a mild effect on floral transition under LD conditions and *PI* and *AP3-1* expression showed no effect. *A. thaliana* expressing the kiwifruit *AG* flowered earlier than the wild-type under SD conditions, and showed reduced height, curled leaves and a loss of inflorescence indeterminancy. Coupled with information on the patterns of expression of these genes in *Actinidia* vegetative tissue and both normal and aberrant floral organs, these studies gave considerable insights into the role of these MADS-box transcription factors in the specification of *Actinidia* floral organs, phase change and flowering time.

Vitamin C is an essential metabolite for plants and animals, and the inability of some animals, including humans, to synthesize vitamin C means that they are dependent upon a dietary source. The *L*-galactose pathway is a significant route for vitamin C production in plants, although the enzyme responsible for the conversion of GDP-*L*-galactose to *L*- galactose-1-phosphate remained elusive until Laing et al. (2007) identified homologous genes from *Arabidopsis* and *A. chinensis* encoding a GDP-*L*-galactose guanyltransferase (GGT) capable of carrying out this function. Transient expression of the *A. chinensis GGT* gene in *N. benthamiana* leaves showed a 3-fold increase in vitamin C levels, and coupled with the biochemical studies, confirmed GGT's role in the *L*-galactose pathway. Further studies of *Actinidia* vitamin C production via the *L*-galactose pathway also made use of *in planta* functional genomics. As qPCR results had indicated GGT and GDP-mannose-3',5'- epimerase (GME) were key enzymes involved in the high vitamin C content of *A. eriantha*, Bulley et al. (2009) expressed the *A. eriantha GGT* in *Arabidopsis*, identifying plants with over four times the amount of vitamin C in leaves. In *N. benthamiana* leaves, transient *GGT* expression increased vitamin C levels 4.2 fold, a 20% increase resulted from transient *GME* expression, and simultaneous expression of *GGT* and *GME* gave an average increase of 8.6-fold in vitamin C.

Biochemical and gene expression studies on the production β-linalool, an acyclic monoterpene alcohol, were supplemented by transient expression of putative (*S*)-linalool synthase cDNAs from *A. eriantha* and *A. polygama*, to further understand their role in floral aroma (X.Y. Chen et al., 2010). The production of large amounts of linalool in *N. benthamiana* leaves transiently expressing these cDNAs confirmed their function as linalool synthases. A biochemical study of *A. deliciosa* 'Hayward' and its male pollinator 'Chieftain' identified the sesquiterpene (*E,E*)-α-farnesene as the major terpene floral volatile, with germacrene D, (*E*)-β-ocimene, (*Z,E*)-α-farnesene, also present (Nieuwenhuizen et al., 2009). Transient expression in *N. benthamiana* leaves of two *A. deliciosa* cDNAs encoding putative terpene synthases (*AFS1* and *GDS1*), followed by dynamic headspace sampling and GC-MS analyses, showed that expression of *AFS1* resulted in the production of large quantities of (*E,E*)-α-farnesene and smaller quantities of (*Z,E*)-α-farnesene and (E)-β-ocimene. *GDS1* expression resulted in production of germacrene D.

Glycosyltransferases are responsible for much of the diversity of anthocyanins, a subgroup of the flavonoids that give much of the red, purple and blue pigmentation to plants. Montefiori et al. (2011) characterized two glycosyltransferases (*F3GT1* and *F3GGT1*) from a red-fleshed *A. chinensis*. Recombinant *F3GT1* produced in *E. coli* catalyzed the addition of galactose to the 3-*OH* position in cyanidin, whilst recombinant *F3GGT1* catalyzed the addition of UDP-xylose to cyanadin-3-galactosidase. Confirmation of the roles of these genes in the red pigmentation of fruit flesh was demonstrated firstly through establishing that transient expression of *Arabidopsis PAP1* and *TT8* transcription factors in *A. eriantha* fruit resulted in red pigmentation, localized mainly around the fruit core, and with the major accumulated anthocyanin being cyanidin 3-*O*-xylo-galactoside. Concomitant transient expression of the two *Arabidopsis* genes with an *F3GT1* RNAi construct resulted in little orno visible red colour in *A. eriantha* fruit, indicating the *F3GT1* gene's critical role in anthocyanin biosynthesis. Concomitant transient expression of *PAP1* and *TT8* with an *F3GGT1* RNAi construct greatly reduced the amount of the major anthocyanin, cyanidin 3- *O*-xylo-galactoside.

Despite the availability of *Actinidia* transformation systems, to date there has been little published research on functional genomics of *Actinidia* genes through over-expression or silencing of genes in *Actinidia.* Such research is ongoing at Plant & Food Research and elsewhere, and the lack of published information is probably because many of these studies are related to fruit characteristics, and the time from initiating transformation to fruiting is at least three years. Of the few reports that have been published, one using gene silencing of ACC oxidase (Atkinson et al., 2011) has been discussed earlier in this chapter. Other studies have involved the over-expression of an *Actinidia Lfy* transcription factor cDNA, in an attempt to enhance early fruit set (Guo et al., 1999) and the silencing of ACC synthetase or ACC oxidase genes in *A. deliciosa* and *A. chinensis* (Li et al., 2003). In both these studies, only the production of transgenic plants was reported, with no analysis of their phenotype. Ledger et al. (2010), however, describe the use of transgenic*A. chinensis* plants to examine the role of *carotenoid cleavage dioxgenase* (*CCD*) genes in branching and vine architecture. The involvement of *CCD* genes, or their orthologs, in branching has been shown through the characterization of branching mutants in a range of annual plant species. Ledger et al. (2010) showed that expression of *A. chinensis CCD7* and *CCD8* cDNAs was able to complement their corresponding *Arabidopsis* branching mutants *max3* and *max4*. In *A. chinensis* plants transformed with a *CCD8* gene silencing construct, a number of plants showed greatly reduced *CCD8* expression levels at eight and 13 months of growth in the glasshouse. The *CCD8*-silenced plants showed significantly more primary and higher order branches, and a higher incidence of short branches, compared with control plants, but no difference in internode length on the main stem. Another finding was that leaves on some *CCD8*-silenced plants were slower to senesce and had a greater chlorophyll content than leaves of control plants. This study confirmed the role of *CCD8* in branching, and identified that *CCD8* plays a role in senescence in a deciduous woody perennial plant.

The studies described above give some valuable insights into the enzymatic or structural function of proteins encoded by these *Actinidia* genes and the roles they may play in the plant's phenotypic characteristics. However, in assigning a definitive function to a gene, it is essential also to understand the temporal and spatial regulation of its expression. In many of the studies described above, microarray and/or qPCR analysis were used to determine the transcriptional level of these genes in various tissues and in some cases at different developmental phases of the plant. Another approach to gain insights into the regulation of a gene's expression has been the analysis of promoter-reporter gene fusions in transgenic systems. Lin et al. (1993) fused an upstream region of an actinidin coding region to the β-glucuronidase (GUS) coding region and observed GUS expression during the later stages of transgenic petunia seed pod development, resembling the induction of actinidin in *Actinidia* fruit tissues. Similar promoter- GUS fusions were used to analyse an *A. chinensis* polygalacturonase promoter, and at the breaker stage of transgenic tomato fruit development, GUS expression was observed throughout the inner and outer pericarp, the columella and seeds, and became restricted to the inner pericarp and seeds at the later stages of ripening (Z.Y. Wang et al., 2000).

Table 2. Plant & Food Research's *Actinidia in planta* cDNA over-expression and gene silencing construct collection. Over-expression constructs of full length cDNAs cloned into pART27-derived vectors (Gleave, 1992) and gene silencing constructs of hairpin cDNA structures cloned into pTKO2 (Snowden et al., 2005). cDNAs are under the control of the CaMV35S promoter.

General Process	Metabolic pathway, process or gene classification	No. of over-expression	No. of silencing constructs
Flavour & Aroma	Terpenoid Biosynthesis	9	3
	Ester Biosynthesis	12	1
	Cytochrome P450	22	3
Sugars & Acids	Sugar metabolism	2	
	Aromatic amino acid	4	
Fruit Ripening	Cell wall structure	5	1
	Ethylene biosynthesis and	2	1
Colour	Chlorophyll degradation	3	1
	Carotenoid biosynthesis	2	2
	Anthocyanin biosynthesis	1	
	Phenylpropenoid biosynthesis	1	3
Vitamin C	Ascorbate biosynthesis	24	6
Protein Degradation	Ubiquitination	2	1
	Actinidin	3	2
Allergenicity	Allergens	2	1
Plant Development	Branching	1	1
	Phase change	2	1
	Hormone response	5	4
	Cell signalling	1	
Defence	Antimicrobial peptides	2	
DNA and Replication	Cell cycle		1
	Nucleotide synthesis & DNA		2
	Chromatin remodelling	1	5
Transport	Transporters	3	1
Unknown	Sex-linked		3

General Process	Metabolic pathway, process or gene classification		No. of over-expression	No. of silencing constructs
Transcription	Transcriptional machinery		2	
	miRNA		4	
	Transcription factors	*Myb* & *Myb-*	35	3
		bZIP	19	2
		MADS-box	15	7
		bHLH	13	1
		C2-C2 *Dof*	9	
		C2-C2 *CO-*	8	
		NAC Domain	5	
		AP2-EREB	2	1
		WRKY	1	1
		BZR	1	1
		WD40	1	
Total			**224**	**59**

Transcription factors play a central role in regulating gene expression, through activation or repression of target promoters, and are able to regulate complex developmental processes or entire metabolic pathways co-ordinately. The dual-luciferase reporter vector system, developed to identify transcription factor activation of promoters (Hellens et al., 2005), has been applied to studying transactivation of promoters of *Actinidia* ripening-related *ACO1* and *XET5* genes, by *EIL2* and *EIL3* (Yin et al., 2010). Exploiting this promoter activation tool is reliant on having a cloned *Actinidia* transcription factors; Plant & Food Research has developed such a resource, which includes *Actinidia* cDNA over-expression and RNAi gene silencing constructs, including 109 *Actinidia* transcription factors for use in many of the *in planta* functional genomics approaches described above (Table 2).

Molecular markers and mapping

Molecular markers have been used to carry out genetic characterization of the *Actinidia* genus. They allow germplasm enhancement through systematic crossing of plants, selected on the basis of their intra- and inter-specific phylogenic relationships and patterns of allelic diversity, and the selection of parent plants with desirable alleles for use in breeding programmes. Markers are also used to determine hybridity, pedigree, and for quality control during crossing. The development and application of molecular markers closely linked to desirable traits has the potential to assist kiwifruit breeding greatly through the early selection of those progeny, with a high probability of carrying the genetic information for the desired trait. Various genetic markers have been developed in *Actinidia* using restriction fragment length polymorphisms (RFLPs) (Crowhurst et al., 1990), amplified fragment length polymorphisms (AFLPs) (Novo et al., 2010; Testolin et al., 2001; X.G. Xiao et al., 1999), random amplified polymorphic DNAs (RAPDs) (Gill et al., 1998; H.W. Huang et al., 2002a; Shirkot et al., 2002), SSRs (Fraser et al., 2004; W.G. Huang et al., 1998; Korkovelos et al., 2008) or SNPs (Zhou et al., 2011). Much of the early molecular marker development was primarily to investigate the molecular phylogeny of *Actinidia* species, which in general was consistent with the traditional morphology-based classification. Studies were aimed also at sex determination, and molecular markers confirmed that the dioecious nature in *Actinidia* was a consequence of sex-determining genes localized on a pair of chromosomes that function like an XX/XY system (Gill et al., 1998; Harvey et al., 1997; Testolin et al., 1995). Although the genetic basis for sex determination in kiwifruit remains unknown, RAPD markers linked to this trait led to the development of sequence-characterized amplified regions (SCARs) (Gill et al., 1998). These are now deployed routinely in marker- assisted kiwifruit breeding, to eliminate male plants from crosses at the seedling stage, to select males when breeding for pollinizers, or to ensure a desirable male-to-female ratio of progeny are planted when characterizing families.

A framework *Actinidia* linkage map was first constructed using SSRs and the pseudo-test cross mapping strategy, often used for mapping out-crossing species, followed by the integration of AFLP markers (Testolin et al., 2001). Markers were screened over 94 individuals from a population generated from an inter-specific cross of a diploid *A. chinensis* female and a diploid *A. callosa* male. Linkage maps were produced for each parent, with the female framework map having 160 loci, 38 linkage groups and covering 46% of the estimated genome length, and the male framework map having 116 loci, 30 linkage groups, and covering 34% of the estimated genome length. The maps were produced with LODscores ≥2 (as an indication of coinheritance of loci). Continued progress in *Actinidia* mapping led to a significant advancement in *Actinidia* genetics, with the generation of a gene-rich linkage map of *A. chinensis,* constructed using 644 SSRs, and defining the 29 chromosomes of the haploid genome (Fraser et al., 2009). Again, SSRs were the marker of choice, owing to their abundance, distribution in coding and non-coding regions, reproducibility, Mendelian mode of inheritance and co-dominant nature. The inherent variability of SSRs, because of the high mutation rate, makes SSRs highly informative genetic markers. The linkage maps were produced using a mapping population of 272 individuals, created through an intra- specific cross of diploid *A. chinensis* parents, selected from two very distinct geographical locations in China, and exhibiting a diversity of fruiting characteristics. Resulting comprehensive genetic linkage maps of the male and female parents were produced and an integrated map of the cross was generated, using co-dominant SSR markers. The female and male linkage maps were composed of 464 and 365 markers, respectively, with markers estimated to be within 10 cM of each other in over 96% of the female genome and 94% of the male genome. The robustness of the maps was reflected by the LOD scores of 4–10. Using sex-linked SCAR markers, linkage group 17 was identified as the putative X and Y chromosomes. The sex-determining locus appeared to be sub-telomeric, occupying only a small portion of the chromosome, with little evidence of recombination in this region. These genetic linkage maps provide a valuable resource for the supply of markers for the breeding of novel cultivars, as tools for comparative and quantitative trait mapping. They will contribute to further investigations on the evolution and function of genetic control mechanisms in kiwifruit. They are an essential part of assembling the genome sequence of *Actinidia*.

Genome sequencing

Although the *Actinidia* EST database is a useful resource, it at best represents only 50–60% of the genes within the *Actinidia* genome, and contains no information on elements such as promoters, terminators and introns that play important roles in controlling gene expression. In addition, EST libraries, by the nature of their construction, under-represent genes that are expressed at relatively low levels and yet could play a critical role in a particular trait. Understanding key traits requires detailed information of not only the transcribed regions of a genome but also the intergenic and intron sequences, information that can be gained from the whole genome sequence (WGS) and its subsequent annotation. The advent of second- generation sequencing (2^{nd}GS) and advances in data handling and assembly software have now made it feasible to determine the WGS of a plant species at a fraction of the cost of the Sanger technology used to generate the WGS of *Arabidopsis,* for example (The Arabidopsis Genome Initiative, 2000). Plant & Food Research has recently initiated a research effort to determine the WGS of a diploid *A. chinensis* genotype, the haploid genome of which is predicted to be 650 Mbp. With no di-haploid or homozygous *Actinidia* genotypes available, the heterozygosity of the diploid *A. chinensis* may create some problems in WGS assembly. To minimize this, a genotype that has undergone two generations of sib-crossing and has an inbreeding coefficient of 0.375 has been selected. Genome sequencing is being carried out using an Illumina sequencing platform, using a variety of libraries and resulting sequencing data assembled into scaffolds. WGS assembly is being complemented by BAC-end sequencing, using an *A. chinensis* BAC library (Hilario et al., 2007), and use of the genetic- linkage map discussed earlier.

SOME FUTURE PERSPECTIVES AND CHALLENGES

Completion of the *A. chinensis* WGS will be the first within the Ericales, a large and diverse order that includes persimmon, blueberry, cranberry and tea, and the benefits to be gained in having the *Actinidia* WGS are enormous. Genome annotation is a key to the utility of any WGS, and the advances in transcriptome sequencing will greatly aid the defining and delineating of genes. Building on the availability of the genome sequence, characterization of the interrelationships between the *Actinidia* genome, transcriptome, proteome and metabolome, and functional genomics of alleles, will greatly aid in the understanding of biological processes, phenotypes and traits of kiwifruit. The annotated *A. chinensis* WGS will also provide a reference genome for the sequencing of genomes of other *Actinidia* species, to examine inter-species variability, and to identify SNPs. The knowledge gained from these efforts will open up greater opportunities for molecular breeding in kiwifruit, allowing the use of molecular markers for selective and accelerated introgression of desirable traits from the diverse *Actinidia* germplasm, to create new and novel cultivars.

As detailed in this chapter, much of the molecular research in *Actinidia* has targeted traits such as fruit flavour, aroma, ripening and colour, which could be exploited in the development of new cultivars with novel fruit characteristics. There has been very little molecular research in *Actinidia* targeted towards pathogens and disease. However, the recent devastating effect on commercial kiwifruit orchards in parts of Italy, due to kiwifruit canker, caused by the bacterium *Pseudomonas syringae* pv. *actinidiae* (Psa), may well change the emphasis of the immediate future of kiwifruit research. Although Psa was identified in Italy in 1992 (Scortichini, 1994), the bacterium caused little problem, until severe disease outbreaks in both *A. deliciosa* and *A. chinensis* cultivars in 2009. The presence of Psa has now been reported in most of the major kiwifruit growing regions of the world, although there appear to be a number of haplotypes, differing in their virulence. Minimizing the impact of Psa on the global kiwifruit industry will require a coordinated effort by pathologists, physiologists, breeders and growers. Many of the molecular tools, the knowledge and the *Actinidia* resources described in this chapter will aid in the understanding of the plant– pathogen interactions, the plant's response to infection, the identification and mapping of *Actinidia* genes offering Psa resistance, and ultimately the development of kiwifruit cultivars resistant to Psa, through breeding or genetic manipulation.

REFERENCES

1. Alt-Moerbe, J.; Neddermann, P.; Lintig, J.v.; Weiler, E.W. & Schroder, J. (1988). Temperature-sensitive step in Ti plasmid vir-region induction and correlation with cytokinin secretion by Agrobacteria. *Molecular and General Genetics*, Vol. 213, Issue 1, pp. 1-8, ISSN 0026-8925.

2. Ampomah-Dwamena, C.; McGhie, T.; Wibisono, R.; Montefiori, M.; Hellens, R.P. & Allan,
3. A.C. (2009). The kiwifruit lycopene beta-cyclase plays a significant role in carotenoid accumulation in fruit. *Journal of Experimental Botany*, Vol. 60, Issue 13, pp. 3765-3779, ISSN 0022-0957.
4. Atkinson, R.G.; Johnston, S.L.; Yauk, Y.K.; Sharma, N.N. & Schroder, R. (2009). Analysis of xyloglucan endotransglucosylase/hydrolase (XTH) gene families in kiwifruit and apple. *Postharvest Biology and Technology*, Vol. 51, Issue 2, pp. 149-157, ISSN 0925- 5214.
5. Atkinson, R.G. & MacRae, E. (2007). Kiwifruit. In Pua, E.C. & Davey, M.R., eds, *Biotechnology in Agriculture and Forestry: Transgenic Crops V*, Vol 60, pp. 329-346, ISBN 978-3-540- 49160-6, Springer-Verlag, Berlin.
6. Bachiri, Y.; Song, G.Q.; Plessis, P.; Shoar-Ghaffari, A.; Rekab, T. & Morisset, C. (2001). Routine cryopreservation of kiwifruit (Actinidia spp) germplasm by encapsulation-dehydration:importanceofplantgrowthregulators. *CryoLetters*, Vol. 22, Issue 1, pp. 61-74, ISSN 0143-2044.
7. Barbieri, C. & Morini, S. (1987). Plant regeneration from Actinidia callus cultures. *Journal of Horticultural Science*, Vol. 62, Issue 1, pp. 107-109, ISSN 0022-1589.
8. Bondt, A.d.; Eggermont, K.; Druart, P.; Vil, M.d.; Goderis, I.; Vanderleyden, J. & Broekaert, W.F. (1994). Agrobacterium-mediated transformation of apple (Malus x domestica Borkh.): an assessment of factors affecting gene transfer efficiency during early transformation steps. *Plant Cell Reports*, Vol. 13, Issue 10, pp. 587-593, ISSN 0721- 7714.
9. Bublin, M.; Pfister, M.; Radauer, C.; Oberhuber, C.; Bulley, S.; DeWitt, A.M.; Lidholm, J.; Reese, G.; Vieths, S.; Breiteneder, H.; Hoffmann-Sommergruber, K. & Ballmer- Weber, B.K. (2010). Component-resolved diagnosis of kiwifruit allergy with purified natural and recombinant kiwifruit allergens. *Journal of Allergy and Clinical Immunology*, Vol. 125, Issue 3, pp. 687-694, ISSN 0091-6749.
10. Bulley, S.M.; Rassam, M.; Hoser, D.; Otto, W.; Schunemann, N.; Wright, M.; MacRae, E.; Gleave, A. & Laing, W. (2009). Gene expression studies in kiwifruit and gene over- expression in Arabidopsis indicates that GDP-L-galactose guanyltransferase is a major control point of vitamin C biosynthesis. *Journal of Experimental Botany*, Vol. 60, Issue 3, pp. 765-778, ISSN 0022-0957.
11. Chen, X.; Li, L.; Zhang, Z.; Cheng, Z.; Huang, H.; Jiang, Z.; Wang, Y. & Wang, S. (2006). Study on ovule culture from interspecific hybrids of Actinidia. *Journal of Fruit Science*, Vol. 23, Issue 4, pp. 620-622, ISSN 1009-9980.

12. Chen, X.Y.; Yauk, Y.K.; Nieuwenhuizen, N.J.; Matich, A.J.; Wang, M.Y.; Perez, R.L.; Atkinson, R.G. & Beuning, L.L. (2010). Characterisation of an (S)-linalool synthase from kiwifruit (Actinidia arguta) that catalyses the first committed step in the production of floral lilac compounds. *Functional Plant Biology*, Vol. 37, Issue 3, pp. 232-243, ISSN 1445-4408.

13. Crowhurst, R.N.; Gleave, A.P.; MacRae, E.A.; Ampomah-Dwamena, C.; Atkinson, R.G.; Beuning, L.L.; Bulley, S.M.; Chagne, D.; Marsh, K.B.; Matich, A.J.; Montefiori, M.; Newcomb, R.D.; Schaffer, R.J.; Usadel, B.; Allan, A.C.; Boldingh, H.L.; Bowen, J.H.; Davy, M.W.; Eckloff, R.; Ferguson, A.R.; Fraser, L.G.; Gera, E.; Hellens, R.P.; Janssen, B.J.; Klages, K.; Lo, K.R.; MacDiarmid, R.M.; Nain, B.; McNeilage, M.A.; Rassam, M.; Richardson, A.C.; Rikkerink, E.H.; Ross, G.S.; Schroder, R.; Snowden, K.C.; Souleyre, E.J.; Templeton, M.D.; Walton, E.F.; Wang, D.; Wang, M.Y.; Wang, Y.Y.; Wood, M.; Wu, R.; Yauk, Y.K. & Laing, W.A. (2008). Analysis of expressed sequence tags from Actinidia: applications of a cross species EST database for gene discovery in the areas of flavor, health, color and ripening. *BMC Genomics*, Vol. 9, pp. 351, ISSN 1471-2164.

14. Crowhurst, R.N.; Lints, R.; Atkinson, R.G. & Gardner, R.C. (1990). Restriction-Fragment-Length-Polymorphisms in the Genus Actinidia (Actinidiaceae). *Plant Systematics and Evolution*, Vol. 172, Issue 1-4, pp. 193-203, ISSN 0378-2697.

15. Ferguson, A.R. (1984). Kiwifruit: A botanical review. *Horticultural Reviews*, Vol. 6, pp. 1-64, ISSN 0163-7851.

16. Ferguson, A.R. (1990). Kiwifruit (Actinidia). *Acta Horticulturae*, Issue 290, pp. 601-653, ISSN 0567-7572.

17. Ferguson, A.R. & Bollard, E.G. (1990). Domestication of the Kiwifruit. In Warrington, I.J. & Weston, G.C., eds, *Kiwifruit Science and Management*, pp. 165-246, ISBN 0-908596-28- 6, Ray Richards, Auckland, New Zealand.

18. Ferguson, A.R. & Ferguson, L.R. (2003). Are kiwifruit really good for you?, *Proceedings of the Fifth International Symposium on Kiwifruit*, Wuhan, China, 15-20 September, 2002., pp. 131-138, ISBN 0567-7572

19. Ferguson, A.R. & Huang, H. (2007). Genetic resources of kiwifruit: domestication and breeding. *Horticultural Reviews*, Vol. 33, pp. 1-122, ISSN 0163-7851.

20. Ferguson, A.R. & MacRae, E.A. (1992). Vitamin C in Actinidia. In Warrington, I.J.G.D.H.S.A.M.W.D.J., ed, *Acta Horticulturae*, pp. 481-487, ISBN 0567-7572,

21. Ferguson, A.R.; O'Brien, I.E.W. & Yan, G.J. (1997). Ploidy in Actinidia. In *Acta Horticulturae*, pp. 67-71, ISBN 0567-7572,

22. Ferguson, A.R.; Seal, A.G.; McNeilage, M.A.; Fraser, L.G.; Harvey, C.F. & Beatson, R.A. (1996). Kiwifruit. In Janick, J. & Moore, J.N., eds, *Fruit breeding: Vine and Small Fruit Crops*, Vol 2, pp. 371-417, ISBN John Wiley and Sons, Inc., New York.

23. Fraser, L.G.; Harvey, C.F.; Crowhurst, R.N. & Silva, H.N.d. (2004). EST-derived microsatellites from Actinidia species and their potential for mapping. *Theoretical and Applied Genetics*, Vol. 108, Issue 6, pp. 1010-1016, ISSN 0040-5752.

24. Fraser, L.G.; Kent, J. & Harvey, C.F. (1995). Transformation studies of Actinidia chinensis Planch. *New Zealand Journal of Crop and Horticultural Science*, Vol. 23, Issue 4, pp. 407-413, ISSN 0114-0671.

25. Fraser, L.G.; Tsang, G.K.; Datson, P.M.; Silva, H.N.d.; Harvey, C.F.; Gill, G.P.; Crowhurst, R.N. & McNeilage, M.A. (2009). A gene-rich linkage map in the dioecious species Actinidia chinensis (kiwifruit) reveals putative X/Y sex-determining chromosomes. *BMC Genomics*, Vol. 10, Issue 102, pp. (10 March 2009), ISSN 1471-2164.

26. Gao, Y.; Bi, J. & Liu, Y. (2007). Study on the optimum technological parameters for genetic transformation of Actinidia latifolia. *Journal of Fruit Science*, Vol. 24, Issue 4, pp. 553- 556, ISSN 1009-9980.

27. Gill, G.P.; Harvey, C.F.; Gardner, R.C. & Fraser, L.G. (1998). Development of sex-linked PCR markers for gender identification in Actinidia. *Theoretical and Applied Genetics*, Vol. 97, Issue 3, pp. 439-445, ISSN 0040-5752.

28. Gleave, A.P. (1992). A Versatile Binary Vector System with a T-DNA Organizational- Structure Conducive to Efficient Integration of Cloned DNA into the Plant Genome. *Plant Molecular Biology*, Vol. 20, Issue 6, pp. 1203-1207, ISSN 0167-4412.

29. Godwin, I.; Todd, G.; Ford-Lloyd, B. & Newbury, H.J. (1991). The effects of acetosyringone and pH on Agrobacterium-mediated transformation vary according to plant species. *Plant Cell Reports*, Vol. 9, Issue 12, pp. 671-675, ISSN 0721-7714.

30. Gui, Y.L. (1979). Callus induction from stem segments and plantlet regeneration of Chinese gooseberry. *Acta Botanica Sinica*, Vol. 21, Issue 4, pp. 339-344, ISSN 0577-7496.

31. Gunther, C.S.; Chervin, C.; Marsh, K.B.; Newcomb, R.D. & Souleyre, E.J.F. (2011). Characterisation of two alcohol acyltransferases from kiwifruit (Actinidia spp.) reveals distinct substrate preferences. *Phytochemistry*, Vol. 72, Issue 8, pp. 700-710, ISSN 0031-9422.

32. Guo, W.; Shen, X.; Li, J. & Zheng, X. (1999). Agrobacterium tumefaciens mediated transformation of kiwifruit with Lfy cDNA. *Acta Horticulturae Sinica*, Vol. 26, Issue 2, pp. 116-117, ISSN 0513-353X.

33. Hakozaki, M.; Yoshida, Y. & Suzuki, M. (1996). Viability of calli from hypocotyl of kiwifruit seedlings exposed to liquid nitrogen. *Environment Control in Biology*, Vol. 34, Issue 2, pp. 147-151, ISSN 1880-554X.

34. Han, M.; Gleave, A.P. & Wang, T. (2010). Efficient transformation of Actinidia arguta by reducing the strength of basal salts in the medium to alleviate callus browning. *Plant Biotechnology Reports*, Vol. 4, Issue 2, pp. 129-138, ISSN 1863-5466.

35. Hao, Y.L.; Huang, X.Y.; Mei, X.H.; Li, R.Y.; Zhai, Z.Y.; Yin, S.; Huang, Y. & Luo, Y.B. (2008). Expression, purification and characterization of pectin methylesterase inhibitor from kiwi fruit in *Escherichia coli. Protein Expression and Purification*, Vol. 60, Issue 2, pp. 221-224, ISSN 1046-5928.

36. Harada, H. (1975). *In vitro* organ culture of Actinidia chinensis Pl. as a technique for vegetative multiplication. *Journal of Horticultural Science*, Vol. 50, Issue 1, pp. 81-83, ISSN 0022-1589.

37. Harvey, C.F.; Fraser, L.G.; Kent, J.; Steinhagen, S.; McNeilage, M.A. & Yan, G.J. (1995). Analysis of Plants Obtained by Embryo Rescue from an Interspecific Actinidia Cross. *Scientia Horticulturae*, Vol. 60, Issue 3-4, pp. 199-212, ISSN 0304-4238.

38. Harvey, C.F.; Gill, G.P.; Fraser, L.G. & McNeilage, M.A. (1997). Sex determination in Actinidia. 1. Sex-linked markers and progeny sex ratio in diploid A. chinensis. *Sexual Plant Reproduction*, Vol. 10, Issue 3, pp. 149-154, ISSN 0934-0882.

39. He, Z.C.; Li, J.Q.; Cai, Q. & Wang, Q. (2005). The cytology of Actinidia, Saurauia and Clematoclethra (Actinidiaceae). *Botanical Journal of the Linnean Society*, Vol. 147, Issue 3, pp. 369-374, ISSN 0024-4074.

40. Hellens, R.P.; Allan, A.C.; Friel, E.N.; Bolitho, K.; Grafton, K.; Templeton, M.D.; Karunairetnam, S.; Gleave, A.P. & Laing, W.A. (2005). Transient expression vectors for functional genomics, quantification of promoter activity and RNA silencing in plants. *Plant Methods*, Vol. 1, pp. 1-14, ISSN 1746-4811.

41. Hilario, E.; Bennell, T.; Crowhurst, R.N.; Fraser, L.G.; McNeilage, M.A.; Rikkerink, E. & MacRae, E.A. (2007). Construction of kiwifruit BAC Contig maps by overgo hybridization and their use for mapping the sex locus. *Proceedings of the 6th International Symposium on Kiwifruit,* Vol. 753, Issue 1, pp. 185-189, ISSN 0567-7572.

42. Hirsch; Testolin; Brown; Chat; Fortune; Bureau & De, N. (2001). Embryo rescue from inter-specific crosses in the genus< SMALL> Actinidia</ SMALL> (kiwifruit). *Plant Cell Reports*, Vol. 20, Issue 6, pp. 508-516, ISSN 0721-7714.

43. Honda, C.; Kusaba, S.; Nishijima, T. & Moriguchi, T. (2011). Transformation of kiwifruit using the ipt gene alters tree architechture. *Plant Cell, Tissue and Organ Culture*, Vol. First published online, ISSN 1573-5044.

44. Hood, E.E.; Gelvin, S.B.; Melchers, L.S. & Hoekema, A. (1993). New Agrobacterium helper plasmids for gene transfer to plants. *Transgenic Research*, Vol. 2, Issue 4, pp. 208-218, ISSN 0962-8819.

45. Hood, E.E.; Helmer, G.L.; Fraley, R.T. & Chilton, M.D. (1986). The hypervirulence of Agrobacterium tumefaciens A281 is encoded in a region of pTiBo542 outside of T- DNA. *Journal of Bacteriology*, Vol. 168, Issue 3, pp. 1291-1301, ISSN 0021-9193.

46. Huang, H.W.; Li, Z.Z.; Li, J.Q.; Kubisiak, T.L. & Layne, D.R. (2002a). Phylogenetic relationships in Actinidia as revealed by RAPD analysis. *Journal of the American Society for Horticultural Science*, Vol. 127, Issue 5, pp. 759-766, ISSN 0003-1062.

47. Huang, H.W.; Wang, S.M.; Huang, R.H.; Jiang, Z.W. & Zhang, Z.H. (2002b). 'Jintao', a novel, hairless, yellow-fleshed kiwifruit. *HortScience*, Vol. 37, Issue 7, pp. 1135-1136, ISSN 0018-5345.

48. Huang, W.G.; Cipriani, G.; Morgante, M. & Testolin, R. (1998). Microsatellite DNA in Actinidia chinensis: isolation, characterisation, and homology in related species. *Theoretical and Applied Genetics*, Vol. 97, Issue 8, pp. 1269-1278, ISSN 0040-5752.

49. James, D.J.; Passey, A.J.; Barbara, D.J. & Bevan, M. (1989). Genetic transformation of apple (Malus pumila Mill.) using a disarmed Ti-binary vector. *Plant Cell Reports*, Vol. 7, Issue 8, pp. 658-661, ISSN 0721-7714.

50. James, D.J.; Uratsu, S.; Cheng, J.S.; Negri, P.; Viss, P. & Dandekar, A.M. (1993). Acetosyringone and osmoprotectants like betaine or proline synergistically enhance Agrobacterium-mediated transformation of apple. *Plant Cell Reports*, Vol. 12, Issue 10, pp. 559-563, ISSN 0721-7714.

51. Janssen, B.J. & Gardner, R.C. (1993). The use of transient GUS expression to develop an Agrobacterium-mediated gene transfer system for kiwifruit. *Plant Cell Reports*, Vol. 13, Issue 1, pp. 28-31, ISSN 0721-7714.

52. Jian, L.C. & Sun, L.H. (1989). Cryopreservation of the Stem Segments in Kiwi Fruit. *Acta Botanica Sinica*, Vol. 31, Issue 1, pp. 66-68, ISSN 0577-7496.

53. Kim, M.; Kim, S.; Song, K.; Kim, H.; Kim, I.; Song, E. & Chun, S. (2010). Transformation of carotenoid biosynthetic genes using a micro-cross section method in kiwifruit (Actinidia deliciosa cv. Hayward). *Plant Cell Reports*, Vol. 29, Issue 12, pp. 1339- 1349, ISSN 0721-7714.

54. Kim, M.; Kim, S.C.; Moon, D.Y. & Song, K.J. (2007). Rapid shoot propagation from micro- cross sections of kiwifruit (Actinidia deliciosa cv. 'Hayward'). *Journal of Plant Biology*, Vol. 50, Issue 6, pp. 681-686, ISSN 1226-9239.

55. Kin, M.S.; Fraser, L.G. & Harvey, C.F. (1990). Rescue of hybrid embryos of Actinidia species. *Scientia Horticulturae*, Vol. 44, Issue 1-2, pp. 97-106, ISSN 0304-4238.

56. Kobayashi, S.; Ding, C.K.; Nakamura, Y.; Nakajima, I. & Matsumoto, R. (2000). Kiwifruits (Actinidia deliciosa) transformed with a Vitis stilbene synthase gene produce piceid (resveratrol-glucoside). *Plant Cell Reports*, Vol. 19, Issue 9, pp. 904-910, ISSN 0721-7714.

57. Kobayashi, S.; Nakamura, Y.; Kaneyoshi, J.; Higo, H. & Higo, K.I. (1996). Transformation of kiwifruit (Actinidia chinensis) and trifoliate orange (Poncirus trifoliata) with a synthetic gene encoding the human epidermal growth factor (hEGF). *Journal of the Japanese Society for Horticultural Science*, Vol. 64, Issue 4, pp. 763-769, ISSN 0013-7626.

58. Korkovelos, A.E.; Mavromatis, A.G.; Huang, W.G.; Hagidimitriou, M.; Giakoundis, A. & Goulas, C.K. (2008). Effectiveness of SSR molecular markers in evaluating the phylogenetic relationships among eight Actinidia species. *Scientia Horticulturae*, Vol. 116, Issue 3, pp. 305-310, ISSN 0304-4238.

59. Kumar, S. & Sharma, D.R. (2002). In vitro propagation of kiwifruit. *Journal of Horticultural Science and Biotechnology*, Vol. 77, Issue 5, pp. 503-508, ISSN 1462-0316.

60. Kusaba, S.; Kano-Murakami, Y.; Matsuoka, M.; Matsuta, N.; Sakamoto, T. & Fukumoto, M. (1999). Expression of the rice homeobox gene, OSH1, causes morphological changes in transgenic kiwifruit. *Journal of the Japanese Society for Horticultural Science*, Vol. 68, Issue 3, pp. 482-486, ISSN 0013-7626.

61. Laing, W.A.; Frearson, N.; Bulley, S. & MacRae, E. (2004). Kiwifruit L-galactose dehydrogenase: molecular, biochemical and physiological aspects of the enzyme. *Functional Plant Biology*, Vol. 31, Issue 10, pp. 1015-1025, ISSN 1445-4408.

62. Laing, W.A.; Wright, M.A.; Cooney, J. & Bulley, S.M. (2007). The missing step of the L- galactose pathway of ascorbate biosynthesis in plants, an L-galactose guanyltransferase, increases leaf ascorbate content. *Proceedings of the National Academy of Sciences of the United States of America*, Vol. 104, Issue 22, pp. 9534-9539, ISSN 0027-8424.

63. Langenkamper, G.; McHale, R.; Gardner, R.C. & MacRae, E. (1998). Sucrose-phosphate synthase steady-state mRNA increases in ripening kiwifruit. *Plant Molecular Biology*, Vol. 36, Issue 6, pp. 857-869, ISSN 0167-4412.

64. Lay, V.J.; Prescott, A.G.; Thomas, P.G. & John, P. (1996). Heterologous expression and site- directed mutagenesis of the 1-aminocyclopropane-1-carboxylate oxidase from kiwi fruit. *European Journal of Biochemistry*, Vol. 242, Issue 2, pp. 228-234, ISSN 0014-2956.

65. Ledger, S.E. & Gardner, R.C. (1994). Cloning and characterization of five cDNAs for genes differentially expressed during fruit development of kiwifruit

(Actinidia deliciosa var. deliciosa). *Plant Molecular Biology*, Vol. 25, Issue 5, pp. 877-886, ISSN 0167-4412.

66. Ledger, S.E.; Janssen, B.J.; Sakuntala, K.; Wang, T. & Snowden, K.C. (2010). Modified CAROTENOID CLEAVAGE DIOXYGENASE8 expression correlates with altered branching in kiwifruit (Actinidia chinensis). *New Phytologist*, Vol. 188, Issue 3, pp. 803-813, ISSN 0028-646X.

67. Li, M.; Huang, Z.G.; Han, L.X.; Zhao, G.R.; Li, Y.H. & Yao, J.L. (2003). A high efficient Agrobacterium tumefaciens-mediated transformation system for kiwifruit. In Huang, H.W., ed, *Proceedings of the Fifth International Symposium on Kiwifruit*, pp. 501-507, ISBN 0567-7572.

68. Li, X., Li, J. & Doejarto, D.D. (2009). Advances in the study of the systematics of Actinidia Lindley. *Frontiers of Biology in China*, Vol. 4, Issue 1, pp.55-61, ISSN 1673-3509.

69. Lin, E.; Burns, D.J.W. & Gardner, R.C. (1993). Fruit developmental regulation of the kiwifruit actinidin promoter is conserved in transgenic petunia plants. *Plant Molecular Biology*, Vol. 23, Issue 3, pp. 489-499, ISSN 0167-4412.

70. Lin, Q.L.; Chen, Z.Q. & Wu, J.S. (1994). Propagation in vitro of some excellent clones of kiwi fruit. *Journal of Fujian Agricultural University*, Vol. 23, Issue 3, pp. 271-274, ISSN 1006-7817.

71. MacRae, E.A. (2007). Can biotechnology help kiwifruit breeders? *Acta Horticulturae* Vol. 753, pp. 129-138, ISSN 0567-7572.

72. Matsuta, N.; Iketani, H. & Hayashi, T. (1990). Effect of acetosyringone on kiwifruit transformation. *Japan J. Breed*, Vol. 40, Issue suppl 2, pp. 184-185, ISSN 0536-3683.

73. McNeilage, M.A. & Considine, J.A. (1989). Chromosome-Studies in Some Actinidia Taxa and Implications for Breeding. *New Zealand Journal of Botany*, Vol. 27, Issue 1, pp. 71-81, ISSN 0028-825X.

74. Mei, X.H.; Hao, Y.L.; Zhu, H.L.; Gao, H.Y. & Luo, Y.B. (2007). Cloning of pectin methylesterase inhibitor from kiwi fruit and its high expression in Pichia pastoris. *Enzyme and Microbial Technology*, Vol. 40, Issue 5, pp. 1001-1005, ISSN 0141-0229.

75. Monette, P.L. (1986). Micropropagation of kiwifruit using non-axenic shoot tips. *Plant Cell, Tissue and Organ Culture*, Vol. 6, Issue 1, pp. 73-82, ISSN 0167-6857.

76. Montefiori, M.; Espley, R.V.; Stevenson, D.; Cooney, J.; Datson, P.M.; Saiz, A.; Atkinson, R.G.; Hellens, R.P. & Allan, A.C. (2011). Identification and characterisation of F3GT1 and F3GGT1, two glycosyltransferases responsible for

anthocyanin biosynthesis in red-fleshed kiwifruit (Actinidia chinensis). *Plant Journal*, Vol. 65, Issue 1, pp. 106-118, ISSN 0960-7412.

77. Mu, S.K.; Fraser, L.G. & Harvey, C.F. (1990). Initiation of callus and regeneration of plantlets from endosperm of Actinidia interspecific hybrids. *Scientia Horticulturae*, Vol. 44, Issue 1-2, pp. 107-117, ISSN 0304-4238.

78. Nakamura, Y.; Sawada, H.; Kobayashi, S.; Nakajima, I. & Yoshikawa, M. (1999). Expression of soybean beta -1,3-endoglucanase cDNA and effect on disease tolerance in kiwifruit plants. *Plant Cell Reports*, Vol. 18, Issue 7/8, pp. 527-532, ISSN 0721-7714.

79. Nieuwenhuizen, N.J.; Wang, M.Y.; Matich, A.J.; Green, S.A.; Chen, X.; Yauk, Y.; Beuning, L.L.; Nagegowda, D.A.; Dudareva, N. & Atkinson, R.G. (2009). Two terpene synthases are responsible for the major sesquiterpenes emitted from the flowers of kiwifruit (Actinidia deliciosa). *Journal of Experimental Botany*, Vol. 60, Issue 11, pp. 3203-3219, ISSN 0022-0957.

80. Novo, M.; Romo, S.; Rey, M.; Prado, M.J. & Gonzalez, M.V. (2010). Identification and sequence characterisation of molecular markers polymorphic between male kiwifruit (Actinidia chinensis var. deliciosa (A. Chev.) A. Chev.) accessions exhibiting different flowering time. *Euphytica*, Vol. 175, Issue 1, pp. 109-121, ISSN 0014-2336.

81. Oberhuber, C.; Bulley, S.M.; Ballmer-Weber, B.K.; Bublin, M.; Gaier, S.; DeWitt, A.M.; Briza, P.; Hofstetter, G.; Lidholm, J.; Vieths, S. & Hoffmann-Sommergruber, K. (2008). Characterization of Bet v 1-related allergens from kiwifruit relevant for patients with combined kiwifruit and birch pollen allergy. *Molecular Nutrition & Food Research*, Vol. 52, pp. S230-S240, ISSN 1613-4125.

82. Oliveira, M.M.; Barroso, J. & Pais, M.S. (1991). Direct gene transfer into Actinidia deliciosa protoplasts: comparative analysis of the transient expression of the gus gene introduced by PEG treatment and electroporation, pp. A34, ISBN 0031-9317,

83. Paul, W.; Amiss, J.; Try, R.; Praekelt, U.; Scott, R. & Smith, H. (1995). Correct processing of the kiwifruit protease actinidin in transgenic tobacco requires the presence of the C- terminal propeptide. *Plant Physiology*, Vol. 108, Issue 1, pp. 261-268, ISSN 0032-0889.

84. Qian, Y.Q. & Yu, D.P. (1991). Advances in Actinidia research in China. *Acta Horticulturae*, Issue 297, I, pp. 51-55, ISSN 0567-7572.

85. Qiu, Q.; Wang, Z.; Cai, Q. & Jiang, R. (2002). Changes of DHN1 expression and subcellular distribution in A. delicisoa cells under osmotic stress. *Science in China Series C - Life Sciences*, Vol. 45, Issue 1, pp. 1-9, ISSN 1006-9305.

86. Raquel, M.H. & Oliveira, M.M. (1996). Kiwifruit leaf protoplasts competent for plant regeneration and direct DNA transfer. *Plant Science (Limerick)*, Vol. 121, Issue 1, pp. 107-114, ISSN 0168-9452.

87. Rugini, E.; Pellegrineschi, A.; Mencuccini, M. & Mariotti, D. (1991). Increase of rooting ability in the woody species kiwi (Actinidia deliciosa A. Chev.) by transformation with Agrobacterium rhizogenes rol genes. *Plant Cell Reports*, Vol. 10, Issue 6/7, pp. 291-295, ISSN 0721-7714.

88. Schroder, R.; Atkinson, R.G.; Langenkamper, G. & Redgwell, R.J. (1998). Biochemical and molecular characterisation of xyloglucan endotransglycosylase from ripe kiwifruit. *Planta*, Vol. 204, Issue 2, pp. 242-251, ISSN 0032-0935.

89. Scortichini, M. (1994). Occurrence of Pseudomonas-Syringae Pv Actinidiae on Kiwifruit in Italy. *Plant Pathology*, Vol. 43, Issue 6, pp. 1035-1038, ISSN 0032-0862.

90. Shang, Z.; Wang, X.; Ma, F. & Liang, D. (2009). Cloning of L-galactose dehyrogenase gene from Actinidia latifolia Merr. and its expression in E. coli. *Acta Horticulturae Sinica*, Vol. 36, Issue 12, pp. 1741-1748, ISSN 0513-353X.

91. Shirkot, P.; Sharma, D.R. & Mohapatra, T. (2002). Molecular identification of sex in Actinidia deliciosa var. deliciosa by RAPD markers. *Scientia Horticulturae*, Vol. 94, Issue 1-2, pp. 33-39, ISSN 0304-4238.

92. Snowden, K.C.; Simkin, A.J.; Janssen, B.J.; Templeton, K.R.; Loucas, H.M.; Simons, J.L.; Karunairetnam, S.; Gleave, A.P.; Clark, D.G. & Klee, H.J. (2005). The Decreased apical dominance 1/petunia hybrida carotenoid cleavage dioxygenase8 gene affects branch production and plays a role in leaf senescence, root growth, and flower development. *Plant Cell*, Vol. 17, Issue 3, pp. 746-759, ISSN 1040-4651.

93. Stachel, S.E.; Messens, E.; Montagu, M.v. & Zambryski, P. (1985). Identification of the signal molecules produced by wounded plant cells that activate T-DNA transfer in Agrobacterium tumefaciens. *Nature*, , Vol. 318, Issue 6047, pp. 624-629, ISSN 0028-0836.

94. Stachel, S.E.; Nester, E.W. & Zambryski, P. (1986). *virA* and *virC* control the plant induced activation of the T-DNA transfer process of *A. tumefaciens*. *Cell*, Vol. 46, pp. 325-333

95. Testolin, R.; Cipriani, G. & Costa, G. (1995). Sex segregation ratio and gender expression in the genus Actinidia. *Sexual Plant Reproduction*, Vol. 8, Issue 3, pp. 129-132, ISSN 0934-0882.

96. Testolin, R.; Huang, W.G.; Lain, O.; Messina, R.; Vecchione, A. & Cipriani, G. (2001). A kiwifruit (Actinidia spp.) linkage map based on microsatellites and

integrated with AFLP markers. *Theoretical and Applied Genetics*, Vol. 103, Issue 1, pp. 30-36, ISSN 0040-5752.

97. The Arabidopsis Genome Intiative (2000). Analysis of the genome sequence of the flowering plant Arabidopsis thaliana. *Nature Genetics*, Vol. 408, Issue 6814, pp. 796-815, ISSN 0028-0836.

98. Tian, N.; Wang, J. & Xu, Z.Q. (2011). Over expression of Na+/H+ antiporter gene AtNHX1 from Arabidopsis thaliana improves the salt tolerance of kiwifruit (Actinidia deliciosa). *South African Journal of Botany*, Vol. 77, Issue 1, pp. 160-169, ISSN 0254- 6299.

99. Tingay, S.; McElroy, D.; Kalla, R.; Fieg, S.; Wang, M.B.; Thornton, S. & Brettell, R. (1997). Agrobacterium tumefaciens-mediated barley transformation. *Plant Journal*, Vol. 11, Issue 6, pp. 1369-1376, ISSN 0960-7412.

100. Uematsu, C.; Murase, M.; Ichikawa, H. & Imamura, J. (1991). Agrobacterium-mediated transformation and regeneration of kiwi fruit. *Plant Cell Reports*, Vol. 10, Issue 6/7, pp. 286-290, ISSN 0721-7714.

101. Varkonyi-Gasic, E.; Moss, S.M.; Voogd, C.; Wu, R.; Lough, R.H.; Wang, Y. & Hellens, R.P. (2011). Identification and characterization of flowering genes in kiwifruit: sequence conservation and role in kiwifruit flower development. *BMC Plant Biology*, Vol. 11, Issue 72, pp. (27 April 2011), ISSN 1471-2229.

102. Walton, E.F.; Wu, R.; Richardson, A.C.; Davy, M.; Hellens, R.P.; Thodey, K.; Janssen, B.J.; Gleave, A.P.; Rae, G.M.; Wood, M. & Schaffer, R.J. (2009). A rapid transcriptional activation is induced by the dormancy-breaking chemical hydrogen cyanamide in kiwifruit (Actinidia deliciosa) buds. *Journal of Experimental Botany*, Vol. 60, Issue 13, pp. 3835-3848, ISSN 0022-0957.

103. Wang, H.; Cao, G.H. & Prior, R.L. (1996). Total antioxidant capacity of fruits. *Journal of Agricultural and Food Chemistry*, Vol. 44, Issue 3, pp. 701-705, ISSN 0021-8561.

104. Wang, L.; Li, P.H. & Brutnell, T.P. (2010). Exploring plant transcriptomes using ultra high- throughput sequencing. *Briefings in Functional Genomics*, Vol. 9, Issue 2, pp. 118-128, ISSN 2041-2649.

105. Wang, M.; Li, M. & Meng, A. (2003). Selection of a new red-fleshed kiwifruit cultivar 'Hongyang'. *Proceedings of the Fifth International Symposium on Kiwifruit*, Issue 610, pp. 115-117, ISSN 0567-7572.

106. Wang, T.; Atkinson, R. & Janssen, B. (2007). The choice of Agrobacterium strain for transformation of kiwifruit. *Acta Horticulturae*, Vol. 753, Issue 1, pp. 227-232, ISSN 0567-7572

107. Wang, T.; Ran, Y.; Atkinson, R.G.; Gleave, A.P. & Cohen, D. (2006). Transformation of Actinidia eriantha: a potential species for functional genomics studies in Actinidia. *Plant Cell Reports*, Vol. 25, Issue 5, pp. 425-431, ISSN 0721-7714.

108. Wang, Z.Y.; MacRae, E.A.; Wright, M.A.; Bolitho, K.M.; Ross, G.S. & Atkinson, R.G. (2000). Polygalacturonase gene expression in kiwifruit: relationship to fruit softening and ethylene production. *Plant Molecular Biology*, Vol. 42, Issue 2, pp. 317-328, ISSN 0167-4412.

109. Williams, M.H.; Boyd, L.M.; McNeilage, M.A.; MacRae, E.A.; Ferguson, A.R.; Beatson, R.A. & Martin, P.J. (2003). Development and commercialization of 'baby kiwi' (Actinidia arguta Planch.). *Proceedings of the Fifth International Symposium on Kiwifruit*, Issue 610, pp. 81-86, ISSN 0567-7572.

110. Wu, H.; Sparks, C.; Amoah, B. & Jones, H.D. (2003). Factors influencing successful Agrobacterium-mediated genetic transformation of wheat. *Plant Cell Reports*, Vol. 21, Issue 7, pp. 659-668, ISSN 0721-7714.

111. Wu, J.; Ferguson, A.R. & Murray, B.G. (2009). In vitro induction of autotetraploid Actinidia plants and their field evaluation for crop improvement, Proceedings of the VI International Symposium on In Vitro Culture and Horticultural Breeding, Brisbane, Australia, 25-29 August 2008., pp. 245-250, ISBN 0567-7572

112. Wu, J.; Ferguson, A.R. & Murray, B.G. (2011). Manipulation of ploidy for kiwifruit breeding: in vitro chromosome doubling in diploid *Actinidia chinensis* Planch. *Plant Cell, Tissue and Organ Culture*, Vol. First published online, ISSN 1573-5044.

113. Wu, Y.; Zhao, Y.; Engelmann, F. & Zhou, M. (2001). Cryopreservation of kiwi shoot tips. *CryoLetters*, Vol. 22, Issue 5, pp. 277-284, ISSN 0143-2044.

114. Xiao, X.G.; Zhang, L.S.; Li, S.H.; Wang, B.; Testolin, R. & Cipriani, G. (1999). First step in the search for AFLP markers linked to sex in Actinidia. *Fourth International Symposium on Kiwifruit, Proceedings*, Issue 498, pp. 99-104, ISSN 0567-7572.

115. Xiao, Z. & Han, B. (1997). Interspecific somatic hybrids in Actinidia. *Acta Botanica Sinica*, Vol. 39, Issue 12, pp. 1110-1117, ISSN 0577-7496.

116. Xiao, Z.; Wan, L. & Han, B. (2004). An interspecific somatic hybrid between Actinidia chinensis and Actinidia kolomikta and its chilling tolerance. *Plant Cell, Tissue and Organ Culture*, Vol. 79, Issue 3, pp. 299-306, ISSN 0167-6857.

117. Xu, X.; Gu, Q.; Cai, Z.; Deng, X. & Zhang, Q. (2006). Cryopreservation of in vitro cultured kiwifruit shoot-tips by vitrification and their regeneration. *Acta Horticulturae Sinica*, Vol. 33, Issue 4, pp. 842-844, ISSN 0513-353X.

118. Xu, Z.C.; Hyodo, H.; Ikoma, Y.; Yano, M. & Ogawa, K. (1998). Biochemical characterization and expression of recombinant ACC oxidase in Escherichia coli and endogenous ACC oxidase from kiwifruit. *Postharvest Biology and Technology*, Vol. 14, Issue 1, pp. 41-50, ISSN 0925-5214.

119. Yin, X.; Allan, A.C.; Chen, K. & Ferguson, I.B. (2010). Kiwifruit EIL and ERF genes involved in regulating fruit ripening. *Plant Physiology*, Vol. 153, Issue 3, pp. 1280-1292, ISSN 0032-0889.

120. Yuwono, T. (2004). The presence of actinidin (Cysteine Protease) and recombinant plasmids carrying the actinidin gene influence the growth of Saccharomyces cerevisiae. *World Journal of Microbiology & Biotechnology*, Vol. 20, Issue 5, pp. 441-447, ISSN 0959-3993.

121. Zhai, Z.; Wu, Y.; Engelmann, F.; Chen, R. & Zhao, Y. (2003). Genetic stability assessments of plantlets regenerated from cryopreserved in vitro cultured grape and kiwi shoot- tips using RAPD. *CryoLetters*, Vol. 24, Issue 5, pp. 315-322, ISSN 0143-2044.

122. Zhang, B.; Chen, K.; Bowen, J.; Allan, A.; Espley, R.; Karunairetnam, S. & Ferguson, I. (2006). Differential expression within the LOX gene family in ripening kiwifruit. *Journal of Experimental Botany*, Vol. 57, Issue 14, pp. 3825-3836, ISSN 0022-0957.

123. Zhang, Y.J.; Qian, Y.Q.; Mu, X.J.; Cai, Q.G.; Zhou, Y.L. & Wei, X.P. (1998). Plant regeneration from in vitro-cultured seedling leaf protoplasts of Actinidia eriantha Benth. *Plant Cell Reports*, Vol. 17, Issue 10, pp. 819-821, ISSN 0721-7714.

124. Zhou, J.; Liu, Y.F. & Huang, H.W. (2011). Characterization of 15 Novel Single Nucleotide Polymorphisms (Snps) in the Actinidia Chinensis Species Complex (Actinidiaceae). *American Journal of Botany*, Vol. 98, Issue 5, pp. E100-E102, ISSN 0002-9122.

125. Zhu, D.; Mi, Y.; Chen, Y.; Wang, J. & Liu, Z. (2003). A preliminary study on the transient expression of GFP gene in callus protoplasts of Actinidia arguta. *Journal of Henan Agricultural University*, Vol. 37, Issue 2, pp. 145-148, ISSN 1000-2340.

CHAPTER-2

BIOTECHNOLOGICAL APPROACHES FOR ENHANCING GARLIC PROPAGATION AND QUALITY

INTRODUCTION: AN OVERVIEW

In recent years, the application of biotechnology in agriculture has revolutionized crop production and quality improvement. One such area of interest is the development of biotechnological tools for enhancing garlic propagation and elevating its quality. Garlic, known for its unique flavour and numerous health benefits, has been a staple in culinary and medicinal traditions across the world. However, traditional methods of propagation and cultivation have limitations in terms of scale, efficiency, and genetic enhancement. Biotechnological approaches offer promising solutions to address these challenges, paving the way for a more sustainable and superior garlic production system.

Garlic (*Allium sativum* L.) is a monocotyledonous herb belonging to the genus *Allium* and the family Alliaceae and it is the second most widely distributed species of this genus throughout the world, (Kamenetsky, 2007). Garlic is used as food flavoring or as a medicinal plant. It can be preserved in oil or vinegar or processed into products such as garlic salt, garlic juice, concentrated garlic or most commonly, dehydrated garlic (Brewster, 1994). Although there have been different hypothesis as to the origin of garlic and it was even thought that it was a Mediterranean plant, Vavilov (1926) and Kazakova (1971) suggested Central Asia as its primary center. Years later, this hypothesis was confirmed by the discovery of fertile clones of a primitive garlic type in the Tien-Shan Mountains in Kyrgyzstan (Etoh, 1986; Kotlinska et al., 1991) and by studies using biochemical and molecular markers (Pooler & Simon, 1993).

Fritsch & Friesen (2002) put forward the idea that *Allium sativum* is a complex made up by three main groups: (a) Sativum, (b) Longicuspis and (c) Ophioscorodon, and two subgroups: Subtropical and Pekinense. The commercial types of garlic can be divided into: (1) violet or Asian, which is cultivated in subtropical regions, (2) pink, which needs long photoperiods and has low requirements for cold, (3) white, which needs long photoperiods, has medium to high requirements for cold, and (4) purple, which needs long photoperiods and periods of cold (Heredia-García, 2000). They can also be classified into hard-neck and soft-neck garlic. Hard-neck garlic forms a floral scape whose flowers normally abort and whose end produces top sets, while soft-necked garlic does not form a scape. The majority of garlic cultivated for commercial purposes is soft-neck type because it is easier to cultivate and it has a longer shelf life (Kamenestsky, 2007).

Garlic is grown all over the world from temperate to subtropical climates (Fritsch & Friesen, 2002). Production and world cultivated area have increased over years. In 2004 production was of 14'071,335 t obtained from an area of 1'129,714 ha; while in 2007 15'799,909 t were produced on 1'220,314 ha. The main producer of garlic is China, with 17'967,857 t, accounting for 80.6% of the world production, followed by India (1'070,000 t) and the Republic of Korea (380,000 t) (FAOSTAT, 2009) (Table 1). Garlic bulbs are composed of aggregate of cloves which have their origin in the axillary buds. The cloves are made up of a protective sheath (which is dry and thin at maturity), a thickened storage sheath leaf (which represents the major part of the clove and it is also the usable part), and sprouting and foliage leaves which protect the apical meristem (Mann, 1952; Purseglove, 1988). The number of cloves per bulb varies with the cultivar, but bulbs with a maximum of 16 cloves are preferred.

Table 1. Main garlic-producing countries in the world (FAOSTAT, 2009).

Rank	Country	Production (t)	Production (%)
1	China	17'967,857	80.64
2	India	1'070,000	4.80
3	Republic of Korea	380,000	1.71
4	Russian Federation	227,270	1.02
5	Myanmar (before Burma)	200,000	0.90
6	United States of America	178,760	0.80
7	Egypt	174,659	0.78
8	Bangladesh	154,831	0.69
9	Spain	154,000	0.69
10	Ukraine	150,100	0.67
	World	22'282,060	

Currently, garlic propagates vegetative through cloves or through top sets that develop in the plant´s inflorescences (which can prevent the plant from producing flowers and seeds). Kamenetsky & Rabinowitch (2001) explain that lack of fertility could be due to the fact that in past the floral scapes were removed and plants with low flowering ability were selected in order to obtain bigger bulbs (Kamenetsky & Rabinowitch, 2001; Etoh & Simon, 2002). Nowadays, in some places the bulbs are harvested before the flowering time to avoid their rotting or to use the scapes as vegetable (Etoh & Simon, 2002). In addition, the sterility of the garlic has been mainly attributed to chromosomal deletions, and also to differences in the length of homologous chromosome, to loss of genes involved in gametogenesis, to hypertrophy of the tapetal layer of the anthers at the post-meiotic stage, to microspore degeneration before or after the tetrade stage, to nutritional competition between the top sets and flowers, and to infestation with microorganisms (rickettsias) (Novak, 1972; Konvicka et al., 1978; Etoh, 1985; Pooler & Simon, 1994).

CHEMICAL COMPOSITION AND MEDICINAL TRAITS OF GARLIC

The main components of the garlic bulb are water (65%) and carbohydrates (26-30%), especially fructose polymers (Table 2). Other components are lipids, proteins, fiber, minerals and saponins (Lawson, 1996). Elements such as selenium (700µg per 100g of fresh weight), sulphur, zinc, magnesium, iron, sodium, calcium, as well as vitamins A, C, E and B-complex vitamins (thiamin, riboflavin, niacin) and phenols are also present in the garlic bulb (Koch & Lawson, 1996; Vinson et al., 1998). Garlic produces organosulphur compounds such as the γ- glutamylcysteines and alliin ((+)-S-(2-propenyl)-L-cysteine sulfoxide) which confers its flavor, odor and biological activity (Block, 1985). The alliin can account for 1.4% of the fresh weight of bulb (Keusgen, 2002). It has been found that the activity of alliinase, the enzymethat hydrolyzes the sulphur compounds in garlic, is 10 times higher in bulbs than in leaves (Rabinkov et al., 1994) (Table 2).

Apart from its use for food flavoring, garlic also has medicinal uses for the relief of various ailments such as those caused by aging, arthritis, cancer, artheroesclerosis, immune deficiencies, blood glucose level, respiratory diseases, etc. (Keusgen, 2002; Raham, 2001). Likewise, it has been observed that garlic has antioxidant properties, it reduces blood cholesterol and triglycerides levels, lowers blood pressure and the possibility of blood clot formation and improves arterial oxygenation (Augusti, 1990; Abrams & Fallon, 1998; Bordia et al., 1998). Garlic's effect on reduction of lipids has been most extensively studied. The properties mentioned above are directly related to the sulphur compounds found in the garlic bulb. Alliin is also attributed the antibiotic effect on microorganisms such as *Helicobacter pylori* (bacterium which is associated with stomach cancer), *Salmonella typhi*, yeasts, *Trypanosoma* and *Staphylococcus epidermis*. Its inhibitory effect has also been observed on pathogenic fungi (*Aspergillus*, *Cryptoccocus neoformis*, dermatophytes) (Keusgen, 2002).

Table 2. Chemical composition of garlic bulb (Lawson, 1996).

Component	Amount (fresh weight; %)
Water	62-68
Carbohydrates	26-30
Protein	1.5-2.1
Amino acids: common	1-1.5
Amino acids: cysteine sulphoxides	0.6-1.9
γ-Glutamylcysteines	0.5-1.6
Lipids	0.1-0.2
Fibre	1.5
Total sulphur compounds*	1.1-3.5
Sulphur	0.23-0.37
Nitrogen	0.6-1.3
Minerals	0.7
Vitamins	0.015
Saponins	0.04-0.11
Total oil-soluble compounds	0.15 (whole) – 0.7 (cut)
Total water-soluble compounds	97.00

*Excluding protein and inorganic sulphate (0.5%)

GARLIC CULTIVATION AND STORAGE: PESTS AND DISEASES

- Garlic plant can be affected by various diseases caused by viruses, fungi and bacteria. The viruses that tend to cause it severe damages are poty viruses, such as Leek Yellow Stripe Virus (LYSV), Garlic Yellow Streak Virus (GYSV) and Onion Yellow Dwarf Virus (OYDV) (Bos, 1982; Walkey, 1987). Some Carla viruses, like Common Latent Virus (GCLV) and Shallot Latent Virus (SLV) can also infect the garlic plant (Messiaen et al., 1994). One of the most widely spread diseases in garlic producing countries is white rot, caused by the fungus *Sclerotium cepivorum,* which provokes wilting of the plant and rotting of the bulb.
- Its sclerotia can survive in soil for up to 20 years, therefore limiting garlic production (Delgadillo-Sánchez, 2000). As far as the fungus *Penicillium corymbiferum* is concerned, this attacks plants weakened by other pathogens, and although infested plants survive the infection, bulbs present symptoms during the storage period. Various bacteria

(*Bacillus* spp., *Erwinia* spp., *Pseudomonas* spp.) can also cause damages on bulbs upon storage. Garlic can also be affected by pests like thrips (*Thrips tabaci*), which are insects that infest plants from early developmental stages and cause severe foliage damages. For this reason, thrips are considered the most noxious pest affecting this crop plant. Mites (*Rhizoglyphus* spp.) are another garlic pest that invade the bulbs and limit their sprouting ability (Bujanos-Muñiz & Marín-Jarillo, 2000). On the other hand, bulb nematode (*Ditylenchus dipsaci*) causes the root knot disease, characterized by yellowing and rolling of leaves, as well as rotting of the bulb's base.

BREEDING

Commercial garlic cultivars only propagate themselves vegetative, the increase of genetic variation through conventional crossing is very low, or even absent. For this reason, clonal selection, induced mutations, somaclonal variation or genetic engineering are the only options for breeding improved cultivars (Robinson, 2007). Clonal selection has been the most widely used method for generating new garlic material. It is based on the variability existing in populations as a result of cross pollination between various garlic types and its ancestors when this plant still had the ability of sexual reproduction (Etoh & Simon, 2002; Koul et al., 1979). On the other hand, although mutations may be a source of variability, they are rather limited; therefore, breeding using this strategy has not resulted in significant progress (Etoh & Simon, 2002). The lack of sexuality in garlic limits the increase of variability that is useful for breeding for economically important traits, such as tolerance to biotic and abiotic stress, earliness, yield and quality (Kamenetsky, 2007). Moreover, vegetative propagation has various disadvantages for the crop: (a) a low multiplication rate (5 to 10 per year), (b) expensive and short-term storage that requires wide spaces, (c) transmission of phytopathogens (fungi, viruses, nematodes) through generations and from one production area to another, which can cause a yield decrease of up to 70%, and (d) loss of product quality (Kamenetsky, 2007; Walkey, 1990; Nagakubo et al., 1993).

BIOTECHNOLOGICAL ADVANCEMENTS IN GARLIC PROPAGATION, PRESERVATION AND BREEDING

Biotechnological tools such as micropropagation, meristems culture (in order to obtain virus-free plants), somaclonal variation, and genetic transformation, have contributed to propagation, preservation and breeding of garlic.

Micropropagation

Studies related to the application of tissue culture techniques such as micropropagation for garlic production started in 1970. This technique proved to be advantageous over clove reproduction, as it only requires cells or small tissue fragments to generate high number of plants. Micropropagation can be carried out via two morphogenetic ways: (1) organogenesis, which results in the formation of organs (shoots or roots), and (2) somatic embryogenesis, which leads to the formation of structures having a similar or equal morphology to that of a zygotic embryo. Both processes can involve (indirect) or not (direct) a previous callus phase. Morphogenetic ability in garlic decreases as the callus grows older and the emergence of abnormal plants increases (Novak, 1990). For this reason, regeneration that does not involve a previous callus phase is preferred. Embryogenesis possesses a series of advantages over organogenesis, such as higher potential for high plant output, lower labour requirement and lower cost (Sata et al., 2001). Several micropropagation protocols have been established using both ways of morphogenesis and different explant types; however, most protocols have been developed following the organogenetic way.

Organogenesis

Meristem culture is a technique used for obtaining virus-free plants, and also for micropropagation. Messiaen et al. (1970) were the first in regenerating garlic plants from meristems. Shoot or bud formation from callus was achieved using a combination of 6- furfurylaminopurine (kinetin), indol-3-acetic acid (IAA) and 2,4-diclorophenoxiacetic acid (2,4-D) (9.28μM, 11.4μM and 4.5μM, respectively). Likewise, Havránek & Novak (1973) obtained numerous growth areas on calli produced from meristems on a culture medium with 2,4-D. The subculture of calli to a medium containing IAA (11.4μM) and kinetin (46.5μM) induced formation of adventitious shoots.

In a different work, calli obtained from meristems of three garlic varieties (*Rose de Lautrec*, *California Early* and *California Late*) cultivated in a medium with 2,4-D (4.5μM) and IAA (5.7 μM) produced adventitious shoots when transferred to a medium with IAA (5.7μM) and kinetin (4.6μM) (Kehr & Shaeffer, 1976). For his part, Abo-El-Nil (1977) started cultures from meristems, stems and leaf discs of the variety *Extra Early White*, on a medium with p- chlorophenoxyacetic acid (p-CPA) (10μM), 2,4-D (2μM) and kinetin (0.5μM), from which callus formation was achieved, which in turn resulted in the formation of adventitious shoots in the presence of kinetin (10μM) and IAA (10μM). In other works, meristems were cultivated on B5 medium (Gamborg et al., 1968) with 2.5μM 2-isopentenyladenine (2ip) and 0.55μM α-naphtalenacetic acid (NAA) (Bhojwani, 1980), or in Linsmaier and Skoog (LS) (Linsmaier & Skoog, 1965) medium with 9μM N^6-benzyladenine (BA) alone or with 11.1μM NAA, and multiple shoots were obtained (Osawa et al., 1981).

There are only a few reports on suspension cell culture in garlic. For instance, Nagasawa & Finer (1988) were the first to establish suspension cell cultures obtained from calli derived from meristems of the cultivar *Howaito-Roppen* grown in presence of NAA (5.5μM) and BA (9μM). Adventitious shoots with leaf primordia started differentiating only after transferring the calli to agar-solidified medium. Likewise, Kim et al. (2003) obtained cells in suspension after cultivating shoots of the cultivar *Danyang* differentiated *in vitro* in Murashige and Skoog (MS) (1962) medium with 2.5μM 2iP. These cultures regenerated an average of 21.5 shoots per explant when they were exposed to a light intensity of 50μmol m^{-2} s^{-1}. Thirty bulbs developed per explant in a medium containing 11% sucrose and 135 bulbs in the presence of 10μM jasmonic acid.

Subsequently, Nagakubo et al. (1993) developed a micropropagation protocol for six varieties (*Isshuwase, Isshu-gokuwase, Shanhai, Santo, Furano* and *Howaito-roppen*) starting from shoot-tips which were cultivated in a medium supplemented with NAA (1μM) and BA (1μM). Regenerated adventitious shoots were subcultured for four generations in presence of NAA (5μM) and BA (10μM) for their multiplication. The application of this protocol enables the production of 256 plants from one shoot-tip in 10 months. A novel regeneration protocol was developed by dissecting and sectioning longitudinally the shoots developed from cloves of the cultivar *Extra Select Sets*. These shoots were cultivated on a medium with BA (8μM) and NAA (0.1μM), and after five weeks they produced eight more shoots compared to the ones that had been kept intact (Mohamed-Yassen et al., 1994).

The roots produced by cloves have proved to be a good explant for plant regeneration. When the root tips are cultivated in a medium with NAA (1μM) and BA (10μM) the shoot formation is achieved for 75% of the explants, without an intermediate callus phase. It is estimated that by using this method up to 380 shoots could be produced starting from a single clove (Haque et al., 1997). Other protocols have been developed, which involve callus formation from this type of explants upon their cultivation in MS or N6 culture media (Chu et al., 1975) supplemented with 2,4-D (4.5μM) alone or combined with kinetin (4.7μM). Transferring the calli to a medium with 4.4μM BA allows the regeneration of 169 plants per gram of callus, which have the ability of forming microbulbs (Robledo-Paz et al., 2000). Khan et al. (2004) also regenerated adventitious shoots from calli developed from root tips of two garlic varieties. The highest callus formation frequency was observed when a combination of 2,4-D (6.8μM) and kinetin (23.8μM) was used. Shoot differentiation rose exponentially with increasing BA concentration, reaching the highest value at 45μM BA, while shoot rooting occurred in the absence of growth regulators. A variation in the number of shoots and their regeneration time was observed depending on the genotype used. Approximately 75% of the regenerated plants established successfully when transferred to greenhouse.

The roots developed from adventitious shoots obtained *in vitro* also allowed garlic micropropagation when cultivated on a medium that induced callus formation and then transferred to a medium with BA (13.3μM) and 4-amino-3,5,6-trichloro-picolinic acid (picloram) (1.4μM). This method enables the regeneration of 5.4 shoots per explant (Myers & Simon, 1998a). A protocol named one-step was developed when the same type of explants was cultivated on a modified B5 medium supplemented with 0.1μM 2,4-D, 11.1μM NAA and 13.6μM BA, under two light conditions (16 hours photoperiod and complete darkness). In general, the root tips cultivated under low light conditions displayed the highest percentage of organogenic calli. The application of this protocol allowed the formation of callus and the regeneration of 250 shoots per gram of callus in the same culture medium and under the same light conditions (Martín-Urdíroz et al., 2004). Zheng et al. (2003) also obtained adventitious shoots by using apical and non-apical root fragments, originating from plants generated *in vitro* of four cultivars grown in Europe (*Messidrome*, *Morado de Cuenca*, *Morasol* and *Printanor*). The explants were cultivated on MS medium with 4.5μM 2,4- D and 0.5μM 2iP in order to induce calli formation, which were then transferred to a medium containing 4.7μM kinetin to promote shoot differentiation. The highest regeneration rate was obtained when non-apical fragments were used, although the difference was not significant.

In a different work, Ayabe & Sumi (1998) used the stem disc (consisting in the apical meristem and the lateral buds of the clove) to regenerate plants of the cultivar *Fukuchi- howaito*. When this was cut into various fragments and then cultivated on a medium with BA (0.4μM), 20-25 adventitious shoots were obtained. The same result was observed when protoplasts isolated from shoot primordia were cultured in the presence of NAA (0.5μM) and 2iP (2.4μM), adenine and coconut milk (Ayabe et al., 1995). Barandiaran et al. (1999) used immature bulbs of 23 accessions as a source of axillary buds, which were cultivated during six weeks on B5 medium with 2.5μM 2iP and 0.55μM NAA (establishment phase). Multiplication of regenerated shoots was done on the same culture medium and 20 weeks later shoot clusters were separated in order to cultivate them individually and to induce bulb formation at a low temperature (4°C). Although plants and bulbs were obtained for all accessions under tested conditions, response depended on genotype (accession). Three months later, 60% of bulbs that were transferred to soil survived and produced shoots. This protocol allowed the use of the same culture medium for all phases of micropropagation (establishment, multiplication and bulb formation) and for all accessions, which enabled the handling of all materials tested at the same time, as only three subcultures were required over a period of seven months. Primordial leaf obtained from cloves are also able to produce adventitious shoots when cultivated on a medium with 2,4-D (4.5μM), and develop into plants when transferred onto a medium containing picloram (1.4μM) and BA (13.3μM) (Myers & Simon, 1999). Haque et al. (2003) developed a protocol for plant regeneration and bulb formation from shoot and root meristems of the cultivar *Bangladesh Local.* Meristems were cultivated on MS medium without growth regulators or containing various concentrations of BA (1-10μM) and NAA (1-5μM). None of the combinations of regulators tested produced a higher response than the one observed in their absence (95.5%). In fact, the presence of these compounds suppressed shoot formation in a directly proportional manner to concentration; 45% of root explants formed adventitious shoots, 60% of which produced bulbs. Although a higher number of buds resulted in shoot formation, the root meristems produced more shoots per explant (20). Bulbs derived from root meristems were smaller than the ones derived from bud meristems.

On the other hand, Luciani et al. (2006) tested different explants for micropropagation of variety *069*, which were cultivated on BDS medium (Dustan & Short, 1977), supplemented with picloram, 2,4-D and BA. The basal plates and meristems resulted in the highest values of shoot regeneration, and 2,4-D proved to be better than picloram for inducing callus and shoot formation. By using a combination of 0.25μM 2,4-D and 4.43μM BA, 100% of explants were able to produce calli, which differentiated into both embryos and shoots. It is worth mentioning that *in vitro* propagation is frequently associated with a process known as hyperhydricity or vitrification, which is a physiological disorder caused by the *in vitro* culture conditions that affects the behavior of regenerated plants. This disorder promotes abnormalities at physiological, anatomical and morphological level, which limit the successful establishment of differentiated plants upon their transfer to greenhouse. Hyperhydric plants have a slow growth rate, thick and deformed stems. Their leaves are translucent, thick and wet (Olmos & Hellin, 1998; Kevers et al., 2004).

A study of biochemical and ultrastructural traits of hyperhydric garlic shoots regenerated *in vitro* was carried out by Wu et al. (2009), who observed that organelles such as mitochondria and chloroplasts were compressed against cell wall, in these shoots. In addition, protein content decreased significantly and O2 and H2O2 generation rate increased 45.3% and 63.9%, respectively. Activity of oxidative stress-related enzymes (lipoxygenase, superoxid dismutase, peroxidase, catalase, ascorbate peroxidase) also increased. Moreover, a rise in the level of electrolytes lixiviation was observed, indicating a damage of membrane lipids. Authors concluded that hyperhydric condition of tissues is closely linked to oxidative stress.

Somatic embryogenesis

Formation of structures called embryoids was reported for the first time in 1977. They differentiated from calli obtained from stem tips, bulb leaf discs cultivated in the presence of kinetin (20μM) and IAA (10μM) (Abo-El-Nil, 1977). This response was observed again after a long time when basal plates and floral receptacles were cultivated on a medium containing NAA (1μM) and BA (10μM) (Xue et al., 1991; Al-Zahim et al., 1999). Likewise, Ali & Metwally (1992) induced embryo formation from calli generated from root segments; however, regeneration rate was low. In a different work, Barrueto-Cid et al. (1994) established cultures in suspension of the variety *Chonan* starting from calli initiated on MS medium with 5μM 2,4-D, 5μM picloram and 10μM kinetin. Calli were subcultured onto a medium with 4.5μM 2,4-D and hydrolyzed casein before using them for cell suspension cultures. Plant regeneration occurred after transferring cells to a medium containing 77-153μM adenine.

Later, Sata et al. (2001) obtained somatic embryos directly from basal sections of cloves of the cultivar *Malepur* grown on White medium (White, 1963) supplemented with 4.5μM 2,4-D and 2.3μM kinetin. Under these conditions, each explant formed 20 to 25 embryos, which in the presence of higher concentrations of 2,4-D and kinetin turned into masses of hyperhydric tissue. In the same way, Fereol et al. (2002) produced somatic embryos and plants of the variety *Rouge de la Réunion* after cultivating calli obtained from root tips on a modified B5 medium supplemented with 2,4-D (0.4μM) and kinetin (2.3μM). Thirty percent of the somatic embryos developed into plants which acclimated successfully to greenhouse conditions. Later, Fereol et al. (2005b) established a protocol for embryo regeneration through suspension cultures by using young leaf sections from cloves of the variety *Morasol.* Embryogenic calli were obtained when explants were grown on B5 medium with 4.5μM 2,4- D and 0.47μM kinetin, then transferred to a modified B5 medium with 2.2, 1.1, 1.1, 0.4μM 2,4-D, IAA, NAA and kinetin, respectively, plus 175mM sucrose and 2mM proline. Calli were maintained on this medium for five months and were later used to initiate suspension cultures in a modified N6 medium supplemented with 1.3μM 2,4-D, 0.4μM BA and 131mM sucrose. Embryo production was induced on a medium with 2.3μM kinetin and 0.4μM 2,4-1 Embryos developed into plants, which could produce microbulbs *in vitro*. By using the same explants type and the same culture conditions described above, induction of embryogenic suspension cultures of four garlic cultivars (*Rouge de la Réunion*, *Morasol*, *Messidrome* and *Printanor*) was achieved. Ninety percent of calli differentiated into embryos at globular stage after two months of culture. Out of the regenerated embryos, 50% developed into plants that were successfully established in greenhouse. The histological analysis of the culture revealed that regenerated somatic embryos had a unicellular origin (Fereol et al., 2005a).

Meristem culture

Meristem culture technique has been widely used for the production of virus-free clones. Virus elimination through meristem culture is based on the fact that these meristematic cells are free or almost free of virus and therefore plants regenerated from them will also be virus-free (Salomon, 2002). For this purpose, it is recommended to isolate explants of maximum 5mm, although sometimes their size may limit their establishment *in vitro*. Meristem culture enabled virus-free plants to be produced in various regions in the world(Walkey, 1987; Bhojwani et al., 1982; Peña-Iglesias & Ayuso, 1982; Bertacinni et al., 1986; Conci & Nome, 1991). In Slovenia, Eastern Europe, a trial was conducted to eliminate the OYDV in plants of the cultivar *Ptujksi-spomladanski* through thermotherapy and meristem culture. Meristems of 0.3-0.6mm were first cultivated on B5 medium with 1µM IAA and 1µM BA, then transferred to a multiplication medium containing 5µM jasmonic acid and 5µM 2iP. Meristems obtained from plants that had undergone thermotherapy regenerated a lower number of shoots (1.0-2.2) than the non-treated plants (9.3); 90 to 100% plants were found to be free of the OYDV (Ucman et al., 1998).

Sidaros et al. (2004) attempted to produce plants of three garlic cultivars (*Chinese, Italian* and *Balady*) through meristem culture and chemotherapy. Chemotherapy was carried out by using virazole [or ribavirin (1-β-D-ribofuranosil-1,2,4-triazole-3-carboxamide)] in culture medium. The highest percentage (100%) of virus-free plants was obtained when meristems of 3mm were cultivated on MS medium containing 50mg L^{-1} virazole. In a different study, thermotherapy, chemotherapy and meristem culture were combined in order to obtain plants of the varieties *Taiwan* and *Chileno* free of potyvirus. Thermotherapy consisted in maintaining regenerated plants from embryos dissected from cloves that showed negative results on an ELISA (Enzyme Linked Immuno Absorbent Assay) for potyvirus during one week at 32°C, followed by two weeks at 36°C, and three weeks at 38°C. Embryos were removed from cloves of these plants and cultivated in presence of 205µM ribavirin. Meristems (0.1-0.5mm) of regenerated plants that showed negative results by ELISA were used to generate new plants. Thermotherapy had a more negative effect on plant survival than meristem culture and chemotherapy. However, thermotherapy proved to be more efficient for virus elimination (60.0 to 70.9%) than meristem culture (64.0%), while chemotherapy was not efficient for potyvirus elimination. On the other hand, 10.7% of plants of the cultivar Taiwan grown in field became reinfected, while the Chileno cultivar showed an 8.9% of reinfection after three consecutive cycles of the crop (Ramírez-Malagón et al., 2006).

The use of stems and scape tips of the variety *Red Six Cloves* allows formation of adventitious shoots when cultivated on a medium with NAA (2.6µM) and kinetin

(2.3μM). These shoots developed into plants free of the garlic mosaic virus (GMV) 65 days after starting the culture (Ma et al., 1994). Alternative protocols have been developed for generating virus-free plants starting from inflorescence meristems, bulbils or roots, as apart from being virus free they are available in higher numbers than the apical meristems (Appiano & D'Agostino, 1983). In this way, Verbeek et al. (1995) cultivated meristems obtained from cloves and bulbils (0.15-1.00mm), 71-71% of which regenerated plants; 38% of explants obtained from cloves and 25% of the explants obtained from bulbils were found to be virus-free. In addition, it was observed that meristems smaller than 0.4mm failed to produce shoots. Similarly, Ebi et al. (2000) established a system for elimination of mite-borne mosaic virus using meristems (0.2-0.4mm) obtained from bulbils. These meristems produced plants after being cultivated on MS medium with 5.4μM NAA. The immunoblot assay indicated that several of regenerated plants were virus-free. Senula et al. (2000) obtained plants of 87 accessions free of the viruses OYDV, LYSV, GCLV, SLV and MbFV by cultivating meristems of 0.3-0.8mm originated from bulbils. OYDV and LYSV were eliminated in 85-95% of the regenerated plants. Addition of ribavirin to culture medium reduced regeneration potential, but increased virus elimination. Later, Xu et al. (2001) regenerated virus-free plants from meristems obtained from inflorescences of nine lines. Explants were cultivated on B5 medium containing 0.22μM BA and 0.3mM adenine. Byusing this protocol it was possible to obtain 50-90% plants free of the OYDV, 70-100% plants free of the LYSV and 60-80% free of the SLV. Production of bulbs from virus-free plants was higher than from infected plants. Four to five years would be necessary to obtain virus-free elite seeds that can be established in the field. The economical analysis indicated a net profit of 50.3 to 244.5% (depending on the genotype) for garlic seed producers.

Somaclonal variants

Tissue culture tools such as *in vitro* selection, embryo rescue, somatic hybridization, genetic transformation and somaclonal variation can be used to generate crop variation. Larkin and Scrowcroft (1981) defined somaclonal variation as the phenotypic variation seen in plants regenerated *in vitro* with respect to the original plant. At genetic level, somaclonal variation can be brought about by various DNA changes that include: (a) chromosomal rearrangements, (b) aneuploidy, (c) poliploidy, (d) modification of gene expression by methylation, amplification, inactivation or reactivation, (e) genetic conversion, (f) somatic recombination, (g) transposons movement, (h) genes mutations, etc. (Scowcroft, 1984; Peschke & Phillips, 1991).

Various experiments have been undertaken in garlic in order to generate somaclonal variants that could be used in its improvement. For instance, Novak (1983) treated meristems 0.5-0.7mm with a solution of colchicine (3g L^{-1}) to induce polyploidy. Meristems were treated in two different ways: (1) cultured for 7 days on a solid medium with colchicine, and (2) cultured for 2 days in a liquid medium with colchicine. The latter treatment proved to be more effective. By using this experimental strategy, 35% of regenerated plants were found to be tetraploid, and 14% chimaeras with diploid and tetraploid cells. Dolezel et al. (1986) pointed out that in garlic probability of generating somaclonal variation is higher when disorganized growth occurs, specially for longer periods of time. Similarly, plants generated from old explants or with a high level of differentiation have an increased possibility of suffering this type of variation. The cultivar *Frankon* is resistant to populations of the nematode *Ditylenchus dispsaci* found in Israel, but produces bulbs and cloves of small size and has low yield. A protocol for *in vitro* regeneration was developed to generate somaclonal variants that can produce bulbs with commercial traits. This protocol consisted in cultivating basal plates on BDS medium supplemented with 2.2μM 2,4-D and 2.3μM kinetin to induce callus formation, which were then transferred to a medium containing 9.5μM kinetin and 11.7μM IAA to develop adventitious shoots. Once shoots produced roots, they were transplanted to soil in greenhouse. Assessment of plant characteristics and bulb development revealed that there was variation in bulb size and color, and also in number and size of cloves per bulb compared to the original plant type, which indicates that this could represent a promising material for generation of improved somaclonal variants (Koch & Salomon, 1994).

In a different study, Madhavi et al. (1991) compared ability of calli (organized and disorganized) and bulbs to produce sulphur compounds like alliin. Apart from finding differences in sulphur level compounds, they also observed changes related to proteins, aminoacids, carbohydrates and enzymes. For instance, specific activity of the enzyme alliin lyase in callus was 50% lower than in bulb, while enzymes amino transferase, malate dehydrogenase, polyphenol oxidase, peroxidase and alkaline phosphatase displayed a higher activity than in bulb. Incorporation of precursors to volatile fraction was also higherin organized calli. The somaclone *118.15*, derived from variety *Rosado,* which is grown commercially in Argentina, possesses agronomically desirable traits. A cytological and phenotypical study revealed that both somaclone and original type had the same chromosome number (2n = 16). In addition, some individuals of this clone contained polyploid, aneuploid and haploid cells, probably derived from processes such as endomitosis, nuclear fusion or homologue chromosome pairing in somatic cells. Binucleate cells and differences in length of chromosome pairs were also observed. Plants of clone*118.15* were taller, had a higher diameter of pseudostem and produced bulbs with less, but bigger cloves than the original type (Ordoñez et al., 2002).

El-Aref (2002) regenerated plants of cultivar *Balady* from leaves and roots cultivated first on BDS medium with 2,4-D (4.5μM) and kinetin (9.5μM) for callus initiation, then in presence of 9μM BA and 5.5μM NAA for plant regeneration. These plants formed bulbs which were planted in soil to generate new plants. Isoenzyme analysis showed that 9 out of 29 regenerated plants were different from original plant with regard to some of the enzymes analyzed (phosphatase acid, alcohol dehydrogenase, malate dehydrogenase, esterase). Parental bands were found in all plants, and 5 new bands were observed in 31% of them. Esterase and acid phosphatase displayed a higher polymorphism than alcohol dehydrogenase and malate dehydrogenase. Later, Mukhopadhyay et al. (2005) studied chromosome stability in plants of cultivar *Rossete* generated from callus. It was observed that plants generated from calli iniciated on solid MS medium with 2,4-D (9μM) and kinetin (0.93μM) and sub-cultured in liquid medium with NAA and kinetin exhibited chromosome stability, while the ones grown solely on the initiation solid medium contained hypo- or hyperdiploid cells along with the diploid ones. Frequency of aneuploid cells (2.2-48.9%) increased with callus age.

Recently, Badria & Ali (1999) identified somaclonal variants regenerated from calli obtained from root meristems cultivated on MS medium with kinetin, 2,4-D and IAA. The somaclonal variants formed bulbs without division in the first generation, and displayed normal phenotype in the following generation. After four cycles in field, somaclonal variants that exhibited significant differences of bulb characteristics were found. Cytogenetical analysis revealed that these somaclonal variants had the same chromosome number as original plants. Quantification of alliicin production showed that some somaclones contained three times more of this compound (14.5mg g^{-1}) than control plants (3.8mg g^{-1}). Authors suggest that this technique could be useful for improving alliicin content in garlic.

Germplasm conservation

Conservation of valuable garlic accessions involves their yearly cultivation, as bulbs cannot be stored for long periods of time (6 months at -3°C). This practice is expensive as it requires land use and manpower. Moreover, germplasm grown in field is exposed to environmental changes, pests and diseases that reduces its quality (Panis & Lambardi, 2006). The majority of plant germplasm is stored in seed repositories at temperatures between -15 and -20°C. However, for species whose seeds are recalcitrant (they cannot be dried to humidity levels low enough for storage) or for species that do not produce seeds, like garlic, slow growth storage and cryopreservation are the only tools for conserving them. Slow growth storage involves a condition that maintains tissue growth at a minimum and it allows the medium term storage of material (Botau et al., 2005). It isbased on using organs cultured *in vitro* at 4°C and 10 to 15°C for plants growing in temperate and tropical areas, respectively (Keller et al., 2006).

Cryopreservation is one of the most commonly used tools for germplasm conservation because it requires minimum amount of space and maintenance. In addition, it reduces loss of accessions by contamination, human errors and somaclonal variation which may occur during slow growth storage (Panis & Lambardi, 2006; Sakai & Engelman, 2007). This technique involves the use of liquid nitrogen (which has a freezing temperature of -196°C) for long-term storage of plant material. At this temperature, the majority of biochemical and physical processes are effectively stopped. Cryopreservation is only useful if formation of intracellular ice crystals does not take place, as they may cause irreversible damage to the cell membrane (Panis & Lambardi, 2006). Ice formation without an extreme reduction of the cell water content can only be avoided by a process known as vitrification, in which an aqueous solution turns into an amorphus and glassy state (Sakai, 2000). This procedure substitutes cellular dehydration that occurs during freezing by a reduction of cell water content that is achieved by treating tissues prior to the cooling process with highly concentrated solutions (PVS2, PVS3) containing glycerol, ethylen glycol, dimethylsulfoxid and sucrose, or by air drying (Sakai & Engelman, 2007).

Some countries are already making use of the previously mentioned techniques to conserve their valuable germplasm. For example, China possesses *in vitro* virus-free germplasm banks and their respective databases. These banks have been established taking into account factors such as genotype, culture medium components, light conditions, temperature of incubation rooms, etc. Moreover, studies have been carried out in order to optimize conservation conditions. In this respect, Xu et al. (2005) studied the behavior of six genotypes during their *in vitro* storage. They found out differences in conservation period, depending on genotype. Two genotypes, namely *Cangshan Zaotai* and *Tianjin Baodi* could be stored for 25 months and had a survival percentage of 100%. They also observed that shoots grown at low temperatures on B5 medium with 1.3-2.2μM BA, 0.5-1.6μM NAA and 38-115μM of abscisic acid could be conserved for a longer time. Evaluation of stored material indicated that it was genetically stable and 0.1-0.2% of it became infected with virus. Similarly, the Institute of Plant Genetics and Crop Plant Research at Gatersleben, Germany, one of the biggest gene banks in Europe, possesses a collection of 3039 accessions of species of the genus *Allium*, including the European garlic core collection. Before storage of germplasm, virus elimination is undertaken through meristem culture, then either slow growth storage is carried out for 12 months at 2 and 19°C or cryopreservation-vitrification using a mix of glycerol and sucrose 1:1 as cryoprotectant inside the aluminum foil (Keller et al., 2006).

In the United States, investigations have been carried out for cryopreservation of garlic accessions at the Western Regional Plant Introduction Station, Pullman, Washington. In this respect, Ellis et al. (2005) tested two vitrification solutions to cryopreserve 12 accessions. Shoot tips excised from cloves were treated with the vitrification solutions 2 (PVS2; 15% DMSO, 15% ethylene glycol, 30% glycerol, 0.4M sucrose) and 3 (PVS3; 50% sucrose, 50% glycerol). Eleven out of the 12 accessions could be successfully cryopreserved by using vitrification solutions 2 and 3 as cryoprotectants. Cryopreservation resulted in better regrowth of 7 and 3 accessions when PVS2 and PVS3, respectively, was used. Only onegenotype displayed good recovery in both solutions, which indicated that response depended on genotype.

In Romania, investigations were carried out to establish conditions for slow growth of Romanian garlic landraces. Shoots regenerated from leaf discs were cultured under a slow growth condition that consisted in a medium lacking sucrose at 16°C and under normal conditions on a medium with 3% of sucrose at 24°C. After four weeks, shoots cultivated under slow growth conditions reduced their growth rate by 42%. Shoots grown on sucrose- free medium displayed shorter internodes, but higher foliar surface, and shorter roots. Some landraces proved to be more sensitive to lack of sucrose than to low temperature. Differences in response to growth conditions tested were also observed among the different genotypes under study. The developed protocol allowed medium-term preservation of landraces under study (Botau et al., 2005).

Other investigation groups have focused their efforts on defining the best conditions for slow growth storage or cryopreservation. For example, Makowska et al. (1999) studied the response of apexes from different sources (cloves and bulbils) to cryopreservation after being treated with vitrification solutions. They found out that after freezing, apexes treated with PVS2 solution (30% glycerol, 15% ethylene glycol, 15% DMSO, 0.4M sucrose) restored their growth in a higher proportion compared to the ones that had been submerged in solution PVS3 (40% glycerol and 40% sucrose). A higher number of apexes excised from big bulbils restored their growth, while the ones dissected from small bulbils failed to regrow. On the other hand, apexes obtained from cloves had higher survival rate than the ones obtained from bulbils.

Later, Sudarmonowati (2001) tested different vitrification solutions in an attempt to define the most suitable method for cryopreservation of embryogenic calli of cultivar *Lumbu Hijau*. Calli with embryos at globular stage were precultured on MS medium with 9μM BA and 0.4M sucrose for 1 to 7 days. Next, they were submerged in MS medium with 2M glycerol and 0.4M sucrose (loading solution) in order to be later exposed to different vitrification solutions for various periods of time (5 to 60 min.). Calli were placed in cryotubes with one drop of vitrification solution, then plunged into liquid nitrogen for 30 minutes. After freezing, calli were first plated on MS medium with 5μM 2iP, 2.3μM kinetin and 0.4M sucrose, and later on the same medium containing a lower amount of sucrose. Of the three vitrification solutions tested, the one that contained a mix of glycerol (22%), ethylene glycol (17%), propylene glycol (17%) and DMSO (7%) proved to be better for calli conservation. The highest calli survival percentage was 30%, which indicated that it was necessary to improve the tested methodology.

Recently, Hassan et al. (2007) established slow growth cultures of two garlic varieties grown in Egypt (*Balady* and *Seds 10*). They cultivated bulblets on MS medium with 0.35M sucrose, 5g L^{-1} charcoal and 0.04μM BA, then on a medium with different concentrations of sorbitol and sucrose (0.1, 0.2 and 0.4M). Cultures were incubated in darkness at 4°C for their conservation. Bulblets did not develop neither shoots nor roots under these conditions during the first three months. Addition of 0.1M sorbitol to culture medium delayed growth of shoots and roots of cultivar *Balady* to 6, 12 and 18 months, while sucrose (0.1 or 0.2M) had the same effect on bulblets of cultivar *Seds* 40. The survival rate was of 100% after 18 months of maintaining cultures under these conditions.

Genetic transformation

- Although several methods are available to introduce DNA into plant cells, most of them have been developed using the bacterium *Agrobacterium tumefaciens* (direct method) or biolistic (indirect method) as a vehicle. *A. tumefaciens* is a soil bacterium having natural ability to transfer part of its DNA to plant cells of various species causing formation of crown gall tumors (Hooykaas & Schilperoot, 1992). This ability is conferred by the Ti plasmid (tumor- inducing), which contains a region called T-DNA, that is transferred to the host cell with helping of virulence genes, also present in Ti plasmid. The T-DNA contains genes that are involved in production of cytokinins (2-isopentyl-AMP) and auxins (IAA), which are responsible for tumor formation (Leemans et al., 1982; Barry et al., 1984). Genetic manipulation of this plasmid has resulted in the replacement of genes contained in the T-DNA of the wild strain by genes that confer desirable traits to transformed plants (Christou, 1996).

- A series of physical, electrical and chemical methods (e.g. electroporation and biolistic) have been generated to introduce DNA directly into plant cells (Songstad et al., 1995). Starting from the successful transformation of monocotyledonous plants, such as maize, and soybean by using the biolistic method (McCabe et al., 1988; Fromm et al., 1990), this has become one of the most used systems for gene transfer. This method consists in bombarding target cells with DNA-coated gold or tungsten microparticles accelerated to very high speeds by a gene gun, which allows them to cross the cell walls. Although there are various types of gene guns, the PDS1000 helium designed by Dupont has been the most widely used, specially for transformation of monocotyledonous plants (Vain et al., 1993; Christou, 1995). A great variety of genetic transformation protocols have been developed, but this technology has not been applied with the same efficiency in every species. The species of the genus *Allium* represent an example in this respect, particularly garlic, for which only a small number of publications are available.

Via biolistic

It was not until 1998 that a protocol for garlic transformation was reported for the first time. Barandiaran et al. (1998) bombarded leaf tissue, immature bulbs, cloves and callus of the cultivar *Morado de Cuenca* with four constructs (pDE4, pCW101, pActl-D and pAHC25). Out of these vectors, the one carrying the reporter *uidA* gene (*gusA*) (coding for β-glucuronidase) under control of the promoter 35S from cauliflower mosaic virus (CaMV35S) and the terminator of the nopaline synthase gene (NOS), allowed expression of the *uidA* gene in 43.3% of leaf explants, 76.7% of bulbs, 23.3% of clove tissue and 13% of calli. Transitory expression of the *uidA* gene could only be detected after treating tissues with a nuclease inhibitor (aurintricarboxylic acid). However, regeneration of transgenic plants could not be achieved by using this protocol. Similarly, Myers & Simon (1998) bombarded cell suspensions of *RAL27* clone with *gusA* and *nptII* (conferring resistance to kanamycin) genes which were under control of CaMV35S and NOS promoters, respectively. After 14-16 weeks on selection medium, shoots were regenerated on calli. Incorporation of *gus* and *nptII* genes into garlic transgenic plants was confirmed by PCR assays.

Later, Ferrer et al. (2000) used biolistics to introduce the reporting gene *uidA* and the selection gene *bar*, which codes for N-acetyl-transpherase, into leaf tissue, basal plate discs and embryogenic calli of cultivar *Moraluz*. The *uidA* and *bar* genes were under control of CaMV35S and maize ubiquitin (Ubi) promoters, respectively. Maximum expression of *uidA*was observed in calli and leaves. In a different investigation, Sawehel (2002) developed a transformation system using calli derived from immature cloves of cultivar *Giza 3*. Calli were bombarded with the plasmid pBI22.23 containing the *hpt* gene (coding for the enzyme hygromycin phosphotranspherase that confers resistance to antibiotic hygromycin), and the reporting gene *gusA*. Calli had been previously treated with aurintricarboxilic acid to inhibit activity of endogenous nucleases. Southern blot assays and histochemical analysis proved that this system allowed the transfer, expression and stable integration of transgenes into the garlic genomic DNA.

At the same time, Park et al. (2002) obtained transgenic plants resistant to herbicide chlorsulfuron after bombarding calli of cultivar *Danyang* with the plasmid pC1301-ALS, which contains *gus, hpt* and *als* (coding for acetolactate synthase) genes, under control of the promoter CaMV35S. Out of 1900 calli, 12 grew and regenerated plants resistant to chlorsulphuron (3mg L^{-1}), which formed bulbs and reached maturity. PCR, Southern blot and Northern blot assays confirmed the expression and integration of transgenes into the genome.

In a different work, Robledo-Paz et al. (2004) established a transformation protocol using embryogenic calli derived from root tips of cultivar *GT96-1*. Calli were bombarded with the plasmid pWRG1515 containing *hpt* and *gusA* genes, both under the control of the promoter CaMV35S, and the 3' region of the *nos* gene. Putative transgenic calli were identified after four months of culturing them on a medium containing hygromycin (20mg L^{-1}), and later developed into plants. Molecular (Southern blot) and histochemical (GUS) analysis confirmed transgenic nature of regenerated plants. Transformation efficiency was of 2.2 clones per fresh weight gram of bombarded callus.

Via *Agrobacterium tumefaciens*

Kondo et al. (2000) were the first to achieve the establishment of a transformation protocol in garlic using *A. tumefaciens* as a vehicle. They infected morphogenetic calli with the strain EHA101 carrying the plasmid pIG121, which in turn contained *nptII, hph* and *uidA* genes under control of the promoter CaMV35S. By using this protocol it was possible to regenerate 15 transgenic plants from 1000 inoculated calli grown on a selective culture medium for five months. Zheng et al. (2004) presented a transformation system that apart from producing plants resistant to antibiotics or herbicides, also enabled introduction of genes for resistance to insects. Inoculation of calli of three European cultivars was undertaken using the strain AGLO carrying four different plasmids containing *gusA* and *hpt* genes, and also *cry1Ca* and *H04* genes from *Bacillus thuringiensis,* which confers resistance to the insect *Spodoptera exigua*. The highest transformation frequency (1.47%) was achieved with the cultivar *Printanor* and the plasmid pPB34. Of regenerated plants, only the ones that integrated the *cry1Ca* gene had a good growth under greenhouse conditions and had the ability to form bulbs. These plants were totally resistant to *Spodoptera exigua* in bioassays carried out *in vitro*. Later, Eady et al. (2005) inoculated immature embryos with the strain LBA4404 carrying the vector pBIN *m-gfp-ER* containing the gen *gfp* (encoding for the green flourescent protein) and the gene *nptII*. Out of the 3200 infected embryos, only two transgenic plants (0.06%) were regenerated. Khar et al. (2005) studied the transitory expression of the reporter gene *gusA* in two garlic cultivars after infecting them with a *A. tumefaciens* strain carrying two plasmids. Plasmid pCAMBIA 1301 induced a higher transformation frequency (7.4%) than plasmid pTOK233 (4.1%). Genes conferring resistanceto fungi are still not being commercially used for fighting diseases caused by these phytopathogens. In garlic, Robledo-Paz (2010, personal communication) incorporated chitinase and glucanase genes in an attempt to confer resistance to the fungus *Sclerotium cepivorum.* These experiments revealed that regenerated transgenic plants were not totally resistant to the fungus, but displayed a delay in the infection speed.

Molecular markers

In order to make a more efficient use of garlic germplasm cultivated in various regions of the world, it is necessary to evaluate and characterize the available genetic diversity (Ordás et al., 1994). As the descriptions based on anatomical and morphological characteristics are incomplete and they can be affected by environmental factors, other methods are required to perform this characterization (García-Lampasona et al., 2003). Polymorphism of molecules such as isozymes and DNA can be used to characterize plant germplasm, specially in cases where morphological and biochemical differences are not conspicuous. Although isozyme analysis represented the first application of molecular markers in the genus *Allium,* its main drawback is the low number of enzymatic systems available in garlic. In addition, these markers may suffer changes induced by the developmental stage of plant material analyzed and by environment (Pooler & Simon, 1993; Klaas & Friesen, 2002).

DNA-based markers are less affected by age, physiological condition of the sample and environmental factors. They are not tissue specific and can be detected in any developmental stage of an organism. DNA markers such as RAPDs (Random Amplified Polymorphic DNA), AFLPs (Amplified Fragment Length Polymorphism), SSR (Simple Sequence Repeats) and DNA fingerprinting have been of great use for various studies in garlic. Isozyme analysis, RAPDs and AFLPs have enabled the study of phylogenetic relationships between different garlic clones and determination of their place of origin (Pooler & Simon, 1993; Maaβ & Klaas, 1995; Bradley et al., 1996; Al-Zahim et al., 1997; Lallemand et al., 1997; Ipek & Simon, 1998; García-Lampasona et al., 2003; Buso et al., 2008; Ipek et al., 2008; Abdoli et al., 2009). The use of molecular markers is indispensable for the establishment of core collections that should contain unique, varied and completely identified accessions in order to reduce costs and labour required for maintenance of collections *in situ* (Ipek et al., 2008). On the other hand, germplasm exchange between garlic producing countries can give rise to that a clone be called in different ways in various countries. If this occurred, the germplasm banks could be constituted by duplicated accessions. In this respect, molecular markers such as DNA fingerprinting and AFLPs have been used to detect duplicated accessions in collections (Bradley et al., 1996;

Ipek et al., 2003). The use of AFLPs revealed that 64% of the U.S. National Plant Germplasm System's garlic collection was duplicated (Volk et al., 2004). Moreover, molecular markers can be used for detection of somaclonal variants generated by *in vitro* culture (Al-Zahim et al., 1999; Saker & Sawahel, 1998; Sánchez-Chiang & Jiménez, 2009), for determination of fertile clones (Hong et al., 2000; Etoh & Hong, 2001), disease resistance (Nabulsi et al., 2001) and clones producing S-amino acids (Ovesná et al., 2007).

AT A GLANCE

- Biotechnological tools such as plant tissue culture can help overcome problems associated with vegetative propagation of garlic, specially the low multiplication rate and disease dispersion. Although plant regeneration has been achieved from different explants types, use of root tips has advantages over other explants due to their virus-free condition and to their availability in a relatively high number (30 or more per clove). Moreover, roots developed from bulbs obtained *in vitro* can also be used for tissue culture (Robledo-Paz et al., 2000). On the other hand, production of virus-free plants via meristem culture combined with thermotherapy and chemotherapy can reduce losses caused by phytopathogens, even when propagation of virus- free material is relatively expensive (Salomon, 2002). Tissue culture has also been applied to the establishment of germplasm banks in various parts of the world where valuable garlic collections are maintained for medium (slow growth) and long term (cryopreservation).
- Techniques such as somaclonal variation and genetic engineering could play an important role in the genetic improvement of garlic because they generate genetic variability. However, the somaclonal variants with commercial potential are scarce, and further experiments are necessary to identify the optimal explant type and the culture conditions that enable formation of somaclones (Novak, 1990). In addition, although there have been advances in the field of genetic transformation in garlic, more investigations are required to establish reproducible and efficient protocols. This task will require selection of suitable target cells for inoculation with *Agrobacterium* or biolistics (e.g. embryogenic calli) (Myers & Simon, 1998b; Robledo-Paz et al., 2004), strategies for transgene expression, a suitable selection method and efficient protocols for plant regeneration (McEloy & Brettell, 1994; Hansen & Wright, 1999). Furthermore, molecular markers will be key pieces in phylogenetic and taxonomic studies (Maa□ & Klaas, 1995) and germplasm conservation (Ipek et al., 2008). Moreover, they will be used for detection of somaclonal variants (Al-Zahim et al., 1999), fertile genotypes (Etoh & Hong, 2001), disease resistant genotypes (Nabulsi

et al., 2001) and clones producing compounds of economical importance (Ovesná et al., 2007) which can be used for improving this important crop plant.

REFERENCES

1. Abdoli, M, Habibi-Khaniani, B., Baghalian, K. S., Shahnazi, Rassouli, H. & Naghdi Badi, H. (2009). Classification of Iranian Garlic (*Allium sativum* L.) Ecotypes Using RAPD Markers. *J. Med. Plants,* Vol. 8, pp. 45-51, ISSN 1684-0240.
2. Abo-El-Nil, M. M. (1977). Organogenesis and Embryogenesis in Callus Culture of Garlic.*Plant Sci. Lett.,* Vol. 9, pp. 259-264, ISSN 0304-4211.
3. Abrams, G. A., & Fallon, M. B. (1998). Treatment of Hepatopulmonary Syndrome with *Allium sativum* L. (garlic): a Pilot Trial. *J. Clin. Gastroenterol.,* Vol. 27, pp. 232-235, ISSN 0192-0790.
4. Ali, A., & Metwally, E. E. (1992). Somatic Embryogenesis and Plant Regeneration as a Tool for Garlic Improvement. *Egypt. J. Appl. Sci.,* Vol. 7, pp. 727-735, ISSN 1110-1571.
5. Al-Zahim, M. A., Ford-Lloyd, B. V., & Newbury, H. J. (1999). Detection of Somaclonal Variation in Garlic (*Allium sativum* L.) Using RAPD and Cytological Analysis. *Plant Cell. Rep.*, Vol. 18, pp. 473-477, ISSN 0721-7714.
6. Al-Zahim, M. A., Newbury, H. J., & Ford-Lloyd, B. V. (1997). Classification of Genetic Variation in Garlic (*Allium sativum* L) Revelated by RAPD. *HortSci.*, Vol. 32, pp. 1102-1104, ISSN 0018-5345.
7. Appiano, A., & D'Agostino, G. (1983). Distribution of Tomato Bushy Stunt Virus in Root Tips of Systemically Infected *Gomphrena globosa. J. Ultrastructural Res.,* Vol. 85, pp. 239-248, ISSN 0022-5320.
8. Augusti, K. T. (1990). Therapeutic and Medicinal Values of Onion and Garlic. In: *Onions and Allied Crops*, Rabinowitch, H. D., & Brewster, J. L., Eds., pp. 94-108, Vol. III, CRC Press, ISBN 0849363020, Boca Raton, Fl., U.S.A.
9. Ayabe, M., & Sumi, S. (1998). Establishment of a Novel Tissue Culture Method, Stem-Disc Culture, and Its Practical Application to Micropropagation of Garlic (*Allium sativum* L.). *Plant Cell. Rep.*, Vol. 17, pp. 773-779, ISSN 0721-7714.
10. Ayabe, M., Taniguchi, K., & Sumi, S. (1995). Regeneration of Whole Plants from Protoplasts Isolated from Tissue-Cultured Shoot Primordial of Garlic (*Allium sativum* L.). *Plant Cell. Rep.*, Vol. 15, pp. 17-21, ISSN 0721-7714.
11. Badria, F. A., & Ali, A. A. (1999). Chemical and Genetic Evaluation of Somaclonal Variants of Egyptian Garlic (*Allium sativum* L.). *J. Med. Food,* Vol. 2, pp. 39-43, ISSN 1096-620X.

12. Barandiaran, X., Di Pietro, A., & Martin, J. (1998). Biolistic Transfer and Expression of a *uidA* Reporter Gene in Different Tissue of *Allium sativum* L. *Plant Cell. Rep.*, Vol. 17, pp. 737-741, ISSN 0721-7714.

13. Barandiaran, X., Martín, N., Alba, C., Rodríguez-Conde, M. F., Di Pietro, A. & Martin, J. (1999). An efficient Method for the *in vitro* Management of Multiple Garlic Accessions. *In Vitro Cell. Dev. Biol.- Plant*, Vol. 35, pp. 466-469, ISSN 1054-5476.

14. Barrueto-Cid, L., Illg, R. D., & Piedrabuena, A. E. (1994). Regeneration of Garlic Plants (*Allium sativum* L. cv. "Chonan") Via Cell Culture in Liquid Medium. *In Vitro Cell. Dev. Biol.-Plant*, Vol. 30, pp. 150-155, ISSN 1054-5476.

15. Barry, G. F., Rogers, S. G., Fraley, R. T., & Brand, L. (1984). Identification of a Cloned Cytokinin Biosynthetic Gene. *Proc. Natl. Acad. Sci. U.S.A.*, Vol. 81, pp. 4776-4780, ISSN 0027-8424.

16. Bertacinni, A., Marani, F., & Borgia, M. (1986). Shoot Tip Culture of Different Garlic Lines for Virus Elimination. *Rivista-della Ortoflorofrutticoltura*, Vol. 70, pp. 97-105, ISSN: 0035-5968.

17. Bhojwani, S. S. (1980). *In vitro* Propagation of Garlic by Shoot Proliferation. *Sci. Hort.*, Vol. 13, pp. 47-52, ISSN 0304-4238.

18. Bhojwani, S. S., Cohen, D., & Fry, P. R. (1982). Production of Virus-Free Garlic and Field Performance of Micropropagated Plants. *Sci. Hort.*, Vol. 18, pp. 39-43, ISSN 0304-4238.

19. Block, E. (1985). The Chemistry of Garlic and Onion. *Sci. Am.*, Vol. 252, pp. 114-119, ISSN 0036-8733.

20. Bordia, T., Verma, S. K., & Srivastava, K. C. (1998). Effect of Garlic (*Allium sativum*) on Blood Lipids, Blood Sugar, Fibrinogen and Fibrinolytic Activity in Patients with Coronary Artery Disease. *Prostaglandins Leukot. Essent. Fatty Acids*, Vol. 58, pp. 257-263, ISSN 0952-3278.

21. Botau, D., Danci, M., & Danci, O. (2005). *In vitro* Medium Term Preservation of Different Romanian Landraces. *Acta Biol. Szegediensis*, Vol. 49, pp. 41-42, Rome, Italy, ISSN 1588-385X.

22. Bradley, K. F., Rieger, M. A., & Collins, G. G. (1996). Classification of Australian Garlic Cultivars by DNA Fingerprinting. *Australian J. Exp. Agric.*, Vol. 36, pp. 613-618, ISSN 0816-1089.

23. Brewster, J. L. (1994). *Onions and Other Vegetable Alliums*. CAB International, ISBN 1845933990, Wallingford, U.K.

24. Bujanos-Muñiz, R., & Marín-Jarillo, A. (2000). Plagas: Descripción, Daños y Control. In: *El Ajo en México*. Heredia-García, E., & Delgadillo-Sánchez, F., Eds.,

pp. 64-67, SAGAR- INIFAP, Campo Experimental Bajío, ISBN 968-800-486-3, Celaya, Gto., México.

25. Buso, G. S. C., Paiva, M. R., Torres, A. C., Resende, F. V., Ferreira, M. A., Buso, J. A. & Dusi, A. N. (2008). Genetic Diversity Studies of Brazilian Garlic Cultivars and Quality Control of Garlic-Clover Production. *Gen. Mol. Res.*, Vol. 7, pp. 534-541, ISSN 1676-5680.
26. Christou, P. (1995). Strategies for Variety Independent Genetic Transformation of Important Cereals, Legumes and Woody Species Utilizing Particle Bombardment. *Euphytica*, Vol. 85, pp. 13-27, ISSN 0014-2336.
27. Christou, P. (1996). Transformation Technology. *Trends Plant Sci.*, Vol. 1, pp. 423-431, ISSN 1360-1385.
28. Chu, C. C., Wang, C., Sun, S. C., Hsú, C., Yin, K. C., Chu, C. Y., & Bi, F. Y. (1975). Establishment of an Efficient Medium for Anther Culture of Rice Through Comparative Experiments on the Nitrogen Sources. *Sci. Sinica*, Vol. 16, pp. 659-688, ISSN 1006-9283.
29. Conci, V., & Nome, S. (1991). Virus Free Garlic (*Allium sativum* L.) Plants Obtained by Thermotherapy and Meristem-Tip Culture. *J. Phytopath.*, Vol. 132, pp. 186-192, ISSN 0931-1785.
30. Delgadillo-Sánchez, F. (2000). Enfermedades: Descripción y Tratamiento. In: *El Ajo en México.* Heredia-García, E., & Delgadillo-Sánchez, F., Eds., pp. 68-77, SAGAR- INIFAP, Campo Experimental Bajío, ISBN 968-800-486-3, Celaya, Gto., México.
31. Dolezel, J., Novak, F. J., & Havel, L. (1986). *Cytogenetics of Garlic (Allium sativum L.) in vitro Culture*. IAAE, ISBN 92-0-010086-4, Vienna, Austria.
32. Dustan, D. I., & Short, K. C. (1977). Improved Growth of Tissue Cultures of The Onion, *Allium cepa. Physiol. Plant.*, Vol. 41, pp. 70-72, ISSN 0031-9317.
33. Eady, C., Davis, S., Catanach, A., Kenel, F., & Hunger, S. (2005). *Agrobacterium tumefaciens*- Mediated Transformation of Leek (*Allium porrum*) and Garlic (*Allium sativum*). *Plant Cell. Rep.*, Vol. 24, pp. 209-215, ISSN 0721-7714.
34. Ebi, M., Kasai, N., & Masuda, K. (2000). Small Inflorescence Bulbils are Best for Micropropagation and Virus Elimination in Garlic. *HortSci.*, Vol. 35, pp. 735-737, ISSN 0081-5345.
35. El-Aref, H. M. (2002). An Effective Method for Generating Somaclonal Variability in Egyptian Garlic (*Allium sativum* L.). Faculty of Agriculture, Assiut University, Assiut, Egypt.
36. Ellis, D., Skogerboe, D., Andre, C., Hellier, B., & Volk, G. (2005). Cryopreservation of 12 *Allium sativum* (garlic) Accessions: a Comparison of Plant Vitrification

Solutions (PVS2 and PVS3). *In Vitro-Cellular and Developmental Biology – Plants,* pp. 11-12, In Vitro Biology Meeting, June 2005, Baltimore, Ma., U.S.A.

37. Etoh, T. (1985). Studies on Sterility in Garlic, *Allium sativum* L. *Memoirs of the Faculty of Agriculture, Kagoshima University, Japan*, Vol. 21, pp. 77-132.
38. Etoh, T. (1986). Fertility of Garlic Clones Collected in Soviet Central Asia. *J. Japan Soc. Hort. Sci.*, Vol. 55, pp. 312-319, ISSN 1832-3351.
39. Etoh, T., & Hong, C. J. (2001). RAPD Markers for Fertile Garlic. *Acta Hort.*, Vol. 555, pp. 209- 212, ISSN 0567-7592.
40. Etoh, T., & Simon, P. W. (2002). Diversity, Fertility and Seed Production of Garlic. In: *Allium Crop Sciences: Recent Advances*, Rabinowitch, H. D., & Currah, L., Eds., pp. 101-117, CAB International, ISBN 0-85199-510-1, Wallingford, U. K.
41. FAOSTAT data. http://apps.fao.org/default.htm (Access June 12th, 2011).
42. Fereol, L., Chovelon, V., Causse, S., Michaux-Ferriere, N. & Kahane, R. (2002). Evidence of a Somatic Embryogenesis Process for Plant Regeneration in Garlic (*Allium sativum* L.). *Plant Cell Rep.*, Vol. 21, pp. 197-203, ISSN 0721-7714.
43. Fereol, L., Chovelon, V., Causse, S., Kalumbueziko, M. L. & Kahane, R. (2005a). Embryogenic Cell Suspension Cultures of Garlic (*Allium sativum* L.) as Method for Mass Propagation and Potential Material for Genetic Improvement. *Acta Hort.*, Vol. 688, pp. 65-74, ISSN 0567-7592.
44. Fereol, L., Chovelon, V., Causse, S., Triaire, D., Arnault, I., Auger, J. & Kahane, R. (2005b). Establishment of Embryogenic Cell Suspension Cultures of Garlic (*Allium sativum* L.), Plant Regeneration and Biochemical Analyses. *Plant Cell Rep.*, Vol. 24, pp. 319- 325, ISSN 0721-7714.
45. Ferrer, E., Linares, C., & Gonzalez, J. M. (2000). Efficient Transient Expression of the β- Glucuronidase Reporter Gene in Garlic (*Allium sativum* L.). *Agronomie*, Vol. 20, pp. 869-874, ISSN 0249-5627.
46. Fritsch, R. M., & Friesen, N. (2002). Evolution, Domestication and Taxonomy. In: *Allium Crop Sciences: Recent Advances.* Rabinowitch, H. D., & Currah, L., Eds., pp. 5-30, CAB International, ISBN 0-35199-510-1, Wallingford, U. K.
47. Fromm, M. E., Morrish, F., Armstrong, C., Williams, R., Thomas, J., & Klein, T. M. (1990). Inheritance and Expression of Chimeric Genes in the Progeny of Transgenic Maize Plants. *Nature Biotechnology*, Vol. 8, pp. 833-839, ISSN 1087-0156.
48. Gamborg, O. L., Miller, R. A., & Ojima, K. (1968). Nutrient Requirements of Suspension Culture of Soybean (*Glycine max* Merril) Root Cells. *Exp. Cell Res.*, Vol. 50, pp. 151- 158, ISSN 0014-4827.

49. García-Lampasona, S., Martínez, L., Burba, J. L. (2003). Genetic Diversity among Selected Argentinean Garlic Clones (*Allium sativum* L.) Using AFLP (Amplified Fragment Length Polymorphism). *Euphytica*, Vol. 132, pp. 115-119, ISSN 0014-2336.

50. Hansen, G., & Wright, M. S. (1999). Recent Advances in the Transformation of Plants. *Trends Plant Sci.*, Vol. 4, pp. 226-231, ISSN 1360-1385.

51. Haque, M. S., Wada, T., & Hattori, K. (1997). High Frequency Shoot Regeneration and Plantlet Formation from Root Tip of Garlic. *Plant Cell Tiss. Org. Cult.*, Vol. 50, pp. 83-89, ISSN 0167-6857.

52. Haque, M. S., Wada, T., & Hattori, K. (2003). Shoot Regeneration and Bulblet Formation from Shoot and Root Meristem of Garlic cv Bangladesh Local. *Asian J. Plant Sci.*, Vol. 2, pp 23-27, ISSN 1682-3974.

53. Hassan, N. A., El-Halwagi, A. A., Gaber, A., El-Awady, M. & Klalaf, A. (2007). Slow-Growth *in vitro* Conservation of Garlic Cultivars Grown in Egypt: Chemical Characterization and Molecular Evaluation. *G. J. Mol. Sci.*, Vol. 2, pp. 65-75, ISSN 1990-9241.

54. Havránek, P., & Novak, F. J. (1973). The Bud Formation in Callus Cultures of *Allium sativum* L. *Z. Pflanzen*, Vol. 68, pp. 308-318, ISSN 0340-8159.

55. Heredia-García, E. (2000). Clasificación Taxonómica. In: *El Ajo en México.*

56. Heredia-García, E., & Delgadillo-Sánchez, F., Eds., pp. 18-22, SAGAR-INIFAP, Campo Experimental Bajío, ISBN 968-800-486-3, Celaya, Gto., México.

57. Hong, C. J., Watanabe, H., Etoh, T., & Iwai, S. (2000). A Search of Pollen Fertile Clones in the Iberian Garlic by RAPD Markers. *Memoirs of the Faculty of Agriculture, Kagoshima University, Japan*, Vol. 36, pp. 11-16.

58. Hooykaas, P. J. J., & Schilperoot, R. A. (1992). *Agrobacterium* and Plant Genetic Engineering.*Plant Mol. Biol.*, Vol. 19, pp. 15-38, ISSN 0167-4412.

59. Ipek, M., Ipek, A., & Simon, P. W. (2003). Comparison of AFLPS, RAPD Markers, and Isozymes for Diversity Assessment of Garlic and Detection of Putative Duplicate in Germplasm Collections. *J. Amer. Soc. Hort. Sci.*, Vol. 128, pp. 246-252, ISSN 0003-1062.

60. Ipek, M., Ipek, A., & Simon, P. W. (2008). Rapid Characterization of Garlic Clones with Locus-Specific DNA Markers. *Turk. J. Agric. For.*, Vol. 32, pp. 357-362, ISSN 1300- 011X.

61. Ipek, M., & Simon, P. W. (1998). Genetic Diversity in Garlic (*Allium sativum* L.) as Assessed by Amplified Fragment Length Polymorphism (AFLP). *1998 National Onion* (*and other Allium*) *Research Conference.* Sacramento, Cal., U.S.A, December, 1998.

62. Kamenetsky, R. (2007). Garlic: Botany and Horticulture. *Hort. Rev.*, Vol. 33, pp. 123-171, ISSN 0069-6986.

63. Kamenetsky, R., & Rabinowitch, H. D. (2001). Floral Development in Bolting Garlic. *Sex.Plant Reprod.*, Vol. 13, pp. 235-241, ISSN 0934-0882.

64. Kazakova, A. A. (1971). Most Common Onion Species, Their Origin and Intraspecific Classification. *Trudy po Prikladnoi Botanike, Genetike I Seleksii*, Vol. 72, pp. 135-136 (In Russian), ISSN: 0372-0586.

65. Kehr, A. E., & Schaeffer, T. (1976). Tissue Culture and Differentiation of Garlic. *HortSci.*, Vol.11, pp. 422-423, ISSN 0018-5345.

66. Keller, E. R. J., Senula, A., Leunufna, S., & Grúbe, M. (2006). Slow Growth Storage and Cryopreservation-Tools to Facilitate Germplasm Maintenance of Vegetatively Propagated Crops in Living Plant Collections. *Int. J. Refrigeration, Vol.*, 29, pp. 411- 417, ISSN 0140-7007.

67. Keusgen, M. (2002). Health and *Alliums*. In: *Allium Crop Sciences: Recent Advances*. Rabinowitch, H. D., & Currah, L., Eds., pp. 357-378, CAB International, ISBN 0-85199-510-1, Wallingford, U.K.

68. Kevers, C., Franck, T., Strasser, R. J., Dommes, J. & Gaspar, T. (2004). Hyperhydricity of Micropropagated Shoots: a Typically Stress-Induced Change of Physiological State. *Plant Cell Tiss. Org. Cult.*, Vol. 77, pp. 181-191, ISSN 0167-6857.

69. Khan, N., Alam, M. S., Nath, U. K. (2004). *In vitro* Regeneration of Garlic Through Callus Culture. *J. Biol. Sci.*, Vol. 4, pp. 189-191, ISSN 1727-3048.

70. Khar, A., Yadav, R. C., Yadav, N., & Bhután, R. D. (2005). Transient *gus* Expression Studies in Onion (*Allium cepa* L.) and Garlic (*Allium sativum* L.). Akdeniz Universitesi Ziraat Fakultesi Dergisi., Vol. 18, pp. 301-304, ISSN 1301-2215.

71. Kim, E. K., Hahn, E. J., Murthy, H. N., & Paek, K. Y. (2003). High Frequency of Shoot Multiplication and Bulblet Formation of Garlic in Liquid Cultures. *Plant Cell Tiss. Org. Cult.*, Vol. 73, pp. 231-236, ISSN 0167-6857.

72. Klaas, M., & Friesen, N. (2002). Molecular Markers in *Allium*. In: *Allium Crop Sciences: Recent Advances*. Rabinowitch, H. D., & Currah, L., Eds., pp. 159-185, CAB International, ISBN 0-85199-510-1, Wallingford, U.K.

73. Koch, H. P., Lawson, L. D. (1996). *Garlic: The Science and Therapeutic Application of Allium sativum L. and related species*. Williams and Wilkins Press, ISBN 0683181475, Baltimore, Ma., U.S.A.

74. Koch, M., & Salomon, R. (1994). Improvement of Garlic Via Somaclonal Variation and Virus Elimination. *Acta Hort.*, Vol. 358, pp. 211-214, ISSN 0567-7572.

75. Kondo, T., Hasegawa, H., & Suzuki, M. (2000). Transformation and Regeneration of Garlic (*Allium sativum* L.) by *Agrobacterium*-Mediated Gene Transfer. *Plant Cell Rep.*, Vol. 19, pp. 989-993, ISSN 0721-7714.

76. Konvicka, O., Nienhaus, F., & Fischbeck, G. (1978). Untersuchungen über die Ursachen der Pollensterilität bei *Allium sativum* L. *Z. Pflanzen*, Vol. 80, pp. 265-276, ISSN 0340-8159.

77. Kotlinska, T., Havranek, P., Navratill, M., Gerasimova, L., Pimakov, A., & Neikov, S. (1991). Collecting Onion, Garlic and Wild Species of *Allium* in Central Asia. *Plant Gen. Res. Newslett.*, Vol. 83/84, pp. 31-32, ISSN 0048-4334.

78. Koul, A. K., Gohil, R. N., & Langer, A. (1979). Prospects of Breeding Improved Garlic in the Light of its Genetic and Breeding Systems. *Euphytica*, Vol. 28, pp. 457-464, ISSN 0014-2336.

79. Lallemand, J., Messian, C. M., Briand, F., & Etoh, T. (1997). Delimitation of Varietal Groups in Garlic (*Allium sativum* L.) by Morphological, Physiological and Biochemical Characters. *Acta Hort.*, Vol. 433, pp. 123-132, ISSN 0567-7592.

80. Larkin, P. J., & Scowcroft, W. R. (1981). Somaclonal Variation-a Novel Source of Variability from Cell Cultures for Plant Improvement. *Theor. Appl. Genet.*, Vol. 60, pp. 197-214, ISSN 0040-5752.

81. Lawson, L. D. (1996). The Composition and Chemistry of Garlic Cloves and Processed Garlic. In: *Garlic: The Science and Therapeutic Application of Allium sativum L. and related species*. Koch, H. P., & Lawson, L. D., Eds., pp. 37-107, Williams and Wilkins Press, ISBN 0683181475, Baltimore, Ma., U.S.A.

82. Leemans, J., Deblaere, R., Willmitzer, L., De Greve, H., Hernalsteens, J. P., Van Montagu, M., & Schell, J. (1982). Genetic Identification of Functions of TL-DNA Transcripts in Octopine Crown Galls. *EMBO J.*, Vol. 1, pp. 147-152, ISSN 0261-4189.

83. Linsmaier, E. M., & Skoog, F. (1965). Organic Growth Factor Requirements of Tobacco Tissue Cultures. *Physiol. Plant.*, Vol. 18, pp. 100-127, ISSN 0032-0889.

84. Luciani, G. F., Mary, A. K., Pellegrini, C., & Curvetto, N. R. (2006). Effects of Explants and Growth Regulators in Garlic Callus Formation and Plant Regeneration. *Plant Cell Tiss. Org. Cult.*, Vol. 87, pp. 139-143, ISSN 0167-6857.

85. Ma, Y., Wang, H. L., Cun-Jin, Z., Zhang, C. J., & Kang, Y. Q. (1994). High Rate of Virus-Free Plantlet Regeneration via Garlic Scape-Tip Culture. *Plant Cell Rep.*, Vol. 14, pp. 65- 68, ISSN 0721-7714.

86. Maaβ, H. I., & Klaas, M. (1995). Intraspecific Differentiation of Garlic (*Allium sativum* L.) by Isozyme and RAPD Markers. *Theor. Appl. Genet.*, Vol. 91, pp. 89-97, ISSN 0040-5752. Madhavi, D. L., Prabha, T. N., Singh, N. S., & Patwarhan, M. V.

(1991). Biochemical Studies with Garlic (*Allium sativum*) Cell Cultures Showing Different Flavour Levels. *J. Sci.Food Agric.*, Vol. 56, pp. 15-24, ISSN 0022-5142.

87. Makowska, Z., Keller, E. R. J., & Engelman, F. (1999). Cryopreservation of Apices Isolated from Garlic (*Allium sativum* L.) Bulbils and Cloves. *Cryoletters*, Vol. 20, pp. 175-182, ISSN 0143-2044.

88. Mann, L. K. (1952). Anatomy of the Garlic Bulb and Factors Affecting Bulb Development.*Hilgardia*, Vol. 21, pp. 195-249, ISSN 0073-2230.

89. Martín-Urdíroz, N., Garrido-Gala, J., Martín, J., & Barandiaran, X. (2004). Effect of Light on the Organogenic Ability of Garlic Roots Using a One-Step *in vitro* System. *Plant Cell Rep.*, Vol. 22, pp. 721-724, ISSN 0721-7714.

90. McCabe, D. E., Martinelli, B. J., & Christou, P. (1988). Stable Transformation of Soybean (*Glycine max*) by Particle Acceleration. *Biotechnology*, Vol. 6, pp. 923-926, ISSN 0733- 222X.

91. McEloy, D., & Brettell, R. S. I. (1994). Foreign Gene Expression in Transgenic Cereals. *Trends Biotechnol.*, Vol. 12, pp. 62-68, ISSN 0167-9430.

92. Messiaen, C. M., Lot, H., & Delecolle, B. (1994). Thirty Years of France's Experience in the Production of Disease-Free Garlic and Shallot Mother Bulbs. *Acta Hort.*, Vol. 358, pp. 275-279, ISSN 0567-7592.

93. Messiaen, C. M., Marrov, J., Quiot, J. B., Leclant, F., & Leroux, J. P. (1970). Etude dans le Sud- est de la France d' un Schéma de Sélection Sanitaire de l' ail et de l' échalote. *Comptes Rendus de la 7 Conf. de Pathologie des Plantes.* pp. 101-103. C.N.R.A. Montfavet, France.

94. Mohamed-Yassen, Y., Splittstoesser, W. E., & Litz, R. E. (1994). *In vitro* Shoot Proliferation and Production Sets from Garlic and Shallot. *Plant Cell Tiss. Org. Cult.*, Vol. 36, pp. 243-247, ISSN 0167-6857.

95. Mukhopadhyay, M. J., Sengupta, P., Mukhopadhyay, S., & Sen, S. (2005). *In vitro* Stable Regeneration of Onion and Garlic from Suspension Culture and Chromosomal Instability in Solid Callus Culture. *Sci Hort.*, Vol. 104, Issue No. 1, pp. 1-9, ISSN 0304-4238.

96. Murashige, T., & Skoog, F. (1962). A Revised Medium for Rapid Growth and Bioassays with Tobacco Tissue Cultures. *Physiol. Plant.*, Vol. 15, pp. 473-497, ISSN 0032-0889.

97. Myers, J. M., & Simon, P. W. (1998a). Continuous Callus Production and Regeneration of Garlic (*Allium sativum* L.) Using Root Segments from Shoot Tip-Derived Plantlets. *Plant Cell Rep.*, Vol. 17, pp. 726-730, ISSN 0721-7714.

98. Myers, J. M., & Simon, P. W. (1998b). Microprojectile Bombardment of Garlic, *Allium sativum* L. *1998 National Onion (and other Allium) Research Conference.* Sacramento, Ca.,U.S.A. December, 1998.

99. Myers, J. M., & Simon, P. W. (1999). Regeneration of Garlic Callus as Affected by Clonal Variation, Plant Growth Regulators and Culture Conditions Over Time. *Plant Cell Rep.*, Vol. 19, pp. 32-36, ISSN 0721-7714.

100. Nabulsi, I., Al-Safadi, B., Mir-Ali, N., & Arabi, M. I. E. (2001). Evaluation of Some Garlic (*Allium sativum* L.) Mutants Resistant to White Rot Disease by RAPD Analysis. *Ann. Appl. Biol.*, Vol. 138, pp. 197-202, ISSN 0970-0153.

101. Nagakubo, T., Nagasawa, A., & Ohkawa, H. (1993). Micropropagation of Garlic Through *in vitro* Bulblet Formation. *Plant Cell, Tiss. Org. Cult.*, Vol. 32, pp. 175-183, ISSN 0167- 6857.

102. Nagasawa, A., & Finer, J. J. (1988). Development of Morphogenic Suspension Cultures of Garlic (*Allium sativum* L.). *Plant Cell. Tiss. Org. Cult.*, Vol. 15, pp. 183-187, ISSN 0167-6857.

103. Novak, F. J. (1972). Tapetal Development the Anthers of *Allium sativum* L. and *Allium longicuspis*. *Regel Experientia*, Vol. 28, pp. 363-364, ISSN 0014-4754.

104. Novak, F. J. (1983). Production of Garlic (*Allium sativum* L.) Tetraploids in Shoot-Tip *in vitro* Culture. *Z. Pflanzen*, Vol. 91, pp. 329-333, ISSN 0340-8159.

105. Novak, F. J. (1990). *Allium* Tissue Culture. In: *Onions and Allied Crops*. Rabinowitch, H. D., & Brewster, J. L., Eds., pp. 233-250, Vol. II, CRC Press, ISBN 0849363012, Boca Raton, Fl., U.S.A.

106. Olmos, E., & Hellin, E. (1998). Ultrastructural differences of Hyperhydric and Normal Leaves from Regenerated Carnation Plants. *Sci. Hort.*, Vol. 75, pp. 91-101, ISSN 0304-4238.

107. Ordás, A., Malvar, R. A., & Ron, A. M. (1994). Relationships among American and Spanish Populations of Maize. Euphytica, Vol. 39, pp. 149-161, ISSN 0014-2336.

108. Ordoñez, A., Torres, L. E., Hidalgo, M. G., & Muñoz, J. O. (2002). Análisis Citológico de una Variante Genética Somática de Ajo (*Allium sativum* L.) Tipo Rosado. *Agriscientia*, Vol. 19, pp. 37-43, ISSN 0327-6244.

109. Osawa, K., Kuriyama, T., & Sugawara, Y. (1981). Clonal Multiplication of Vegetatively Propagated Crops Through Tissue Culture. I. Effective Balance of Auxin and Cytokinin in the Medium and Suitable Explants Part for Mass Propagation of Plantlets in Strawberry, Garlic, Scallion, Welsh Onion, Yam and Taro. *Bull. Veg. Ornamental Crops Res. Stn. Ano Mie (Japan)*, Vol. 9, pp. 1-46, ISSN 0387-5407.

110. Ovesná, J., Kucera, L., Králová, J., Leisová, L., Staveliková, H., & Velisek, J. (2007). Genetic Diversity Among Garlic Clones as Revealed by AFLP, Phenotypic Descriptors and S-Amino Acid Level. *Veg. Crops Res. Bull.*, Vol. 66, pp. 105-116, ISSN 1506-9427.

111. Panis, B., & Lambardi, M. (2006). Status of Cryopreservation Technologies in Plants (Crops and Forest Trees). In: *The Role of Biotechnology in Exploring and Protecting Agricultural Genetic Resources.* Ruane, J., & Sonnino, A., Eds., pp. 68-78, FAO, ISBN 978-92-5- 105480-2, Rome, Italy.

112. Park, M. Y., Yi, N. R., Lee, H. Y., Kim, Z. T., Kim, M., Park, J. H., Kim, J. K., Lee, J. S.,Cheong, J. J., & Choi, Y. D. (2002). Generation of Chlorsulfuron Resistant Transgenic Garlic Plant (*Allium sativum* L.) by Particle Bombardment. *Mol. Breed.*, Vol. 9, pp. 171-181, ISSN 1380-3743.

113. Peña-Iglesias, A., & Ayuso, P. (1982). Characterization of Spanish Garlic Viruses and Their Elimination by *in vitro* Shoot Apex Culture. *Acta Hort.*, Vol. 127, pp. 183-193, ISSN 0567-7592.

114. Peschke, V. M., & Phillips, R. L. (1991). Activation of the Maize Transposable Element Suppressor-Mutator (Spm) in Tissue Culture. *Theor. Appl. Genet.*, Vol. 81, pp. 90-97, ISSN 0040-5752.

115. Pooler, M. R., & Simon, P. W. (1993). Characterization and Classification of Isozyme and Morphological Variation in a Diverse Collection of Garlic Clones. *Euphytica*, Vol. 68, pp. 121-130, ISSN 0014-2336.

116. Pooler, M. R., & Simon, P. W. (1994). True Seed Production in Garlic. *Sex. Plant Repr.0od.*, Vol.7, pp. 282-286, ISSN 0934-0882.

117. Purseglove, J. W. (1988). *Tropical Crops. Monocotyledons.* I. Longman, ISBN 0470205687, New York, N.Y., U.S.A.

118. Rabinkov, A., Zhu, X. Z., Grafi, G., & Mirelman, D. (1994). Allin Lyase (Alliinase) from Garlic (*Allium sativum*), Biochemical Characterization and cDNA Cloning. *Appl. Biochem. Biotechnol.*, Vol. 48, pp. 149-171, ISSN 0273-2289.

119. Raham, K. (2001). Historical Perspective on Garlic and Cardiovascular Disease. *J. Nutr.*, Vol.131, pp. 977-979, ISSN 0022-3166.

120. Ramírez-Malagón, R., Pérez-Moreno, L., Borodanenko, A., Salinas-González, G. J., & Ochoa- Alejo, N. (2006). Differential Organ Infection Studies, Potyvirus Elimination, and Field Performance of Virus-Free Garlic Plants Produced by Tissue Culture. *Plant Cell. Tiss. Org. Cult.*, Vol. 86, pp. 103-110, ISSN 0167-6857.

121. Robinson, R. A. (2007). *Self-Organizing Agro-ecosystems.* Sharebooks Publishing, ISBN 698-0- 9783634-1-3, Available: Sharebooks e-book.

122. Robledo-Paz, A., Villalobos-Arámbula, V. M., & Jofre-Garfias, A. E. (2000). Efficient Plant Regeneration of Garlic (*Allium sativum* L.) by Root Tip Culture. *In Vitro Cell Dev. Biol.- Plant*, Vol. 36, pp. 416-419, ISSN 1054-5476.

123. Robledo-Paz, A., Cabrera-Ponce, J. L., Villalobos-Arámbula, V. M., Herrera-Estrella, L., & Jofre-Garfias, A. E. (2004). Genetic Transformation of Garlic

(*Allium sativum* L.) by Particle Bombardment. HortSci., Vol. 39, pp. 1208-1211, ISSN 0018-5345.

124. Sakai, A. (2000). Development of Cryopreservation Techniques. In: *Cryopreservation of Tropical Plant Germplasm—Current Research Progress and Applications*. Engelman, F.,& Takagi, H., Eds., pp. 1-7, International Plant Genetic Resources Institute, ISBN 9290434287, Rome, Italy.

125. Sakai, A., & Engelman, F. (2007). Vitrification, Encapsulation-Vitrification and Droplet- Vitrification: a Review. *Cryoletters*, Vol. 25, pp. 219-226, ISSN 0143-2044.

126. Saker, M. S., & Sawahel, W. A. (1998). Cultivar Identification and Detection of Somaclonal Variations Using RAPD Fingerprinting in Garlic. *Arab. J. Biotech.*, Vol. 1, pp. 69-75, ISSN 1110-6875.

127. Salomon, R. (2002). Virus Diseases in Garlic and Propagation of Virus-Free Plants. In: *Allium Crop Sciences: Recent Advances*. Rabinowitch, H.D., & Currah, L., Eds., pp. 311-328, CAB Internacional, ISBN 0-85199-510-1, Wallingford, U.K.

128. Sánchez-Chiang, N., & Jiménez, V. M. (2009). Técnicas Moleculares para la Detección de Variantes Somaclonales. *Agricultura Mesoamericana*, Vol. 20, pp. 135-151, ISSN 1021- 7444.

129. Sata, S. J., Bagatharia, S. B., & Thaker, V. S. (2001). Induction of Direct Somatic Embryogenesis in Garlic (*Allium sativum*). *Meth. Cell. Sci.*, Vol. 22, pp. 299-304, ISSN 1381-5741.

130. Sawehel, W. A. (2002). Stable Genetic Transformation of Garlic Planting Using Particle Bombardment. *Cell. Mol. Biol. Lett.*, Vol. 7, pp. 49-59, ISSN 1425-8153.

131. Scowcroft, W. R. (1984). *Genetic Variability in Tissue Culture: Impact on Germplasm Conservation and Utilization*, IBPGR, Bars Code: 0161793, Rome, Italy.

132. Senula, A., Keller, E. R. J., & Leseman, D. E. (2000). Elimination of Viruses Through Meristem Culture and Thermotherapy for the Establishment of an *in vitro* Collection of Garlic (*Allium sativum*). *Acta Hort.*, Vol. 530, pp. 121-128, ISSN 0567-7592.

133. Sidaros, S. A., Omar, R. A., El-Kewey, S. A., & El-Khalik, S. A. (2004). Virus Elimination from Infected Garlic Plants Using Different Techniques. *Egyptian J. Virol.*, Vol. 1, pp. 333- 341, ISSN 0022-538X.

134. Songstad, D. D., Somers, D. A., & Griesbach, R. J. (1995). Advances in Alternative DNA Delivery Techniques. *Plant Cell Tiss. Org. Cult.*, Vol. 40, pp. 1-15, ISSN 0167-6857.

135. Sudarmonowati, E. (2001). Cryopreservation of Garlic (*Allium sativum*) cv Lumbu Hijau Using Vitrification Techniques. *Annales Bogorienses*, Vol. 8, pp. 39-46, ISSN 0517- 8452.

136. Ucman, R., Zel, J., & Ravnikar, M. (1998). Thermotherapy in Virus Elimination from Garlic: Influences on Shoot Multiplication from Meristems and Bulb Formation *in vitro*. *Sci. Hort.*, Vol. 73, pp. 193-202, ISSN 0304-4238.

137. Vain, P., McMullen, M. D., & Finer, J. J. (1993). Osmotic Treatment Enhances Particle Bombardment–Mediated Transient and Stable Transformation of Maize. *Plant Cell Rep.*, Vol. 12, pp. 84-88, ISSN 0721-7714.

138. Vavilov, N. I. (1926). Studies on the Origin of Cultivated Plants. *Bull. Appl. Bot.*, Vol. 16, pp.1-248.

139. Verbeek, M., Van Dijk, P., & Van Well, P. M. A. (1995). Efficiency of Four Viruses From Garlic (*Allium sativum*) by Meristem-Tip Culture. *Eur. J. Plant Pathol.*, Vol. 101, pp. 231-239, ISSN 0929-1813.

140. Vinson, J. A., Hao, Y., Su, X., & Zubik, L. (1998). Phenol Antioxidant Quality in Foods: Vegetables. *J. Agric. Food Chem.*, Vol. 46, pp. 3630-3634, ISSN 0021-8561.

141. Volk, G. M., Henk, A. D., & Richards, C. M. (2004). Genetic Diversity Among U.S. Garlic Clones as Detected Using AFLP Methods. *J. Amer. Soc. Hort. Sci.*, Vol. 129, pp. 559- 569, ISSN 0003-1062.

142. Walkey, D. G. A. (1987). Production of Virus-Free Garlic (*Allium sativum* L.) and Shallot (*A. ascalonicum* L.) by Meristem Tip Culture. *J. Hort. Sci.*, Vol. 62, pp. 211-220, ISSN 0970-2873.

143. Walkey, D. G. A. (1990). Virus Diseases. In: *Onions and Allied Crops*. Rabinowitch, H. D., & Brewster, J. L., Eds., pp. 191-212, Vol. II, CRC Press, ISBN 0849363012, Boca Raton, Fl., U.S.A.

144. White, P. R. (1963). *The Cultivation of Animal and Plant Cells*. The Ronald Press Company, New York, N.Y., U.S.A.

145. Wu, Z., Chen, L. J., & Long, Y. J. (2009). Analysis of Ultrastructure and Reactive Oxygen Species of Hyperhydric Garlic (*Allium sativum* L.) Shoots. *In Vitro Cell Dev. Biol.- Plant*, Vol. 45, pp. 483-490, ISSN 1054-5476.

146. Xu, P., Yan, C., Qu, S., Yan, C.Y., & Srinives, P. (2001). Inflorescence Meristem Culture and Economic Analysis of Virus–Free Garlic (*Allium sativum* L.) in Commercial Production. *Acta Hort.*, Vol. 555, pp. 283-288, ISSN 0567-7592.

147. Xu, P., Yan, C., & Yan, C. Y. (2005). Biotechnology Applied to Garlic and Onion. *Acta Hort.*, Vol. 688, pp. 59-75, ISSN 0567-7572.

148. Xue, H. E., Araki, H., Shi, L., & Yakuwa, T. (1991). Somatic Embryogenesis and Plant Regeneration in Basal Plate and Receptacle Derived-Callus Culture of Garlic (*Allium sativum* L.). *J. Japan Soc. Hort. Sci.*, Vol. 60, pp. 627-634, ISSN 1882-3351.

149. Zheng, S. J., Henken, B., Ahn, Y. K., Krens, F. A., & Kik, C. (2004). The Development of a Reproducible *Agrobacterium tumefaciens* Transformation System for Garlic (*Allium sativum* L.) and the Production of Transgenic Resistant to Beet Armyworm (*Spodoptera exigua* Hübner) *Mol. Breed.*, Vol. 14, pp. 293-307, ISSN 1380-3743.

150. Zheng, S. J., Henken, B., Krens, F. A., & Kik, C. (2003). The Development of an Efficient Cultivar Independent Plant Regeneration System From Callus Derived From Both Apical and Non-Apical Root Segments of Garlic (*Allium sativum* L.). *In Vitro Cell Dev. Biol.- Plant*, Vol. 39, pp. 288-292, ISSN 1054-5476.

CHAPTER-3

PLANT BENEFICIAL MICROBES AND THEIR APPLICATION IN PLANT BIOTECHNOLOGY

INTRODUCTION: AN OVERVIEW

The utilization of plant beneficial microbes represents a transformative aspect of modern plant biotechnology. These microorganisms, including bacteria, fungi, and other beneficial agents, play a pivotal role in promoting plant growth, health, and resilience. Their applications encompass a wide array of agricultural practices, from enhancing nutrient uptake and disease resistance to reducing the need for synthetic chemicals in crop production. Harnessing the potential of these plant-microbe interactions has become a central theme in sustainable agriculture and holds promise for addressing various agricultural challenges, thereby significantly contributing to the field of plant biotechnology.

Plants are involved in a complex network of interactions with microorganisms; some of those are beneficial, others are detrimental, but the former are by far the largest and still widely unexplored part. This chapter reviews the status of development and application of beneficial microbes that provide an option for future prospects.

There is a growing worldwide demand for sound and ecologically compatible environmentally friendly techniques in agriculture, capable of providing adequate nourishment for the increasing human population and of improving the quality and quantity of certain agricultural products. For these reasons, the application of beneficial microorganisms is an important alternative to some of the traditional agricultural techniques which, as it has been well documented, very often severely alter the agro- ecosystem balance and cause serious damage to health. For example, contamination of groundwater by leaching of nitrogen fertilizers, accumulation of nitrates and persistence of chemicals used in crop protection in edible portion of foods are cause of grave concern.

The use of beneficial microorganisms in the replacement or the reduction of chemicals has been so far attested (Dobbelaere et al., 2003; Burdman et al., 2000). Beneficial microorganisms such as diazotrophs bacteria, biological control agents (BCAs), plant growth promoting rhizobacteria (PGPRs) and fungi (PGPFs), can play a key role in this major challenge, as they fulfil important ecosystem functions for plants and soil (Whipps, 1997; Raaijmakers et al., 2009; Hermosa et al., 2011). Moreover, modern agriculture, based on the cultivation of a very limited number of crop species and cultivars, is susceptible to epidemic diseases traditionally contrasted through the use of chemicals. With most crops, no effective fungicides are available against a lot of fungal diseases. Plant growth stimulation and crop protection may be improved by the direct application of a number of microorganisms known to act as bio-fertilizers and/or bio-protectors. How beneficialmicroorganisms really do act to improve plant rooting is only partially known, as several aspects have to be considered, including (i) the production of metabolites related to root development growth and pathogen control (phytohormones, antimicrobials, antibiotics), and (ii) the difficulty to discriminate the direct effects on the specific/total activities and the indirect effects due to the enhanced availability of nutrients and growth regulators.Though over the past 150 years bacteria and fungi have been repeatedly demonstrated to promote plant growth and suppress plant pathogens, this knowledge has yet to be extensively exploited in agricultural biotechnology (Berg, 2009).

INTERACTIONS BETWEEN PLANTS AND MICROORGANISMS: ECOLOGICAL RAMIFICATIONS

Soil-borne microorganisms interact with plant roots and soil constituents at the root-soil interface, where root exudates and decaying plant material provide sources of carbon compounds for the heterotrophic biota (Barea et al., 2005; Bisseling et al., 2009). The number of bacteria in the rhizosphere (the narrow region of soil that is directly influenced by root secretions and associated soil microorganisms) and rhizoplane (the external surface of roots together with closely adhering soil particles and debris) is higher than in the soil devoid of plants; this happens because soils devoid of plants are poor in many attractive substances secreted from the roots. As soon as a seed starts to germinate, a relatively large amount of carbon and nitrogen compounds i.e., sugars, organic acid, aminoacids, and vitamins are excreted into the surrounding environment. This attracts a large population of microorganisms inducing vigorous competition between the different species (Okon, 1994). Moreover, rhizosphere microbiomes typically differ between plant species (Bisseling et al., 2009).

Beneficial microorganisms are known to be biocontrol agents and/or growth promoters. There are several modes of action by which they can be beneficial to plant health, which can be related to an indirect or a direct positive effect. Microorganisms have indirect positive effects on plants, affecting adversely the population density, dynamics and metabolic activities of soil-borne pathogens, mainly through competition, antibiosis, lysis, and hyperparasitism. Competition takes place for space and nutrients at the root surface; competitive colonization of the rhizosphere and successful establishment in the root zone is a prerequisite for effective biocontrol. Antagonistic microorganisms can often produce a range of different antimicrobial secondary metabolites, and/or extracellular lytic enzymes. Hyperparasitism is well documented for *Trichoderma*; it involves secretion of chitinases and cellulases, contact with the pathogen, coiling of hyphae around the hyphae of the pathogen, enzymatic digestion of its cell wall, and penetration. Direct positive effects on plants are exerted by rhizosphere microorganisms through a phytostimulation and a biofertilization of plants; these processes involve production of phytohormones, non-symbiotic nitrogen fixation, and the increase of availability of phosphate and other nutrients in the soil (Burdman et al., 2000). Numerous compounds that are toxic to pathogens, such as HCN, phenazines, pyrrolnitrin, and pyoluteorin as well as, enzymes, antibiotics, metabolites and phytohormones are the means by which PGPRs act; similarly other phenomena such as quorum sensing and chemotaxis, are vital for rhizosphere colonization (Castro-Sowinski et al., 2007; Ramette et al., 2011; Jousset et al., 2011).

Under iron-limiting conditions of soil habitats and plant surfaces, PGPRs can produce low- molecular weight compounds called siderophores that sequester iron in a competitive way thus depriving pathogenic fungi of this essential and often scarcely bioavailable element (Pedraza et al., 2007).

Many rhizosphere microorganisms can induce a systemic response in plants, activating plant defence mechanisms. Inoculation with non-pathogenic root zone bacteria can trigger signalling pathways that lead to higher pathogen resistance of the host, the so-called induced systemic resistance (ISR). Several of the bacteria that have been used to study beneficial effects under abiotic stress conditions, such as *Bacillus* sp., have been shown to induce ISR (Chakraborty et al., 2006). Some PGPRs elicit physical or chemical changes related to plant defense, a process often referred to as ISR, and/or tolerance to abiotic stress, such as drought, salt and nutrient excess or deficiency. For the latter PGPR-induced changes in plants, it has been proposed the term "induced systemic tolerance" (IST). IST relates to an enhanced tolerance to abiotic stresses (Yang et al., 2009). The metabolic pathways for signal transduction in plant defense responses can intercommunicate with other plant stress responses. In addition, the genes that are involved in plant responses to biotic and abiotic stresses can be co-regulated (Dimkpa et al., 2009).

The effect of the growth promotion exerted by PGPRs is mainly related to the release of metabolites and nitrogen fixation processes, the provision of bioavailable phosphorus for plant uptake, sequestration of iron by siderophores, production of plant hormones like auxins, cytochinins and gibberellins, and lowering of plant ethylene levels (Glick, 1995; Glick et al., 1999; Tortora et al., 2011). On the contrary, biocontrol occurs through an indirect action of the BCAs that interact with soil pathogens through several mechanisms such as antibiosis (production of antimicrobial compounds), competition for iron and nutrients or for colonization sites, predation and parasitism, induction of resistance factors (for example the plant is strongly stimulated to synthesize substance called phytoalexins, small molecules with antibiotic activity, which can inhibit the growth of many pathogenic microorganisms), production of enzymes such as chitinase, glucanase, protease and lipase (Whipps, 2001). Growth promotion and biocontrol can be due to the same microorganism that positively influences the development of the plant through different mechanisms, for instance the increased availability and assimilation of the mineral nutritional components, the release of growth factors and the suppression of pathogenic microorganisms. This is translated in more resistant and healthy plants. In addition, PGPR species are able to metabolize numerous and varying carbon sources, to multiply quickly and above all to show a greater competence in colonizing the rhizosphere in comparison to deleterious microorganisms.

The beneficial bacteria are widely studied by microbiologists and agronomists because of their potential in increasing plant production (Somers et al., 2004). The research involving the use of PGPRs were made mainly on herbaceous plants in open field environments and in horticultural crops. Moreover, their application has recently expanded both in forestry and in phytoremediation of contaminated soils. Strains belonging to the genera *Azospirillum* (Okon & Labandera-Gonzalez, 1994; Okon & Itzigshon, 1995; Dobbelaere et al., 2001), *Bacillus* (Reddy & Rahe, 1989; Kokalis-Bourelle et al., 2002; Kokalis-Burelle et al., 2006) and *Pseudomonas* (McCullaugh et al., 1996; Meyer et al., 2010) have been used in experimental tests on a wide range of economically important crops.

Endophytic bacteria, those bacteria that dwell intercellularly in association with plants for most, if not all, of their life cycles (Bacon & Hinton, 2007), have been used for biological control of various plant diseases, as well as for enhanced plant agronomic characteristics, such as increased drought tolerance and nitrogen efficiency.

These bacteria, that include anaerobic, aerobic, and microaerobic species, live within the intercellular spaces of plant, where they feed on apoplastic nutrients, as non-pathogens. They can be found within a wide variety of plant tissue, including seeds, fruit, stems, roots and tubers (Surette et al., 2003). Among them are comprised bacterial diazotrophs that do not form nodules on hosts, such as *Azospirillum* species, and some *Rhizobium* species. Isolated from a large diversity of plants (Rosenblueth & Martínez-Romero, 2006), in general they occur at lower population density than rizospheric bacteria or bacterial pathogens and can positively affect host plant growth (Long et al., 2008). What makes bacterial endophytes suitable as biocontrol agents is their colonization of an ecological niche similar to that of phytopathogens (Ryan et al., 2008).

Endophytes can be strictly dependent on the host plant for their growth and survival ("obligate endophytes"); alternatively, "facultative endophytes" have a stage in their life cycle in which they exist outside host plants (Hardoim et al., 2008). The latter group probably comprises the vast majority of the microorganisms that can thrive inside plants. These endophytes often originate from the soil, initially infecting the host plant by colonizing, for instance, the cracks formed in lateral root junctions and then quickly spreading to the intercellular spaces in the root. Hence, to be ecologically successful, endophytes that infect plants from soil must be competent root colonizers. Endophytic colonization of the plant interior is presumably similar, at least in the initial phases, to colonization of plant roots by rhizobacteria. Competitive rhizosphere bacteria, for example members of the genera *Pseudomonas* (e.g. *P. fluorescens*), *Azospirillum* (e.g. *A. brasilense*) and *Bacillus* (Pedraza et al., 2007; Mano & Morisaki, 2008), are often also found as colonizers of the internal tissue of plants. A suite of environmental and genetic factors is presumed to have a role in enabling a specific bacterium to become endophytic. Inside the plant tissues, modulation of plant physiology by tinkering with the plant ethylene levels has emerged as a major strategy, because any effect on this plant stress signal has major impacts on the bacterial niche. How bacteria modulate plant ethylene concentrations is the key to their ecological success or competence as endophytes. The concept of "competent endophytes" has been proposed as a way to characterize those bacteria that possess key genetic machinery required to colonize the endosphere and to persist in it. This is in contrast to "opportunistic endophytes", which are competent rhizosphere colonizers that might become endophytic by coincidentally entering root tissue, but lack genes that are a key to their ecological success inside the plant. Moreover, it is possible to distinguish "passenger endophytes" that, in the absence of any machinery for efficient root colonization or entry, might enter plants purely as a result of chance events (Rosenblueth & Martínez-Romero, 2006; Mercado-Blanco & Bakker, 2007).

Bacterial endophytes, used for biological control of various plant diseases and for improved plant agronomic characteristics, may be of particular interest as they have the advantage of being relatively protected from the competitive soil environment; moreover, they usually grow in the same plant tissue where bacterial plant pathogens are detected (Bulgari et al., 2009). Their importance to crop production systems is only just beginning to be appreciated: so far, they have been shown to promote growth in potatoes, tomatoes, and rice, and they have been shown to be capable of inducing both biotic and abiotic stress resistance (Surette et al., 2003). A large number of mechanisms are being proposed to explain this effect: production of antimicrobial compounds, macronutrient competition, siderophore production, induced systemic resistance. This array of proposed mechanisms reflects the high diversity of endophytic bacteria.

Tolerance to salinity

Soil salinity in arid regions is frequently an important limiting factor for cultivating agricultural crops. PGPR-elicited plant tolerance against salt stress has been intensively studied, showing that inoculation with endophytic bacteria can mitigate the effects of salt stress in different plant species. High K^+/Na^+ ratios were found in salt-stressed maize in which selectivity for Na^+, K^+ and Ca^{2+} was altered upon inoculation with *Azospirillum* (Hamdia et al., 2004).

Similarly, inoculation of pepper with *Bacillus* sp.TW4 led to relief from osmotic stress, which is often manifested as salinity (and/or drought) stress. In these plants, genes linked with ethylene metabolism under abiotic stress were down-regulated (Sziderics et al., 2007). Because *Bacillus* sp. TW4 showed ACC deaminase activity, the authors speculated that the enzyme may be involved in the lower expression of these genes. Salt stress has also been shown to affect nodulation during *Phaseolus–Rhizobium* interaction. However, secondary inoculation of the salt-stressed plants with *Azospirillum* caused an extended exudation of plant flavonoids compared to *Rhizobium* alone, implying an induction of flavonoid genes in the presence of *Azospirillum* (Dardanelli et al., 2008). Thus, the co-inoculation of plants with different bacterial species may contribute to relieving abiotic stress.

IST to salt stress was also noted with *Arabidopsis* (Zhang et al., 2008) using *Bacillus subtilis* GB03, a species that has previously been used as a commercial biological control agent. Interestingly, some of the volatile organic compounds (VOCs) that are emitted from *B. subtilis* GB03 (Ryu et al., 2004) are bacterial determinants involved in IST. The response to saline stress has also been evidenced in barley seedlings where inoculation with *Azospirillum* seemed to mitigate NaCl stress (Zawoznik et al., 2011).

Drought tolerance

Land surface becoming arid or semi-arid has been rising progressively in these last decades; water use efficiency is a current priority for the United Nations policy and a key issue for plant research. Under water stress conditions, leaf transpiration and leaf conductance decrease, and the water use efficiency rises; this mechanism keeps plant growth under water-limited environments (Aroca & Ruiz-Lozano, 2009).

Plant responses to drought include an increase in abscisic acid (ABA) levels, that cause stomatal closure to minimize water loss; these events involve production of activated oxygen species (Cho et al., 2008). Other plant-signalling compounds are involved in regulating stomatal closure, such as methyl jasmonate, salicylic acid and ethylene.

The mechanisms that allow plants to cope with drought stress are regulated by changes in gene expression; drought regulated genes can be divided in two groups: functional genes (encoding for transporters, detoxification enzymes, osmolyte biosynthesis enzymes etc.) and regulatory genes, that encode for transcription factors (Aroca & Ruiz-Lozano, 2009). On the whole, the beneficial effects of PGPR on plant drought tolerance is caused by changes in hormonal contents, mainly that of ABA, ethylene and cytokinins.ABA is involved in the enhancement of plant drought tolerance by PGPR; *Arabidopsis* plants inoculated with *A. brasilense* Sp245 showed more ABA content than non-inoculated ones (Cohen et al., 2008).

Different strains of *A. lipoferum* were used to inoculate wheat seedlings under drought stress. Inoculation alleviated the plant drought stress, increasing wheat growth and yield; different strains performed differently (Arzanesh et al., 2011).

Exactly how a beneficial bacterium induces changes in plant root morphology is not yet clear. Bacterial production of hormone-like substances and their ability to stimulate endogenous hormone levels were believed to play the key role in this process (Dobbelaere et al., 1999). However, more recently, it has been found that, under aerobic conditions, *A. brasilense* produces significant amounts of the small diffusible gas, nitric oxide, which has been shown to act as a signalling molecule in an IAA-induced pathway involved in adventitious root development (Creus et al., 2005; Molina-Favero et al., 2008).

At the transcriptional level, the bacterium *P. polymyxa* caused the induction of a drought- responsive gene, *ERD15*, isolated from drought-stressed *A. thaliana* (Timmusk & Wagner, 1999). The inoculated plants were more tolerant to drought stress than non-inoculated ones; that could be caused by a mild biotic stress that could help plants cope with subsequent drought stress.

EXPERIMENTAL CONSIDERATIONS ABOUT PLANT-BENEFICIAL BACTERIA

In the Mediterranean area the use of microorganisms became indeed widespread in the ’80s, in coincidence with the sudden spread of soybean crop, that required the inoculation of the nitrogen fixing *Bradyrhizobium japonicum*, mainly applied to seeds as peat based or liquid inocula at sowing. Operators became familiar with the use microorganisms in agriculture. After that, despite the huge potentiality of beneficial microorganisms, a relative low diffusion must be highlighted, owing to "inconsistent" results in field experiments, but also owing to prejudices derived from the easy and large availability of chemicals. At the moment, as a consequence of (i) a growing interest towards low input agriculture systems (organic farming, biodynamic, natural farming), (ii) a favourable opinion of consumers for food with no chemicals, and (iii) the increased difficulties in the employment of chemicals according to the most recent laws, we are assisting at a "microbiological revolution", and the use of microorganisms is increasing.

The use of beneficial microorganims is mostly oriented to improve plant growth and protection in an agricultural context, nevertheless several applications in a wider environmental sense could be prospected, as reported by our group in scientific literature. *Pseudomonas fluorescent* (Russo et al., 1996; 2001; 2005), *Bacillus subtilis* (Felici et al., 2008), *Rhizobium* spp (Toffanin et al., 2000; Casella et al., 2006), are some of beneficial bacteria applied in our experimental/scientific work as biofertilizers and/or biocontrol agents in agriculture. Other potential applications currently include micropropagation, bioremediation and phytoremediation, phosphate solubilization, soil aggregation, sewage treatment, bioleaching, oil recovery, coal scrubbing and biogas production.

Azospirillum brasilense is a free-living, aerobic Gram-negative bacterium,that fixes N2 under microaerobic conditions, highly motile, displaying a mixed pattern of flagellation, which offers these microrganisms the advantage of moving towards favorable nutrient conditions. These bacteria have been isolated in particular from the rizosphere of cereals and grasses, in soils with low organic content and low doses of nitrogen fertilization (Dobbelaere et al., 2001; 2003). They are able to penetrate the roots and grow endophytically in the intercellular spaces; they have been isolated from a large variety of soils and locations worldwide, in tropical and temperate regions (Steenhoudt & Vanderleyden, 2000). The *Azospirillum* species are plant growth promoting rhizobacteria, which positively affect the growth and the yield of many plants of agricultural and ecological importance (Bashan et al., 2004). Since the '80s, *Azospirillum* species have been extensively studied for their potential in improving the growth and yield of cereal crops, particularly in sub-tropical regions, firstly speculating that their ability in freely fixing nitrogen could improve soil fertility and increase nutrient uptake of plants. In the last years, much evidence has arisen that the beneficial effects of *Azospirillum* species depend on an array of contributions, such as production of phyto-hormones and other bioactive substances, rather than the bacterial nitrogen fixation. Multiple mechanisms are currently suggested to explain the beneficial effects on plant growth (Bashan &

de-Bashan, 2010). *Azospirillum* is used in many Countries as bacterial inoculant, alone or together with other bacteria and vesicular arbuscular mycorrhizal (VAM) fungi, for many crops (Bashan et al., 2004). The effects on the yield have not been reported to be always positive; they depend on the bacterial strain, the inoculated plant cultivar, and the environmental conditions (Pandey et al., 1998). In particular, temperature has turned out to be of crucial importance when this bacterium is inoculated in winter crops, where responses can be low or non-significant (Kaushik et al., 2001). Hence, the growth response of inoculated crops is not completely predictable (Hartmann & Bashan, 2009); nevertheless, much progress has been made in this field, and the practical field application of *Azospirillum* is expanding worldwide, especially in Central and South America. It has been estimated that there were 300,000 ha inoculated fields in Mexico in 2007, while in Argentina over 220,000 ha of wheat and corn were inoculated in 2008.

Azospirillum brasilense has been proposed in our studies in different fields, ranging from agriculture (crops, micropropagation, grape and olive propagation, ornamental plants nursery) to environmental sciences (bioremediation, environmental engineering), as reported below.

Micropropagation

Micropropagation is an efficient method of propagating large numbers of genetically uniform plants (Honda & Kobayashi, 2004), although serious problems concerning specific steps including explant sterilization, media manipulation, and acclimatization phase, can often invalidate its success, making the plantlets production a cost-intensive process.

In vitro bacterization of potato plantlets has been shown to enhance their transplant stress tolerance thereby eliminating the need of an expensive greenhouse hardening step, which even now is commonly used by pre-elite seed potato producers. Plants bacterized *in vitro* with *Pseudomonas fluorescens* strains CHA0 and IP10 were found to have a significantly higher fresh shoot weight compared to non-bacterized plants in the same system. Arbuscular mycorrhizal (AM) fungi have also been shown to reduce drought stress and increase disease resistance.

We have investigated the possibility of using the PGPR *Azospirillum brasilense* Sp245 to improve the micropropagation of cherry plum (*Prunus cerasifera*) trees (Russo et al., 2008). We have examined the ability of *A. brasilense* to promote rooting of explants during *in vitro* culture in growth-chamber tests and to promote plant growth and plant health during *ex vitro* acclimatization in greenhouse. In the presence of indolebutyric acid (IBA), both rooting and growth of *P. cerasifera* cuttings were significantly improved by bacteria inoculation. During the acclimatation phase, the main positive effect of inoculation was an increased biomass production, as compared with uninoculated control, suggesting that during acclimatation the rhizobacterium produced phytohormones, increased the nutrient uptake of the roots, and caused an improvement of overall plant performance. An ability to protect plants against pathogen attack was also observed, with a plant survival rate of nearly 100% in inoculated plants as compared to 0% in the negative control. This protective effect was demonstrated both *in vitro* and *in vivo* against the pathogenic fungus *Rhizoctonia* spp.

The effectiveness of *A. brasilense* cells application to micropropagated plantlets at the time of transplanting from *in vitro* culture to acclimatization conditions, has been furthermore assessed on three different fruit tree rootstocks: Mr.S2/5 (*Prunus cerasifera* X *P. spinosa*), GF 677 (*Prunus persica* X *P. amygdalus*), MM 106 (Northen spy X M1). This is a critical phase, in which plantlets are subjected to numerous environmental stresses that may lead to significant plant loss. Plant growth and plant health at the end of *post-vitrum* acclimatization, both in growth-chamber and greenhouse trials, respectively, were evaluated.

After 60 days, growth parameters were positively affected by Sp245 inoculum. In the case of Mr.S 2/5, an increase in rootstock stem length and node number by 37% and 42%, respectively, compared to the control was noted. In the case of GF 677, the bacterial inoculum increased stem length and node number by up to the 75% and 65%, respectively, compared to the control. The inoculum did not exert on MM 106 for both parameters suggesting that the effects of Sp245 could depend on a specific clone-microbe association. In all the cases, however, a higher vigor, consistent with a wider leaf area, was present in the inoculated plantlets demonstrating that the use of *Azospirillum* can significantly contribute to optimize plant performance during the phase of adaptation of plants to *post-vitrum* conditions (Vettori et al., 2010).

Considering that the main obstacles to an intensive and widespread use of beneficial microorganisms, at the commercial level, have been so far the so called "inconsistent" results in field application, mainly related to biotic and abiotic adverse environmental factors, the *in vitro* and *post vitro* inoculation with bacteria may be a way of overcoming a part of these difficulties.

Co-inoculation strategies

In recent years, a number of studies on co-inoculation of two or more beneficial microorganisms for better crop productivity have been reported. Positive effects, such as increase in biomass parameters, nitrogen-content and yield have been found in legumes inoculated with *Rhizobium* and *Azospirillum*. These positive effects may be attributed to early and increased nodulation, enhanced N_2 fixation rates, and a general improvement of root development. Stimulation of nodulation following the inoculation with *Azospirillum* may be derived from an increase in production of lateral roots, root hair density and branching, butalso from the differentiation of a greater number of epidermal cells into root hairs susceptible for infection by rhizobia. Nodulation by rhizobia co-inoculated with *Azospirillum* may also be enhanced by an increased secretion of root flavonoid substances that are involved in the activation of the nodulation genes in *Rhizobium* (Dobbelaere et al., 2001). Considerable results have also been obtained on grain yield, N, P, K content in wheat co- inoculated with *Azospirillum brasilense* and *Rhizobium meliloti* (Askary et al., 2009).

The effects of co-inoculation of bacteria and fungi has also been reported. *Azospirillum*–AM fungus combination seems suitable for sustainable agriculture practices, since both types of microorganisms are compatible with each other. The stimulatory effect of the *Azospirillum* inocula on root growth did not significantly influence the mycorrhization, regardless of the AM fungus involved, either in wheat or in maize plants, grown in the greenhouse and/or under field conditions. The effect of *Azospirillum brasilense* Sp245 was assessed at greenhouse level in three different cultivars of durum wheat, in the presence of indigenous AM fungi, and in maize plants artificially inoculated with *Glomus mosseae* and *Glomus macrocarpum*. At field level, the establishment of natural AM fungal symbiosis was evaluated with the commercial strain *Azospirillum lipoferum* CRT1 in maize plants (Russo et al., 2005). Positive effects of *Azospirillum brasilense* and arbuscular mycorrhizal colonization on rice growth and drought resistance have also been attested (Ruíz-Sánchez et al., 2011).

On the other hand, the combination of two rhizobacteria had no synergistic or comparable effects on plant biomass, with respect to their single applications. Indeed, individual inoculation of *B. subtilis* and *A. brasilense* Sp245 positively affected the growth in dry weight of both shoots and roots of tomato plants, but the combination of the two rhizobacteria had no synergistic or comparable effects on plant biomass. *In vitro* tests and cellular analysis of root tips revealed a growth inhibition of the primary root, which is not related to a reduced persistence in the rhizosphere of one or both bacteria (Felici et al., 2008). Moreover co- inoculation with mycorrhiza and rhizobia of different bean genotypes resulted in the reduction in the trehalose content and the authors concluded supporting the idea of using rhizobial or mycorrhizal inoculation separately (Ballesteros-Almanza et al., 2010).

These results suggest that mixing different microorganisms in the same inocula/treatment can cause interferences and consequent bad or lower than expected performances. Interactions and antagonist phenomena in contaminant species against *Bradyrhizobium japonicum*, obtained from the same soybean inocula preparation, had already been evidenced in the '80s.

Bioremediation and phytoremediation

Phytoextraction, actuated by hyperaccumulating or non-hyperaccumulating species, could be improved by using a plant-microbe system (Zhuang et al. 2007), thus contributing to novel promising methods for the cleaning-up of soils contaminated by heavy metals. Rhizobacteria of the genus *Azospirillum* have been extensively used for crop phytostimulation as above stated, thanks to the positive interaction between bacteria and plants at root level (Dobbeleare et al., 2001; Dobbeleare et al., 2003; Russo et al., 2005; Russo et al., 2008).

The implementation of lead phytoextraction in contaminated industrial soils by applying*A. brasilense* Sp245 to plants of indigenous species belonging to Mediterranean forestry was investigated. The possible phytoextraction ability was evaluated in *Myrtus communis*L. and *Laurus nobilis* L., previously selected among other plant species that were found able to grow in the contaminated areas, on the basis of the Pb content (EmissionSpectrophotometer Atomic Plasma, ICP-AES), the growing speed and the vegetative *habitus*. By trials carried out in greenhouse, it was shown that *A. brasilense* Sp245 can enhance the plant growth in Pb contaminated soil and affect the plant total lead content. Greenhouse trials were performed for 2 and 9 months, and plants were grown in pot in the presence of two level of Pb (312 and 4345 ppm).

The presence of Sp245 positively affected the total amount of Pb that was removed by plants, either as total biomass produced (Figure 1) or as specific Pb concentration, as a consequence of the incremented root growth, attesting the synergic effect of plants and microorganisms in a bioremediation system, and as higher specific Pb concentration (Table 1 and Table 2). Moreover the bioconcentration factor (Pb in plant tissues/Pb in soil) and translocation factor (Pb in leaves and shoots/Pb in roots) were significantly affected by the presence of *A. brasilense* Sp245, attesting the synergetic effect of plants and microorganisms in a rhizoremediation system.

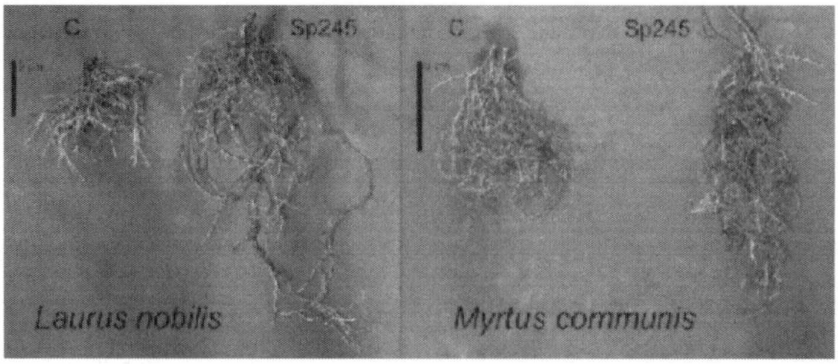

Fig. 1. Effect of *Azospirillum brasilense* Sp245 inoculation on root system in *Laurus nobilis* and*Myrtus communis* after nine months of pot cultivation with Pb polluted soil.

Table 1. Effect of *A. brasilense* Sp245 inoculation on Pb uptake (mg/Kg d.m.) by each plant of *Laurus nobilis* and *Myrtus communis* after two and nine months of pot cultivation with soils at different Pb concentration (Pb- : 312 mg/Kg and Pb+ : 4345 mg/Kg). Within each shrub species and each period means with the same letter were not significantly different according to the SNK'test (P ≤0.05).

Shrub species	Pb uptake (mg/Kg dry matter)		Pb uptake Effect of *A. brasilense* Sp 245	
Laurus nobilis	Pb-	Pb+	Pb-	Pb+
Two months	63b	430a	77b	438a
Nine months	79c	466b	94c	534a
Myrtus communis	Pb-	Pb+	Pb-	Pb+
Two months	103c	806b	116c	954a
Nine months	191d	1176b	305c	1324a

Table 2. Effect of *A. brasilense* Sp245 inoculation on biomass produced (g/plant d.m.) by *Laurus nobilis* and *Myrtus communis* after two and nine months of pot cultivation with Pb polluted soil. Within shrub species means with the same letter were not significantly different according to the SNK'test (P ≤0.05).

Shrub species	Control	Effect of *A. brasilense* Sp 245
Laurus nobilis		
Two months	3.1c	5.6b
Nine months	6.6a	7.5a
Myrtus communis		
Two months	2.9d	5.2b
Nine months	4.1c	6.3a

Engineering Rooted in Natural Principles and End-of-Therapy Approaches

- The increased root density and branching, the improving in rooting of cuttings and the better adaptation to biotic and abiotic stresses derived from beneficial microorganisms, may represent an advantage in soil bioengineering and landscape construction. The role of vegetation in slope stability and restoration of steep rock faces with shrubs and trees is difficult due to extreme microclimatic and edaphic conditions (Beikircher et al., 2010). Inoculation with *A. brasilense* Sp245 of plant material used for restoration of drought-prone sites during preconditioning, can increase the drought tolerance and can play a synergetic and pivotal role in that phase. Work in progress with some angiosperm species, known for their vulnerability, gives us good expectation.

- Another promising field could be the use of selected beneficial microorganisms in endotherapy, or trunk injection. This is an alternative method of treatment of urban woodland plant, with many advantages compared to traditional air treatments, including the absence of spraying of chemicals, and hence the complete harmlessness for the health of citizens, birds and other animals (Sánchez-Zamora & Fernández-Escobar, 2004; Hubbard & Potter, 2006).

AT A GLANCE

- Agriculture is the oldest economic sector in the world, and is more dependent on fertile soils and a stable climate than any other trade. At the same time, it has a huge influence on the ecological balance, water and soil quality, and on the preservation of biological diversity. Since the middle of the last century, agricultural techniques and economic framework conditions worldwide have undergone such a radical transformation that agriculture has became a major source of environmental pollution. The investigation about ecologically compatible techniques in agriculture and environmental sciences can take essential advantage from the use of beneficial microorganisms as plant- microbe interactions fulfil important ecosystem functions.
- Plant diseases are a major cause of yield losses and ecosystem instability worldwide. Use of agrochemicals to protect crop against plant pathogens has been increasing along with the intensification of agricultural production over the last few decades. New biotechnological methods for crop protection are based on the use of beneficial microorganisms applied as biofertilizers and/or biocontrol agents; this approach represents an important tool for plant disease control, and could lead to a substantial reduction of chemical fertilizer use, which is an important source of environmental pollution. Nevertheless, despite dedicated efforts to study beneficial microorganisms, relatively few products have been registered for agricultural use and they count for a very small fraction of the potential market. This is a consequence of several limitations affecting their commercial expansion, which are mainly related to the survival capability of microorganisms under various environmental conditions.
- Moreover, nowadays, microbial inoculants, some of which have a historical record for safe use since 1896 (the well-known rhizobia, for the inoculation of legumes) or since the 1930s (e.g. *Bacillus thuringiensis*, for the biological control of invertebrate pests) are being widely applied in modern agriculture as biofertilizers and biocontrol agents. Other interesting applications include micropropagation, bioremediation and phytoremediation,

phosphate solubilization, soil aggregation, sewage treatment, bioleaching, oil recovery, coal scrubbing and biogas production, and represent incoming fields of application.

- In short, from the examples and references cited above, it is manifest that useful microorganisms of agricultural importance represent an alternative and ecological strategy for disease management, in order to reduce the use of chemicals in agriculture and to improve cultivar performance. At the same time, their application is a highly efficient way to resolve environmental problems, for example through bioremediation and bioengineering. However, although beneficial microorganisms hold a great promise for dealing with different environmental problems, it is important to acknowledge that much of this promise has yet to be realized. Indeed, much needs to be learned about how microorganisms interact with each other and with the environments. For the future development of biotechnology in this field, the contribution of a combination of scientific disciplines is of primary importance to promote sustainable practices in plant production system, as well as in conservation and ecosystem restoration.

REFERENCES

1. Aroca, R. & Ruíz-Lozano, J.M. (2009). Induction of plant tsolerance to semi-arid environments by beneficial soil microorganisms (a review) En: *Climate Change, Intercropping, Pest Control and Beneficial Microorganisms, sustainable Agriculture* Reviews 2. (Lichtfouse, E., Ed.) Springer, Netherlands pp.121-135.

2. Arzanesh, M.; Alikhani, H.; Khavazi, K.; Rahimian, H. & Miransari, M. (2011). Wheat (*Triticum aestivum* L.) growth enhancement by *Azospirillum* sp. under drought stress. *World Journal of Microbiology and Biotechnology*, Vol.27, pp.197-205.

3. Askary, M.; Mostajeran, A.; Amooaghaei, R.; Mostajeran, M. (2009). Influence of the co- inoculation *Azospirillum brasilense* and *Rhizobium meliloti* plus 2,4-D on grain yield and N, P, K content of *Triticum aestivum* (cv. Baccros and Mahdavi). American- Eurasian Journal of Agricultural and Environmental *Science*, Vol.5, No.3 pp.296-307.

4. Barea, J.M.; Pozo, M.J.; Azcón, R. & Azcón-Aguilar, C. (2005). Microbial cooperation in the rhizosphere. *Journal of experimental botany*, Vol.56, pp.1761-1778

5. Ballesteros-Almanza, L.; Altamirano-Hernandez, J.; Peña-Cabriales, J.J.; Santoyo, G.; Sanchez-Yañez, J.M.; Valencia-Cantero, E.; Macias-Rodriguez, L.; Lopez-Bucio, J.;

6. Cardenas-Navarro, R. & Farias-Rodriguez, R. (2010). Effect of co-inoculation with mycorrhiza and rhizobia on the nodule trehalose content of different bean genotypes. *Open Microbiol J.*, Vol.17, No.4, pp.83-92.

7. Bacon, C.W. & Hinton, D.M. (2007). Isolation, in planta detection, and uses of endophytic bacteria for plant protection. In "Manual of environmental microbiology" ed. ByC.J. Hurst *et al.* ASM Press Washington D.C.

8. Bashan, Y.; Holguin, G. & De-Bashan, L.E. (2004). *Azospirillum*-plant relationships: physiological, molecular, agricultural, and environmental advances (1997-2003). *Canadian journal of microbiology* Vol.50, pp.521-577

9. Bashan, Y. & De-Bashan, L.E. (2010). How the Plant Growth-Promoting Bacterium *Azospirillum* Promotes Plant Growth-a Critical Assessment. *Advances in Agronomy,* Vol. 108, pp.77-136

10. Beikircher, B.; Florineth, F. & Mayr, S. (2010). Restoration of rocky slopes based on planted gabions and use of drought-preconditioned woody species. *Ecol Eng.*, Vol.36, No.4, pp.421-426

11. Berg, G. (2009). Plant–microbe interactions promoting plant growth and health: perspectives for controlled use of microorganisms in agriculture. *Applied Microbiology and Biotechnology*, Vol.84, pp.11-18

12. Bisseling, T.; Dangl, J.L. & Schulze-Lefert, P. (2009). Next-Generation Communication.*Science,* Vol.324, pp.691-691

13. Bulgari, D.; Casati, P.; Brusetti, L.; Quaglino, F.; Brasca, M.; Daffonchio, D. & Bianco, P.A. (2009). Endophytic bacterial diversity in grapevine (*Vitis vinifera* L.) leaves described by 16S rRNA gene sequence analysis and length heterogeneity-PCR. *Journal of microbiology (Seoul,Korea),* Vol.47, pp.393-401

14. Burdman, S.; Jurkevitch, E. & Okon, Y. (2000). Recent advance in the use of plant growth promoting rhizobacteria (PGPR) in agriculture. In: *Microbial Interaction In Agriculture Forestry*, Vol. II, Subba Rao NS & Dommergues YR pp.229-250

15. Casella, S; Shapleigh, J.P.; Toffanin, A. & Basaglia, M. (2006) Investigation into the role of the truncated denitrification chain in *Rhizobium sullae* strain HCNT1, *Biochemical Society Transactions*, Vol.34, pp.130-132

16. Castro-Sowinski, S.; Herschkovitz, Y.; Okon, Y. & Jurkevitch, E. (2007). Effects of inoculation with plant growth-promoting rhizobacteria on resident rhizosphere microorganisms. *FEMS Microbiol Lett.*, Vol.276, pp.1-11

17. Chakraborty, U.; Chakraborty, B. & Basnet, M. (2006). Plant growth promotion and induction of resistance in *Camellia sinensis* by *Bacillus megaterium*. *Journal of Basic Microbiology*, Vol.46, pp.186-195

18. Cho, S.M.; Kang, B.R.; Han, S.H.; Anderson, A.J.; Park, J.-Y.; Lee, Y.-H.; Cho, B.H.; Yang, K.- Y.; Ryu, C.-M. & Kim, Y.C. (2008). 2R,3R-butanediol, a bacterial volatile produced by *Pseudomonas chlororaphis* O6, is involved in induction of systemic tolerance to drought in *Arabidopsis thaliana. Molecular plant-microbe interactions.* 21 pp.1067-1075

19. Cohen, A.; Bottini, R. & Piccoli, P. (2008). *Azospirillum brasilense* Sp produces ABA in chemically-defined culture medium and increases ABA content in *Arabidopsis* plants. *Plant Growth Regulation*, Vol.54, pp.97-103

20. Creus, C.M.; Graziano, M.; Casanovas, E.M.; Pereyra, M.A.; Simontacchi, M.; Puntarulo, S.; Barassi, C.A. & Lamattina, L. (2005). Nitric oxide is involved in the *Azospirillum brasilense*-induced lateral root formation in tomato. *Planta,* Vol.22, pp.297-303

21. Dardanelli, M.S.; De Cordoba, F.J.F.; Espuny, M.R.; Carvajal, M.A.R.; Diaz, M.E.S.; Serrano, A.M.G.; Okon, Y. & Megias, M. (2008). Effect of *Azospirillum brasilense* coinoculated with *Rhizobium* on *Phaseolus vulgaris* flavonoids and Nod factor production under salt stress. *Soil Biology & Biochemistry,* Vol.40, pp. 2713-2721

22. Dimkpa, C.; Weinand, T. & Asch, F. (2009). Plant-rhizobacteria interactions alleviate abiotic stress conditions. *Plant Cell and Environment*, Vol.32, pp. 1682-1694

23. Dobbelaere, S.; Croonenborghs, A.; Thys, A.; Vande Broek, A. & Vanderleyden, J. (1999). Phytostimulatory effect of *Azospirillum brasilense* wild type and mutant strains altered in IAA production on wheat. *Plant and Soil,* Vol.212, pp. 153-162

24. Dobbelaere, S.; Croonenborghs, A.; Thys A.; Ptacek, D.; Vanderleyden, J.; Dutto, P.; Labandera-Gonzalez, C.; Caballero-Mellado, J.; Aguirre, J.F.; Kapulnik, Y.; Brener, S.; Burdman, S.; Kadouri, D.; Sarig, S. & Okon, Y. (2001). Responses of agronomically important crops to inoculation with *Azospirillum. Australian Journal of Plant Physiology,* Vol.28, pp.871-879

25. Dobbelaere, S.; Vanderleyden, J. & Okon, Y. (2003). Plant growth-promoting effects diazotrophs in the rhizosphere. *Crit Rev Plant Sci,* Vol.22, pp.107–149.

26. Felici, C.; Vettori, L.; Toffanin, A. & Nuti, M. (2008). Development of a strain-specific genomic marker for monitoring a *Bacillus subtilis* biocontrol strain in the rhizosphere of tomato. *FEMS Microbiol Ecol.*, Vol.65, No.2, pp.289-98.

27. Glick, B.R. (1995). The enhancement of plant growth by free-living bacteria. *Can. J. Microbiol.* Vol.41, pp. 109-117

28. Glick, B.R.; Patten, C.L.; Holguin, G. & Penrose, D.M. (1999). Biochemical and genetic mechanisms used by plant growth-promoting bacteria. Imperial College Press, London, UK.

29. Hamdia, A.B.E.; Shaddad, M.A.K. & Doaa M.M. (2004). Mechanisms of salt tolerance and interactive effects of *Azospirillum brasilense* inoculation on maize cultivars grown under salt stress conditions. *Plant Growth Regulation*, Vol.44, pp.165-174

30. Hardoim, P.R.; Van Overbeek, L.S. & Elsas, J.D.V. (2008). Properties of bacterial endophytes and their proposed role in plant growth. *Trends in Microbiology*, Vol.16, pp. 463-471

31. Hermosa, R.; Botella, L.; Alonso-Ramírez, A.; Arbona, V.; Gómez-Cadenas, A.; Monte, E. & Nicolás, C. (2011). Biotechnological applications of the gene transfer from the beneficial fungus *Trichoderma harzianum* spp. to plants. *Plant Signal Behav.*, Vol.6, No.8.

32. Honda, H. & Kobayashi, T. (2004). Large scale micropropagation system of plant cells. *Adv.Biochem. Eng. Biotechnol.*, Vol.91, pp.105-34

33. Hubbard,J.L. & Potter A.D. (2006). Managing Calico Scale (Hemiptera: Coccidae) Infestations on Landscape Trees. *Arboriculture & Urban Forestry*, Vol.32, No.4, pp138-147

34. Jousset A, Rochat L, Lanoue A, Bonkowski M, Keel C, Scheu S. (2011). Plants respond to pathogen infection by enhancing the antifungal gene expression of root-associated bacteria. *Mol Plant Microbe Interact.*, Vol.24, pp.352-358

35. Kokalis-Bourelle, N.; Vavrina, E.N.: Rosskopf, E.N. & Shelby, R.A. (2002). Field evaluation of plant growth-promoting rhizobacteria amended transplant mixes and soil solarization for tomato and pepper production in florida. *Plant Soil*, Vol.238, pp.257-266

36. Kokalis-Burelle, N.; Kloepper, J.W. & Reddy, M.S. (2006). Plant growth-promoting rhizobacteria as transplant amendments and their effects on indigenous rhizosphere microorganisms. *Applied Soil Ecology*, Vol. 31, No.1-2, pp.91-100

37. Long, H.H.; Schmidt, D.D. & Baldwin, I.T. (2008). Native Bacterial Endophytes Promote Host Growth in a Species-Specific Manner; Phytohormone Manipulations Do Not Result in Common Growth Responses. PLoS One, Vol. 3, p.2702

38. Mano, H. & Morisaki, H. (2008). Endophytic Bacteria in the Rice Plant. *Microbes and Environments*, Vol.23, pp. 109-117

39. Mercado-Blanco, J. & Bakker, P.A.H.M. (2007). Interactions between plants and beneficial *Pseudomonas* spp.: exploiting bacterial traits for crop protection. *Antonie van Leeuwenhoek,* Vol.92, pp.367-389

40. McCullaugh, M.; Utkhede, R.; Menzies, J.G.; Punja, Z.K. & Paulits, T.C. (1996). Evaluation of plant growth promoting rhizobacteria for biological control of *Pythium* root rot of cucumbers grown in rockwool and effects on yield. *Europ. J. Plant Pathol.*, Vol.102, pp.747-755

41. Meyer, J.B.; Lutz M.P.; Frapolli, M.; Péchy-Tarr, M.; Rochat, L.; Keel, C.; Défago, G. & Maurhofer, M. (2010). Interplay between wheat cultivars, biocontrol pseudomonads, and soil. *Appl Environ Microbiol.*, Vol.76, pp.6196-204

42. Molina-Favero, C.; Creus, C.M.; Simontacchi, M.; Puntarulo, S. & Lamattina, L. (2008). Aerobic nitric oxide production by *Azospirillum brasilense* Sp245 and its influence on root architecture in tomato. *Molecular plant-microbe interactions*, Vol.21, pp.1001-1009

43. Okon, Y. (1984). *Azospirillum*/Plant Associations. Y. Okon, Ed, CRC Press, Boca Raton, FL Okon, Y.; Labandera-Gonzales, C.A. (1994). Agronomic application of *Azospirillum*: an evaluation of 20 years worldwide field inoculation. *Soil Biol. Biochem.*, Vol.26, pp.1591-1601

44. Okon, Y. & Itzigshon, R. (1995). The development of *Azospirillum* as a commercial inoculant for improving crop yields. *Biotech. Advances*, Vol.13, pp.415-424

45. Pedraza, R.; Motok, J.; Tortora, M.; Salazar, S. & Díaz-Ricci J. (2007). Natural occurrence of *Azospirillum brasilense* in strawberry plants. *Plant and Soil*, Vol.295, pp.169-178 Raaijmakers, J.M.; Paulitz, T.C.; Steinberg, C.; Alabouvette, C.; Moënne-Loccoz, Y. (2009). The rhizosphere: a playground and battlefield for soilborne pathogens and beneficial microorganisms. *Plant and Soil*, Vol.321

46. Ramette A, Frapolli M, Fischer-Le Saux M, Gruffaz C, Meyer JM, Défago G, Sutra L, Moënne-Loccoz Y. (2011). *Pseudomonas protegens* sp. nov., widespread plant- protecting bacteria producing the biocontrol compounds 2,4-diacetylphloroglucinol and pyoluteorin. *Syst Appl Microbiol.*, Vol.34, pp.180-8

47. Reddy, M.S. & Rahe, J.E. (1989). Growth effects associated with seed bacterization not correlated with populations of *Bacillus subtilis* inoculant in onion seedling rhizospheres. *Soil Biol. Biochem.*, Vol.21, pp.373-378

48. Rosenblueth, M. & Martínez-Romero, E. (2006). Bacterial endophytes and their interactions with hosts. *Molecular plant-microbe interactions*, Vol.19, pp.827-837

49. Ruíz-Sánchez, M.; Armada, E.; Muñoz, Y.; García de Salamone, I.E.; Aroca, R.; Ruíz-Lozano, J.M. & Azcón, R. (2011). *Azospirillum* and arbuscular mycorrhizal colonization enhance rice growth and physiological traits under well-watered and drought conditions. *J. Plant Physiol.*, Vol.169, No.10, pp1031-7

50. Russo, A.; Moenne-Loccoz, Y.; Fedi, S.; Higgins, P.; Fenton,A.; Dowling, D.N.; O'Regan,M.&O'Gara,F.(1996).Improveddeliveryofbiocontrol *Pseudomonas* and their antifungal metabolites using alginate polymers *Appl Microbiol Biotechnol*, Vol.44, pp.740-745

51. Russo, A.; Basaglia, M.; Tola, E. & Casella, S. (2001). Survival root colonisation and biocontrol capacities of *Pseudomonas fluorescens* F113 LACZY in dry alginate micro- beads *J. Ind. Microbiol. and Biotchnol.*, Vol.27, No.6, pp.337-342

52. Russo, A.; Basaglia, M.; Casella, S. & Nuti, M.P. (2005). Biocontrol activity of *Pseudomonas fluorescens* encapsulated in alginate beads towards *Rhizoctonia solani* on cotton plants *Biotechn Progress*, Vol.21, pp.309-314

53. Russo, A.; Vettori, L.; Felici, C.; Fiaschi, G.; Morini, S. & Toffanin, A. (2008). Enhanced micropropagation response and biocontrol effect of *Azospirillum* Sp245 on *Prunus cerasifera* L. clone Mr.S 2/5 plants. *J. Biotechnol.*, Vol.134, pp.312–319

54. Ryan, R.P.; Germaine, K.; Franks, A.; Ryan, D.J. & Dowling D.N. (2008). Bacterial endophytes: recent developments and applications. *FEMS Microbiology Letters*, Vol.278, pp.1-9

55. Ryu, C.M.; Farag, M.A.; Hu, C.H.; Reddy, M.S.; Kloepper, J.W. & Pare, P.W. (2004). Bacterial volatiles induce systemic resistance in *Arabidopsis*. *Plant Physiology*, Vol.134 pp.1017-1026

56. Sánchez-Zamora M.A. and R. Fernández-Escobar (2004). Uptake and distribution of trunk injections in conifers. *Journal of Arboriculture*, Vol.30, No.2, pp.73-79

57. Somers, E.; Vanderleyden, J. & Srinivasan, M. (2004). Rhizosphere bacterial signalling: A love parade beneath our feet. *Critical Reviews in Microbiology*, Vol.30, pp.205-240

58. Steenhoudt, O. & Vanderleyden, J. (2000). *Azospirillum*, a free-living nitrogenfixing bacterium closely associated with grasses: genetic, biochemical and. *FEMS Microbiology Reviews, Vol.*24, pp.487-506

59. Surette, M.A.; Sturz, A.V.; Lada, R.R. & Nowak, J. (2003). Bacterial endophytes in processing carrots (Daucus carota L. var. sativus): their localization, population density, biodiversity and their effects on plant growth. *Plant and Soil*, Vol.253, pp.381-390

60. Sziderics, A.H.; Rasche, F.; Trognitz, F.; Wilhelm, E. & Sessitsch, A. (2007). Bacterial endophytes contribute to abiotic stress adaptation in pepper plants (Capsicum annuum L.). *Canadian journal of microbiology*, Vol.53, pp.1195-1202

61. Timmusk, S. & Wagner, E.G. (1999). The plant-growth-promoting rhizobacterium *Paenibacillus polymyxa* induces changes in *Arabidopsis thaliana* gene expression: a possible connection between biotic and abiotic stress responses. *Mol. Plant Microbe Interact*, Vol.12, pp.951-9

62. Toffanin, A.; Basaglia, M.; Ciardi, C.; Vian, P.; Povolo, S. & Casella, S. (2000). Energy content decrease and viable not culturable (VNC) status induced by oxygen limitation coupled to the presence of nitrogen oxides in *Rhizobium "hedysari"*. *Biol. Fert. Soil.*, Vol.31, No.6, pp.484-488

63. Tortora, M.L.; Díaz-Ricci, J.C. & Pedraza, R.O. (2011). *Azospirillum brasilense* siderophores with antifungal activity against *Colletotrichum acutatum*. *Arch Microbiol.*, Vol.193, No.4, pp.275-86

64. Yang, J.; Kloepper, J.W. & Ryu, C.M. (2009). Rhizosphere bacteria help plants tolerate abiotic stress. *Trends in Plant Science*, Vol.14, pp.1-4

65. Vettori, L.; Russo, A.; Felici, C.; Fiaschi, G.; Morini, S. & Toffanin, A. (2010). Improving micropropagation: effect of *Azospirillum brasilense* Sp245 on acclimatization of rootstocks of fruit tree. *Journal of Plant Interactions*, Vol.5, Par.4, pp.249-259

66. Whipps, J.M. (1997). Ecological considerations involved commercial development of biological control agents for soil-borne disease. *Modern Soil Microbiology*. Van Elsas JD, Trevors JT, Wellington EMH pp.525-533

67. Whipps, J.M. (2001). Microbial interactions and biocontrol in the rhizosphere. *J. Exp. Bot.*, Vol.52, pp.487-511

68. Zawoznik, M. S.; Ameneiros, M.; Benavides M. P.; Vázquez, S. & Groppa, M. D. (2011). Response to saline stress and aquaporin expression in Azospirillum-inoculated barley seedlings. *Applied Microbiology and Biotechnology*, Vol.90, No.4, pp.1389

69. Zhang, H.; Kim, M.S.; Sun, Y.; Dowd, S.E.; Shi, H.Z. & Pare P.W. (2008). Soil confer plant salt tolerance by tissue-specific regulation of the sodium transporter HKT1. *Molecular Plant-Microbe Interactions*, Vol.21, pp.737-744

70. Zhuang, X.; Chen, J.; Shim H. & Bai Z. (2007). New advances in plant growth-promoting rhizobacteria for bioremediation. *Environment International*, Vol.33, No.3, pp 406- 413

PART-II
Medical Biotechnology

CHAPTER-4

ASSESSMENT OF CIRCULAR RNA EXPRESSION *IN VIVO* USING THE PERMUTED INTRON-EXON TECHNIQUE

INTRODUCTION: AN OVERVIEW

The study of circular RNA (circRNA) expression has emerged as a significant area of research in molecular biology and genomics. CircRNAs, characterized by their covalently closed loop structure, have been found to play essential roles in gene regulation and various biological processes. To understand their expression patterns and functions, innovative techniques are continually being developed. One such method that has garnered attention is the "Permuted Intron-Exon" technique, which allows for the in vivo assessment of circRNA expression. This approach offers a unique perspective on circRNA dynamics, shedding light on their potential significance in cellular processes and disease mechanisms.

Functional RNAs, *e.g.*, aptamers (Lee *et al.*, 2005; Que-Gewirth & Sullenger, 2007), ribozymes (Malhbacher *et al.*, 2010), antisense oligonucleotides (Hnik *et al.*, 2009), and double-stranded RNA (dsRNA) (Watts & Corey, 2010), hold promise for use as RNA drugs in the near future. However, the linear form of RNA without chemical modifications is rapidly degraded in both human serum and cell extracts due to endogenous nucleases. Therefore, it will likely be necessary to chemically modify these RNA drugs (Pestourie *et al.*, 2005; Watts *et al.*, 2008) to protect them from nuclease-dependent degradation. In fact, the recently developed aptamer drug pegaptanib sodium (Macugen®; Pfizer) for use against macular degradation consists of 2'-F- or 2'-OCH3-substituted nucleotides, thus preventing its rapid degradation in the ocular environment. Although at present it is the only commercially available RNA drug, we infer from the selling price of Macugen® that similar novel chemically modified RNA drugs are likely to be expensive because production of a chemically modified RNA molecule and scaling up the production yield of the RNA are expensive in principle. Therefore, the development of not only inexpensive but also durable RNA drugs will facilitate the widespread use of easily administered RNA drugs. To address the problems outlined above, our research group has considered *in vivo* circular RNA expression as a model for inexpensive RNA drug production because circular RNA molecules are resistant to exoribonucleases without any chemical modifications under cellular conditions. Therefore, the circular form of RNA would be a promising RNA drug candidate without requiring chemical modification.

Circular RNA can be produced both *in vitro* and *in vivo* using two methodologies. The first makes use of ligase to ligate both ends of the linear form of RNA transcripts (Chen & Sarnow, 1998; Beaudry & Perreault, 1995), while the second uses a spontaneous group I intron self- splicing system, designated as the permuted intron-exon (PIE) method (Puttaraju & Been, 1992). The latter technique is the only methodology available for *in vivo* circular RNA production because it has no requirement for proteinaceous components, such as ligases. Therefore, the PIE method is a promising economical methodology for producing circular RNA drugs. In this chapter, we describe our circular streptavidin RNA aptamer expression by the PIE method as a model for RNA drug production (Umekage & Kikuchi, 2006, 2007, 2009a, 2009b). Then, we discuss our recent improvements in the circular RNA expression technique, *i.e.*, the **t**andem **o**ne-way transcription of **P**IE (TOP) method, to achieve higher yields of *in vivo*circular RNA expression. In this system, we achieved production of approximately 0.19 mg of circular RNA from a 1-L culture of the *Escherichia coli* strain JM101Tr. To our knowledge, this is the highest circular RNA expression yield reported to date. Finally, we will discuss *in vivo* circular RNA expression by the marine phototrophic bacterium *Rhodovulum sulfidophilum*. This bacterium produces RNA both within the cell and in the culture medium in nature and produces no RNases in the culture medium (Suzuki *et al.*, 2010), whereas strong RNase activity is observed in the culture medium of a conventional *E. coli* strain that can be used for RNA production. Therefore, we speculated that *Rdv. sulfidophilum* would be a suitable strain for RNA production in the culture medium bypassing the total RNA extraction procedure to break the cell membrane, such as the acid guanidinium thiocyanate phenol chloroform (AGPC) method (Chomczynski & Sacchi, 1987).

SELF-SPLICING OF GROUP I INTRONS AND THE PERMUTED INTRON-EXON (PIE) APPROACH

Group I intron self-splicing RNA from the ciliate *Tetrahymena* was the first discovered ribozyme (Cech *et al.*, 1981). The group I intron sequence has been widely detected in eukaryotes (Cech *et al.*, 1981), prokaryotes (Xu *et al.*, 1990) and some bacteriophages (Ehrenman *et al.*, 1986). This self-splicing does not require any proteinaceous components but does require the presence of Mg^{2+} and guanosine nucleotides (Cech *et al.*, 1981). After self-splicing, the concomitant ligation of the two exons takes place (Fig. 1). This self-splicing mechanism consists of a well-defined two-step transesterification mechanism, and the sequential self-splicing steps take place after formation of the higher-order intron architecture. In the first step, a guanosine nucleotide attacks the phosphate at the 5' splicing site and scission occurs between the upstream exon and the intron, and the guanosine nucleotide is then ligated to the 5' side of the intron. Next, the hydroxyl group of the 3' end of the upstream exon shows nucleophilic attack of the downstream splicing site of the phosphorus, and intron circularisation and exon ligation occur. Therefore, it is assumed that both the 5' end of the 5' half exon and 3' end of the 3' half exon are somehow ligated before self-splicing occurs, and the resulting spliced exon product has a circular conformation. Several biochemical (Galloway-Salbo *et al.*, 1990) and structural investigations of group I intron self-splicing (Stahley & Atrobel, 2006) indicated that the peripheral region of the intron architecture and internal open reading frame (ORF) sequence does not participate in formation of the intron architecture and the self-splicing event mentioned above. Theses investigations allowed us to permute the order of the intron and exon sequence without distorting the tertiary structure of the permuted intron architecture. Puttaraju & Been (1992) first reported that circular permutation of the group I intron from both the *Anabena* pre- tRNA intron and the *Tetrahymena* intron generated a circular RNA exon *in vitro*. Another PIE from the T4 phage group I intron was later shown to be applicable for generating the circular exon (Ford & Ares, 1994). As the exon sequence does not participate in the self- splicing reaction, the exon sequence in the PIE sequence is replaced with another foreign sequence. Based on this concept, several circular RNAs have been developed by the PIE method, *i.e.*, the tat-activated response (TAR) RNA (Puttaraju & Been, 1995; Bohjanen *et al.*, 1996; Bohjanen *et al.*, 1997), rev responsive element (Puttaraju & Been, 1995), HDV ribozyme (Puttaraju *et al.*, 1993; Puttaraju & Been, 1996), tRNA (Puttaraju & Been, 1992), *Bacillus subtilis* PRNA (Puttaraju & Been; 1996), mRNA encoding GFP (Perriman & Ares, 1998), yeast actin exon (Ford & Ares, 1994), hammerhead ribozyme (Ochi *et al.*, 2009), and streptavidin RNA aptamer (Umekage & Kikuchi, 2006, 2007, 2009a, 2009b) (Table 1).

These results confirmed the availability of the PIE method to yield a wide variety of circular RNAs. As circularisation is driven only by magnesium ions and guanosine nucleotides, the circularisation of RNA in E. coli (Puttaraju & Been, 1996; Perriman & Ares, 1998; Umekage & Kikuchi, 2007, 2009a, 2009b) and Saccharomyces cerevisiae (Ford & Ares, 1994; Puttaraju & Been, 1996) have been demonstrated, and the in vivo expressed circular RNAs reported above are functionary active.

Our group also showed that the circular streptavidin RNA aptamer produced both in vitro and in E. coli (Umekage & Kikuchi, 2006, 20007, 2009a, 2009b), and the expressed circular streptavidin RNA aptamer was purified from the total RNA fraction by the solid-phase DNA probe method (Suzuki et al., 2002). This is the first evidence that both in vitro and in vivo circularisation of an RNA aptamer and the in vivo circularised RNA generated by the PIE method can be purified (Umekage & Kikuchi, 2009a).

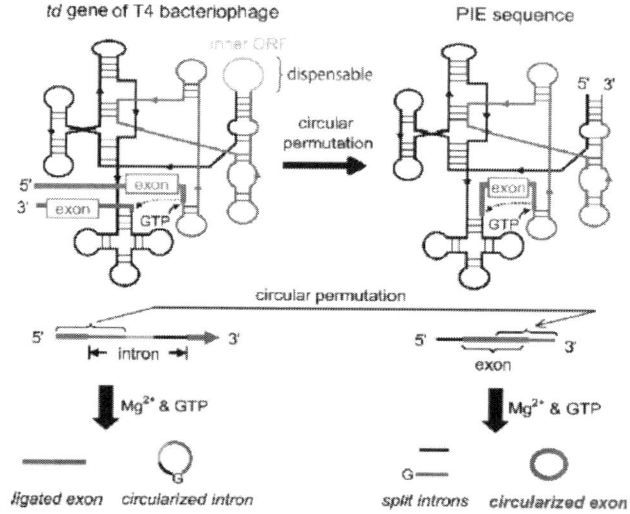

Fig. 1. Group I intron self-splicing RNA and the permuted intron-exon (PIE). Predicted secondary structure of the *td* group I intron sequence (upper left side) and the PIE sequence (upper right side). The pink and black lines show the intron sequence and the green line indicates the internal ORF sequence. Coloured horizontal lines shown in the middle of this figure illustrate the circular permutation of the *td* intron. In the normal *td* intron (left side), after transesterification, the exon sequence is ligated. In the permuted *td* intron (right side), the exon sequence is circularised and a split intron sequence appears. Circularisation of the exon sequence by the PIE method also requires magnesium ions and guanosine nucleotides.

On the other hand, we found that this circularisation affected the original activity of the linear form of functional RNAs. The dissociation constant (*K*d) of the circular streptavidin RNA aptamer increased (Umekage & Kikuchi, 2009a) and ribozyme activity of the hammerhead ribozyme decreased (Ochi *et al.*, 2009). These observations suggest that structural constraints were induced by circularisation. Although it is also important to take into consideration the circular RNA structure before constructing the PIE sequence, it is difficult to predict the tertiary structure of the circularised RNA molecule.Therefore, optimisation of the circularised sequence would be required involving randomising the spacer sequence, inserting the poly(A) sequence, *etc.* We succeeded in recovering the functional activity of the circular hammerhead ribozyme by adding a poly(A) spacer between the ribozyme sequence and the indispensable linkage sequence derived from the exon sequence for circularisation by the PIE method (Ochi *et al.*, 2009), and the recovered ribozyme activity of the circular hammerhead ribozyme was dependent on the length of the poly(A) spacer (Ochi *et al.*, 2009).

Table 1. Summary of circular RNA production. "Category of circularised RNA": Source of the circularised exon sequence. "Intron": Source of the intron sequence used for constructing the PIE sequence. "Expression": Circular RNA production *in vitro* or *in vivo* (the expression host strain is listed).

Category of circularised RNA	Intron	Expression	Reference
tRNA exon from *Anabaena* PCC7120	Anabaena & Tetrahymena	*in vitro*	Puttaraju & Been, 1992
HDV ribozyme	Anabaena	*in vitro*	Puttaraju *et al.*, 1993; Puttaraju & Been, 1996
td exon from T4 phage	T4 phage	*in vitro* & *E. coli* DH5α & *S. cerevisiae* IH1097	Ford & Ares, 1994
actin ORF from yeast	T4 phage	*in vitro* & *E. coli* DH5α& *S. cerevisiae* IH1097	Ford & Ares, 1994
Rev-responsive element RNA	Anabaena	*in vitro*	Puttaraju & Been, 1995
tat-activated response RNA	Anabaena	*in vitro*	Puttaraju & Been, 1995 Bohjanen *et al.*, 1996; Bohjanen *et al.*, 1997
B. subtilis PRNA	Anabaena	*in vitro* & *E. coli* BL21(DE3)	Puttaraju & Been, 1996
GFP ORF	T4 phage	*in vitro* & *E. coli* BL21(DE3)	Perriman & Ares, 1998
streptavidin aptamer	T4 phage	*in vitro* & *E. coli* JM109(DE3) & *E. coli* JM101Tr	Umekage & Kikuchi, 2006, 2007, 2009a, 2009b
hammerhead ribozyme	T4 phage	*in vitro*	Ochi *et al.*, 2009
streptavidin aptamer	T4 phage	*Rdv. sulfidophilum* DSM 1374^T	this study (see 2.4)

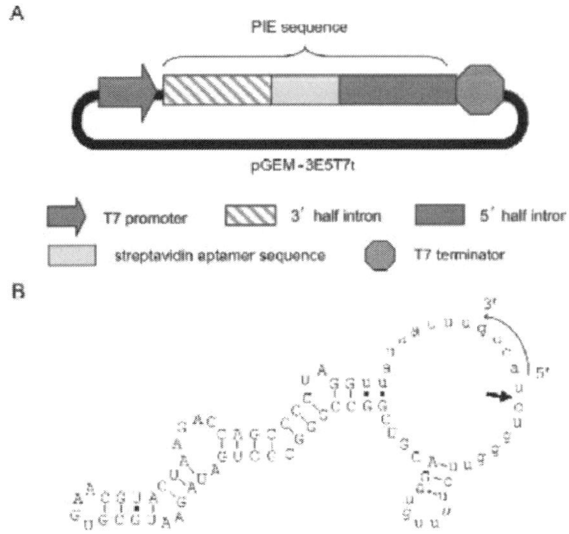

Fig. 2. Schematic representation of the plasmid pGEM-3E5T7t, and the predicted secondary structure of the circular streptavidin RNA aptamer produced from the plasmid. (A) The figure shows pGEM-3E5T7t. The PIE sequence is located between the T7 promoter sequence and the T7 terminator sequence. The PIE sequence consists of the 3' half intron, streptavidin aptamer sequence and 5' half intron sequence. The intron sequence is derived from the *td* intron of bacteriophage T4. (B) Predicted secondary structure of the circular streptavidin RNA aptamer. Upper and lower case letters indicate the aptamer sequence and exon sequence derived from the original exon sequence of the *td* gene, respectively. Streptavidin RNA is derived from the S1 aptamer reported by Srisawat & Engelke (2001). The thick arrow represents the self-ligated junction. The thin arrow (anticlockwise) indicates the orientation of the circular RNA from 5' to 3'. The thin line and black dot represent Watson– Crick base pairing and G-U Wobble base pairing, respectively.

In vivo circular streptavidin RNA aptamer expression by the PIE method

To demonstrate *in vivo* circular RNA expression, our group designed the circular streptavidin RNA aptamer as a model RNA dug and the PIE sequence for production of the circular streptavidin RNA aptamer. The PIE sequence consists of the 3' half intron, aptamer sequence and 5' half intron sequence in this order (Fig. 2A), with omission of the internal ORF sequence (Fig. 1, shown as the green-coloured line) in the *td* intron for this PIE construction that does not participate in the self-splicing reaction of the intron. We constructed a circular streptavidin RNA aptamer expression vector, pGEM-3E5T7t, which consists of three parts: the T7 RNA promoter, the PIE sequence and the T7 terminatorsequence in the multicloning site in the standard cloning vector, pGEM-3Z (Promega) (Fig. 2A). The streptavidin RNA aptamer sequence in the PIE sequence was derived from the S1 aptamer sequence reported by Srisawat & Engelke (2001), and the intron sequence derived from the *td* intron of bacteriophage T4. Both ends of the resulting circular streptavidin RNA aptamer sequence were ligated by an indispensable linker sequence derived from a partial exon sequence of the *td* gene of bacteriophage T4 (Fig. 2b, lower case).

To express the circular streptavidin RNA aptamer *in vivo*, pGEM-3E5T7t was transformed into JM109(DE3) (*endA*1 *recA*1, *gyrA*96, *thi*-1, *hsdR*17 (rK^-, mK^+), *relA*1, *supE*44, ?(*lac-proAB*), [F', *traD*36, *proAB*, *lacI*qZ, M15], (DE3)) (Promega), which encodes an isopropyl-β-D- thiogalactopyranoside (IPTG)-inducible T7 RNA polymerase on its genomic sequence. The full-growth culture of JM109(DE3) harbouring pGEM-3E5T7t was transferred into 1 L of fresh LB broth and cultured until the optical density at 600 nm (OD600) reached 0.7 at 30?C. Then, IPTG was added to a final concentration of 0.4 mM and circular RNA expression was induced by cultivation for 2 h at 30?C with vigorous shaking. After extraction of the total RNA, which included the circular RNA, using the AGPC method to break the cell membrane (Chomczynski &Sacchi, 1987), the circular RNA expression in the total RNA fraction was monitored by ethidium bromide staining and Northern blotting analysis.

We performed two-dimensional (2D) denaturing polyacrylamide gel electrophoresis (2D PAGE) (Schumacher, 1983; Feldstein, 2007) to monitor whether the circular RNA was present in the recovered total RNA fraction. This 2D-electrophoresis is based on the differences in migration behaviour between linear and circular RNA under denaturing gel conditions. Unlike linear RNA, migration of a circular RNA molecule of the same length varies with the acrylamide and/or bis-acrylamide concentration on denaturing polyacrylamide electrophoresis (PAGE). Therefore, after 2D denaturing gel- electrophoresis, the linear RNA fraction has migrated in the diagonal direction, whereas the circular RNA appears beside the diagonal line of the linear molecule. Both ethidium bromide staining and Northern blotting analysis showed a single spot beside the diagonal migration line, indicating that the circular RNA was present in the total RNA fraction. This spot was eluted and subjected to partial alkaline digestion. The partially digested nicked circular RNA migrated faster than the intact circular molecule on denaturing 10% PAGE, confirming that the eluted RNA was a circular form. The degradation products of circular RNA were not detected by Northern blotting analysis. This clearly showed that the circular RNA expressed in *E. coli* cells was protected against exonuclease-induced degradation, such as that induced by ribonuclease II (RNase II) (Frazão *et al.*, 2006). The expression level of the circular RNA aptamer was determined to be 2.5 ± 0.46 ng per 1 μg of total RNA by Northern blotting analysis. The yield of the circular RNA aptamer in that total RNA was estimated to be approximately 24 µg.

Next, we developed a circular RNA purification method for future inexpensive and economical RNA production and purification. We showed that the circular streptavidin RNA aptamer was successfully purified with the solid-phase DNA probe technique (Suzuki *et al.*, 2002). To purify the circular RNA produced by the PIE method, we designed a 5' biotinylated DNA probe that can hybridise with the circular RNA and the circular RNA- DNA hybrid can be easily trapped using a streptavidin-coated column. The trapped circular RNA-DNA hybrid is also denatured with a high concentration of urea (7 M) solution and the circular RNA is eluted. We can also choose the elution buffer such that the DNA probestill binds to the solid-phase and it can be reused for another round of RNA purification (data not shown). Using a streptavidin-coated column (GE Healthcare), the circular streptavidin RNA aptamer was eluted under denaturing conditions and yielded 21 µg of the circular RNA (about 88% recovery) from 1 L of *E. coli* cell culture. Electrophoretic mobility shift assay (EMSA) also showed that the purified circular streptavidin RNA aptamer from JM109(DE3) retained its binding properties toward streptavidin.

To verify the suitability of the circular RNA for future RNA therapeutic uses, we measured the half-life of the purified circular RNA aptamer in HeLa cell extracts as a model of intracellular conditions. The estimated half-life of the purified circular streptavidin RNA aptamer was at least 1,386 min, while that of the S1 aptamer, which is the linear form of the streptavidin RNA aptamer, was 43 min. These observations suggested that the circular RNA escapes exoribonuclease-dependent RNA degradation under intracellular conditions. However, the circular RNA degraded completely within 15 s in 25% human serum. This is reasonable because human serum contains the RNaseA family ribonucleases (Haupenthal *et al.*, 2006; Haupenthal *et al.*, 2007; Turner *et al.*, 2007). These findings indicated that the circular RNA would be useful under cellular conditions only when delivered into the cell in a precise manner, *e.g.*, by using cationic liposomes (Sioud & Sorensen, 2003; Sorensen *et al.*, 2003) or virus vector systems (Mi *et al.*, 2006), to prevent RNaseA family ribonuclease-dependent degradation.

Expressing a Circular Streptavidin RNA Aptamer in Vivo Using the Pie Method on a Constitutive Basis

We then considered the constitutive circular RNA expression, as the previous expression procedure requires monitoring of the optical density for optimal IPTG induction (see 2.1). For constitutive expression of the RNA sequence in *E. coli*, we followed the procedure of Ponchon & Dardel (2004). They reported that the M3 vector containing the strong constitutively active lipoprotein (*lpp*) promoter, which is one of the strongest promoters in *E. coli* (Movva *et al.*, 1978; Inoue *et al.*, 1985), is applicable for *in vivo* RNA expression in the *E. coli* strain JM101Tr (Δ(*lac pro*), *supE*, *thi*, *recA56*, *srl-300.*:Tn*10*, (*F'*, *traD36*, *proAB*, *lacIq*, *lacZ*, Δ*M15*)). In addition, total RNA expression in JM101Tr is higher than that of JM109(DE3) (our unpublished observation).

Before constructing the constitutive PIE expression plasmid, we replaced the original tRNA^{Met} sequence between the *lpp* promoter and *rrn*C terminator sequence in the M3 vector with the PIE sequence from pGEM-3E5T7t. The resulting expression vector is designated as pM3-3E5. The PIE sequence was amplified from the PIE sequence in pGEM-3E5. After transformation of pM3-3E5 into the JM101Tr strain, cell density (OD600) was measured at several time points during cultivation and 1-mL aliquots were collected from 200 mL of 2×YT medium. Total RNA was recovered by ISOGEN (Nippon Gene) and Northern blotting analysis was performed. At various time points in culture from early logarithmic phase to stationary phase, circular RNA was visible in each lane on electrophoretic analysis even with ethidium bromide staining. The presence of circular RNA, but not the nicked form, was clearly detected on Northern blotting analysis and the amount of circular RNA increased with cell growth. These results suggested that the *lpp* promoter was active and drove expression of the PIE sequence without any induction. The stain JM101Tr is positive for ribonucleases, such as ribonuclease II (Frazão *et al.*, 2006). Therefore, these observationsindicated that the circular RNA also accumulated in the *E. coli* JM101Tr strain, escaping degradation by exonucleases as seen in the previous expression system described in Section 2.1. The resulting yield of circular RNA after 18 h of cultivation at 30°C was estimated to be3.6 ± 0.15 ng per 1 µg of total RNA, which was approximately 1.5-fold higher than that of the previous method (Umekage & Kikuchi, 2009a) (see 2.1). These observations indicated effective constitutive circular RNA expression in this system.

Enhancing Circular RNA Expression through the Utilization of the Tandem One-Way Transcription of Pie (Top) Technique

To augment the circular RNA expression in *E. coli*, we developed the TOP (**t**andem **o**ne-way transcription of **P**IE) technique, which is a simple methodology for increasing the copy number of the PIE sequence in a single plasmid. The TOP technique is shown schematically in Fig. 3A. With this technique, it is easy to amplify the copy number by sequential insertion of the transcriptional unit in a single plasmid (Fig. 3B). First, we amplified the transcriptional unit, which consists of the *lpp* promoter, PIE sequence and *rrn*C terminator in pM3-3E5 (see Section 2.1) with both the 5' flanking sequence containing *Kpn*I–*Xho*I sites and the 3' flanking sequence containing a *Sal*I site. Next, we digested the amplified sequence with *Kpn*I and *Sal*I, and the resulting fragment was inserted into the M3 plasmid double- digested with *Kpn*I and *Xho*I. The digested *Xho*I site on the M3 plasmid and the *Sal*I site on the amplified fragment can hybridise with mutual 3' protruding ends of the palindromic TCGA sequence, and the resulting ligated fragment forms the sequence GTCGAG, which can be digested with neither *Xho*I nor *Sal*I (Fig. 3B). Therefore, the inserted sequence is as follows: 5'-*Kpn*I-*Xho*I-*lpp* promoter-PIE sequence-*rrnC* terminator sequence-GTCGAG site-3' (Fig. 3C). Thus, the subsequent transcriptional unit can be inserted at the *Kpn*I–*Xho*I site. We constructed four series of pTOP vectors using M3 designated as pTOP(I), pTOP(II), pTOP(III) and pTOP(IV) in parallel with the number of inserted transcriptional units.

This pTOP plasmid has a constitutive *lpp* promoter and therefore the constitutive expression of the PIE sequence in JM101Tr is expected, similar to that using the constitutive expression plasmid pM3-3E5 described in Section 2.2. To demonstrate the availability of the TOP technique, we then analysed the circular streptavidin RNA aptamer expression in *E. coli* by Northern blotting analysis and we detected that the circular RNA expression was expressed in all pTOP vectors (pTOP(I), (II), (III) and (IV)) (Fig. 3D).

As shown in the Fig. 3D., the circular RNA expression increased until two tandem insertions of the PIE, and the expression yields were almost the same using pTOP(II) and pTOP(III) (Table 2). These results indicated that the TOP system is a potentially useful and simple methodology for increasing circular RNA expression in *E. coli*. The circular RNA expression using pTOP(II) was estimated to be about 9.7 ± 1.0 ng per 1 µg of total RNA after 18 h of cultivation and this yield was approximately 2.7-fold higher than that of the expression procedure using the pM3-3E5 system as described in Section 2.2. In addition, the circular RNA expression in 1 L of culture medium was estimated to be approximately 0.19 mg, which is the highest yield of circular RNA expression in *E. coli* reported to date. In contrast, expression of the circular RNA dropped dramatically when using pTOP(IV); the reason for this drop in expression level is not yet clear. To address this problem, we collected pTOP(IV) after 18 h of cultivation in JM101Tr and the plasmid was single-digested with *Hind*III and then subjected to 1% agarose gel electrophoresis. A few single-digested pTOP(IV) fragments

Showed unexpected migration behaviour (data not shown), suggesting that it was difficult for pTOP(IV) to undergo replication in JM101Tr during 18 h of cultivation. Although the expressional host strain JM101tr has the *recA56* mutant, which results in defects in recombination, this genetic mutation is not sufficient to confer stability on pTOP(IV). This instability of pTOP(IV) in JM101Tr indicates the necessity for optimisation of the TOP technique for further augmentation of circular RNA expression; *e.g.*, optimisation of the intervening sequence between the two transcriptional units, considering the direction of transcription, changing the expressional host to a strain lacking another gene that results in defective recombination, such as *sbcB*, *C* or another *rec* gene (Palmer *et al.*, 1995), and optimising the copy number of PIE sequences in the single transcriptional unit to avoid accumulation of *lpp* promoter in the single plasmid.

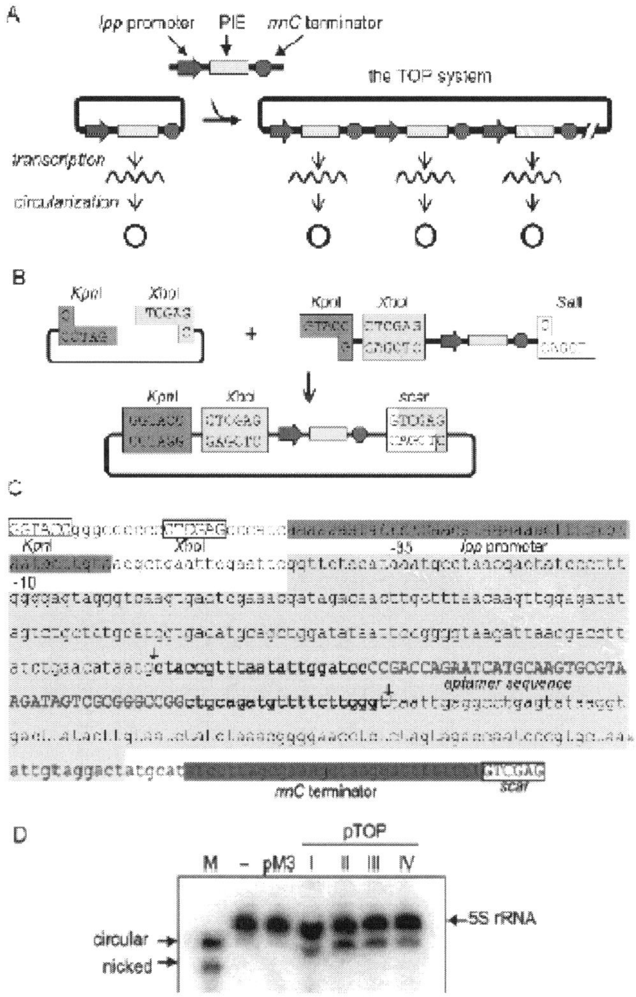

Fig. 3. Construction of the pTOP vectors, and the availability of the TOP method for generating circular RNA in JM101Tr. (A) Outline of the TOP method. (B) Illustration of sequential insertion of the PIE sequence into the same plasmid. First, *Kpn*I and *Xho*I double digested plasmid and *Kpn*I and *Sal*I double digested insertion sequence were prepared. Both the *Kpn*I site from the plasmid and the insertion sequence are ligated and the *Xho*I-digested site in the plasmid and the *Sal*I-digested site in the insertion sequence are ligated, resulting in the sequence GTCGAG at the 3' side of the inserted site. (C) Nucleotide sequence of one unit of the TOP system. Arrows represent splicing positions of this PIE sequence: yellow, the PIE sequence; blue box, *lpp* promoter sequence; italicised sequence in the blue box, −35 and −10 regions of the *lpp* promoter; red upper case letters,

aptamer sequence and *rrn*C terminator sequence; lower case letters in the yellow region, intron sequence of the *td* gene; bold lower case letters, exon sequence of the *td* gene; bold, circularised sequence; boxed sequence, ligated sites. (D) Northern blotting analysis of the circular RNA expression by each pTOP series. Total RNA derived from JM101Tr containing the *in vivo* expressed circular streptavidin RNA aptamer was fractionated by 10% denaturing PAGE. In addition, the circular RNA expression monitored using the ^{32}P-labelled complementary oligo-DNA probe of the aptamer sequence (5'-CCAATATTAAACGGTAGACCCAAGAAAACATC-3'). 5S rRNA was monitored as an internal control using the ^{32}P-labelled complementary oligo- DNA probe sequence (5'- GCGCTACGGCGTTCACTTC-3'). Arrows indicate the migration positions of the circular RNA (circular), nicked RNA (nicked) and 5S rRNA. Circular RNA control marker (M) was prepared by *in vitro* transcription (Umekage & Kikuchi, 2009a). "-", Total RNA from JM101Tr; "M3", negative control of the TOP system lacking the PIE sequence. Roman numerals I, II, III and IV represent the total RNA from JM101Tr harbouring pTOP(I), pTOP(II), pTOP(III) and pTOP(IV), respectively.

Expression of circular RNA in the marine phototrophic bacterium *Rhodovulum sulfidophilum*

Finally, we would like to discuss our new project to develop an economical and efficient method for RNA production using the marine phototrophic bacterium *Rdv. sulfidophilum* (Fig. 4), taking advantage of its unique characteristics in that nucleic acids are produced extracellularly (Suzuki *et al.*, 2010). In addition this bacterium produces no RNases in the culture medium (Suzuki *et al.*, 2010). Although the mechanism of extracellular RNA production by this bacterium has not been fully characterised, this extracellular RNA expression system represents an economical and efficient methodology for RNA production as it is only necessary to collect the culture medium containing extracellularly produced RNA and purify the RNA of interest with a column bypassing the need for a cell extraction procedure using phenol or various other extraction reagents to rupture the cell membrane.

We began by constructing the engineered circular RNA expression plasmid, pRCSA, based on the broad-host range plasmid pCF1010 (Lee & Kaplan, 1995). The PIE sequence was amplified from pGEM-3E5T7t, and the *rrn*A promoter and *puf* terminator sequence were amplified from the genomic DNA of *Rdv. sulfidophilum* DSM 1374^{T} (Hansen & Veldkamp, 1973; Hiraishi & Ueda, 1994). The resulting amplified DNA fragments were inserted into pCF1010 to give pRCSA, which was then transformed into *Rdv. sulfidophilum* DSM 1374^{T} byconjugation using the mobilising *E. coli* strain S-17 as a plasmid donor (Simon *et al.*, 1983). The heat shock transformation method can also be used (unpublished observation) (Fornari& Kaplan, 1982). The transformed *Rdv. sulfidophilum* DSM 1374^{T} was cultured under anaerobic conditions under incandescent illumination (about 5,000 lx) for 12 – 16 h at 25°C in PYS-M medium (Nagashima *et al.*, 1997, Suzuki *et al.*, 2010). Cultured cells were harvested and the total intracellular RNA was extracted with the AGPC method. The estimated yield of the intracellular circular RNA was approximately 1.3 ng per 1 L of culture medium by Northern blotting analysis. On the other hand, the circular RNA expression in the culture medium was barely detected by Northern blotting analysis; however, RT-PCR analysis demonstrated the existence of circular RNA in the cultured medium (data not shown). At present, neither intracellular nor extracellular expression of the circular RNA aptamer can be achieved at practical levels for economic and efficient circular RNA expression, and the overall improvement of RNA expression using this bacterium is strongly promoted.

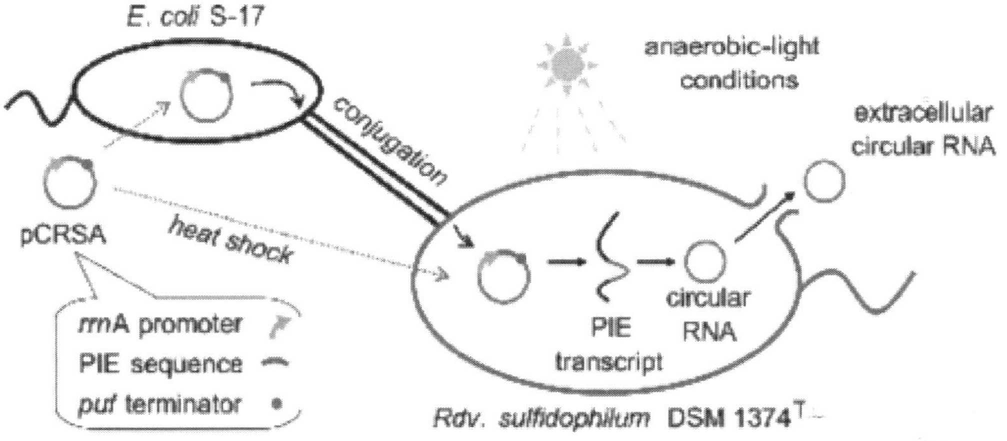

Fig. 4. Overview for circular RNA expression using *Rdv. sulfidophilum* DSM 1374T. Circular RNA expression plasmid, pRCSA, was transformed into *Rdv. sulfidophilum* DSM1374T byconjugationusingthemobilising *E. coli* strain S-17 (Simon *et al.*, 1983) or by direct transformation using the heat shock method (Fornari & Kaplan, 1982). The transformed *Rdv. sulfidophilum* was grown under anaerobic-light conditions. The PIE sequence in pRCSA was transcribed with the endogenous RNA polymerase and circular RNA was generated from the PIE sequence. The circular RNA produced inside the cell was released extracellularly into the culture medium.

AT A GLANCE

Our circular streptavidin RNA aptamer expression system described in Sections 2.1, 2.2 and 2.3 is summarised in Table 2. To our knowledge, the TOP method is the most effective means of circular RNA expression, and the *in vivo* constitutive RNA expression is suitable for circular RNA expression, as the spontaneously expressed circular RNA can exist stably within the cell avoiding endogenous exoribonuclease-dependent degradation. By using the circular streptavidin RNA aptamer expression plasmid pTOP(II) and *E. coli* JM101Tr as a host stain, the expression yield of the circular RNA was estimated to be approximately 0.19 mg per 1 L of culture. Although the TOP method requires further improvement to augment circular RNA expression, it is notable that this method easily increased the level of circular RNA expression by simple multiplying the copy number of transcription units in the singleplasmid. Therefore, we assumed that the TOP strategy will be more effective especially using a low copy number plasmid, because increasing the plasmid copy number by genetic engineering is not easy. We also presented the solid-phase DNA probe method as a simple purification procedure for *in vivo* expressed circular RNA, because this technique does not require electrophoresis for purifying the circular RNA (Umekage & Kikuchi, 2009a).

The most remarkable advantage of circularising functional RNAs is protection from exoribonuclease-induced degradation without the need for chemical modifications, such as use of 2'-protected nucleotides (*e.g.*, 2'-fluoro, 2'-*O*-methyl, LNA) (Schmidt *et al.*, 2004; Burmeister *et al.*, 2005; Di Primo *et al.*, 2007; Pieken *et al.*, 1991) or phosphorothioate linkages (Kang *et al.*, 2007). Although chemical synthesis of RNA molecules is currently the main methodology used for synthetic RNA production, the *in vivo* circular RNA production technique described in this chapter is a promising method for future RNA drug production because it is both economical and the product can be purified simply. In addition, circular RNA without any chemical modification would be safer than chemically modified RNA for therapeutic human use.

This PIE method can be applied in any species because it requires only magnesium ions and guanosine nucleotides. However, the expression of circular RNA inside human cells or other mammalian cells in culture has not been examined. Therefore, we are currently examining circular RNA expression in human cells based on this method for future development of gene therapy methodologies. We assume that PIE transcription and concomitant RNA circularisation take place in the nucleus, and therefore the circular functional RNA (including aptamers, ribozymes, dsRNA *etc.*) expression within the nucleus will represent a novel gene regulation method targeting nuclear events, such as transcription (Battaglia *et al.*, 2010), RNA splicing (van Alphen *et al.*, 2009), telomere repairing (Folini *et al.*, 2009) and chromatin modification (Tsai *et al.*, 2011).

Table 2. Summary of circular RNA expression. "IPTG" and "constitutive" indicate that the circular RNA expression was induced by the addition of IPTG and constitutive expression of the circular RNA by the constitutive *lpp* promoter, respectively. "Yield" represents the circular RNA expression yield (ng) per 1 μg of total RNA recovered from the harvested cells. The data include standard deviations (±), which were derived from three independent experiments (*n* = 3).

Plasmid	Host strain	Expression	Yield (ng/μg)	Reference
pGEM-3E5T7t	JM109(DE3)	IPTG	2.5 ± 0.46	Umekage & Kikuchi, 2009a
pM3-3E5	JM101Tr	constitutive	3.6 ± 0.15	Umekage & Kikuchi, 2009b
pTOP(I)	JM101Tr	constitutive	5.0 ± 1.5	this study
pTOP(II)	JM101Tr	constitutive	9.7 ± 1.0	this study
pTOP(III)	JM101Tr	constitutive	9.0 ± 1.8	this study
pTOP(IV)	JM101Tr	constitutive	1.8 ± 0.70	this study

REFERENCES

1. Battaglia, S.; Maguire, O. & Campbell, M.J. (2010). Transcription factor co-repressors in cancer biology; roles and targeting. *Int. J. Cancer*, Vol. 126, No. 11, pp. 2511-2519

2. Beaudry, D. & Perreault, J.P. (1995). An efficient strategy for the synthesis of circular RNA molecules. *Nucleic Acids Res.*, Vol. 23, pp. 3064-3066

3. Bohjanen, P.R.; Liu, Y. & Garcia-Blanco, M.A. (1997). TAR RNA decoys inhibit tat-activated HIV-1 transcription after preinitiation complex formation. *Nucleic Acids Res.*, Vol. 25, pp. 4481-4486

4. Bohjanen, P.R.; Colvin, R.A.; Puttaraju, M.; Been, M.D. & Garcia-Blanco, M.A. (1996). A small circular TAR RNA decoy specifically inhibits Tat-activated HIV-1 transcription. *Nucleic Acids Res.*, Vol. 24, pp. 3733-3738

5. Burmeister, P.E.; Lewis, S.D.; Silva, R.F.; Preiss, J.R.; Horwitz, L.R.; Pendergrast, P.S.; McCauley, T.G.; Kurz, J.C.; Epstein, D.M.; Wilson, C. & Keefe, A.D. (2005). Direct in vitro selection of a 2'-O-methyl aptamer to VEGF. *Chem. Biol.*, Vol. 12, pp. 25-33

6. Cech, T.R.; Zaug, A.J. & Grabowski, P.J. (1981). *In vitro* splicing of the ribosomal RNA precursor of Tetrahymena: involvement of a guanosine nucleotide in the excision of the intervening sequence. Cell, Vol. 27, 3 Pt 2, pp. 487-496

7. Chen, C.Y. & Sarnow, P. (1998). Internal ribosome entry sites tests with circular mRNAs.*Methods Mol. Biol.*, Vol. 77, pp. 355-363

8. Chomczynski, P. & Sacchi, N. (1987). Single-step method of RNA isolation by acid guanidinium thiocyanate-phenol-chloroform extraction. *Anal. Biochem.* Vol. 162, pp. 156-159

9. Di Primo, C.; Rudloff, I.; Reigadas, S.; Arzumanov, A.A.; Gait, M.J. & Toulme, J.J. (2007). Systematic screening of LNA/2'-O-methyl chimeric derivatives of a TAR RNA aptamer. *FEBS Lett.*, Vol. 581, pp. 771-774.

10. Ehrenman, K.; Pedersen-Lane, J.; West, D., Herman, R.; Maley, F. & Belfort, M. (1986). Processing of phage T4 td-encoded RNA is analogous to the eukaryotic group I splicing pathway. *Proc. Natl. Acad. Sci. USA*, Vol. 83, No. 16, pp. 5875-5879

11. Feldstein, P.A.; Levy, L.; Randles, J.W. & Owens, R.A. (1997). Synthesis andtwo-dimensional electrophoretic analysis of mixed populations of circular and linear RNAs. *Nucleic Acids Res.*, Vol. 25, pp. 4850-4854

12. Folini, M.; Gandellini, P. & Zaffaroni, N. (2009). Targeting the telosome: Therapeutic implications. *Biochim. Biophys. Acta*, Vol. 1972, pp. 309-316.

13. Ford, E. & Ares, M. Jr. (1994). Synthesis of circular RNA in bacteria and yeast using RNA cyclase ribozymes derived from a group I intron of phage T4. *Proc. Natl. Acad. Sci. USA*, Vol. 91, pp. 3117-3121

14. Fornari, C.S. & Kaplan, S. (1982). Genetic transformation of *Rhodopseudomonas sphaeroides* by plasmid DNA. *J. Bacteriol.*, Vol. 152, pp. 89-97

15. Frazão, C.; McVey ,C.E.; Amblar, M.; Barbas, A.; Vonrhein, C.; Arraiano, C.M. & Carrondo,(2006). Unravelling the dynamics of RNA degradation by ribonuclease II and its RNA-bound complex. *Nature*, Vol. 443, pp. 110-114

16. Galloway-Salbo, J.L.; Coetzee, T. & Belfort, M. (1990). Deletion-tolerance and trans-splicing of the bacteriophage T4 td inton. Analysis of the P6-L6a region. *J. Mol. Biol.*, Vol. 211, No. 3, pp. 537-549

17. Hansen, T.A. & Veldkamp, H. (1973) *Rhodopseudomonas sulfidophila*, nov. spec., a new species of the purple nonsulfur bacteria., *Arch. Mikrobiol.*, Vol. 92, pp. 45-48

18. Haupenthal, J., Baehr, C., Kiermayer, S., Zeuzem, S. and Piiper, A. (2006). Inhibition of RNAse A family enzymes prevents degradation and loss of silencing activity of siRNAs in serum. *Biochem. Pharmacol.*, Vol. 71, pp. 702-710

19. Haupenthal, J.; Baehr, C.; Zeuzem, S. & Piiper, A. (2007). RNase A-like enzymes in serum inhibit the anti-neoplastic activity of siRNA targeting polo-like kinase 1. *Int. J. Cancer*, Vol. 121, pp. 206-210

20. Hiraishi, A. & Ueda, Y. (1994). Intrageneric structure of the genus *Rhodobacter*: transfer of *Rhodobacter sulfidophilus* and related marine species to the genus *Rhodovulum* gen. nov., *Int. J. Syst. Bacteriol.*, Vol. 44, pp. 15-23

21. Hnik, P.; Boyer, D.S.; Grillone, L.R.; Clement, J.G.; Henry, S.P. & Green, E.A. (2009). Antisense oligonucleotide therapy in diabetic retinopathy. *J. Diabetes Sci. Technol.*, Vol. 3, No. 4, pp. 924-930

22. Inouye S. & Inouye, M., (1985). Up-promoter mutations in the *lpp* gene of *Escherichia coli. Nucleic Acids Res.*, Vol. 13, pp. 3101-3110

23. Kang, J.; Lee, M.S.; Watowich, S.J. & Gorenstein, D.G . (2007). Combinatorial selection of a RNA thioaptamer that binds to Venezuelan equine encephalitis virus capsid protein. *FEBS Lett.* Vol. 581, pp. 2497-2502.

24. Lee, J.H.; Canny, M.D.; De Erkenez, A.; Krilleke, D.; Ng, Y.S.; Shima, D.T.; Pardi, A. &Jucker, F. (2005). A therapeutic aptamer inhibits angiogenesis by specifically targeting the heparin binding domain of VEGF165. *Proc. Natl. Acad. Sci. USA*, Vol. 102, pp. 18902- 18907

25. Lee, J. K. & Kaplan, S. (1995). Transcriptional regulation of puc operon expression in *Rhodobacter sphaeroides, J. Biol. Chem*, Vol. 270, pp. 20453-20458

26. Mulhbacher, J.; St-Pierre, P. & Lafontaine, D.A., (2010). Therapeutic applications of ribozymes and riboswitches. *Curr. Opin. Pharmacol.*, Vol. 10, No. 5, pp. 551-556

27. Movva, N.R.; Katz, E.; Asdourian, P.L.; Hirota, Y. & Inouye M., (1978). Gene dosage effects of the structural gene for a lipoprotein of the *Escherichia coli* outer membrane. *J. Bacteriol.*, Vol. 133, pp. 81-84

28. Nagashima, K.V.P.; Hiraishi, A.; Shimada, K. & Matsuura, K., (1997). Horizontal transferof genes coding for the photosynthetic reaction centers of purple bacteria. *J. Mol. Evol.*, Vol. 45, pp. 131-136

29. Ochi, A.; Umekage, S. & Kikuchi, Y. (2009). Non-enzymatic *in vitro* production of circular hammerhead ribozyme targeting the template region of human telomerase RNA. *Nucleic Acids Symp. Ser.*, Vol. 53, pp. 275-276

30. Palmer, E.L.; Gewiess, A.; Harp, J.M.; York, M.H. & Bunick, G.J. (1995). Large-scale production of palindrome DNA fragments. *Anal. Biochem.*, Vol. 231, No. 1, pp. 109- 114

31. Perriman, R. & Ares, M. Jr. (1998). Circular mRNA can direct translation of extremely long repeating-sequence proteins in vivo. *RNA*, Vol. 4, pp. 1047-1054

32. Pestourie, C.; Tavitian, B. & Duconge, F. (2005). Aptamers against extracellular targets for in vivo applications, *Biochimie*, Vol. 87, pp. 921-930

33. Pieken, W.A., Olsen, D.B., Benseler, F., Aurup, H. and Eckstein, F. (1991). Kinetic characterization of ribonuclease-resistant 2'-modified hammerhead ribozymes. *Science*, Vol. 253, pp. 314-317

34. Ponchon, L. & Dardel, F. (2007). Recombinant RNA technology: the tRNA scaffold. *Nat.Methods* , Vol. 4, pp. 571-576

35. Puttaraju, M. & Been, M.D. (1992). Group I permuted intron-exon (PIE) sequences self-splice to produce circular exons. *Nucleic Acids Res.*, Vol. 20, 5357-5364
36. Puttaraju, M.; Perrotta, A.T. & Been, M.D. (1993). A circular trans-acting hepatitis delta virus ribozyme. *Nucleic Acids Res.*, Vol. 21, pp. 4253-4258
37. Puttaraju, M & Been, M.D. (1995). Generation of nuclease resistant circular RNA decoys for HIV-Tat and HIV-Rev by autocatalytic splicing. *Nucleic Acids Symp. Ser.*, Vol. 33, pp. 152-155
38. Puttaraju, M. & Been, M.D. (1996). Circular ribozymes generated in Escherichia coli using group I self-splicing permuted intron-exon sequences. *J. Biol. Chem.*, Vol. 271, pp. 26081-26087
39. Que-Gewirth, N.S. & Sullenger, B.A. (2007). Gene therapy progress and prospects: RNA aptamers. Gene Ther., Vol. 14, No. 4, pp. 283-291
40. Schmidt, K.S.; Borkowski, S.; Kurreck, J.; Stephens, A.W.; Bald, R.; Hecht, M.; Friebe, M.; Dinkelborg, L. & Erdmann, V.A. (2004). Application of locked nucleic acids to improve aptamer *in vivo* stability and targeting function. *Nucleic Acids Res.*, Vol. 32, pp. 5757-5765
41. Schumacher, J.; Randles, J.W. & Riesner, D. (1983). A two-dimensional electrophoretic technique for the detection of circular viroids and virosoids. *Anal. Biochem.*, Vol. 135, pp. 288-295
42. Simon, R.; U. Priefer, U. & Pühler, A. (1983). A broad host range mobilization system for *in vivo* genetic engineering: transposon mutagenesis in gram negative bacteria. *Biotechnology*, Vol. 1, pp. 37–45.
43. Sioud, M. & Sorensen, D.R. (2003). Cationic liposome-mediated delivery of siRNAs in adult mice. *Biochem. Biophys. Res. Commun.* Vol. 312, pp. 1220–1225.
44. Sorensen, D.R.; Leirdal, M. & Sioud, M. (2003). Gene silencing by systemic delivery of synthetic siRNAs in adult mice. *J. Mol. Biol.* Vol. 327, pp. 761-766
45. Srisawat, C. and Engelke, Q.R. (2001). Streptavidin aptamers: affinity tags for the study of RNAs and ribonucleoproteins. *RNA*, Vol. 7, pp. 632-641
46. Stahley, M.R. & Strobel, S.A. (2006), RNA splicing: group I intron crystal structures reveal the basis of splice site selection and metal ion catalysis. *Curr. Opin. Struct. Biol.* Vol. 16, No. 3, pp. 319-326
47. Suzuki, H.; Ando, T.; Umekage, S.; Tanaka, T. & Kikuchi, Y. (2010). Extracellular production of an RNA aptamer by ribonuclease-free marine bacteria harboring engineered plasmids: a proposal for industrial RNA drug production. *Appl. Environ. Microbiol.*, Vol. 76, No. 3, pp. 786-793

48. Suzuki, T.; Suzuki, T.; Wada, T.; Saigo, K. & Watanabe, K. (2002). Taurine as a constituent of mitochondrial tRNAs: new insights into the functions of taurine and human mitochondrial diseases. *EMBO J.*, Vol. 21, pp. 6581-6589

49. Tsai, M.C.; Spitale, R.C. & Chang, H.Y. (2011). Long intergenic non-coding RNAs-New links in cancer progression. *Cancer Res.* Vol. 71, No. 1, pp. 3-7

50. Turner, J.J.; Jones, S.W.; Moschos, S.A.; Lindsay, M.A. & Gait, M.J. (2007). MALDI-TOF mass spectral analysis of siRNA degradation in serum confirms an RNase A-like activity. *Mol. BioSystems*, Vol. 3, pp. 43-50

51. Umekage, S. & Kikuchi, Y., (2006). Production of circular form of streptavidin RNA aptamer*in vitro*. Nucleic Acids Symp. Ser. (Oxf), Vol. 50, pp. 323-324

52. Umekage, S. & Kikuchi, Y., (2007). Production of circular streptavidin RNA aptamer in vivo. Nucleic Acids Symp. Ser. (Oxf), Vol. 51, pp. 391-392

53. Umekage, S. & Kikuchi, Y., (2009). *In vitro* and *in vivo* production and purification of circular RNA aptamer. J. Biotechnol., Vol. 139, No. 4, pp. 265-272, (2008 Epub ahead of print)

54. Umekage, S. & Kikuchi, Y., (2009). In vivo circular RNA production using a constitutive promoter for high-level expression. J. Biosci. Bioeng., Vol. 108, No. 4, pp. 354-356

55. van Alphen R.J., Wiemer, E.A., Burger, H. & Eskens, F.A. (2009). The spliceosome as target for anticancer treatment. *Br. J. Cancer*, Vol. 100, No. 2, pp. 228-232

56. Watts, J.K. & Corey, D.R., *Bioorg. Med. Chem. Lett.*, (2010). Clinical status of duplex RNA. Vol. 20, No. 11, pp. 3203-3207

57. Watts, J.K.; Deleavey, G.F. & Damha, M.J., *Drug Discov. Today*, (2008), Chemically modified siRNA: tools and applications. Vol 13, No. 19-20, pp842-855

58. Xu, M.Q.; Kathe, S.D.; Goodrich-Blair, H.; Nierzwicki-Bauer, S.A. & Shub, D.A. (1990). Bacterial origin of a chloroplast intron: conserved self-splicing group I introns in cyanobacteria. *Science*, Vol. 250, No. 4987, pp. 1566-1570

CHAPTER-5

MIMICKING DNA THROUGH THE USE OF ANTI-RESTRICTION & PENTAPEPTIDE REPEAT (PPR) PROTEINS

INTRODUCTION: AN OVERVIEW

In the realm of molecular biology and genetic engineering, a fascinating and pivotal area of research involves the art of mimicking DNA. This intricate endeavor relies on the utilization of two intriguing components: anti-restriction proteins and pentapeptide repeat (PPR) proteins. These biological agents possess the unique ability to imitate the structure and function of DNA, opening up exciting possibilities for applications in gene regulation, genome editing, and beyond.

Protein mimicry of DNA is a recently discovered direct mechanism of regulation of DNA- dependent enzyme activity by means of proteins that mimic DNA structure and interact with a target enzyme and completely inhibit (or modulate) its activity. DNA-mimicking inhibitor proteins bind directly to the enzyme and thus blocks or alters the activity of the latter. Protein mimicry of DNA was first described in Ugi derived from PBS2 bacteriophage of *Bacillus subtilis* (Mol et al, 1995). This protein of 84 amino acid residues with a total charge of (–12) inhibits uracil-DNA glycosylase (UDG), an enzyme involved in DNA repair (Mol et al, 1995; Putnam & Tainer, 2005). Subsequently, this type of protein mimicry was found in the ribosomal elongation factor EF-G (tRNA-like motif), and in the dTAFII 230 component of eukaryotic transcription factor TFIID (DNA-like domain) (Liu et al., 1998). The family of DNA mimetics further includes DinI, a negative SOS response regulator in *E. coli* (Ramirez et al., 2000), and a nucleosome forming protein HI1450 of *Haemophilus influenzae* (Parsons et al., 2004). However, in most of these cases, only a part of the protein molecule is DNA-like, in contrast to antirestriction and pentapeptide repeat (PPR) proteins, whose entire structure mimics the B-form of DNA. For instance, the X-ray structure of Ugi reveals a domain similar to the B-form of DNA, but the molecule as a whole is globular. Note that, in Ugi, the crucial negative charges are those of E20, E28, and E31 in the N domain (Mol et al.,1995).

Horizontal gene transfer is a fundamental mechanism for driving diversity and evolution. Transmission of DNA to bacterial cells that are not direct descendants of the donor is often achieved via mobile genetic elements such as plasmids, conjugative transposons and bacteriophages. Mobilization of these elements can lead in the spread of antimicrobial resistance in clinical environments and in the wider community.

Over 50% of eubacteria and archaea contain the genes for one or more of the four classes of known DNA restriction and restriction-modification (RM) systems (Roberts et al., 2005). RM systems work by recognizing specific DNA sequences and triggering an endonuclease activity which rapidly cleaves the foreign DNA allowing facile destruction by exonucleases (Bickle & Kruger,1993; Murray, 2000; Loenen, 2003).

Mobile genetic elements such as plasmids, transposons and bacteriophage contain the specific genes encoding anti-RM systems. Activation of anti-RM system weakens or negatesthe RM defence system allowing further horizontal gene transfer (Wilkins, 1995; Zavilgelsky, 2000; Murray, 2002; Tock & Dryden, 2005).

The genes encoding antirestriction proteins are situated on conjugational plasmids (*ard*A gene) and some bacteriophages (*ocr* and *dar*A genes). Antirestriction proteins inhibit the type I restriction-modification enzymes and thus protect unmodified DNA of plasmids and bacteriophages from degradation. Genes *ard* (*a*lleviation of *r*estriction of DNA) facilitate the natural DNA transfer between various types of bacteria ensuring overcoming intercellular restriction barriers (horizontal genes transfer). Genes *ocr* (bacteriophage T7) and *dar*A (bacteriophage P1) significantly increase the infection efficiency by phages of the bacterial cells.

Antirestriction proteins ArdA and Ocr belong to the group of very acidic proteins and contain a characteristic sequence of negative charges (Asp and Glu). X-ray diffraction study of proteins ArdA and Ocr carried out demonstrated that these proteins were like the B-form of DNA (Walkinshaw et al., 2002; McMahon et al., 2009). Therefore the antirestriction proteins operate on the principle of concurrent inhibition replacing DNA in the complex with the enzyme (DNA mimicry).

DNA-mimetic antirestriction proteins ArdA and Ocr can be electroporated into cells along with transforming DNA and protect unmodified DNA from degradation. As a result the antirestriction proteins improve transformation efficiency. The highly charged, very acidic proteins Ocr and ArdA can be used as a purification handle similar to other fusion tags. A monomeric mutant of the Ocr protein was used as a novel fusion tag which displayed solubilizing activity with a variety of different passenger proteins (DelProposto et al., 2009).

The pentapeptide repeat is a recently discovered protein fold. MfpA and Qnr (A,B,C,D,S) are two newly characterized pentapeptide repeat proteins (PPRs) that interact with type II topoisomerase (DNA gyrase) and confer bacterial resistance to the drugs quinolone and fluoroquinolone [Hegde et al., 2005; Hedge et al., 2011). The *mfp*A gene is chromosome borne in *Mycobacterium tuberculosis (*Hegde et al., 2005; Montero et al., 2001), while *qnr* genes are plasmid borne in Gram-negative enterobacteria (Martinez-Martinez, L. et al.,1998; Tran et al., 2005; Cattoin & Nordmann, 2009; Rodriguez-Martinez et al. 2011). The size, shape, and surface potential of MfpA and Qnr proteins mimics duplex DNA (Hegde et al., 2005; Vetting et al., 2009; Hegde et al., 2011).

TYPE I RESTRICTION-MODIFICATION SYSTEMS

Restriction–modification (RM) systems form a barrier protecting a cell from the penetration by foreign DNA (Murray, 2000; Loenen, 2003). In the modern understanding, RM enzymes are a part of the "immigration control system", which discriminates between its own and foreign DNA entering the cell (Murray, 2002). The system is based on two conjugated enzymatic activities: those of restriction endonucleases and DNA methyltransferases. RM enzymes recognize a specific nucleotide sequence in the DNA, and the restriction endonuclease cleaves the double strand of unmodified DNA. The host DNA is protected from enzymatic cleavage by specific methylation of the recognition sites produced by DNA methyltransferases. RM enzymes are classified in four types. We shall now discuss the features of type I RM systems, since it is these systems that are efficiently inhibited by antirestriction proteins. Figure 1 schematically represents the activity of a type I enzyme, e.g., EcoKI. EcoKI comprises five subunits (R_2M_2S): two R subunits are restrictionendonucleases that cleave the double helix of unmodified DNA, two M subunits are methyltransferases that methylate adenine residues at the recognition site, and an S subunit recognizes a specific DNA site (sK) and forms a stable complex with it.

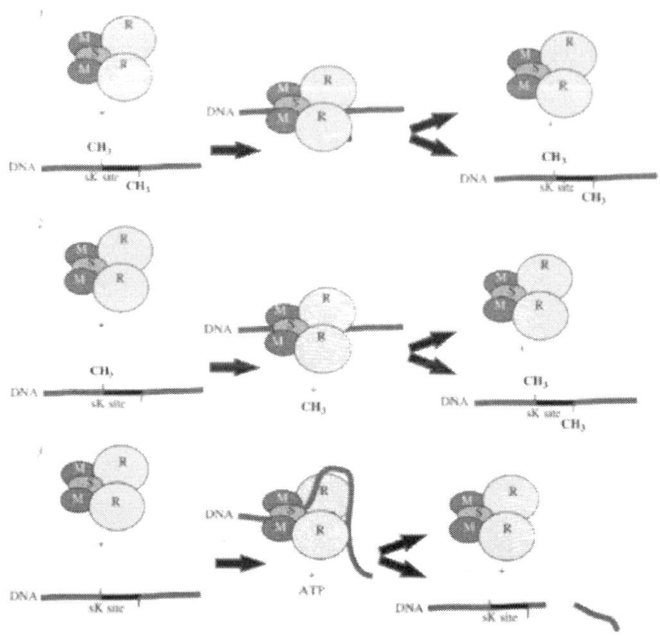

Fig. 1. Activity of a type I restriction–modification enzyme. 1, Both DNA strands at the sK site are methylated. The enzyme–DNA complex dissociates. 2, One of DNA strands at the sK site is methylated. The methylase (M) methylates the adenyl residue of the other strand, and the complex dissociates. 3, Both DNA strands at the sK site are unmethylated. The enzyme initiates DNA translocation through the R subunits accompanied by the formation of a supercoiled loop and subsequent double-stranded DNA break.

The sK site is "hyphenated", i.e., only seven outmost nucleotides of the 13 bp long recognition sequence are conserved (e.g. EcoKI recognizes 5'-AACNNNNNNGTGC-3'). According to the footprinting data, EcoKI covers 66 bp of the DNA sequence. Further events depend on the sK status. If both DNA strands at the site are methylated, the complex dissociates.If only one strand is methylated, the methylase M methylates the respective adenyl residue, and the complex dissociates. If both DNA strands are unmethylated, the DNA helix is translocated through the R subunits, while the S subunit remains bound to the sK site. The endonuclease R randomly cleaves the DNA strands at a considerable distance from the sK site. This is the principal difference between the type I RM enzymes and type II restrictionendonucleases, which introduce a double-strand DNA break directly at the recognition site or at a specific distance from it. The translocation process itself is associated with considerable energy expenditure in the form of ATP. As a result, type I RM enzymes are ATP-dependent, whereas type II enzymes are not. Another characterizing feature of the EcoKI–sK complex is that the S subunit binds only to the outmost conserved nucleotides of the site. As a result, the double stranded DNA undergoes significant deformation, acquiring a kink of approximately 34°, which sets additional energy demands. Nucleotide sequences of the recognition sites vary and are specific for each type I enzyme (EcoK, EcoB, EcoA, EcoD, Eco124, StyLT, StySP, CfrAI, and many others). Based on their homology and the possibility of subunit exchange, type I RM systems are classified into four families: IA, IB, IC, ID. Restriction is efficient against foreign DNA irrespective of the way it is introduced into the cell: by injection from a phage, transformation, or conjugative transmission. Thus, type I RM systems constitute a socalled restriction barrier that prevents interspecies horizontal gene transfer.

CONJUGATIVE PLASMIDS & TRANSPOSONS, BACTERIOPHAGES, AND ANTI-RESTRICTION MECHANISMS

Natural horizontal gene transfer between bacteria is mediated primarily by transmissible plasmids, conjugative transposones, and bacteriophages (Wilkins, 1995). Evolution of all transmissible plasmids, conjugative transposones and some bacteriophages gave rise to systems enabling them to overcome restriction barriers. This phenomenon has been termed antirestriction (Zavilgelsky, 2000; Tock & Dryden, 2005). An investigation of antirestriction mechanisms employed by transmissible plasmids showed that the process involves a specialized antirestriction protein encoded by the *ard*A gene (*a*lleviation of *r*estriction of *D*NA). *ard*A genes were first discovered in plasmids of the incompatibility group N in 1984– 1985 (Belogurov et al.,1985), and later in other types of plasmids (Kotova et al., 1988; Delver et al., 1991). In 1991–1995, *ard*A genes were sequenced and the primary structure of ArdA proteins was determined (Delver et al., 1991; Chilley & Wilkins, 1995). Genes *ard*A are located in the leader region of the plasmid sequence, which lies next to *ori*T and is the first to enter the host cell in the course of conjugative transfer. The *ori*T site, the origin of plasmid conjugative replication, is located at the boundary of the *tra* operon with the rest of plasmid.

The conjugative transposon Tn916 of the bacterial pathogen *Enterococcus faecalis* contains *orf*18 gene, which is located within position region and encodes an ArdA antirestriction protein (Serfiotis-Mitsa et al., 2008). Genes of the *ard*A family encode small, very acidic proteins comprised of 160–170 amino acid residues and bearing a characteristic total negative charge of (–20 to –30) which act as specific highly efficient inhibitors of cellular type I RM enzymes. ArdA proteins inhibit restriction endonucleases of different families (IA, IB, IC, and ID) and with different recognition site sequences with nearly the same efficiency. Thanks to this property of ArdA, transmissible plasmids can overcome the restriction barriers through horizontal transmission from the donor cell into bacteria of various species and genera.

Some bacteriophages also possess genes encoding antirestriction proteins, such as *0.3(ocr)* (phage T7) and *dar*A (phage P1) (Dunn et al., 1981; Kruger et al., 1983; Iida et al., 1988). These genes increase the efficiency of phage infection.

Antirestriction proteins, both of plasmid (ArdA) and phage origin (Ocr), inhibit only type I RM enzymes, whose genes (*hsd*RMS) are usually located on the bacterial chromosome, but not type II restriction endonucleases, the genesof which are normally located on plasmids.

DNA MIMICRY USING ANTI-RESTRICTION PROTEINS

It has been supposed that antirestriction proteins of the ArdA family, as well as Ocr are modulator proteins with a structure similar to that of the B-form DNA, and the characteristic surface distribution of negatively charged D and E residues (aspartic and glutamic acids) imitates the distribution of negatively charged phosphate groups along the DNA double helix (Zavilgelsky, 2000). That is, antirestriction proteins imitate the DNA structure, which is currently termed "protein mimicry of DNA". The spatial structure of the smallest antirestriction protein, Ocr of phage T7 (116 amino acids), was published in 2002 (Walkinshaw et al., 2002). As shown by X-ray crystallography, the spatial structure of Ocr was similar to the B-form of DNA (Fig. 2). The major stem of the Ocr monomer is constituted by three α-helices: A (residues 7–24), B (residues 34– 44), and a long, somewhat bent one, D (residues 73–106); the helices form a tightly packed bunch with strictly regularly positioned negatively charged D and E carboxyls along the stem axis, nearly reproducing the distribution of negatively charged phosphate groups along DNA double helix. The short α-helix C (residues 49–57) is a part of the interface determining the contact of monomers and stable dimer formation.

The structure of the Ocr dimer, both in solution and in crystal form, is similar in length and charge distribution to 24 bp of DNA double helix. The contact of monomers is established by a Van der Waals interaction between hydrophobic clusters within the C α-helices in the middle of the polypeptide: A50, F53, S54, M56, A57, and V77.

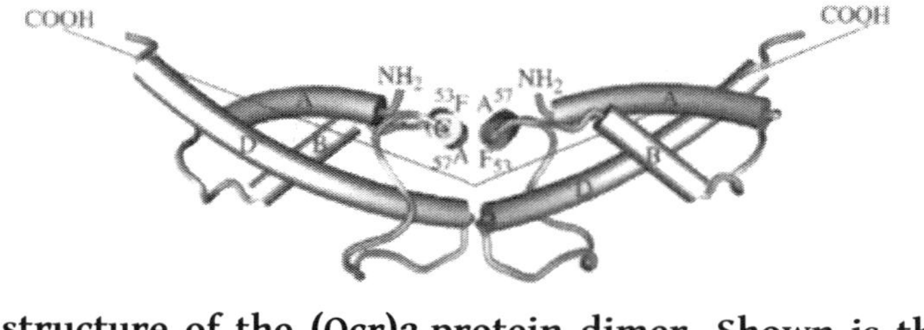

Fig. 2. Spatial structure of the (Ocr)2 protein dimer. Shown is the positioning of α-helices A, B, C, D, and amino acid residues 53F and 57A in the hydrophobic cluster 52IFSVMAS, which determines the Van der Waals attraction of the monomers.

The spatial structure of the ArdA protein from the conjugative transposon Tn916 (166 amino acids), was published in 2009 (McMahon et al., 2009). As was shown by X-ray crystallography, ArdA protein has a extremely elongated curved cylindrical structure witn defined helical groowes. The high density of Asp and Glu residues on the surface follow a helical pattern and the whole protein mimics a 42-base pair stretch of B-form DNA making ArdA dimer by far the largest DNA mimic known (Fig. 3). Each monomer of this dimeric structure can be decomposed into three domains: the N-terminal domain 1 (residues 3-61), the central domain 2 (residues 62-103) and the C-terminal domain 3 (residues 104-165). The N-terminal domain 1 consists of a three-stranded anti-parallel β-sheet and one short α -helix interspersed with three large loops of 10 or more residues. The central domain 2 of ArdA is a four α–helix bundle. The C-terminal domain 3 has a three-stranded β -sheet and three α- helices packed together in a manner that creates a groove in the structure 11 angstrem wide. Analysis of the electrostatic surface of ArdA shows that 2 and 3 domains have a profoundly negative potential (the pI of ArdA is 4). The ArdA dimer, like the monomer, is highlyelongated and curved (Fig. 3). The chord that connects the extreme ends has the length of140 angstrem. The pattern of negative charge even extends across the dimer interface through the conserved residues D109, D111, D112, D115, E122, E123 and E129.

This distribution and conservation of charged residues is evidencc for the necessity of dimer formation for protein function and suggests that ArdA across all species will have similar structural requirements. The dimer interface contains the anti-restriction motif (amino acids 126-140 in the Tn916 ArdA protein) identified previously (Belogurov & Delver, 1995) conserved as well.

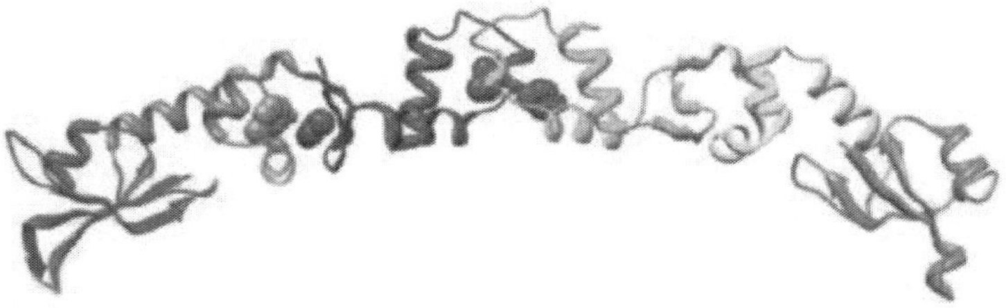

Fig. 3. Spatial structure of the (ArdA)2 protein dimer.

The ArdA dimer appears to mimic about 42 bp of bent B-form DNA. This is comparable in length to the footprint of the EcoKI Type IA RM enzyme, without its cofactors, on DNA. In comparison, the Ocr dimer from phage T7 mimics only about 24 bp., similar in length to the 30 bp footprint of the Type I RM enzyme in the presence of its cofactors and to the footprint of the MTase core, M.EcoKI, of the Type I RM enzyme. The typical DNA target for a Type I RM enzyme is 14 bp long and bipartite, e.g. EcoKI recognizes 5'-AACNNNNNNGTGC-3', and lies centrally in the experimental DNA footprint. It was built the M.EcoKI-ArdA model: domain 3 overlaps the EcoKI target sequence, domain 2 contacts the extremites of the DNA-binding groove in M.EcoKI and domain 3 projects beyond the M.EcoKI structure. Domain 1 is not essential for antirestriction as it can be deleted (Delver et al., 1991) indicating that the key aspect of antirestriction by ArdA is the binding to the MTase core using domains 2 and 3.

The mimicry of DNA enables antirestriction proteins to compete with DNA for binding with the RM enzyme and thus to inhibit DNA degradation (restriction) and methylation (modification). From the point of view of classical enzymatic catalysis, antirestriction is a case of competitive inhibition based on structural similarity between the enzyme substrate and the inhibitor molecule. The relative positioning of monomers in the (Ocr)2 dimer is typical: the angle between their longitudinal axes is approximately 34° (Fig. 2). This dimer structure is nearly equivalent to the kinked DNA double helix structure that is formed at the recognition site of the type I RM enzyme–DNA complex (Murray, 2000). Consequently, Ocr does not require additional energy to bind to EcoKI, and efficiently displaces double- stranded DNA from the complex (the complex formation constant for Ocr–EcoKI is approximately 100 times higher than for DNA–EcoKI) (Atanasiu et al., 2002).

INHIBITION OF RESTRICTION AND MODIFICATION PROCESSES BY ARDA AND OCR PROTEINS

Both ArdA and OCR inhibit ATP-dependent type I RM enzymes. However, the great difference between the life cycles of transmissible plasmids (symbiosis with a bacterial cell) andbacteriophages (infection and lysis of bacteria) makes it interesting to compare the inhibition efficiencies of these proteins. For this purpose, we cloned *ard*A and *ocr* under a strictly regulated promoter. To quantify the intracellular concentration of the antirestriction proteins, we developed a bioluminescence method that utilizes the *Photorhabdus luminescens luxCDABE* genes as reporters. The *luxCDABE* genes were cloned in the pZE21 and pZS33 vectors under the control of the PltetO_1 promoter. The hybrid plasmids were introduced in MG1655Z1 cells. Expression of the *lux* genes was induced by adding anhydrotetracycline in the medium, and the bioluminescence intensity was measured. Since the bioluminescence intensity is directly proportional to the luciferase concentration and the sensitivity of the bioluminescence method is high, it is possible to estimate the enzyme concentration in the cell within a broad range, starting with extremely low concentrations. A calibration plot was constructed to characterize the intracellular content of the enzyme (in relative units (RU)) as a function of the inductor (anhydrotetracycline) concentration (Fig. 4). The luciferase content in MG1655Z1 cells varied from 1 (in the absence of anhydrotetracycline) to 5000 (20 ng/ml anhydrotetracycline or more) RU. It is natural to assume that the relative contents of the proteins synthesized from the *ardA* and *0.3(ocr)* genes cloned in the pZE21 and pZS33 vectors vary within the same range as the luciferase content under the same expression conditions.

To measure the antirestriction activities of the ArdA and Ocr proteins, titration with phage λ.0 was performed for MG1655Z1 cells carrying a hybrid plasmid with the *ardA* or *0.3(ocr)* gene; cells without the hybrid plasmid were used as a control. Since the genome of strain MG1655Z1 contains the *hsdRMS* genes, which code for the *Eco*KI restriction–modification enzyme, the phage λ.0 seeding efficiency was approximately four orders of magnitude lower than in the case of control strain TG_1. However, when MG1655Z1 cells contained a plasmid with the cloned *ardA* or *0.3(ocr)* gene, the phage seeding efficiency changed depending on the production of the antirestriction protein. As the protein production increased, the phage seeding efficiency grew from 10^{-4} (no inhibition) to 1 (complete inhibition of restriction–modification enzymes).

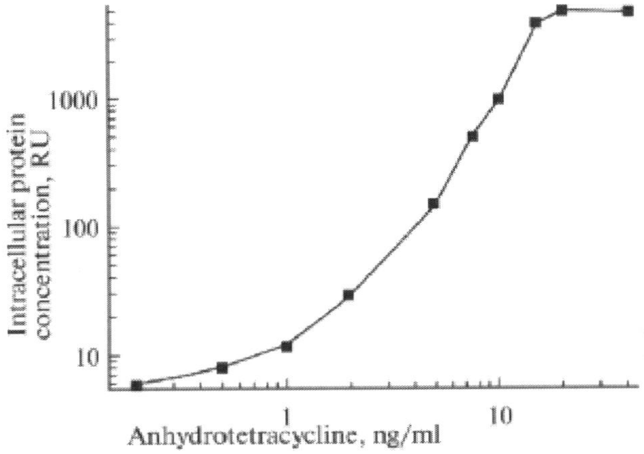

Fig. 4. Luciferase content (relative units, RU) in *E. coli* MG1655Z1 cells containing the pZS33_*lux* or pZE21_*lux* plasmid as a function of anhydrotetracycline content. The *P. luminescens luxCDABE* genes were cloned in the pZS33 and pZE21 vectors under the control of the P1tetO_1 promoter. The luciferase content in the presence of the pZS33_*lux* plasmid and the absence of the inductor anhydrotetracycline was taken as unity.

ArdA ColIb-P9, Ocr T7 antirestriction and antimodification activities were avaluated as a function of the inhibitor concentration, that enabled us to estimate the relative difference in dissociation constants (K d) that describe the interaction efficiency for ArdA or Ocr and EcoKI (Fig. 5) (Zavilgelsky et al., 2008).

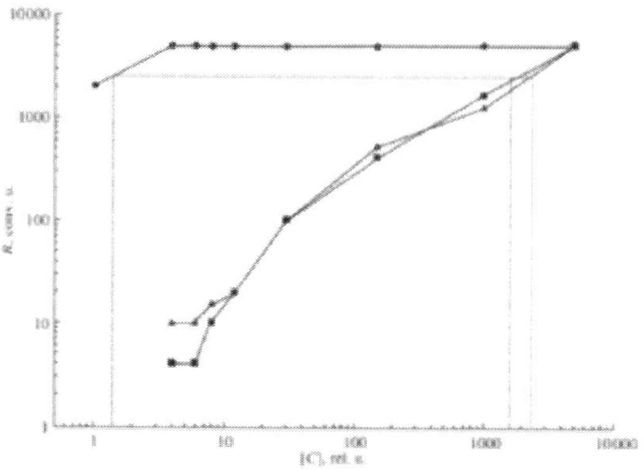

Fig. 5. Antirestriction activity of ArdA ColIb-P9, Ocr T7, and Ocr mutant F53D A57E as a function of their intracellular levels. X-axis: intracellular antirestriction protein concentration (relative units). Y-axis: Antirestriction activity (unmodified λ DNA was used as an EcoKI target). Dotted lines indicate the Kd points. Circles, native Ocr; squares, Ocr F53D A57E; triangles, ArdA.

The antimodification activity of the ArdA and Ocr proteins was inferred from the seeding efficiency of phage λMG1655Z1 (phage λ.0 propagated for one cycle in MG1655Z1 cells carrying a plasmid with the *ardA* or *0.3(ocr)* gene) on strains AB1167 r+m+ and TG1 r−m−. The ratio between the phage titers on these strains reflected the extent of phage DNA modification (methylation). The *ardA* and *0.3 (ocr)* genes were cloned in the pZE21 and pZS33 vectors with the strongly regulated PltetO-1 promoter; the results are summarized in Tables 1 and 2. The intracellular concentrations of the ArdA and Ocr proteins were estimated from the calibration plot constructed by the bioluminescence method (Fig. 1). The ArdA and Ocr proteins substantially differed in the capability of inhibiting the *Eco*KI enzyme. The Ocr protein almost completely inhibited the *Eco*KI restriction–modification system, affecting both restriction and modification activities of the enzyme in a broad Ocr concentration range. The effect was already detectable when Ocr was present at several tens of molecules per cell (1 RU corresponds approximately to ten molecules of the inhibitor protein per cell) .

In the case of the ColIb_P9 ArdA protein, the efficiency of inhibition of the restriction activity of the *Eco*KI enzymes started to decrease when the protein concentration was approximately half its threshold value (which corresponded to complete inhibition of *Eco*KI activity), that is, when ArdA occurred at 10000– 15000 molecules per cell. Inhibition of modification activity of the *Eco*KI enzyme started at higher intracellular ArdA concentrations, at approximately 45000–50000 ArdA molecules per cell (Table 2).

The antirestriction and antimodification activities of the ArdA and Ocr proteins as functions of their intracellular concentrations (in RU) are shown in Fig. 6. While the Ocr proteininhibited both activities of the *Eco*KI enzyme with similar efficiencies and acted already at extremely low concentrations in the cell, the antirestriction and antimodification activity curves substantially differed in the case of the ArdA protein. As estimations showed, the dissociation constant *K*d(met) characteristic of ArdA_dependent inhibition of methylase activity of the *Eco*KI enzyme was tenfold higher than Kd(rest).

The difference in inhibitory properties of the Ocr and ArdA proteins toward type I restriction–modification enzymes is probably determined by the difference in life cycle between phages and transmissible plasmids; i.e., a phage kills the cell, while a plasmid becomes part of cell genetic material.

The ArdA proteins lose their capability of inhibiting modification activity of *Eco*KI_like proteins relatively easy. For instance, the ArdA antirestriction proteins encoded by the R16 (incB) and R64 (incI1) transmissible plasmids inhibit restriction activity of the *Eco*KI enzyme, but do not affect its modification activity [25, 26]. Yet the proteins are highly homologous to the Collb_P9 ArdA protein. In the 166 amino acid residues, differences are observed only in four positions with R64 ArdA and in nine positions with R16 ArdA. We have earlier found that certain single or double substitutions of hydrophobic amino acid resdues for negatively charged residues (D and E) in the region of the antirestriction motif abolish antimodification activity of ArdA encoded by the pKM101(incN) transmissible plasmid, while its its antirestriction activity is still preserved [17].

In this work, we used site_directed mutagenesis and constructed the Collb_P9 ArdA mutant that contained three amino acid substitutions in the C_terminal domain; hydrophobic residues were replaced with a more hydrophobic one: F156I, F158I, and V163I.Activities of the mutant protein are characterized in Table 3. As is seen, the mutant protein inhibited antirestriction activity of the *Eco*KI enzyme, but lost the inhibitory effect on its modification activity.

Likewise, certain amino acid substitutions transform the Ocr protein into an antirestriction protein that inhibits only antirestriction activity of the *Eco*KI enzyme. X_ray analysis of the Ocr protein in crystal demonstrates that a contact of the monomers in the (Ocr)2 homodimer is due to hydrophobic interactions between F53 and A57, which are in the hydrophobic fragment 52_IFSVMAS_ in a short α_helix [11]. We constructed an Ocr mutant with two substitutions, F53D and A57E, assuming that repulsion of negative charges (D...E) would lead to dissociation of the dimer. The *0.3(ocr)* gene with a single or double mutation was cloned in the pUC18 vector. The Ocr F53D A 57E double mutant was tested for functional activity and proved to efficiently inhibit only *Eco*KI restriction activity without affecting methylase activity of the enzyme (Table 4, data on the antirestriction activity of the proteins are omitted). Note that the single amino acid substitutions of the interface region did not affect the antimodification activity of the Ocr protein (Table 4). Like the Ocr protein, the ArdA proteins are active in a homodimeric form. This is true for both the native ColIb_P9 ArdA protein and the R64 ArdA mutant, which is incapable of inhibiting methylase activity of the enzymes.

Based on the data obtained for the Ocr and ArdA mutant proteins, we assume that the antirestriction proteins form complexes of two types with a type I restriction–modification type, which consists of five subunits (R2M2S) [27]. When an antirestriction protein interacts with the S subunit, which recognizes a specific site in DNA, the DNA strand is displaced, and both restriction and modification activities of the enzyme are inhibited. When an antirestriction protein interacts with the R subunit, which is responsible for ATP_dependent translocation and endonucleolytic cleavage of nonmethylated DNA, only restriction activity ofthe enzyme is inhibited. To check this hypothesis, it was important to construct the Ocr mutants that were incapable of inhibiting methylase activity of the enzymes and preserved the effect on their restriction activity. Such properties were observed for the Ocr F53D A57E mutant, which was constructed in this work and had two substitutions of negatively charged amino acid residues for hydrophobic residues in the interface region of the (Ocr)2 homodimer. Thus, the model of type I restriction–modification enzymes with two different binding sites for antirestriction proteins is applicable not only to the ArdA proteins, whose genes are in transmissive plasmids, but also to the Ocr proteins, whose genes are in bacteriophage genomes.

Table 1. Antimodification and antirestriction activities of the Ocr protein as dependent on its intracellular concentration*

Anhydrotetracyclin, ng/ml	Ocr concentration in the cell, RU	EcoKI modification alleviation factor (R) for Ocr**	EcoKI restriction alleviation factor (R) for Ocr
0.0 (vector pZS33)	1	2000	2000
0.0 (vector pZE21)	4	5000	5000
0.2	6	5000	5000
0.5	8	5000	5000
1.0	12	5000	5000
2.0	30	5000	5000
5.0	150	5000	5000
10.0	1000	5000	5000
20.0	5000	5000	5000
40.0	5000	5000	5000

Notes: * The *0.3* (*ocr*) gene was cloned either in the pZE33 vector (row 1) or in the pZE21 vector (other rows) under the control of the P1tetO_1 promoter.

**** Here and in Table 3: Restriction or modification alleviation factor *R* = *K*+ / *K*–, where *K*– is the coefficient of restriction for MG1655Z1 cells without the plasmid containing the *0.3* (*ocr*) gene and *K*+ is the coefficient of restriction for MG1655Z1 cells carrying the plasmid.**

Table 2. Antimodification and antirestriction activities of the ColIb_P9 ArdA protein as dependent on its intracellular concentration*

Anhydrotetracyclin, ng/ml	Ard concentration in the cell, RU	EcoKI modification alleviation factor (R) for ArdA	EcoKI restriction alleviation factor (R) for ArdA**
0.0 (vector pZS33)	1	Not determined	Not determined
0.0 (vector pZE21)	4	1	5
0.2	6	1	6
0.5	8	1	10
1.0	12	1	20
2.0	30	1	120
5.0	150	4	400
7.5	500	10	1000
10.0	1000	100	2500
15.0	4000	400	5000
20.0	5000	1000	5000

We conclude that the dimeric form of an antirestriction protein is essential for inhibiting both activities of a type I restriction–modification system, while the monomeric form is sufficient for inhibition of its restriction activity.

The *ardA* gene was cloned in the pZE21 vector under the control of the P1 tetO_1 promoter.

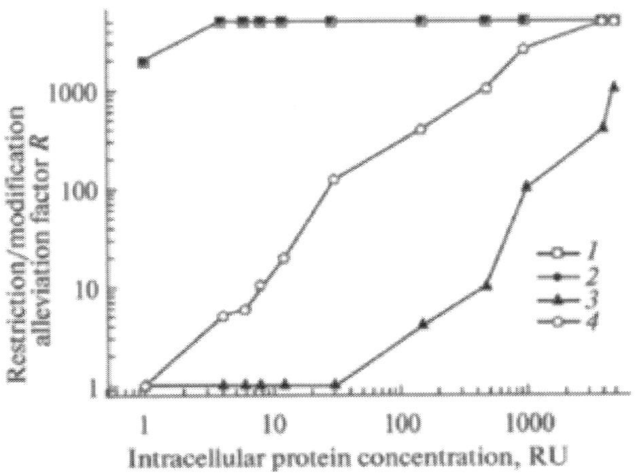

Fig. 6. Antirestriction and antimodification activities of the ColIb_P9 ArdA and T7 Ocr proteins as functions of their intracellular concentrations. Curves: *1*, antimodifi cation activity of Ocr; *2*, antirestriction activity of Ocr;*3*, antimodification activity of ArdA; *4*, antirestriction activity of ArdA.

ArdA and OCR differ considerably in their ability to inhibit the methylase (modification) activity of EcoKI-like enzymes. As a rule, if *ard*A and *0.3(ocr)* genes are governed by a strong promoter, antirestriction and antimodification activities of ArdA and Ocr are established simultaneously (Delver et al., 1991; Chilley & Wilkins, 1995; Atanasiu et al., 2002). Some data suggest, however, that the inhibition of endonuclease and methylase activities depends on different interactions of ArdA proteins with type I RM enzymes. For instance, some natural ArdA proteins inhibit only the endonuclease activity of EcoKI. The respective genes are located in transmissible plasmids R16 (incB) (Thomas et al., 2003) and R64 (incI1) (Zavilgelsky et al., 2004). Furthermore, *in vitro* quantification of the ArdA– EcoKI complex showed that ArdA interacts more efficiently with the complete enzyme R2M2S than with its methylase form M2S, which can only modify DNA (Nekrasov et al., 2007). In contrast to ArdA proteins, Ocr from phage T7 binds to the entire EcoKI enzyme and to its methylase form with nearly equal affinities (Atanasiu et al., 2002), and, therefore, even in very low concentrations it inhibits both the endonuclease and the methylase activities of the enzyme (Fig. 5). This property of OCR is probably related to the difference between the life cycles of a phage and of a transmissible plasmid: a phage kills the host cell, whereas a plasmid becomes part of its genetic material. However, a double amino acid substitution in the 52IFSVMAS hydrophobic cluster of the Ocr interface (an Ocr homodimer is formed by a Van der Waals interaction between these clusters), that is, a substitution of acidic 53D and 57E for hydrophobic 53F and 57A (Fig. 6), causes the mutant protein Ocr F53D A57E to lose the antimethylation while retaining the antirestriction activity against EcoKI. In addition, the mutant protein Ocr F53D A57E has a *K*d of 10^{-7}M, which is 1000 times higher than the *K*d of the native Ocr form (Fig. 6) (Zavilgelsky et al., 2009).

Table 3. Effects of the ArdA (ColIb_P9) protein and its F156I F158I V163I mutant on *Eco*KI restriction and *Eco*KI modification in *E. coli* K_12 AB1157 r+ m+ and MJ109 r–m+* cells upon the cloning of the corresponding genes in the pUC18 vector

Plasmid	Protein	Coefficient of restriction (K) of phage λ.0 on AB1157 r^+m^+(**)	Coefficient of restriction (K) of phage λjm109 on AB1157 r^+m^+
pUC18	Absent	2.0×10^{-4}	1
pVB2(pUC18)	ArdA F1561 F1581 V1631	1	1
pSR3(pUC18)	ArdA native	1	2.0×10^{-4}
pUC18	Absent	1	1
pSR8	Ocr native	1	2.0×10^{-4}
pSR9	Ocr F53D	1	2.0×10^{-4}

Plasmid	Protein	Coefficient of restriction (K) of phage λ.0 on AB1157 r^+m^+(**)	Coefficient of restriction (K) of phage λjm109 on AB1157 r^+m^+
pSR10	Ocr A57E	1	2.0×10^{-4}
pSR11	Ocr F53D A57E	1	1
pSR10	Ocr A57E	1	2.0×10^{-4}
pSR11	Ocr F53D A57E	1	1

Notes: * Phage λ.0 was used to infect E. coli JM109 r–m+ cells. A phage lysate obtained after one reproduction cycle (λjm109) was titrated on strains TG_1 and AB1157.
** The coefficient of restriction K (column 3), which was used to estimate the antirestriction activity of the ArdA proteins, was determined as the ratio of the titer of phage λ.0 on strain AB1157 to the titer of the same phage on strain TG_1 r–m–.
* Phage λ.0 was used to infect E. coli JM109 r– m+ cells. A phage lysate obtained after one reproduction cycle (λjm109) was titrated on strains TG_1 and AB1157. The results were averaged over five replicate experiments.
Table 4. Effects of the native and mutant T7 Ocr proteins on EcoKI_dependent modification in E. coli K_12 JM109 r– m+ cells*
Kd is determined by intracellular protein concentration characterized with a 50% decrease in the inhibition of EcoKI endonuclease activity. For Ocr, this level was approximately 1700 times lower than for ArdA. According to in vitro data, the Ocr–EcoKI complex formation had a Kd of 10^{-10} M (Atanasiu et al., 2002). Therefore, the Kd for ArdA–EcoKI complex formation is 1.7×10^{-7}M.

The fact that endonuclease and methylase EcoKI activities are inhibited by ArdA or Ocr separately suggests that antirestriction proteins can bind type I enzymes in two ways: the complex formation of the first type inhibits both endonuclease and methylase activity of the enzyme, whereas in the complex of the second type, endonuclease activity is blocked while methylase activity is retained. As a working hypothesis, we propose the following model of interaction between antirestriction proteins (ArdA and OCR) and type I RM enzymes (Fig. 7). ArdA and OCR can form a complex both with the S-subunit that contacts with the *s*K site onDNA, and with the R-subunit responsible for the translocation and cleavage of unmodified DNA. The binding of ArdA or Ocr to the S-subunit simultaneously inhibits both endonuclease and methylase activity by displacing DNA from its complex with the R2M2S enzyme (Fig. 7, *1*). However, the binding can be easily disrupted if, as a result of amino acid substitutions, the protein is not in the dimeric form, or if the angle between the longitudinal axes of the monomers differs from the critical 34°. As a consequence, it becomes energetically unfavorable for a DNA-mimic protein to displace kinked DNA from its complex with the S-subunit. On the other hand, the interaction of ArdA or OCR with the R- subunit probably does not depend on the particular dimer structure, since the R-subunit is responsible for DNA strand translocation and the respective complex is not site-specific. Thus ArdA and OCR inhibit only the endonuclease activity of the enzyme, while its methylase activity is preserved: DNA can still bind to the S-subunit, and the M-subunit specifically methylates adenyl residues at the sK site (Fig. 7, *2*).

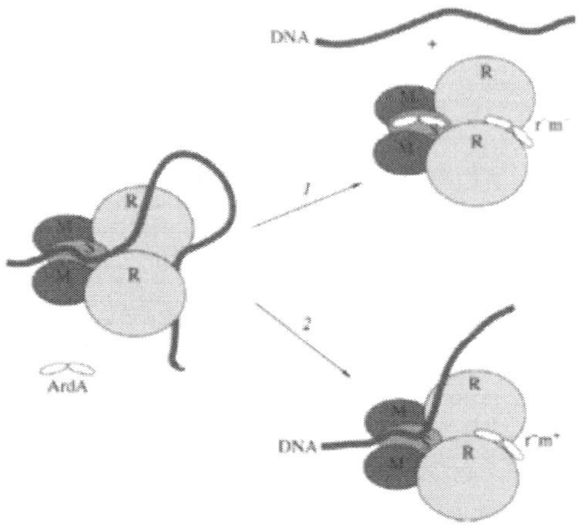

Fig. 7. Putative scheme of ArdA or OCR interaction with a type I RM enzyme (R2M2S). *1*, An ArdA/Ocr complex with the S subunit: unmodified DNA is entirely displaced. Both endonuclease and methylase activities are inhibited (r^- m^- - phenotype). *2*, An ArdA/Ocr complex with the R subunit: a DNA strand is displaced from the translocation center. Only endonuclease activity is inhibited (r– m+ phenotype). Endonuclease and methylase activities of a type I RM enzyme are designated as "r" and "m" respectively.

In vitro experiments showed that the R2M2S form of EcoKI binds two (Ocr)2 dimers, while the methylase form M2 S binds only one (Atanasiu et al., 2002). This result fits well into the above model of inhibition by antirestriction proteins. As the ArdA binding constant is higher for M2S than for R2M2S, moderate levels of ArdA synthesized under natural conditions inhibit only the endonuclease activity of type I RM enzymes so as to protect the plasmid DNA in transmission, but do not affect the methylase activity which is crucial for maintaining the integrity of the plasmid and the host chromosome. The native Ocr form from phage T7 binds to RM enzymes, simultaneously inhibiting both the endonuclease and the methylase activity, and, therefore, interacts with the S-subunit. There is an obviousreason for the Ocr activity being so high (*K*d = 10^{-10} M): in the course of infection, the phage DNA is immediately attacked by cellular endonucleases.

PENTAPEPTIDE REPEAT PROTEINS (PPR PROTEINS)

Inhibitors of DNA gyrase

Quinolones and also fluoroquinolones are synthetic derivatives of nalidixic acid; they belong to a group of antibiotics with wide spectrum of action and high activity and inhibit DNA gyrase. Quinolones bind to the gyrase-DNA complex.

This results in stabilization of the covalent enzyme tyrosyl-DNA phosphate ester (a transient reaction intermediate) and causes death of bacteria. Quinolones have been successfully used for inactivation of *Mycobacterium tuberculosis* cells. During the first years of clinical use of quinolones, findings of *M. tuberculosis* strains resistant to quinolones were rather rare events. Studies of the nature of resistance to quinolones in the laboratory strains of *M. tuberculosis* and the related strain *M. smegmotis* have shown that this effect is determined by missense mutations (amino acid substitutions) in A-chain of DNA gyrase, or it represents the result of regulatory mutation potentiating expression of a protein pump responsible for the extracellular efflux of toxic compounds. However, the wide use of quinolones in medical practice resulted in the discovery of a new type of quinolone resistance. It was shown that the gene determining such type of resistance in *M. smegmotis* and *M. tuberculosis* encodes the MfpA protein, a specific inhibitor of DNA gyrase (Hegde et al., 2005; Montero et al., 2001). The MfpA proteins of *M. tuberculosis* and *M. smegmotis* consist of 183 and 192 residues correspondently; they share 67% identity. In 1998, the resistance to quinolones found in *Klebsiella pneumoniae* was shown to be encoded by the *qnr*A gene and transferred by the conjugated plasmid (Martinez-Martinez et al., 1998). Subsequent investigations have established that *qnr* genes have a worldwide distribution in a range of bacterial pathogens, mainly Gram-negative opportunist (particularly *Enterobacteriaceae*)(Robicsek et al., 2006). Sequence comparison of plasmids isolated from clinical Gram-negative strains differentiates five distinct *qnr* subfamilies *qnr*A, *qnr*B, *qnr*S (Jacoby et al., 2008), and most recently *qnr*C and *qnr*D (Wang et al., 2009; Cavaco et al., 2009). The proteins encoded by these genes exhibit the same function of DNA gyrase inhibition.

MfpA and QnrABCDS proteins belong to the pentapeptide repeat protein (PRP) family. Amino acid sequences of these proteins contain a repeated pentapeptide with the consensus [S, T, A, V][D, N][L, F][S, T, R][G]. MfpA consists of 183 amino acid residues and in these pentapeptides each second amino acid is D or N and each third amino acid is L or F. Table1 shows that MfpA protein consists of 30 pentapeptides, which determine characteristic features of its spatial structure. Figure 6a (taken from (Vetting et al., 2006) shows the spatial structure of the MfpA protein; it consists of a righ-handed β-helix, which corresponds to B- form DNA in size, shape, and electrostatics. In solutions, MfpA forms a dimer due to hydrophobic contact of several amino acids located at the C-end of an α-helical site. The monomeric MfpA consists of eight coils, and four repeated pentapeptides form four sides of a quadrant (1-4) (Table 5). Such spatial structure was named RHQBH (right-handed quadrilateral beta-helix) or "Rfr" (Repeated five-residues). The dimer (MfpA)2 has a rod-like shape 100 angstrem in length and 27 angstrem in diameter. The total charge of the dimer is (−10), but the negative charges are distributed non-randomly. This results in (MfpA)2 dimer, which mimicks a 30 bp segment of B-form duplex DNA. Docking analysis revealed theexistence of tight contact between (MfpA)2 dimer and A2 dimer of the DNA gyrase A subunit (Fig. 8) due to electrostatic complementation between strongly cationic "seat" of the A2 dimer interface and a strongly anionic surface of the (MfpA)2 dimer.

Structural analysis of the *Aeromonas hydrophila,* AhQnr protein is shown that it contain two prominent loops (1 and 2) that project from the PRP structure (Xiong et al., 2011). Deletion mutagenesis demonstrates that both contribute to the protection of *Escherichia coli* DNA gyrase from quinolones. A model for the Qnr:DNA gyrase interaction was suggested, where loop1 interacts with the gyrase A "tower" and loop2 with the gyrase B TOPRIM domains.

Structural similarity between MfpA and Qnr proteins and DNA duplex of the gyrase substrate determines the effectiveness of competitive inhibition of the gyrase; this represents the molecular basis of bacterial resistance to quinolone antibiotics. It should be noted that in contrast to gyrase inhibition by quinolones, the inhibition of gyrase by MfpA and Qnr proteins is not accompanied by cell chromosome degradation. Consequently, the presence of the genes *mfp*A or *qnr* in the bacterial genome is very important because the "fee" for the rescue from the inactivating effect of antibiotics is delayed development of the cell. It is possible that the main function of DNA mimic inhibitors of gyrase consists in modulation of DNA supercoiling, which may potentiate supercoiling at the stage of DNA replication and decrease the rate of supercoiling when the level of chromosome compactness becomes optimal in a particular cell.

Another PRP family proteins

The first protein of the PRP family was originally found in *Anabaena* cyanobacteria (Black et al., 1995). The HglK protein (encoded by the *hgl*K gene and consisting of 727 residues) contains a series of 36 tandem pentapeptides with the consensus sequence ADLSG. Using methods of bioinfor matics, a group of proteins belonging to PRP family has been identified in *Synechocystis cyanobacteria*; there are 15 proteins with series of tandem pentapeptide repeats varying from 13 to 44 (Bateman et al., 1998). By now the proteins of the PRP family have been found in almost all living organisms excluding yeasts. According to data analysis (Vetting et al., 2006), 525 proteins (484 prokaryotic and 41 eukaryotic) with the pentapep- tide motif have been identified. Sequencing of the genome of the cyanobacterium *Cyanothece sp.* PCC 51142 revealed 35 pentapeptide- containing proteins. It was determined (Buchko et al., 2006a) the spatial structure of the Rfr32 protein, which consists of 167 residues. The authors demonstrated that the 21 tandem pentapeptide repeats (with the consensus motif A(N/D)LXX) fold into a right-handed quadrilateral β- helix, or Rfr-fold (as in the case of the MfpA protein); this structure imitates the rod-like structure of B-form DNA. The Rfr structure is also typical for another protein, Rfr23, encoded by a gene that has also been found in the genome of *Cyanothece sp.* PCC 51142 (Buchko et al., 2006b). The real functions of the pentapeptide-containing proteins found in cyanobacteria remain unknown. Some proteins determining immunity of bacteria to their own synthesized antibiotics also belong to the PRP family. These include the McbG protein (encoded by a *mcbG* gene located in the operon responsible for biosynthesis of microcin B17 (Pierrat & Maxwell, 2005) and the OxrA protein, which determines the resistance of *Bacillus megatherium* to oxetanocin A (Morita et al., 1999). In contrast to quinolones, microcin B17 interacts with B-subunit of DNA gyrase. A significant group of pentapeptide repeat family proteins has complex structure and containsseveral domains, including those with catalytic functions. However, the functional role of the pentapeptide repeats in this group remains unknown. But if the putative catalytic function of such protein consists of posttranslational modification of some DNA-binding protein (e.g. histone acetylation), one can suggest that binding of the target protein to the pentapeptide domain would significantly increase selectivity of such a modification reaction.

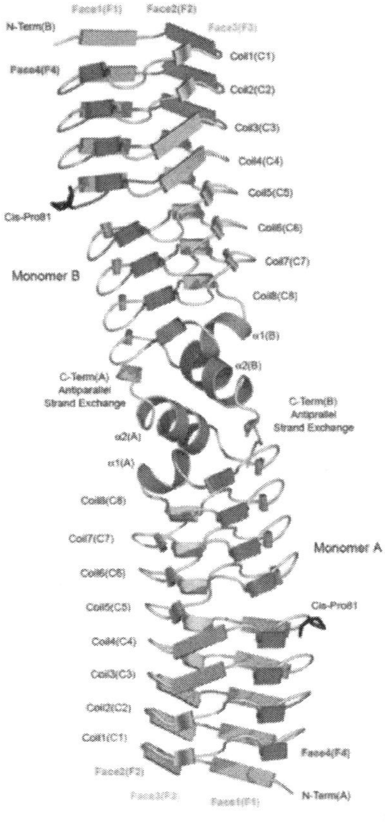

Fig. 8. Ribbon diagram of the *Mycobacterium tuberculosis* MfpA dimer. The four faced of the quadrilateral β-helix are colored green (face 1), blue (face 2), yellow (face 3) and red (face 4).

Table 5. Position of pentapeptides along the axis of the MfpA protein molecule.

Coil	Quadrant sides				Amino acid position in the protein chain
	1	2	3	4	
1	QQWVD	CEFTG	RDFRD	EDLSR	21
2	LHTER	AMFSE	CDFSG	VNLAE	41
3	SQHRG	SAFRN	CTFER	TTLWH	61
4	STFAQ	CSMLG	SVFVA	CRLRP	81
5	LTLDD	VDFTL	AVLGG	NDLRG	101
6	LNLTC	CRLRE	TSLVD	TDLRK	121
7	CVLRG	ADLSG	ARTTG	ARLDD	141
8	ADLRG	ATVDP	VLWRT	ASLVG	161
	ARVDV	*DQAVA*	*FAAAH*	*GLCLA*	181

UTILIZING NOVEL DNA MIMICS FOR VARIOUS APPLICATIONS

Since genes encoding DNA mimics (e.g. *ard*A and *qnr*ABCDS) are located on transmission elements, transposons, and plasmids, this promotes their wide distribution among bacteria of various species and genera.

Thus it is important to investigate in detail the structure of such proteins and the mechanisms of their action. The most illustrative example is the distribution of qnrABCDS genes responsible for the resistance to quinolone antibiotics among clinical bacterial strains. The search for and analysis of genes encoding DNA mimics and representing constituents of transmission elements are important tasks. Below we consider some putative variants of use of DNA-mimicking proteins. The DNA mimics may be successfully used for substitution of DNA during elucidation of spatial structure of the DNA-dependent enzymes by means of X- ray analysis (Dryden, 2006). In some cases, it is difficult to obtain crystals of the complexes of the DNA dependent enzymes and DNA and it is possible that substitution of DNA by the DNA mimics may solve this problem. There are examples illustrating successful use of such substitutions: Ugi–UDG (Putnam &Tainer, 2005). It is suggested that substitution of DNA by the Ocr protein might be used for crystallization of Ocr in its complex with an S-subunit of EcoKI. Therefore, it should be noted that spatial structure of S-subunit of two type I restriction–modification enzymes has been determined (Kim et al., 2005; Calisto et al., 2005). The DNA mimics can be used in affinity chromatography. Affinity columns with a DNA mimics can be used with high effectiveness for detection and purification of various types of DNA-dependent enzymes. Use of radioactive or fluorescent labels will increase thesensitivity of such method. The perspectives of in vitro construction of new types of DNA mimics (i.e. generation of proteins with different "design" and new functions) may be quite wide. These include potential tasks of constructing of DNA mimics, inhibiting or modulating activity of specific groups of DNA-dependent enzymes and tasks related to site directed changes in the structure of already known DNA mimics. Such works are rather successful. For example, using site-directed mutagenesis we have modified the structure of ArdA and Ocr; the modified proteins selectively inhibit the endonuclease (restriction) activity of type I restriction–modification enzymes without any influence on their methylase (modification) activity (Zavilgelsky et al., 2011). The use of such type of antirestriction proteins in gene engineering gives an opportunity to develop stable strains with hybrid plasmids because the process of specific modification of chromosome DNA remains unimpaired. The protein Ocr has already been used as an effective factor promoting significant increase of bacterial transformation by plasmids. Adding a small amount of the Ocr protein to solution with plasmid DNA causes significant (by several orders of magnitude) increase in effectiveness of cell transformation during electroporation. In this case unmethylated DNA and host bacteria with active type I restriction–modification system are used. The Ocr protein (as well as plasmid DNA) easily penetrates inside cells and immediately protects unmethylated DNA against degradation (EPICENTRE Forum 9, 8, htpp//www.epibio.com/forum.asp).

The highly charged, very acidic proteins OCR and ArdA may be used as a purification handle similar to other fusion tags. A monomeric mutant of the OCR protein (13.8 kDa, very acidic, pI = 3.8) was used as a novel fusion tag whith displays solubilizing activity with a variety of different passenger proteins (DelProposto et al., 2009).

In general, perspectives of the use of the DNA mimics might be related diagnostics and therapy of various diseases (e.g. for inhibition of specific enzymes and corresponding biochemical processes in cells).

AT A GLANCE

Modern data on the mechanisms of the modulation of the DNA - binding enzymes by protein mimicry of DNA are reviewed. It has recently been demonstrated that DNA-binding enzymes can be controlled by the direct binding of a control protein to the DNA-binding site on the enzyme. The structures of these control proteins have been discovered to mimic the structure and electrostatics of DNA. Such DNA-mimics might be able to target bacterial restriction systems (Ocr, ArdA), drug resistance systems (MfpA, QnrABCS), as well as replication, recombination, and repair. It puts forward a range of potential uses of new DNA mimics in applied biotechnology.

Figure 9 shows structures of Ocr and MfpA monomers and B-form DNA. Their comparison emphasizes the extraordinary capacities of living nature to develop unique forms crucial for adaptation. The most surprising thing is that nature has chosen different ways for design of proteins mimicking the DNA duplex. In one case (e.g. Ocr) these are tightly packed α- helices, in the other it is a right-handed β-helix (MfpA). Existence of significant negative charge (of the whole macromolecule or particular domain) required for similarity with the DNA polyanion is a common feature of DNA mimics. However, this is a necessary but not sufficient precondition. At the moment the only reliable method for detection of DNA mimics is X-ray analysis.

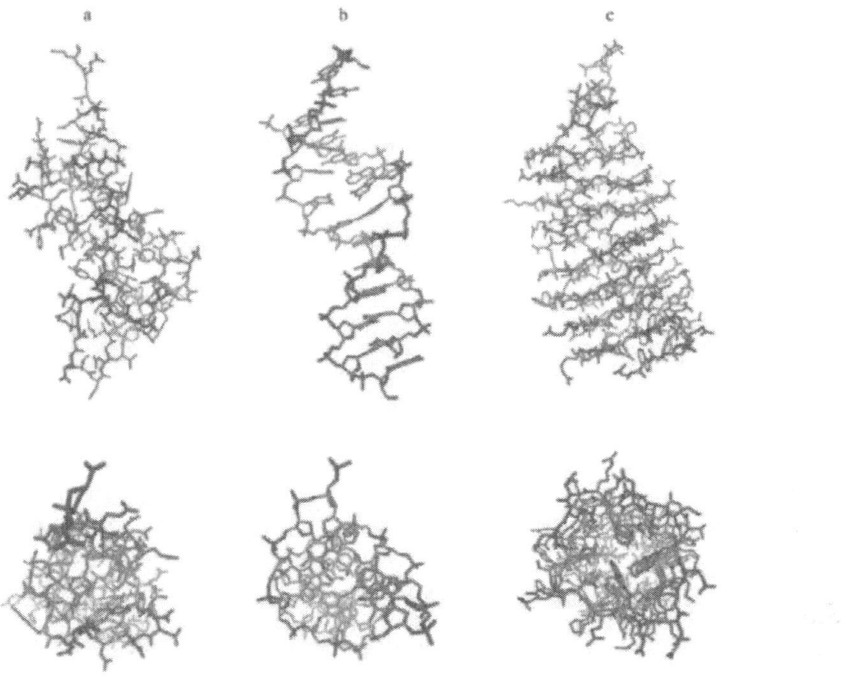

Fig. 9. The structures of Ocr(a) and MfpA(c) monomers and B-form of DNA(b)

REFERENCES

1. Atanasiu, C., Su, T.-J., Sturrock, S., & Dryden, D. (2002). Interaction of the Ocr gene 0.3 protein of bacteriophage T7 with EcoKI restriction-modification enzymes. *Nucleic Acids Res.* Vol.30, pp. 3936-3944

2. Bateman, A., Murzin, A., & Teichman, S. (1998). Structure and distribution of pentapeptide repeats in bacteria. *Protein Sci.* Vol.7, pp. 1477-1480

3. Bickle, T., & Kruger, D. (1993). Biology of DNA restriction. *Microbiol. Rev.* Vol.57, pp. 434–450 Belogurov, A., Yussifov,T., Kotova, V., & Zavilgelsky G. (1985).The novel gene *ard* of plasmidpKM101: alleviation of EcoKI restriction. *Mol. Gen. Genet.* Vol.198, pp. 509–513.

4. Belogurov, A., & Delver, E. (1995). A motif conserved among the Type I restriction-modification enzymes and antirestriction proteins: a possible basis for mechanism of action of plasmid encoded anti-restriction function. *Nucleic Acids Res.* Vol.23, pp. 784-787

5. Black, K., Buikema, W., & Haselkorn, R. (1995). The *hgl*K gene is required for localization of heterocyst-specific glycolipids in the cyanobacterium *Anabaena sp.* strain PCC7120. *J. Bacteriol.*, Vol.171, pp. 6440-6448

6. Buchko, G., Ni, S., Robinson, H., Welsh, E., Pakrasi, H., & Kennedy, M. (2006a). Characterization of two potentially universalturn motifs that shape the repeated five residues fold-crystal structure of a luminal pentapeptide repeat protein from *Cyanothece* 51142. *Protein Sci.*, Vol.15, pp. 2579-2595

7. Buchko, G., Robinson, H., Ni, S., Pakrasi, H., & Kennedy, M. (2006b). Cloning, expression , crystallization and preliminary crystallographic analysis of a pentapeptide-repeat protein (Rfr23) from the bacterium *Cyanothece* 51142. *Acta Crystallogr. Sect. F. Struct. Biol. Cryst. Commun.* Vol.62 (Pt. 12), pp. 1251-1254

8. Calisto, B., Pich, O., Pifiol, J., Fita, I., Querol, E., & Carpena, X. (2005). Crystal structure of a putative type I restriction-modification S subunit from *Mycoplasma genitalium*. *J. Mol. Biol.*, Vol.351, pp. 749-762

9. Cattoin, V., & Nordmann, P. (2009). Plasmid-mediated quinolone resistance in gram- negative bacterial species. *Curr.Med.Chem.* Vol.16, pp. 1028-1046

10. Cavaco, L., Hasman, H., Xia, S., & Aarestrup, F. (2009). *qnr*D, a novel gene conferring transferable quinolone resistance in *Salmonella enterica* serovar Kentucky and Bovismorbificans strains of human origin. *Antimicrob.Agents Chemother.* Vol.53. pp. 603-608

11. Chilley, P., & Wilkins, B. (1995). Distribution of the *ard*A family of antirestriction genes on conjugative plasmids. *Microbiology*. Vol.141, pp. 2157–2164

12. DelProposto, J., Majmudar, C., Smith, J., & Brown, W. (2009). Mocr: A novel fusion tag for enhancing solubility that is compatible with structural biology applications. *Protein Expression and Purification.* Vol.63, pp. 40-49

13. Delver, E., Kotova, V., Zavilgelsky, G., & Belogurov, A. (1991). Nucleotide sequence of the gene (*ard*) encoding the antirestriction protein of plasmid ColIb-P9. *J. Bacteriol.* Vol.173, pp. 5887–5892

14. Dryden, D.. (2006). DNA mimicry by proteins and the control of enzymatic activity on DNA.*Trends Biotech.* Vol.24, pp. 378-382

15. Dunn, J., Elzinga, M., Mark, K., & Studier, F. (1981). Amino acid sequence of the gene *0.3* protein of bacteriophage T7 and nucleotide sequence of its messenger RNA. *J. Biol. Chem.*Vol.256, pp. 2579–2585

16. Hegde, S., Vetting, M., Roderick, S., Mitchenall, L., Maxwell, A., Takiff, H., & Blanchard, J. (2005). A fluoroquinolone resistance protein from *Mycobacterium tuberculosis* that mimics DNA. *Science*. Vol.308, pp. 1480-1483.

17. Hegde, S., Vetting, M., Mitchenal, L., Maxwell, A., & Blanchard, J. (2011). Structural and biochemical analysis of the Pentapeptide-Repeat Protein, EfsQnr, a potent DNA gyrase inhibitor. *Antimicrob. Agent Chemother.* Vol.55, pp. 110-117.

18. Iida, S., Hiestand-Nauer, R., Sandmeier, H., Lehnherr, H., & Arber, W. (1988). Accessory genes in the *dar*A operon of bacteriophage P1 affect antirestriction function generalized transduction, head morphogenesis, and host-all lysis. *Virology*. Vol.251, pp. 49–58

19. Jacoby, G., Cattoir, V., Hooper, D., Martinez-Martinez, L., Nordmann, P., Pascual, A., Poirel, L., & Wang, M. (2008). *qnr* gene nomenclature. *Antimicrob. Agents Chemother.*Vol.52, pp. 2297-2299

20. Kim, J.-S., DeGiovanni, A., Jancarik, J., Adams, P., Yokota, H., Kim, R., & Kim, S.-H. (2005). Crystal structure of DNA sequence specificity subunit of a type I restriction- modification enzyme and its functional implications. *Proc. Natl. Acad. Sci. USA*. Vol.102, pp. 3248-3253

21. Kotova, V., Zavilgelsky, G., & Belogurov, A. (1988). Alleviation of type I restriction in the presence of Incl plasmids: General characterization and molecular cloning of the *ard* gene. *Mol. Biol. (Moscow)*. Vol.22, PP. 270–276

22. Kruger, D., Hansen, S., & Reuter, M. (1983). The *ocr* gene function of bacteriophages T3 and T7 counteracts the *Salmonella typhimurium* DNA restriction systems SA and SB. *J. Virol.* Vol.45, pp. 1147–1149

23. Liu, D., Ishima, R., Tong, K., Bagby, S., Kokubo, T., Muhandiram, D., Kay, L., Nakatani, Y.,& Ikura, M. (1998) Solution structure of a TBP-TAF(II)230 complex: protein mimicry of the minor groove surface of the TATA box unwoind by TBP. *Cell,* Vol.94, pp. 573-583

24. Loenen, W. (2003). Tracking *Eco*KI and DNA fifty years on: A golden story full of surprises.*Nucleic Acids Res.* Vol.31, pp. 7059–7069

25. Martinez-Martinez, L., Pascual, A., & Jacoby, G. (1998). Quinolone resistance from a transferable plasmid. *Lancet*, Vol.351, pp. 797-799

26. McMahon, S., Roberts, G., Johnson, K., Cooper, L., Liu, H., White, J., Carter, L., Sanghvi, B., Oke, M., Walkinshaw, M., Blakely, G., Naismith, J., & Dryden, D. (2009). Extensive DNA mimicry by the ArdA antirestriction protein and its role in the spread of antibiotic resistance. *Nucleic Acids Res.* Vol.37, pp. 4887-4897

27. Mol, C., Arval, A., Sandersen, R., Slupphaug, G., Kavil, B., Krokan, H., Mosbaugh, D., & Tainer, J. (1995). Crystal structure of human uracil-DNA-glycosylase in complex with a protein inhibitor: protein mimicry of DNA. *Cell.* Vol.82, pp. 701–708

28. Montero, C., Mateu, G., Rodrigez, R., & Takiff, H. (2001). Intrinsic resistance of *Mycobacterium smegmatis* to fluoroquinolone may be influenced by new pentapeptide protein MfpA. *Antimicrob. Agents Chemother.* Vol.45, pp. 3387-3393

29. Morita, M., Tomita, K., Ishizawa, M., Tagaki, K., Kawamura, F., Takahashi, H., & Morino, T. (1999). Cloning of oxetanocin A biosynthetic and resistance genes that reside on a plasmid of *Bacillus megaterium* strain NK84-0128. *Biosci. Biotechnol. Biochem.* Vol.63, pp. 563-566

30. Murray, N. (2000).Type I restriction systems: Sophisticated molecular machines (a legacy of Bertani and Weigle). *Microbiol. Rev.* Vol.57, pp.412–434

31. Murray, N. (2002). Immigration control of DNA in bacteria: Self versus non-self.*Microbiology.* Vol.148, pp. 3–20

32. Nekrasov, S., Agafonova, O., Belogurova, N., Delver, E., & Belogurov, A. (2007). Plasmid- encoded antirestriction protein ArdA can discriminate between type I methyltransferase and complete restriction-modification system. *J. Mol. Biol.* Vol.365, pp. 284–297.

33. Parsons, L., Yeh, D., & Orban, J. (2004). Solution structure of the highly acidicprotein HI1450 from *Haemophilus influenzae*, a putative double-stranded DNA mimic. *Protein Struct. Func. Bioinform.* Vol.54, pp. 375-383

34. Pierrat, O., & Maxwell, A. (2005). Evidence for the role of DNA strand passage in the mechanism of action of microcin B17 on DNA gyrase. *Biochemistry.* Vol.44, pp. 4202-4215

35. Putnam, C., & Tainer, J. (2005). Protein mimicry of DNA and pathway regulation. *DNA repair.* Vol.4, pp. 1410–1420

36. Ramirez, B., Voloshin, O., Camerini-Otero, R., & Bax, A. (2000) Solution structure of DinI provides insight into mode of RecA inactivation. *Protein Sci.* Vol.9, pp. 2161-2169

37. Roberts, R., Vincze,T., Posfai, J., & Macelis, D. (2005). REBASE: Restriction enzymes and DNA methyltransferases. *Nucleic Acids Res.* Vol. 33, D230-D232

38. Robicsek, A., Jacoby, G. & Hooper, D. (2006). The worldwide emergence of plasmid - mediated quinolone resistance. *Lancet Infect Dis.* Vol.6, pp. 629-640

39. Rodriguez-Martinez, J., Cano, M., Velasco, C., Martinez-Martinez, L., & Pascual, A. (2011). Plasmid-mediated quinolone resistance: an update. *J. Infrct. Chemother.* Vol.17, pp. 149-182

40. Serfiotis-Mitsa, D., Roberts, G., Cooper, L., White, J., Nutley, M., Cooper, A., Blakely, G., & Dryden, D. (2008). The *orf*18 gene product from conjugative transposon Tn916 is an ArdA antirestriction protein that inhibits type I DNA restriction-modification systems. *J. Mol. Bio*l. Vol.383, pp. 970-981

41. Thomas, A., Brammar, W. & Wilkins, B. (2003). Plasmid R16 ArdA protein preferentially targets restriction activity of the type I restriction-modification system *Eco*KI. *J. Bacteriol.* Vol.185, pp. 2022–2025

42. Tock, M. & Dryden, D. (2005). The biology of restriction and antirestriction. *Curr. Opin.Microbiol.* Vol.8, pp. 466–472

43. Tran, J., Jacoby, J. & Hooper, D. (2005). Interaction of the plasmid-encoded quinolone resistance protein Qnr with *Escherichia coli* DNA gyrase. *Antimicrob. Agent Chemother.* Vol.49, pp. 118-125

44. Vetting, M., Hegde, S., Fajardo, E., Fiser, A., Roderick, S., Takiff, H. & Blanchard, J. (2006). Pentapeptide repeat proteins. *Biochemistry.*Vol.45, pp. 1-10

45. Vetting, M., Hegde, S. & Blanchard, J. (2009). Crystallization of a pentapeptide-repeat protein by reductive cycle pentylation of free amines with glutaraldehyde. *Acta Crystallogr. D Biol. Crystallogr.* Vol.65, pp. 462-469

46. Walkinshaw, M., Taylor, P., Sturrock, S., Atanasiu, C., Berge, T., Henderson, R., Edwardson,J. & Dryden, D. (2002). Structure of Ocr from bacteriophage T7, a protein that mimics B-form DNA. *Mol. Cell.* Vol.9, pp. 187–194

47. Wang, M., Guo, Q., Xu, X., Wang, X., Ye, X., Wu, S., Hooper, D. & Wang, M. (2009). New plasmid-mediated quinolone resistance gene, *qnr*C, found in a clinical isolate of *Proteus mirabilis*. *Antimicrob. Agents Chemother.* Vol.53. pp. 1892-1897

48. Wilkins, B. (1995). Gene transfer by bacterial conjugation: Diversity of systems and functional specializations. In: *Population Genetics of Bacteria. (Soc. Gen. Microbiol.Symp. 5.* pp. 59-88. S.Baumberg et al. Cambridge: Cambridge Univ. Press.

49. Xiong, X., Bromley, E., Oelschlaeger, P., Woolfson, D. & Spencer, J. (2011). Structural insights into quinolonw antibiotic resistance mediated by pentapeptide repeat proteins: conserved surface loops direct the activity of a Qnr protein from a Gram-negative bacterium. *Nucleic Acids Res.* Vol.39, pp. 3917-3927

50. Zavilgelsky, G. (2000). Antirestriction. *Mol. Biol. (Moscow).* Vol.34, pp. 854–862

51. Zavilgelsky, G., Letuchaja, T. & Rastorguev, S. (2004). Antirestriction activity of ArdA protein encoded by the IncI1 R64 transmissible plasmid. *Mol. Biol. (Moscow)*. Vol.38, pp. 901–906

52. Zavilgelsky, G., Kotova, V. & Rastorguev, S. (2008). Comparative analysis of antirestriction activities of ArdA (ColIb-P9) and Ocr (T7) proteins. *Biochemistry (Moscow).* Vol.73, pp. 1124–1130

53. Zavilgelsky, G., Kotova, V. & Rastorguev, S. (2009). Antirestriction and antimodification activities of the T7 Ocr protein: Effect of mutations in interface. *Mol. Biol. (Moscow).* Vol.43, pp. 103–108

54. Zavilgelsky, G., Kotova, V. & Rastorguev, S. (2011). Antimodification activity of the ArdA and Ocr proteins. Russian. J. Genetics. Vol.47, pp. 139-146

CHAPTER 6

BIOTECHNOLOGY INVOLVING PLATELET-RICH PLASMA (PRP): IDEAS AND CLINICAL USES IN ORTHOPEDICS AND SPORTS MEDICINE

INTRODUCTION: AN OVERVIEW

In the dynamic fields of orthopedics and sports medicine, Platelet-Rich Plasma (PRP) biotechnology has emerged as a groundbreaking approach with significant potential for enhancing patient care and performance. This innovative biotechnological advancement harnesses the healing power of the body's own platelets, offering a multitude of clinical applications that extend far beyond traditional medical practices. In this exploration, we delve into the multifaceted world of PRP biotechnology, uncovering its diverse applications, therapeutic possibilities, and the promising future it holds in the realms of orthopedics and sports medicine.

Regenerative medicine is the augmentation or substitution of diseased or injured cells or tissues by one of two means: (1) an improvement in the ability of endogenous cells to reform damaged tissue or (2) the use of exogenous cells or tissues to replace damaged cells or tissues. Advances in regenerative medicine essentially depend on improving our understanding of cell biology and molecular signaling. Cell signaling is complex and incompletely understood due to the multiple interactions and cross-talk among system components. The human body has some 100 trillion cells, which in the healthy state coordinate their actions through an exchange of chemical signals to maintain body homeostasis. Every cell phenotype secretes signaling proteins that influence their own behavior (autocrine) or the behavior of other neighboring cells (paracrine) through interactions with specific transmembrane receptors located in the cellular membrane. Currently, a great deal of research is directed towards improving our understanding of intercellular communication and the intracellular transduction of these signals; in the field of regenerative medicine, this knowledge will help to disentangle the mysteries of tissue repair and to achieve proper tissue repair and regeneration. Moreover, to reach this goal we must integrate all the information and understanding derived from basic research into novel therapies that yield quicker and more efficient tissue regeneration.

Within the last decade, the development of platelet-rich plasma (PRP) technology has emerged. The impact of the discoveries regarding the potential of PRP healing has fueled the optimism about autologous regenerative medicine. Indeed, the emergence and application of PRP technology, i.e., autologous molecular pool, has revolutionized the field of regenerative medicine in part due to the repair capacities of growth factors (GFs) and cytokines secreted by platelets. The easy preparation protocols, biosafety and versatility of PRP preparations have stimulated translational research and interest by both the scientific and medical communities. PRP therapies represent a major breakthrough in the treatment of many medical conditions and are currently one of the hottest topics in regenerative medicine because of their important implications for our future health. Discovery and contributions in the field have not only improved the clinical treatment of many patientswith different clinical conditions but, from a multimolecular perspective, have opened the field of PRP science to cellular and molecular exploration of healing mechanisms. This technology provides the opportunity of moving molecular knowledge off the shelves and into practice, making it relevant in a clinical context, and achieves a true marriage between what we have learned through research and clinical applications.

This chapter will provide an overview of the potential therapeutic use of platelets and plasma for the release of signaling proteins in regenerative medicine. For the purposes of this chapter, the basic principles of healing and the role of platelets as molecular reservoirs will be discussed. A detailed description of the potential technological relevance of PRP biotechnology is followed by a section on applications of PRP therapies in numerous clinical conditions and medical fields with a special emphasis on orthopedics and sports medicine. There is no question that the key to both future advances in PRP science and its application in the treatment of disease and trauma lies in a better understanding of repair processes.

A PICTURE OF HEALING MECHANISMS

The most effective way to improve tissue repair is to understand normal healing mechanisms after a perturbation due to disease, which then becomes the basis for improving patient care and health. Healing mechanisms are, to a great extent, shared by the different tissues of the body and can be depicted by overlapping and successive phases characterized by a preponderance of cell signaling from various systems. The spatially and temporally dynamic nature of healing mechanisms presents a challenge to the identification of critical mechanisms. Firstly, hemostasis is accomplished through a network of processes that include the platelet system and the coagulation cascade; such processes arrest bleeding and set in motion the inflammatory response.

Early inflammatory response

Inflammation and blood coagulation are intimately linked. Acute inflammation, the complex systemic early defense system, is the first reaction of the innate immune system (platelet, leukocytes and macrophages) to injury. Direct exposure of cells to physical, mechanical or chemical trauma has immunological consequences relative to the degree of injury, i.e., the apoptotic or necrotic condition of resident fibroblasts. Accordingly, local regulatory mechanisms adjust the magnitude of the response so that inflammatory processes are localized to areas of damage, and the amount and duration of immune cell infiltration are adequate to phagocyte apoptotic/necrotic cells. In addition, endothelial cells, which are actively involved in healing, limit clot formation to the sites of injury. Activated platelets and leukocytes within this clot then release growth factors and numerous cytokines, establishing the onset of inflammation.

Eventually, spatially and temporally changing patterns of various leukocyte subsets transmigrate across the endothelium. Circulating neutrophils are rapidly captured by selectins that are presented by endothelial cells; they then invade the wounded tissue in response to chemical signals. The lifespan of neutrophils in the injured tissue is about two days, during which they perceive signals from the environment and respond by secreting cytokines (Borregaard et al., 2007). Furthermore, neutrophils release stored substances carried in different granule subsets, including reactive oxygen species, cationic peptides orproteases. The key role of neutrophils is to clear the early rush of contaminating bacteria; in a sterile wound, such as surgical incisions that are experimentally induced, neutrophil absence does not perturb the healing process.

Monocyte recruitment and infiltration at the injury site happens days later and is highly regulated by adhesion molecules expressed by endothelial cells and by chemokines and other substances released by platelets, neutrophils (Soehnlein et al., 2009) and apoptotic/necrotic cells (Nathan, 2006). Commanded by signals present in the environment, monocytes turn into macrophages (the dedicated phagocytes) and induce major changes in gene expression and cell function. Indeed, the severity of tissue injury may determine the different states of macrophage activation. "Innate" activation occurs through lipolysaccharide or interferon-γ (IFN-γ) and is associated with a pro-inflammatory state [the production of interleukin-6 (IL-6), interleukin-1β (IL-1β) and tumor necrosis factor-α (TNF- α). Alternatively, "classical" activation occurs through IL-4/IL-23 and is associated with the synthesis of healing factors including transforming growth factors (TGF-β and TGF–α), basic fibroblastic growth factor (bFGF), platelet-derived growth factor (PDGF), and vascular endothelial growth factor (VEGF) (Krysko et al., 2006).

Recent research suggests that these features of inflammation may determine the difference between efficient repair and the failure to repair. For example, in animal experiments, neutropenia accelerated the closure of incision wounds (Dovi et al., 2003) but did not affect the healing of surgically repaired tendons. Macrophage depletion impaired skin wounding by reducing collagen deposition and angiogenesis and also impaired the response to wounding in diabetic mice. Other studies suggest that targeting macrophage activation may provide a new therapeutic approach to protect tissues from ischemia and promote repair. Notwithstanding, macrophage depletion significantly improved the morphology and biomechanical properties of the tendon-bone interface after experimental anterior cruciate ligament (ACL) surgery. Thus, there are large gaps in the understanding of how neutrophils and macrophages influence repair. The difficulty in understanding the inflammatory response stems, in part, from biological redundancy: one molecule may have several functional roles, and different molecules may perform overlapping functions.

Trophic phase

New tissue formation occurs 2-10 days after injury and is characterized by cellular proliferation and the migration of different cell types. New blood vessels are formed by a process known as angiogenesis, and later, the sprouts of capillaries along with fibroblasts and macrophages replace the fibrin matrix with granulation tissue that forms the new substrate for cell migration.

Cell proliferation and migration

The proliferative phase begins with the formation of a fibrin, fibronectin glycosiaminoglycan, and hyaluronic acid matrix that is initially populated with macrophages and platelets. The various cytokines secreted by these cells enhance cell migration into the site using the fibrin and fibronectin matrix as a scaffold. Progenitors of differentiated cell types, such as bone, cartilage, muscle, nerve sheath and connective tissue cells, are thought to contribute to a collection of proliferating progenitor cells. Alternatively, progenitor stem-cell-like for tissue niches migrate, divide and differentiate into tissue fibroblasts. Fibroblasts move through the extracellular matrix by binding fibronectin, vitronectin and fibrin via their arginine-glycine- aspartic acid amino acid sequence recognized by their integrin receptors. The fibroblasts proliferate in response to GFs and cytokines and become the predominant cell type by the third to fifth day following injury. Fibroblasts also secrete extracellular zinc dependent endopeptidases called metalloproteinases (MMPs), which facilitate their movement through the matrix and help with the removal of damaged matrix components. Once the fibroblasts have entered the wound, they produce collagen, proteoglycans and other components. Fibroblast activities are predominantly regulated by GFs such as PDGF and TGF-β. PDGF secreted by platelets and macrophages stimulates fibroblast proliferation, chemotaxis and collagenase expression. TGF-β has pleiotropic actions that are context-dependent.

Angiogenesis

Angiogenesis occurs with the formation of new capillary networks through endothelial cell migration and division. Endothelial cells are activated to initiate angiogenesis such that new blood vessels are initiated to promote blood flow to support the high metabolic activity in the newly deposited tissue. Angiogenesis is regulated by a combination of local stimulatory factors such as VEGF and anti-angiogenic factors such as angiostatin, endostatin, and thrombospondin. Local factors that stimulate angiogenesis include low oxygen tension, low pH and high lactate levels. Soluble mediators such as bFGF, HGF, TGF- β and VEGF also stimulate endothelial cells to produce vessels. Tissue oxygen levels directly regulate angiogenesis through hypoxia inducible factor (HIF), which binds oxygen. When there is a decrease in oxygen levels surrounding capillary endothelial cells, HIF-1 levels increase and stimulate VEGF transcription to promote angiogenesis. Animal studies have shed some light on the natural pattern of GF expression during this stage. For example, signaling of VEGF-A via the endothelial receptors VEGFR1 and VEGFR2 is present at the healing site early after tissue injury. Other growth factors including TGF-β, PDGF-BB, and angiopoietin-1, which are important for vessel stabilization, are expressed later at the healing site.

The new vasculature allows the delivery of nutrients and the removal of by-products. As noted above, granulation tissue consists of a dense network of blood vessels and capillaries, elevated cellular density of fibroblasts and macrophages and randomly organized collagen fibers. The metabolic rate of this tissue is high and reflects the activity required for cell migration, division and protein synthesis, which emphasizes the importance of adequate nutrition and oxygen to properly heal the wound. Granulation tissue is particularly abundant and accompanies the process of wound healing by secondary intention.

Synthesis of the extracellular matrix

The high concentration of growth factors and cytokines initially secreted by platelets and leukocytes and later amplified by macrophages induces a rapid increase in specific cell populations, including migrating fibroblasts and resident cells. The number of stromal cells increases in parallel with angiogenesis, which is readily evident in the hypoxic environment produced by the injury. So, the production of extracellular matrix molecules grows in proportion with increasing cell number.Growth factors, including TGF-β1, PDGF, BDNF, bFGF and type-I insulin-like growth factor (IGF-I), function at various stages during the healing process and produce differentoutcomes depending on the conditions. For instance, PDGF, a chemotactic and mitotic factor for fibroblasts, also induces the synthesis of collagen type I. TGF-β1, which peaks early in injuries, is essential for the recruitment and maintenance of progenitor cells during neo-tissue formation, and its function might be necessary during healing. Additionally, the interactions of TGF-β1 with other TGF-β isoforms, namely TGF-β2 and TGF-β3, mediate which type of collagen is synthesized in the healing tissue. IGF-I anabolic and anti-apoptotic activities are regulated by IGF-I binding proteins (BP), IGFBP- 2, IGFBP-3, and IGFBP-4, which are also present in the early healing response. In both humans and animals, the expression of IGF-I and TGF-β1 preceded the stimulation of collagen synthesis, a relevant issue in tissue healing. The bioactivity of these growth factors is regulated not only at the receptor level but through activation of TGF-β complex and IGF-binding proteins.

Tissue remodeling and scarring

- Finally, the tissue enters into the last phase of healing, a long remodeling phase in which granulation tissue matures into a scar. Collagen accumulation reaches a maximum at 2-3 weeks after injury, and the transition to remodeling begins. There is a balance between synthesis, deposition and degradation during this phase. Small capillaries aggregate into larger blood vessels, and there is an overall decrease in the water content of the wound. Similarly, cell density and the overall metabolic activity of the wound decrease. The most dramatic change occurs in the overall type, amount and organization of the collagen fibers, resulting in an increased tensile strength of the tissue. Initially, there is increased deposition of collagen type III, also referred to as reticular collagen, that is gradually replaced by collagen type I. Collagen fibers are cross-linked by the enzyme lysyl oxidase, which is secreted by fibroblasts in the extracellular matrix. The normal adult 4:1 ratio of type I to type III collagen is restored during remodeling. Equilibrium is established as new collagen is formed and collagen type III is degraded. The MMPs, collagenases, gelatinases and stromelysins, control the degradation of extracellular matrix components to facilitate cell migration into the wound, angiogenesis and overall tissue remodeling.

- In each of the described healing phases, the specific signaling activity is silenced or counterbalanced by other endogenous signals that serve to limit the duration and to promote progression to a new stage. During all these stages, local and migratory cells synthesize different patterns of GFs and cytokines in an attempt to cope with the temporal demands of the healing tissue. Consequently, therapeutic approaches to manipulate healing may need to integrate multiple cell types and large signaling networks that are necessary for the dynamic communication between cells. The need to target various signaling pathways simultaneously demands the administration of a balanced combination of mediators instead of administering a purified isolated protein, which could not cope with the multiple requirements of the injured tissue. Therefore, the ability to release signaling molecules in a spatiotemporal manner that mimics the needs of the injured tissue has become a challenge in the scientific and medical fields.

Pathologic tissue healing

- There are many categories of impediments for wound healing. These include local or systemic impediments. The former include tissue viability, seroma and/or hematoma,infection, insufficient blood supply and/or mechanical factors. For example, adequate blood supply must exist to provide nourishment and oxygenation to healing tissues. A lack of blood supply may lead to tissue ischemia and an increased risk of infection. Tissues do not heal if there are more than 10^5 bacteria per gram of tissue. Hence, necrotic tissue must be debrided to avoid the risk of infection and because it interferes with normal healing.
- Mostly clinical differences between chronic and acute healing tissues are thought to be explained in part by alterations in the local biochemical environment. The observation that increased amounts of TGF-β1 were present in hypertrophic scars led to clinical efforts to block scar formation through administration of antibodies against TGF-β1 and other pro-inflammatory mediators. Recent evidence also suggests that changes in the molecular environment of the cells in the wound may change the fate of healing tissues or organs. It is becoming increasingly apparent that growth factors and cytokines play a variety of key roles during normal tissue repair, and many have also been shown to act therapeutically in situations where normal healing is impaired. Although tremendous strides have been made in delineating the myriad of factors involved in normal and pathological healing, it has become clear that single-agent therapies,

such as administration of growth factors, have only a moderate impact on tissue repair in the clinical setting, probably due to the redundancy and plasticity of the components of tissue repair or their rapid degradation at the injured site.

- In this context, the emergence of PRP biotechnology as a way to harness tissue regeneration for medical needs has fueled the optimism surrounding cell-signaling based regenerative medicine. A deeper understanding will accelerate the development of PRP therapies.

PLATELETS: MOLECULAR CONTRIBUTION TO HEALING

The ultimate solution to tissue healing is likely to be the administration of multimolecular preparations with the ability to elaborate the full complexity of biological signaling, including all the environmental cues that are needed to regulate the biological mechanisms described above. Platelets are a natural source of growth factors and cytokines involved in tissue healing. Until now, it has not been possible to provide a pool of molecular signals and the temporary cell scaffold necessary to initiate healing in the same therapeutic agent. Substantial progress in the understanding of platelet biology has revealed much about the complexity of PRP therapies. Additional insights come from combining the information from the plasma proteome.

Platelet biology

- Our understanding of fundamental aspects of platelet biology and function has been enriched in the last decades. Platelets are discoid cellular elements that are heterogeneous in size and have the smallest density of all blood cells, at 2 μm in diameter (a leukocyte is about 20 μm in diameter). They are anucleate and originate in the bone marrow as bulges along the length of pseudopodial extensions of megakaryocytes. The so termed proplatelets are then fragmented into individual platelets and released into the blood stream where they travel for about 7-10 days before removal from circulation after senescence and are replaced with younger platelets possessing greater functional capabilities. Platelets are replete with secretory granules, which are critical to platelet function. Among the three types of granules,dense granules, alpha-granules and lysosomes, the alpha-granule is the most abundant. There are approximately 50-80 alpha-granules per platelet, although they are heterogeneous with regard to cargo (Villeneuve et al., 2009).
- For example, anti-angiogenic proteins are packaged in different alpha-granules subpopulations than pro-angiogenic proteins. Moreover, there is some evidence that secretion of pro- versus anti-angiogenic stores may be

agonist-specific (Italiano et al., 2008). The total protein content of platelets includes not only soluble proteins that are released into the extracellular space but also membrane bound proteins that become expressed on the platelet surface. Many of the proteins found in α- granules are also present in plasma. In a recent survey of the platelet membrane proteome, 629 membrane proteins were detected (Maynard et al., 2007). Overall proteomic studies suggest that more than 1048 soluble proteins are present in the supernatant of platelets.

Platelet function

- Not long ago, platelets were merely considered to function as haemostatic agents. However, as researchers broadened their understanding of platelets, many more facets were identified. Around 1980, platelets were recognized for their healing function. More than a decade later, the involvement of platelets in angiogenesis was discovered. Subsequently, Folkman showed that angiogenesis regulating proteins were selectively pumped into the budding pro-platelets from the mother megakaryocyte and that PF-4 is captured by platelets in tumor-bearing animals. Further developments using PRPs as a therapeutic biotechnology in the past few years have allowed the direct observation of platelet secretomes, not within, but outside the blood stream that interact with various injured tissues and organs.
- In the physiological process of wound healing, platelets embedded within blood clots serve as a primary source of biologically active factors. Therefore, the PRP concept is straightforward. As platelets are a major source of healing factors within blood clots, the idea that concentrating platelets at the injured site could accelerate and optimize healing mechanisms opens the door for the development of PRP therapies. For example, typically after muscle strains or contusion, the hematoma that originates as a consequence of vessel disruption contains about 94% red blood cells, a small amount of platelets (4%) and less than 1% leukocytes. The rational for the use of PRPs involves replacing the blood clot with adhesive PRP, thus minimizing the presence of red blood cells (about 95% in volume) while increasing platelet concentration at the injury site. In doing so, we would achieve a supra-physiological concentration of platelet and plasma proteins that accelerates the repair process by direct or indirect mechanisms, i.e., by attracting immune cells via chemotaxis or enhancing further synthesis of healing proteins by local cells. Moreover, the ability to release these signaling factors in a spatiotemporal manner using the fibrin scaffold perfectly meets the needs of the injured tissue over time.However, the present knowledge of both PRP therapies and healing mechanisms

needs to be better explored to translate such knowledge into improved biological therapies.

BIOTECHNOLOGICAL INNOVATIONS IN PLATELET-RICH PLASMA FOR TISSUE REGENERATION

The history: A three decades perspective

From a historical point of view, the first blood bank PRP preparations began during the 1960s and become routine preparations through the 1970s. In the 1980s, the advent of regenerative medicine aiming to rapidly translate the science into patient care using the patient's ownresources opened the door to the use of platelets as vehicles for the delivery of a balanced pool of healing factors. At that time, platelets were found to release wound healing substances that initiated the repair of injured tissues and vessels in cutaneous ulcers (Margolis et al., 2001). Later in the 1990s, platelets were introduced into maxillofacial surgery as autologous modifications of potent adhesives known as fibrin glues. The use of platelets was particularly fortuitous given that the main initial interest was to take advantage of the adhesive and haemostatic properties of the homologous fibrin during bone surgery. A realization of the clinical potential of PRP-therapies followed the positive clinical observations, such as enhanced bone formation and anti-inflammatory functions, during oral and maxillofacial applications (Whitman et al., 1997; Marx et al., 1998; Anitua E, 1999). At the beginning of the millennium, PRP was used for the first time to treat knee injuries in arthroscopic surgery (Sánchez et al., 2003 a and b), and later it was extended to the treatment of tendons (Sánchez et al., 2007), muscles injuries (Sánchez et al., 2005), osteoarthritic knees (Sánchez et al., 2008) and hips (Sánchez et al., 2011) and for use in chondropathies (Kon et al., 2010). Below we show the temporal sequence of the development of PRP therapies.

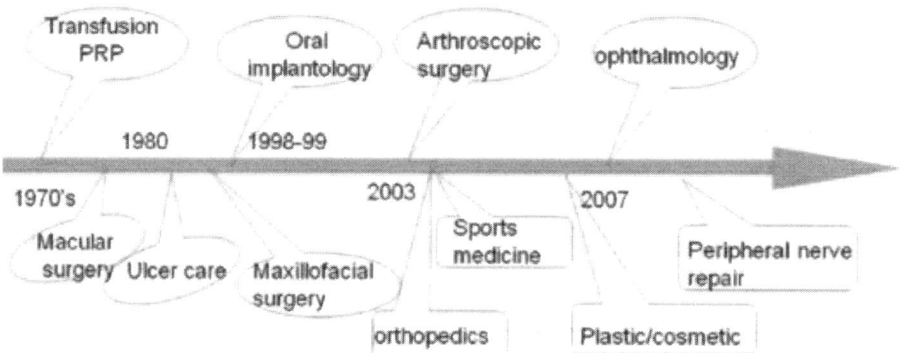

Fig. 1. Temporal sequence of the development of PRP technologies

Terminology

Long before any therapeutic application was imagined, the term PRP, which described plasma with a platelet count above the peripheral blood, was coined by hematologists. In 2007, the novel connotation of PRP was introduced to the Medical Subject Heading database (MeSH): PRP refers to a product consisting of PLATELETS concentrated in a limited volume of PLASMA used in various surgical tissue regeneration procedures where the GROWTH FACTORS in the platelets enhance wound healing and regeneration. At present (2011), the field is growing more complex, and the primacy of growth factors is now shared by new classes of platelet released biomolecules, which are also critical in healing.

PRP preparation

Peripheral blood is the supply source for the preparation of PRPs; the mean number of circulating platelets is 200,000 plt/μl. For PRP preparation, peripheral blood is drawn from the patient under sterile conditions, with or without anticoagulants, and the plasma is prepared by centrifugation or filtration. The volume can be adapted to the clinical needs, ranging from 10 to 100 mL. Essentially, the methods of producing PRPs determine the composition and concentration of leukocytes, erythrocytes and platelets in a given plasma volume. There are three methods: 1) the double spinning method using automated machines and commercial kits, 2) the single spinning method using conventional laboratory centrifuges followed by manual PRP separation, and 3) selective blood filtration using commercially available technology. Single spinning yields a 1-3 fold change in platelet concentration over baseline levels, and double spinning yields a 4-8 fold change in platelet concentration over baseline levels. Double spinning also concentrates leukocytes. Accordingly, platelet concentrates have been categorized as pure platelet-rich plasma (P-PRP), in which leukocytes are purposely eliminated from the PRP, and leukocyte and platelet-rich plasma (L-PRP), which contains a high concentration of leukocytes (Dohan et al., 2009).

At present, there is much debate surrounding four central questions of clinicians: (1) is the number of platelets important, (2) is the presence of leukocytes important, (3) when should PRP be activated, and (4) how should PRP be activated. The clinical variability observed throughout the studies points out that some techniques might not produce a sufficient number of functional platelets to produce the expected outcome. Similarly, there is no consistency in the methods of application of this therapy, the timing of treatment, the number of injections per series or the volume of injections. This has precluded the establishment of the standards necessary to integrate the extensive relevant literature in basic and clinical science. For example, double spinning techniques yield a PRP concentrate with a volume of about 10% of the volume of blood withdrawn (i.e., 20 mL of whole blood would result in 2 mL of PRP). In contrast, 40-50% of the blood volume is obtained after single spinning. Also, each method leads to a different product with differing biological properties and potential uses. Currently, it is unclear whether these differences have any clinical relevance. Some authors have suggested that PRP preparations containing only moderately elevated platelet concentrations induce optimal biological benefit, whereas lower platelet concentrations produce suboptimal effects and higher concentrations produce inhibitory effects. According to others, the 'therapeutic dose' of PRP is at least 4-6 times higher than the normal platelet count. To add to the discussion, the actual growth factor content does not correlate with the platelet count in whole blood or in PRP when leukocytes are present in the preparation, and there is no evidence that gender or age affects platelet count or growth factor concentrations. However, age may influence the number of receptors on local cells interacting with the plasma signals.

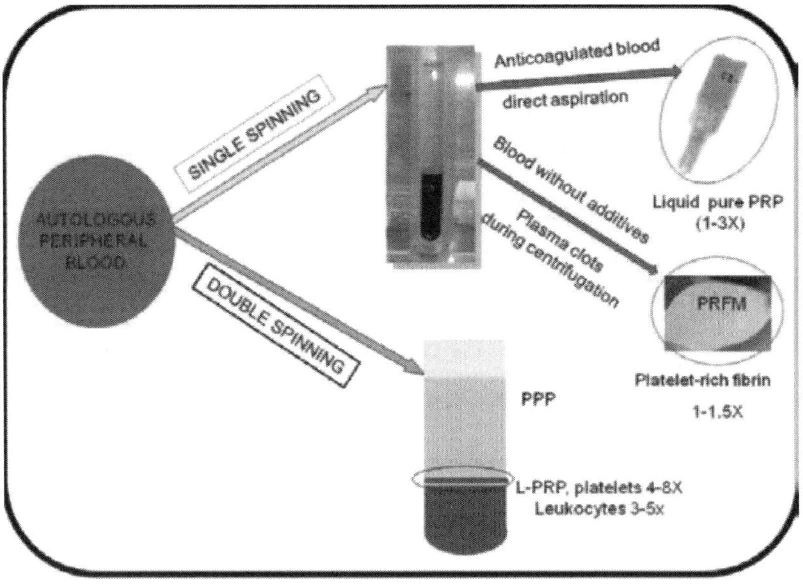

Fig. 2. Methods of producing PRP determine the composition and concentration of leukocytes, and platelets in a given plasma volume

PRP activation and fibrin delivery

Because these procedures are considered to be an autograft by the regulatory authorities of most countries, the plasma should be prepared and immediately used at the point of care, and the plasma should not be stored. Prior to application, platelets can be slowly activated by setting in motion the coagulation cascade with the addition of calcium chloride, a necessary cofactor for prothrombin conversion to thrombin. Alternatively, coagulation and platelets can be instantly activated by adding a standard solution of bovine or human thrombin with 10% calcium chloride to the PRP. After plasma activation, the fibrin scaffold can be formed *in vivo* or *ex vivo*: the latter is suitable for implantation in surgery or in ulcer care and provides a gradual release of growth factors in the area where it has been applied. Depending upon the activation mechanism, induced by CaCl2, collagen or thrombin can achieve a sudden burst of GFs or a gradual release. Indeed, a central question in biology and cell signaling is how extracellular factors elicit a complex set of signaling events to achieve specific cellular functions.

Figure 3 shows fibrin which is a natural biopolymer involved in the coagulation cascade formed upon fibrinogen cleavage by thrombin. It acts as a reservoir for growth factors, cells and enzymes during wound healing and provides a scaffold for the synthesis of the extracellular matrix. Fibrin scaffolds provide nature's cues for tissue regeneration. Fibrin is a key scaffold material for the delivery of biomolecules, and it mimics natural processes and provides adequate exposure time to maximize biological interactions.

The kinetics of signaling may be influenced not only by distinct cell surface receptors but also by the method that their cognate ligands are secreted or delivered. A receptor may be acutely activated by an immediate increase in ligand concentration, a process mimicked in most pharmacological studies. In many cellular processes in vivo, however, cells encounter a gradual increase in the concentration of extracellular factors, i.e., constitutively secreted factors need to accumulate over time to reach a threshold set by the affinity of the receptor.

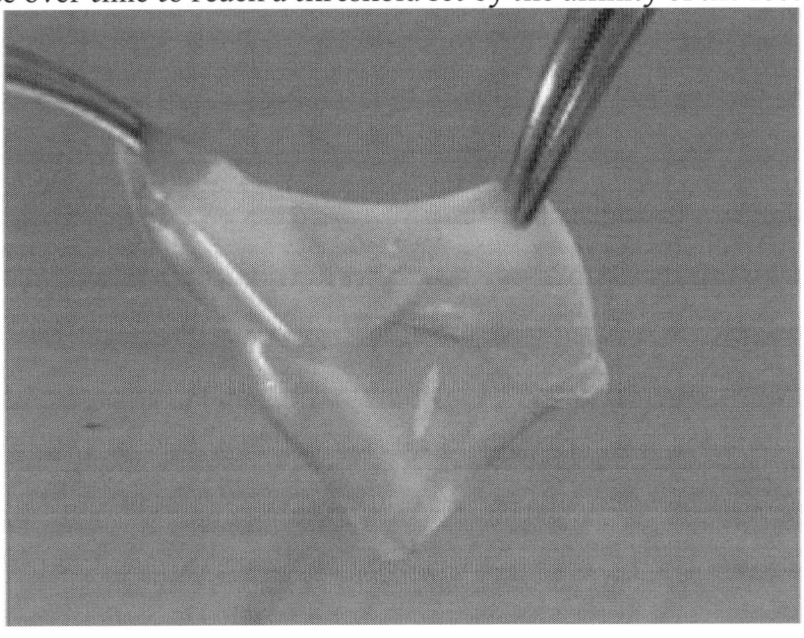

Fig. 3. Fibrin is a temporary scaffold for cell adhesion at the injured site and also functions as a vehicle for the delivery of growth factors and cytokines

Questions about safety still linger regarding the routine use of PRP. Any concerns regarding the transmission of diseases such as HIV, hepatitis, or Creutzfeldt-Jakob disease, or of the development of immunogenic reactions, a concern with the use of allografts or xenografts, are by definition not applicable due to the autologous nature of PRP. However, some systems use purified bovine thrombin to activate the platelets. This may produce coagulopathies, and most commercial systems now use recombinant human thrombin.

Some authors have raised the issue of genetic instability and have hypothesized that the use of PRP may lead to the development of neoplasms. Growth factors act on receptors located on the cell membranes rather than on the cell nucleus and activate normal gene expression via intracellular signaling proteins, which promote normal, not abnormal, gene expression. Growth factors are not directly mutagenic, and their activities in normal wound healing are highly regulated by various feedback control mechanisms. Furthermore, up to now, no systemic effect on circulating growth factors has been shown after PRP application.

Some antimicrobial activity of PRP (platelet-leukocyte gel) against *Staphylococcus aureus* has been shown in vitro and in vivo, although it is not comparable to systemic antibiotic treatment.

Technological relevance of PRP biotechnology

The medical industry is benefiting from a robust demand for technologically advanced products that accommodate the increasingly active baby boomer (people born between the 1940s and 1960s) life style as well as the sedentary lifestyle that accompanies the escalating levels of obesity. Expansion is rapidly occurring in the bone growth factor and protein segments, termed orthobiologics. In fact, PRP technologies are now very important in at least two main market segments: (1) bone repair, which includes the use of regular PRP or PRP composites made of PRP mixed with structural biomaterials or bone grafts and (2) soft tissue repair, which includes the development of techniques for applying PRPs and the surgical tools needed to correctly apply the different physical- chemical configurations of the biomaterial. An estimated 30% of new products under development are "combo products" that involve medical devices embedded with pharmaceutical or biologics components. Ambulatory PRP treatments benefit from point- of-care ultrasonography; ultrasound guidance improves success in PRP per-cutaneous and intra-articular procedures.

Recently, the medical industry has realized the potential of autologous products. Thus, although not fully developed yet, autologous technologies are readily available, and the present leading firms that control the orthopedic industry and market, Zimmer Holdings, Stryker, Biomet, Arthrex, DePuy, Smith & Nephew and Synthes, have introduced PRP devices. In the last few years, several semi-automatic machines have been developed for the centrifugal separation of PRP for therapeutic use. The process of PRP preparation is relatively straight forward and can be performed in the clinic or in the operating room. In most cases, it can be completed within minutes. The cost to both medical practitioners and patients varies widely depending on the method used to produce the PRP.

THERAPEUTIC APPLICATIONS

The versatility and biocompatibility of PRP biotechnology has stimulated its therapeutic use in many different fields (see Figure 4), including orthopedics, sports medicine, ophthalmology, dentistry, and cosmetic, plastic and maxillofacial surgery. Here we present some of the most interesting therapeutic applications with a special emphasis on musculoskeletal applications.

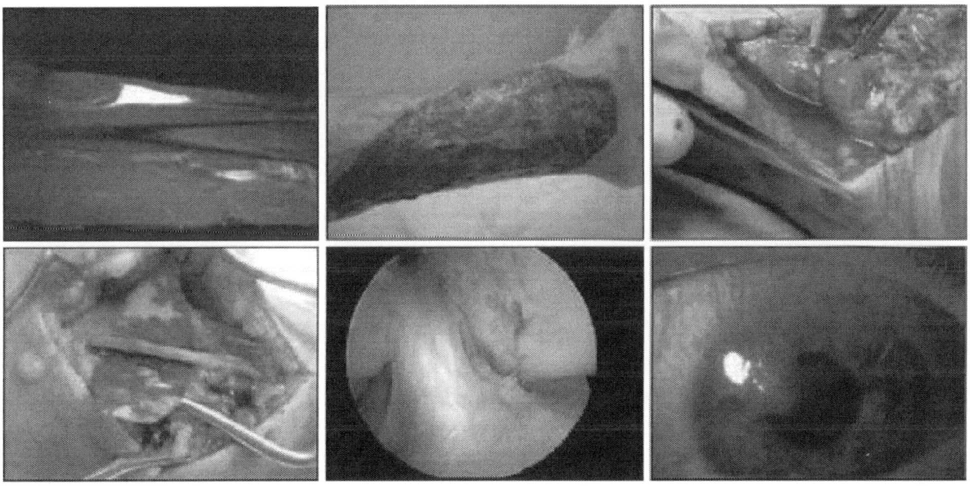

Fig. 4. Application of PRP therapies in the different medical fields: management of muscle injuries in sports medicine, open orthopedic surgery, arthroscopic surgery, ulcer care, peripheral nerve repair or the treatment of corneal ulcers in ophthalmology

Demand for musculoskeletal care

Musculoskeletal disorders, which affect millions of people worldwide, can lead to chronic pain and physical disability. According to leading pain experts, more people around the world experience musculoskeletal pain than any other type of pain (Global year against musculoskeletal pain, Oct 2009-Oct 2010, www.iasp-pain.org). In the United States alone, 2.5 million orthopedic reconstructions, including bone, cartilage, ligament and tendon reconstructions, are performed annually. With an aging population and a prolonged life expectancy, an increase in the number of patients suffering from musculoskeletal disorders such as osteoporosis and arthritis are expected in the future. The former is currently suffered by 10 million Americans over the age of 50, and the latter is a chronic musculoskeletal disease that affects 1 in 3 adult Americans. In addition, every year sporting activities result in a variety of injuries to cartilage, ligaments and especially muscles. Taken together, these musculoskeletal diseases increase patient morbidity and disability and the social and economical consequences are staggering. As a result, the period of 2000-2010 has been named "the bone and joint decade" in an effort to increase the attention of scientists regarding the problems related to these disorders and to promote advancement in these fields [Anitua et al., 2006].

Orthopedic Applications

Bone repair

More than 6 million bone fractures are reported annually in the USA, of which 5-10% have impaired healing that causes pain and disability. To improve patient-care results, scientists are making great efforts to create bone substitutes and to develop ways of improving bone healing. The use of platelet rich preparations may help to fulfill some of these requirements, particularly as an aid to bone regeneration. In fact, in vitro studies have clearly demonstrated that platelet derived growth factors stimulate the proliferation of human trabecular bone cells and the differentiation of human osteoblast-like cells. Studies have confirmed that the local application of PRPs is especially important in pathological conditions in which bone healing is weakened due to an inadequate blood supply, such as that observed in atrophic nonunion fractures. Both percutaneous injection and surgical augmentation with freshly prepared PRP have been shown to normalize fracture callus (Sánchez et al., 2008). The hundreds of soluble proteins released from both plasma and platelets include VEGF-A, PDGF, FGF, EGF, HGF, and IGF. These angiogenic activators collectively promote vessel wall permeability and promote the growth and proliferation of endothelial cells (Nurden et al., 2008).These findings are consistent with those seen in diabetic patients with a Charcot foot who showed improved healing and fewer complications after ankle fusion treated with fresh PRP. In contrast, previously frozen and thawed PRP supplementation in long bone nonunions treated with external fixation failed to provide clinical usefulness. In orthopedic trauma to date, there are not enough clinical studies to make definite conclusions. However, in some clinical conditions, the development of newly grown bone may be a realistic target if PRP is applied with cells or scaffolds. In fact, the effectiveness of bone grafting can be enhanced by creating custom-made biomaterials that will meet specific structural and biological tissue requirements in different anatomical locations. In this context, a wide array of composite biomaterials can be created by mixing PRP with either artificial or natural biomaterials. Moreover, the use of PRPimproves the handling, adhesion and adaptation of the composite graft. This is in part because these biological products may act as a biologic glue to hold together the matrix particles. Apart from facilitating the handling and manipulation, the combination of both materials may have synergistic effects on bone regeneration. For example, when patients with solitary bone cysts were treated with allogenic grafts and PRP, the cysts were filled with newly formed bone after 12 months (Pedzizs et al., 2010). In a randomized control trial among people undergoing a medial, opening-wedge osteotomy of the proximal tibia, the use of an allograft with PRP showed better radiographic osseointegration at all stages of follow-up (Dallari D et al., 2007)Developing engineered tissue is another interesting approach for bone regeneration. This may be feasible after combining mesenchymal stem cells (MSCs) and scaffold-like platelet rich plasma preparations. In fact, isolated cells, growth factors and biocompatible supporting scaffolds have generally been considered essential prerequisites to tissue engineering approaches. In the last few years, several attempts have been reported especially for bone regeneration but also for cartilage and periodontal tissue engineering. For example, the potential bone regeneration capacity of an MSC and platelet rich plasma mixture (MSC/PRP) was analyzed and compared with other approaches, including a natural deproteinized bovine bone, an autologous bone and the platelet rich product alone. Compared with the other treatments, the results of histology and mechanical properties showed that the MSC/PRP combination provided greater bone maturation and early stage bone regeneration. This mixed preparation has also been successfully used for bone regeneration in several patients. Encouraging results were observed in clinical studies exclusively concerning children. For instance, in the distraction of long bones, Kitoh et al. (2007) reported less complications in children treated with PRP plus MSCs than in children that did not receive PRP and MSC augmentation. The same authors reported an enhanced healing index in a controlled series of children with achondroplasia or hypochondroplasia undergoing limb-lengthening procedures. Even so, achieving control of bone healing is difficult, and the challenges associated with PRP therapies are enormous, extending beyond the present knowledge.

Joint repair

Arthroscopy is a minimally invasive procedure that allows doctors to treat joint injuries and disease through small incisions in the skin. The concept of not having to perform extensive soft tissue dissection is appealing because the recovery is quicker and less painful than open techniques. The use of PRP in arthroscopic surgery was initially introduced by Sanchez et al. (2003) in the treatment of a cartilage avulsion in the knee of a young athlete and in the reconstruction of the anterior cruciate ligament, as explained below (Sánchez et al., 2003). In recent work, Guadilla et al. (2011) showed how the arthroscopic management of the femoral head may be enhanced by the application of PRP in several ways. First, by grafting the necrotic area with trabecular bone mixed with PRP to induce angiogenesis and to enhance cell survival and function. Second, platelet-rich plasma can be applied within the intra-articular space to improve the conditions of synovial cells, chondrocytes, and subchondral osteoblasts [Andia et al, 2011].

Other authors have shown that the perioperative application of platelet rich plasma and fibrin sealant in arthroplasties reduces blood transfusion requirements, the length of thehospital stay, and the incidence of blood leakage and arthrofibrosis while it improves the range of motion. Another step forward would be to explore the analgesic and anti- inflammatory effects of PRP. Additional potential benefits, including blood loss, shorter hospital stay, and faster recovery time, should also be investigated.

PRP and cartilage engineering ex-vivo: Evidence of the effects of PRP on cellular proliferation and differentiation comes mainly from studies of tissue engineering. For example, chondral lesions represent a clinical challenge due to the limited capacity of chondrocytes to proliferate in vivo. Thus, autologous cells can be harvested from a small tissue biopsy and sufficiently expanded ex-vivo for re-implantation. When articular chondrocytes are the cellular source, PRP improves ex-vivo proliferation but also causes de-differentiation. Importantly, PRP-expanded cells retain their capacity to re-differentiate and synthesize cartilage-specific proteins when transferred to a 3D environment.

The cultivation of stem cells is another alternative that is under clinical investigation for the treatment of osteoarthritis; given their capacity to differentiate into chondrocytes and secrete a wide array of biologically active factors that support cell proliferation and tissue formation. The sources of these stem cells include the bone marrow and the synovial fluid. In addition, the Hoffa fat pad contains stem cells with chondrogenic potential. Stem cells derived from the meniscus, synovium, Hoffa fat, synovial fat and ACL share similar gene expression profiles. Culturing these cells under hypoxic conditions has been shown to enhance their differentiation into cartilage-like tissue.

To avoid contact of the cells with bovine products and to implement GMP-compatible protocols, PRP releasates or lysates provide a feasible alternative to fetal calf or bovine serum in the expansion of these cells for cartilage engineering purposes. The addition of PRP (compared to fetal calf serum) improves cellular expansion and imparts a differentiation capacity towards the osteogenic, chondrogenic and adipogenic lineage. In addition, PRPs can be used as carriers for chondrocyte delivery during re-implantation.

Osteoarthritis

The dramatic increase in the incidence and prevalence of joint pathology over the past two decades has focused attention on therapeutic interventions that can reverse or ameliorate progressive joint damage and pathology. Degenerative osteoarthritis (OA) is the most common form of arthritis and affects nearly 27 million adults in the US (Lawrence et al., 2008). Despite the vast amount of molecular knowledge accrued during the last few years, a major breakthrough in OA therapy has not emerged. A large part of the problem is that researchers do not know enough about the biology of OA to identify the right targets. The disease is the result of a long chain of events, but some of the links in that chain are still a mystery; nobody is certain which link to cut in order to stop disease progression. Limiting factors in the current efforts are to some extent attributed to a poor understanding of the molecular basis of the disease progression and the lack of dynamic biomarkers that reflect specific biological or pathological processes. Hence, with the exception of surgery, all approaches are merely palliative. The conservative management of OA and chondropathies with PRP biotechnologies is becoming increasingly popular, but clinical evidence is preliminary and modest and is limited mostly to observational case studies that have used patient-reported outcomes as end points. Our preliminary clinical results in a retrospective cohort study of knee OA showed that intra-articular injection of PRP decreased pain andenhanced function compared to hyaluronic acid (HA) injections (Sánchez et al., 2008). In a case series study that involved 115 young patients with low degrees of articular degeneration in the knee, Kon et al. (2010) reported reduced pain and improved function that was maintained at 12 months but not 24 months after treatment (Filardo et al., 2011). Sampson et al. (2010), also in a small case series (n=13), reported significant pain and symptom relief but did not find any significant change in the daily activities or quality of life of the patients treated. PRP injections for hip OA produced clinically significant reductions in pain and function, although this was only seen in 40% of the patients studied (Sánchez et al., 2011). When discussing PRP therapies, differences between the preparations and the re- administration procedures used should be acknowledged. Although pure PRP and leukocyte PRP formulations are not comparable in terms of leukocyte content, platelet count or plasma volume, the resulting improvements in pain and function were not exclusive to any one formulation. The pursuit to identify a unifying therapy for OA would be enhanced by refining the end points in future clinical studies.

Sport medicine

Sports related soft tissue injuries cause athletes to lose a significant amount of time from their sport and represent a significant burden to society in terms of health care resources, personal disability and activity restriction. In 2002, an estimated

15.8 billion dollars in total health care expenditures was used for the medical management of these injuries (Yu WW 2005). Soft tissue disorders, including muscle, tendon, ligament and joint capsular injuries, represent more than 50% of all the musculoskeletal injuries reported each year in the USA. Primary care studies have shown that 16% of the general population suffers from shoulder pain, whereas elbow tendinopathy affects 1-2% of the population. The importance of this problem is substantial because the field of sports medicine influences millions of people from athletes to those who participate in recreational sports or simply exercise to stay healthy and active.

Muscle injuries

Muscle injuries resulting from extrinsic or intrinsic mechanisms are extremely common in sports, accounting for about 35-45% of all injuries. Contact sports and sports that require the generation of large eccentric forces present the highest risk. The vulnerability of soccer players to strains and contusions is a substantial problem for professional players and their clubs; such injuries involve significant time lost from training and competition. Due to the increasing demands of the competitive soccer season, muscle treatments able to accelerate the recovery time without adversely affecting the recurrence rate (i.e., those that can minimize the scarring response) are of paramount importance [Andia et al, 2011].

At present, no drugs have been developed that hasten the restoration of muscle function after injury. Therefore, in the absence of any available evidence-based treatments, injection therapies may be an important option to help professional athletes. At the 2nd World Congress of Regenerative Medicine, Sanchez (2005) reported for the first time the application of leukocyte-free PRP to 21 muscle injuries of different severities and different anatomical locations. Small tears progressed well with a single application, whereas more severe tears required 2-3 ultrasound-guided injections. The injected volume depended on tear severity. These athletes, who played in first division teams of the Spanish SoccerLeague, resumed normal training activities in half the time needed by matched historical controls. Using the same leukocyte-free PRP preparation, Wee (2009) reported good outcomes (1 week to return to pre-injury activities) after three weekly US-guided injections to treat an adductor longus strain in a professional bodybuilder.

Another autologous blood derived biotechnology is named ACS (Autologous Conditioned Serum). This technology consists of an autologous liquid serum conditioned by the incubation of whole blood with glass beads. It contains signaling proteins that include interleukin-1b (IL- 1b), tumor necrosis factor-alpha (TNF-a), IL-7, FGF-2, interleukin 1 receptor antagonist (IL- 1Ra), HGF, platelet derived growth factor (PDGF-AB), transforming growth factor (TGF-β1) and IGF-1. Wright-Carpenter (2004) assessed the effects of ACS injections in a non-blinded, non-randomized case control study. The experimental group was treated with ACS, and the control group, which was analyzed retrospectively, included patients who had received Traumeel®/Actovegin®. Traumeel is a homeopathic formulation that contains both botanical and mineral ingredients in homeopathic concentrations. It is purported to suppress the release of inflammatory mediators and to stimulate the release of anti-inflammatory cytokines. Actovegin is a deproteinized calve blood hemodialysate that consists of a physiological mix of amino acids. The RICE principle was employed for initial care in both groups. The primary measured outcome was the time needed to resume full sporting activities. The experimental group returned to competition after 16.6 days, whereas the control group took 22.3 days. In addition, MRI scans taken at 16 days in both groups confirmed that regression of the edema/bleeding was faster in the ACS group. Both treatments were safe.

Tendon pathology

Chronic pain in tendons is very common and studies show that overuse, underloading and overloading, all contribute to tendon injuries and pain. More than 30-50% of the injuries among professional and recreational athletes are overuse tendon injuries resulting in the onset of pain and discomfort. Data collected from sedentary people showed that tendinosis is not necessarily a consequence of overuse. Nevertheless, the odds of having tendinopathy among elite endurance athletes are one in two (Kujala et al., 2005). Thus, the development of innovative strategies to treat tendon injuries is an essential task, but it requires a more thorough understanding of the underlying cellular and molecular mechanisms. The use of platelet rich preparations in this context may be focused on restoring the normal tissue composition while avoiding further degeneration. When we evaluated the effects of the pool of growth factors released from PRP on tendon cells, the results showed that human tendon cells increased their proliferation rate and were stimulated to release VEGF and HGF. The former promotes angiogenesis, which is directly related with tendon healing capability; the latter is a potent antifibrotic agent that can reduce scar formation around tendon tissues. Other studies have reported that injections of platelet rich plasma one week postoperatively increased tendon regenerate strength. The clinical translation of this approach was assayed in a pioneer study involving professional and recreational athletes. PRP was injected into the tendon fibers after the tendon was sutured. After closing the paratenon and before closing the overlying skin, the affected area was covered with the fibrin scaffold. The results showed that those receiving the PRP-therapy experienced a significant acceleration in functional recovery compared with a matched group that underwent conventional surgery. Moreover, the effects induced by PRP therapies had long-term consequences such asdecreased cross-sectional area of the

Achilles tendon after 18 months [Sánchez et al., 2007]. The feasibility and biosafety of PRP therapies made their application possible not only in surgeries but in the conservative management of tendon problems.

Currently, conservative management with PRP injections and its research attention are increasing [Andia et al, 2011]. Recently, three studies on PRP injection, of which two were on patients with chronic patellar tendinopathy [62,65] receiving three injections of leukocyte-platelet concentrate (double centrifugation), were reported. Significant improvements in the Tegner scores were described in one of the two studies. In addition, improvement in pain and function was reported after a single PRP injection in patients with epycondylitis [Mishra et al., 2006]. More recently, two double-blind, randomized clinical trials were performed on patients with lateral epycondylitis [Peerbooms et al., 2010] and chronic Achilles tendinopathy [De Vos et al., 2010], respectively. In both studies, the experimental treatment consisted of a single injection of an identical buffered PRP. The clinical results were significant for patients with lateral epycondylitis, for which PRP reduced pain and improved function. In contrast, in patients with Achilles tendinopathy, PRP injection did not reduce pain or improve activity [De Vos et al., 2010]. It seems improbable that a single injection could stop or reverse an ongoing degenerative process. Instead repeated injections appear to be more efficient in degenerative pathologies. No complications were reported after PRP treatments.

ACL reconstruction

Finally, a great deal of effort has been paid to the development of novel medical tools for the repair of injured anterior cruciate ligaments (ACL). The ACL is one of the four major ligaments connecting the bones of the human knee. A torn ACL is a common injury and is typical among the active younger population. The injury requires surgical intervention to stabilize the knee and to prevent cartilage and meniscal injuries, which lead to degenerative joint disease. ACL reconstruction, namely ACL tissue engineering, involves the manipulation of cells and tissues to replace the injured ligament; this process is a complex undertaking and involves many mechanical and biological challenges. It requires both the application of mechanical knowledge and an understanding of how cells are maintained and grow into functioning tissues to replace defective or injured ligaments. At present, the most common options in ACL replacement are allografts or autografts. A novel approach using PRP technologies seeks to facilitate ACL healing by mimicking the native tissue and improving tissue function with the appropriate cues (see Figure 5), ultimately leading to better patient care.

Cell cultures and animal research, in addition to human clinical studies, drive the main hypotheses for the application of PRP biotechnology in ACL reconstruction. These applications involve first promoting bone-bone and bone-tendon healing, and second, influencing the pattern of change within the autograft body (ligamentization). Finally, the application of PRP-therapies will help in donor site healing. Graft fixation is the weakest link in ACL reconstruction because knee laxity develops during the immediate postoperative period until biologic fixation occurs within the bone tunnel. Classically, graft stabilization is achieved more rapidly with a bone plug-patella tendon-bone (BPTB) graft than with the hamstring. The BPTB graft becomes anchored to the bone wall via appositional bone formation, and in these circumstances, the use of PRP may aid in the formation of the callus and may accelerate bone fusion (Sánchez et al., 2010). In apreliminary study, Sánchez et al. (2003) described a procedure for treating bone tunnels and for conditioning the graft prior to implantation with PRP. They compared a group of 50 patients treated with surgery and pure PRP with another group of 50 patients who underwent surgery alone. The two groups were matched for age and graft type. The authors reported better integration of PRP-treated grafts within the tunnels, as assessed by X-rays, and a larger number of completely stable knees in the PRP group. Other authors have explored the influence of autologous bone plugs, either alone or combined with PRP therapies, on the promotion of femoral bone-tendon healing. They reported that bone plugs, but not PRP-therapies, significantly prevented femoral tunnel widening.

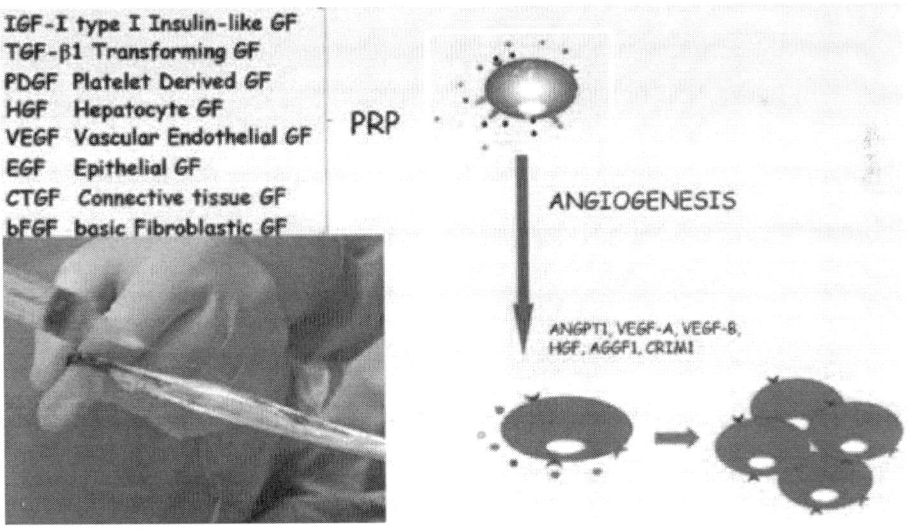

Fig. 5. Transfer of autologous GFs and cytokines to the tendon graft, applying the principles of tissue engineering and using PRP biotechnology to estimulate biological mechanisms such as angiogenesis.

The appropriate function of ACL grafts, essential for normal knee biomechanical functioning, entails a successful intra-articular graft ligamentization. One exciting option to enhance ligamentization is to simultaneously transfer multiple cytokines and growth factors (including PDGF, TGF-β1 and VEGF, among others) to the graft by applying an endogenous PRP. Autografts could be loaded in situ with a balanced pool of signaling molecules. These molecules would have the potential to not only activate the graft tenocytes but also to attract cells, such as endothelial or stem cells, from adjacent niches (such as the synovium and/or the intrapatellar pad) to the graft structures using the synovial fluid for passage. The corroboration and clinical translation of this notion may be enhanced healing and intrasynovial adaptation of the tendon graft to the synovial milieu. Recently, we have compared the gross appearance and microscopic qualities of the PRP-treated and untreated grafts during the remodeling period (6-24 months). Gross morphology was evaluated using second-look arthroscopy focusing on graft thickness, apparent tension and synovium coverage. The overall arthroscopic evaluation provided evidence that a higher percentage of the grafts rated as excellent in the PRP group (57% versus 33%). No grafts were scored as poor in the PRP group, but 20% of the controls showed poor morphology. At the same time, PRP treatment influenced the histological characteristics of the tendon graft, which resulted in tissue that was more mature than in the controls. Histology displayed newly formed connective tissue enveloping the graft in 77.3% of the PRP-treated grafts and in 40% of the controls (Sanchez et al., 2010). Other authors have used a compressed gelatin sponge soaked with leukocyte and platelet-rich concentrate (GPS system by Biomet Biologic, Warsaw, USA) sutured to the intra-articular part of the graft, which confirmed the acceleration of the maturation of the grafts treated with PRP as assessed by magnetic resonance imaging.

Cutaneous ulcers

Clinical differences between acute and chronic wounds are in part explained by alterations in the local biochemical environment. For example, acute wounds are associated with a greater mitogenic activity than chronic wounds.

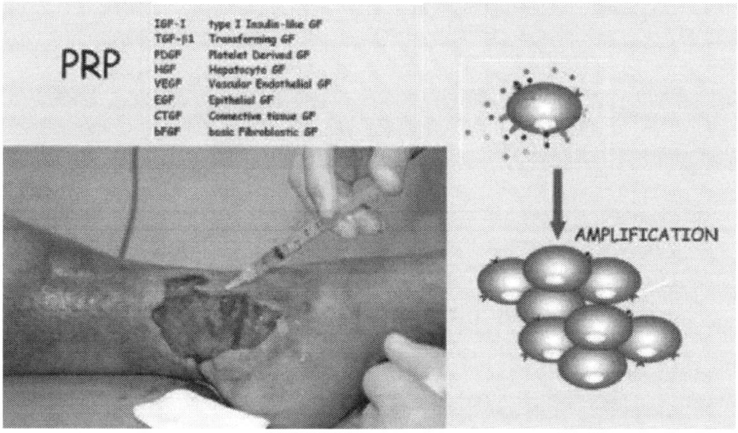

Fig. 6. Chronic ulcers are treated with several applications of PRP in order to enhance cell proliferation, and the formation of granulation tissue

Chronic wounds are associated with a higher level of pro-inflammatory cytokines than acute wounds. As chronic wounds begin to heal, they progress to a less inflammatory state. Elevated protease activities in chronic wounds may directly contribute to poor healing by degrading the proteins necessary for normal wound healing. Chronic wounds can be defined as those failing to proceed through an orderly and timely process to produce anatomic and functional integrity. Practically, a chronic wound is one that has failed to heal within 3 months. The cellular, biochemical and molecular events that characterize chronic wounds have been well defined, including a prolonged inflammatory phase, cellular senescence, deficiency of growth factors and/or their receptors, deficient fibrin production andhigh levels of proteases. In normally healing wounds, acute inflammation with neutrophil infiltration brings neutrophil-derived matrix protease enzymes that debride the wound and pave the way for new tissue deposition and remodeling. In chronic wounds, the orderliness of the healing process is disrupted by some underlying abnormality that prolongs the inflammatory phase and produces a cascade of tissue responses that perpetuates the non- healing state. Repeated trauma, foreign bodies, pressure necrosis, infection, ischemia and tissue hypoxia also amplify the chronic inflammatory state, which is characterized by excess neutrophils, macrophages and lymphocytes. Fragments of dead tissue, bacterial products and foreign bodies are powerful chemoattractants that sustain a continuous influx of inflammatory cells, which in turn produce a variety of growth factors, cytokines, and matrix-degrading enzymes. Among the most potent of these enzymes are elastase and MMPs, which are present in large quantities in chronic wounds. Given the low levels of TIMPs, the MMP/TIMP balance is distorted; thus, the excess of proteolytic enzymes shifts the balance towards ECM destruction and the degradation of signaling proteins. Therefore, any effective intervention must include a strategy for disrupting this cycle and setting the wound on a permanent path towards healing. Historically, the first clinical application of platelet derived preparations was conducted in chronic leg ulcers in which wounds were filled with collagen embedded in platelet secreted proteins. This initial product, known as PDWHF (platelet-derived wound healing factors) stimulated the formation of the vascularized connective tissue found in healing wounds. Thereafter, various other types of platelet products have been assayed in several pilot studies, case series and clinical trials.

Growth factors are crucial for timely wound healing; inadequate levels of growth factors may be an important factor contributing to the chronicity of the wound, which may be degraded in excess by cellular or bacterial proteases. Initially, Margolis et al. (2001) showed that platelet releasates were more effective than standard therapy. Subsequently, PRP formulations were refined and primarily applied as fibrin membranes for the treatment of non-healing ulcers. More recently, the use of PRP in the management of chronic diabetic foot ulcers has been successful (Setta HS 2011). Moreover, PRP provides advantages in skin grafting for recalcitrant ulcers (Chem Tim et al., 2010). Allogenic platelet preparations have been used recently to treat recalcitrant ulcers in very elderly hypomobile patients for whom autologous blood processing may be difficult (Greppi et al., 2011). Finally, the use of PRP gel resulted in an improved quality of life and a lower cost of care over a 5-year period than other treatment modalities for patients with non-healing diabetic foot ulcers. Although actual treatment outcomes may differ from those modeled, PRP gel represents a potentially attractive treatment alternative for insurers and health care providers to address the cost burden and health effects of non-healing diabetic foot ulcers (Dougherty EJ 2008)

Alternative Therapeutic Strategies

The potential therapeutic value and versatility of platelet rich products has stimulated research in additional medical fields. PRP biotechnology holds promise as a healing preparation in surgical procedures and in the treatment of many different diseases. The use of PRPs for cell delivery and tissue engineering permit insights into the development of novel therapies. For example, autologous fat grafting, also known as fat transfer or fat injection, has long been a staple of cosmetic and reconstructive surgery. Fat grafts have proven very effective in the reconstruction of soft tissue defects, particularly for facial plastic and reconstructiveprocedures. However, there has always been one significant disadvantage associated with autologous fat grafting: the unpredictable and often inconsistent graft survival rate. Promising new evidence has shown that PRP can enhance the fat graft survival rate. Moreover, nasolabial folds, superficial rhytids and acne scars have been successfully treated with injections of autologous PR fibrin matrices (Sclafani AP 2010). Additionally, PRP can be associated with novel dermatologic procedures as an aid in healing. For example, PRP is an effective method for enhancing wound healing and reducing transient adverse effects after fractional carbon dioxide laser resurfacing (Na et al., 2011).

Another remarkable application of PRP is in ophthalmology. Several successful examples include the use of PRP releasates as eye drops for the treatment of a broad spectrum of corneal persistent epithelial defects (Lopez-plandolit et al., 2010). Furthermore, the use of autologous platelet rich plasma was shown to be very effective in the treatment of patients suffering from dry eye symptoms; it improved both patient symptoms and major clinical signs [Alio et al., 2007]. Platelet rich plasma also promotes healing of dormant corneal ulcers even in eyes that are threatened by corneal perforation, and it is a reliable and effective therapeutic tool for the enhancement of epithelial wound healing on the ocular surface.

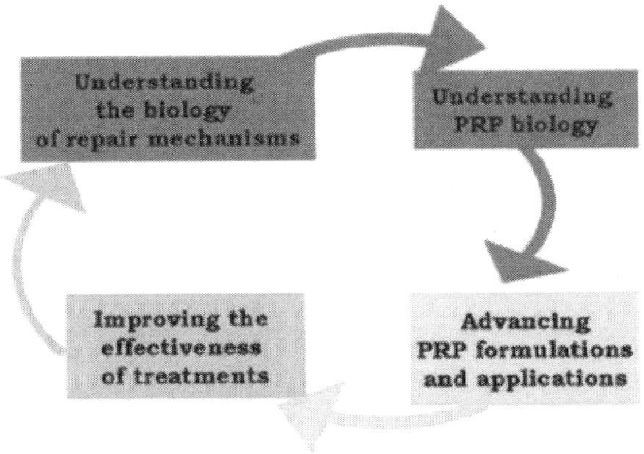

Fig. 7. The four domains of PRP science. Improved understanding of the biology of PRPs and repair mechanisms have emerged as a potential way of improving PRP formulations and applications. The identification of critical molecules that interact with healing will be critical in developing new approaches to treatments.

- Other interesting recent approaches using PRP biotechnology include the successful application of platelet rich plasma in peripheral nerve regeneration (Sariguney et al., 2008) and the use of PRP biotechnology to treat damaged myocardial tissue. Utilizing a murine myocardial permanent ligation and ischemia/reperfusion model, a proprietary PRP formulation demonstrated a positive effect in left ventricular cardiac function. The use of PRP for skin rejuvenation is another application of PRP biotechnology.

AT A GLANCE

Realistically, a substantial amount of research is needed to bring PRP technologies to the bedside, as clinical and laboratory findings that indicate its potential benefits must be followed by comprehensive clinical studies to demonstrate efficacy. Demonstrating effectiveness in different pathologies will be critical for the widespread adoption of PRP technology, including re-imbursement. Below is a schematic representation that illustrates the four domains of PRP science.

Because of the safety of these products, basic science, clinical discovery and patient-oriented research should be interdependent rather than successive steps. The substantial challenges of incorporating such research into clinical care must be pursued if the potential of PRPs is to be realized. Although PRP therapies have many compositions and procedures for application, they all try to maximize the cell signals that may enhance tissue healing. Our increased understanding of the healing mechanisms that result in tissue repair is paving the way towards the optimization of healing therapies

REFERENCES

1. Alio JL, Pastor S, Ruiz-Colecha J, Rodriguez A, Artola A. (2007). Treatment of ocular surface syndrome after LASIK with autologous platelet-rich plasma. *Journal of Refractive Surgery* Jun 23(6): 617-9
2. Andia I, Sánchez N, Maffulli N. Joint Pathology and PRP therapies. Expert Opin Biol Ther 2011;12(01):1-16
3. Andia I, Sánchez N, Maffulli N. Tendon healing and platelet-rich plasma therapies. Expert Opin Biol Ther 2010;10(10):1415-26
4. Andia I, Sánchez N, Maffulli N. Platelet rich plasma therapies for sports muscle injuries: any evidence behind clinical practice? Experte Opin Biol Ther 2011; 11(4):509-18
5. Anitua E. (1999). Plasma rich in growth factors: preliminary results of use in the preparation of future sites for implants. *International journal of Oral and maxillofacial Implants* Jul- Aug;14(4):529-35
6. Anitua E, Andia I, Ardanza B, Nurden P, Nurden AT. (2004). Autologous platelets as a source of proteins for healing and tissue regeneration. *Thrombosis and Haemostasis* 91(1):4-15
7. Anitua E, Sanchez M, Nurden AT, Nurden P, Orive G, Andia I. (2006). New insights into and novel applications for platelet-rich fibrin therapies. *Trends in Biotechnology* 24(5):227-34
8. Anitua E, Sanchez M, Orive G, Andia I. (2007). The potential impact of the preparation rich in growth factors (PRGF) in different medical fields. *Biomaterials* 28:4551-60
9. Borregaard N, Sorensen OE, Theilgaard-Mönch K. (2007). Neutrophil granules: a library of innate immunity proteins. *Trends in Immunology* 28(8): 340-345
10. Dallari D, Savarino L, Stagni C, Cenni E, Cenacchi A, Fornasari PM, Albisinni U, Rimondi E, Baldini N, Giunti A. (2007). Enhanced tibial osteotomy healing with use of bone grafts supplemented with platelet gel or platelet gel and bone marrow stromal cells. *J Bone Joint Surg Am*. Nov ;89(11):2413-20
11. de Vos RJ, Weir A, van Schie HTM, et al. (2010). Platelet-Rich Plasma Injection for Chronic Achilles Tendinopathy A Randomized Controlled Trial. *Jama-Journal of the American Medical Association* 303(2):144-149.
12. Dhollander AAM, De Neve F, Almqvist KF, et al. (2011). Autologous matrix-induced chondrogenesis combined with platelet-rich plasma gel: technical description and a five pilot patients report. *Knee Surgery Sports Traumatology Arthroscopy* 19(4):536-42

13. Dohan Ehrenfest DM, Rasmusson L, Albrektsson T. (2009). Classification of platelet concentrates: from pure platelet-rich plasma (P-PRP) to leucocyte- and platelet-rich fibrin (L-PRF). *Trends in Biotechnology* 27(3):158-67.

14. Dougherty, EJ. (2008). An evidence-based model comparing the cost-effectiveness of platelet-rich plasma gel to alternative therapies for patients with nonhealing diabetic foot ulcers. *Advanced Skin Wound Care* 21: 568-575.

15. Dovi JV, DiPietro LA. (2003). Accelerated wound closure in neutrophil depleted mice.*Journal of Leukocyte Biology* 7:448-455

16. Filardo G, Kon E, Della Villa S, Vincentelli F, Fornasari PM, Marcacci M. (2009) Use of platelet-rich plasma for the treatment of refractory jumper's knee. *International Orthopaedics* 34(6):909-915.

17. Greppi, N, Mazzucco, L, Galetti, G, et al. (2011) Treatment of recalcitrant ulcers with allogeneic platelet gel from pooled platelets in aged hypomobile patients. *Biologicals* 39: 73-80.

18. Guadilla J, Fiz N, Andia I, Sánchez M. (2011) Arthroscopic management and platelet-rich plasma therapy for avascular necrosis of the hip. *Knee Surg Sports Traumatol Arthrosc* Jun 22.

19. Italiano JE, Richardson JL, Patel-Hett S, et al. Angiogenesis is regulated by a novel mechanism: pro- and antiangiogenic proteins are organized into separate platelet alpha granules and differentially released. Blood 2008;111:1227-33

20. Kitoh H, Kitakoji T, Tsuchiya H, Katoh M, Ishiguro N. (2007). Transplantation of culture expanded bone marrow cells and platelet rich plasma in distraction osteogenesis of the long bones. *Bone* Feb;40(2):522-8. Epub 2006 Oct 27.

21. Kitoh H, Kitakoji T, Tsuchiya H, Katoh M, Ishiguro N. (2007). Distraction osteogenesis of the lower extremity in patients with achondroplasia/hypochondroplasia treated with transplantation of culture-expanded bone marrow cells and platelet-rich plasma. *J Pediatr Orthop* Sep; 27(6):629-34.

22. Kon E, Buda R, Filardo G, et al. (2010) Platelet-rich plasma: intra-articular knee injections produced favorable results on degenerative cartilage lesions. *Knee Surgery Sports Traumatology Arthroscopy*. 18(4):472-479.

23. Kon E, Filardo G, Delcogliano M, et al. (2009). Platelet-rich plasma: New clinical application A pilot study for treatment of jumper's knee. *Injury-International Journal of the Care of the Injured*. 40(6):598-603.

24. Krysko DV, D'Herde K, Vandenabeele P. (2006) Clearance of apoptotic and necrotic cells and its immunological consequences. *Apoptosis* 11(10):1709-26

25. Lawrence RC, Feltson DT, Helmick CG, et al. (2008) Estimated of the prevalence of arthritis and other rheumatic conditions in the United States. *Arthritis Rheumatism* 58(1):26-35

26. Lopez-plandolit S, Morales MC, Freire V, Echevarría J, Durán JA. (2010). Plasma rich in growth factors as a therapeutic agent for persistent corneal epithelial defects. *Cornea* 29(8):843-8

27. Margolis DJ, Kantor J, Santanna J, Strom BL, Berlin JA. (2001). Effectiveness of platelet releasate for the treatment at diabetic neuropathic foot ulcers. *Diabetes Care* 24(3):483-88

28. Marx RE, Carlson ER, Eichstaedt RM, Schimmele SR, Strauss JE, Georgeff KR. (1998). Platelet-rich plasma: Growth factor enhancement for bone grafts Oral Surg Oral Med Oral Pathol Oral Radiol Endod Jun;85(6):638-46.

29. Maynard DM, Heijnen HFG, Horme MK, et al. (2007). Proteomic analysis of platelet alpha- granules using mass spectrometry. *Journal of Thrombosis and Haemostasis* 5:1945-1955

30. Mishra A, Pavelko T. (2006). Treatment of chronic elbow tendinosis with buffered platelet- rich plasma. *American Journal of Sports Medicine.* 34(11):1774-8.

31. Mummery WK, Schofield G, Spence JC. (2002). The epidemiology of medically attended sport and recreational Injuries in Queensland. *Journal of Science and Medicine in Sport.* Dec;5(4):307-320.

32. Na, JI, Choi, JW, Choi, HR, et al. (2011). Rapid Healing and Reduced Erythema after Ablative Fractional Carbon Dioxide Laser Resurfacing Combined with the Application of Autologous Platelet-Rich Plasma. *Dermatologic Surgery* 37: 463-468.

33. Nathan C. (2006). Neutrophils and immunity: challenges and opportunities. *Nature Reviews Immunology* Mar;6(3):173-82.

34. Nurden AT, Nurden P, Sanchez M, Andia I, Anitua E. (2008). Platelets and wound healing.*Frontiers in Bioscience.* 13:3532-48.

35. Pedzisz P, Zgoda M, Kocon H, Benke G, Górecki A. (2010). Treatment of solitary bone cysts with allogenic bone graft and platelet-rich plasma. A preliminary report. Acta Orthop Belg. Jun;76(3):374-9

36. Peerbooms JC, Sluimer J, Bruijn DJ, Gosens T. (2010) Positive Effect of an Autologous Platelet Concentrate in Lateral Epicondylitis in a Double-Blind Randomized Controlled Trial Platelet-Rich Plasma Versus Corticosteroid Injection With a 1-Year Follow-up. *American Journal of Sports Medicine.* 38(2):255-262.

37. Poon IK, Hulett MD, Parish CR. (2010). Molecular mechanisms of late apoptotic/ necrotic cell clearance. *Cell Death and Differerentiation* 17(3):381-397

38. Randelli PS, Arrigoni P, Cabitza P, Volpi P, Maffulli N. (2008) Autologous platelet rich plasma for arthroscopic rotator cuff repair. A pilot study. *Disability and Rehabilitation*. 30(20-22):1584-1589.

39. Sampson S, Reed M, Silvers H, et al. (2010) Injection of Platelet-Rich Plasma in Patients with Primary and Secondary Knee Osteoarthritis A Pilot Study. *American Journal of Physical Medicine & Rehabilitation*. 89(12):961-9

40. Sánchez M, Azofra J, Aizpurua B, et al. (2003). Use of autologous plasma rich in growth factors in arthroscopic surgery. *Cuadernos de Artroscopia* 10:12-19.

41. Sánchez M, Azofra J, Anitua E, et al. (2003). Plasma rich in growth factors to treat an articular cartilage avulsion: a case report. *Medicine Sciences Sports Exercise* 35:1648-52

42. Sánchez M, Anitua E, Andia I. Application of autologous growth factors on skeletal muscle healing. Second International Congress on Regenerative Medicine. http://www.plateletrichplasma.com /pdf/Orthopedic- PRP/ Sports%20Medicine/66-SanchezRegMed2005.pdf

43. Sanchez M, Anitua E, Azofra J, Andia I, Padilla S, Mujika I. (2007). Comparison of surgically repaired achilles tendon tears using platelet-rich fibrin matrices. *American Journal of Sports Medicine* 35(2):245-51.

44. Sanchez M, Anitua E, Orive G, Mujika I, Andia I. (2009) Platelet-rich therapies in the treatment of orthopaedic sport injuries. *Sports Medicine* 39(5):345-54.

45. Sanchez, M, Anitua, E, Azofra, J, et al. (2008). Intra-articular injection of an autologous preparation rich in growth factors for the treatment of knee OA: a retrospective cohort study. *Clinical Experimental Rheumatology* 26:910-3

46. Sánchez, M; Anitua, E; Cugat, R; Azofra J, Guadilla J, Seijas R, Andia I. (2009). Nonunions Treated With Autologous Preparation Rich in Growth Factors. *Journal of Orthopedic Trauma* 23 (1): 52-59.

47. Sánchez M, Anitua E, Cole A, et al. (2009). Management of post-surgical Achilles tendon complications with a Preparation Rich in Growth Factors: A study of two-cases. *Injury EXTRA* 40:11-15.

48. Sanchez M, Anitua E, Azofra J, Prado R, Muruzabal F, Andia I. (2010). Ligamentization of Tendon Grafts Treated With an Endogenous Preparation Rich in Growth Factors: Gross Morphology and Histology. *Arthroscopy-the Journal of Arthroscopic and Related Surgery*. 26(4):470-480

49. Sanchez, M, Anitua, E, Lopez-Vidriero, E, Andia, I. (2010). The Future: Optimizing the Healing Environment in Anterior Cruciate Ligament Reconstruction. *Sports Medicine and Arthroscopy Review* 18: 48-53.

50. Sanchez M, Guadilla J, Fiz N, Andia I. (2011). Ultrasound-guided platelet rich plasma injections for the treatment of osteoarthritis of the

hip. *Rheumatology* doi:10.1093/rheumatology/ker303

51. Sariguney, Y, Yavuzer, R, Elmas, C, et al. (2008). Effect of platelet-rich plasma on peripheral nerve regeneration. *Journal of Reconstructive Microsurgery* 24: 159-167.

52. Schmikli SL, Backx FJG, Kemler HJ, van Mechelen W. (2009). National Survey on Sports Injuries in the Netherlands: Target Populations for Sports Injury Prevention Programs. *Clinical Journal Sport Medicine* 19(2):101-6

53. Sclafani AP. Platelet-rich fibrin matrix for improvement of deep nasolabial folds. (2010).*Journal of Cosmetic Dermatology* Mar;9(1):66-71

54. Semple JW, Freedman J. (2010). Platelets and innate immunity. *Cell Molecular Life Sciences*67:499-511

55. Setta, HS, Elshahat, A, Elsherbiny, K, et al. (2011). Platelet-rich plasma versus platelet-poor plasma in the management of chronic diabetic foot ulcers: a comparative study. *International Wound Journal* 8: 307-312.

56. Soehnlein O, Zernecke A, Weber C. (2009). Neutrophils launch monocyte extravasation by release granule proteins. *Thrombosis and Haemostasia* 02:198-205

57. Villeneuve J, Block A, Le Bousse-Kerdiles MC, et al. Tissue inhibitors of matrix metalloproteinases in platelets and megakaryocytes: A novel organization for these secreted proteins. Exp Hematology 2009;37: 849-56

58. Wee LL, Lee DYH, Soon MYH. (2009). Plasma Rich in Growth Factors to Treat Adductor Longus Tear. *Annals Academy of Medicine Singapore.* Aug 38(8):733-734.

59. Whitman DH, Berry RL, Green DM. (1997). Platelet gel: an autologous alternative to fibrin glue with applications in oral and maxillofacial surgery. J Oral Maxillofac Surg. Nov;55(11):1294-9

60. Wright-Carpenter T, Klein P, Schaferhoff P, Appell HJ, Mir LM, Wehling P. (2004). Treatment of muscle injuries by local administration of autologous conditioned serum: A pilot study on sportsmen with muscle strains. *International Journal of Sports Medicine.* Nov 25(8):588-593

61. Yard EE, Schroeder MJ, Fields SK, Collins CL, Comstock RD. (2008).The epidemiology of United States high school soccer injuries, 2005-2007. *American Journal of Sports* Medicine 36(10):1930-37

CHAPTER-7

THE UTILIZATION OF POLYMERS IN PHARMACEUTICAL CONTEXTS: UTILIZING NATURAL BIOACTIVE COMPOUNDS AS INITIATORS AND CATALYSTS IN BIODEGRADABLE POLYMER SYNTHESIS

INTRODUCTION: AN OVERVIEW

In the realm of pharmaceutical science, the strategic integration of polymers has revolutionized drug delivery systems, formulation techniques, and biomedical applications. One of the pioneering frontiers within this domain is the incorporation of natural bioactive compounds as initiators and catalysts in the synthesis of biodegradable polymers. This innovative approach not only enhances the eco-friendliness of pharmaceutical materials but also leverages the potency of nature's toolbox for designing advanced drug delivery platforms.

Biopolymers, synthetic polymers and their derivatives are commonly used in medicine and pharmacy. Significant progress attained in the polymer chemistry and technology has boosted the dynamic development of the medicinal engineering.

Recently, particular interest of scientists has been focused on biomedical polymers, especially those used for drug delivery systems, therapeutic systems and macromolecular prodrugs. The aforementioned applications have opened new exciting prospects for medicine, because specially designed polymers are capable of delivering medicinal substances to the target diseased tissues and cells together with dosing those drugs according to controlled specified pharmacodynamics. Particular attention has recently been paid to chemistry of biocompatiable and biodegradable polymers, because they have an advantage of being readily hydrolyzed into removable and non-toxic products, which can be subsequently eliminated by metabolic pathways. Furthermore, the biomedical polymers have to be synthesized now using friendly for the environment and safe for human health, effective natural initiators, co-initiators and/or catalysts.

Therefore, the main objective of this work is to discuss various polymers recommended for the pharmaceutical applications and then to describe natural compounds used as initiators, catalysts and co-initiators in the synthesis of biodegradable and bioresorbable polyesters and polycarbonates.

POLYMERS IN THE PHARMACEUTICAL APPLICATIONS

Macromolecules are applied in pharmacy as the pharmacological substances, blood substitutes, drug delivery and therapeutic systems, in the synthesis of macromolecular prodrugs and in the technology of prolonged release drug formulations.

Polymers with the pharmacological effects and polymeric blood substitutes

One of the most interesting polymers used in pharmacy, are those exerting a pharmacological effect. DIVEMA, copolymer of divinyl ether-maleic anhydride (Florjanczyk & Penczek, 1998; Papamatheakis et al., 1978) is an example of such compound with antitumoral and antiviral properties. Its action probably includes the stimulation of the glycoprotein production, which suppresses viral RNA translocation in cells and division of cancer cells.

Furthermore, the polymers are often applied as swelling, relaxation and sliding agents. Methylcellulose taken orally is not absorbed from the alimentary tract. However, it detains water on swelling and in consequence causes relaxation of the stercorous mass (Tonnesen & Karlsen, 2002; Zejc & Gorczyca, 2002).

A copolymer of ethylene and propylene glycols has found an application in the therapy of constipations (Tonnesen & Karlsen, 2002; Zejc & Gorczyca, 2002). This non-ionic, surface- active polymer is unable to penetrate through the gut walls because of large average molecular weight. However, it causes relaxation and hydration of the stercorous mass by the reduction of the surface tension.

A linear polymer of uronic acids - alginic acid (mannuronic acid conjugated β-1,4 and L- guluronoic acid glycosidically conjugated α-1,4) is mainly obtained from the *Laminaria algae*. This polymer neutralizes hydrochloric acid (Janicki et al., 2002; Zejc & Gorczyca, 2002). Its action relies on detaining of water in stomach followed by reduction of irritations and pain.

A polyvinylpyrrolidone has found an application as anti-diarrhoeal drug (Tonnesen & Karlsen, 2002; Zejc & Gorczyca, 2002). Its amphoteric properties normalize pH in stomach and intestines through acids or bases adsorption, which are usually raised as result of fermentation or putrefaction.

The synthetic hormones with the protein structure play an important role in the modern pharmacology (Zejc & Gorczyca, 2002). The Buserelin, Goserelin, Leuprorelin and Triptorelin are known as synthetic analogues of Gonadoliberin (the hormone of hypothalamus). These oligopeptides are obtained by exchanging of some amino acids in Gonadoliberin molecule and then are used to treat prostate and breast cancers or endometriosis. Another example is synthetic analogue of Somatoliberin used for treating children with some forms of GH deficiency. The synthetic analogue of Somatostatin - Octreotide is applied to treat the alimentary tract (Zejc & Gorczyca, 2002).

Corticotrophins are examples of synthetic hormones of the anterior pituitary, often applied in the therapy of rheumatoid diseases and severe asthma. Thus, Oxitocin, Vasopressin and Ornipressin are belong to the group of hormones of posterior lobe of the hypophysis (Zejc & Gorczyca, 2002). First of them causes uterine contractions, second can contract the smooth muscles of the blood vessels while Ornipressin is often added to the anaesthetics. Moreover causes the vessels contraction.

The peptide antibiotics are the relatively numerous group of the natural oligomers. They are composed of peptide-bounded amino acids to form cyclic, linear or cyclic-linear structures (Markiewicz & Kwiatkowski, 2001; Patrick, 2003; Zejc & Gorczyca, 2002). They may act the Gram-negative (Polymyxin) and Gram-positive (Gramicidin, Prostinamycin) bacteria as well as fungi and protozoa.

A cyclosporine A – branched and cyclic oligopeptide composed of 11 amino acids is an important macromolecular immunosuppressive drug (Markiewicz & Kwiatkowski, 2001; Zejca & Gorczyca, 2002). Cyclosporine A selectively inhibits lymphocytes T function, thus is widely used as an immune barrier tolerance agent in the transplantology.

Macromolecular inhibitors that absorb the cholesterol from the intestines are also known; form them insoluble in water polymers, which produce complexes with the bile acids. To this polymer group belongs: copolymer of divinylbenzene and styrene substituted with quaternary trimethylammonium group and copolymer of diethyltriamine and epichlorohydrin (Zejc & Gorczyca, 2002).

A heparin, obtained from the animal tissues (mainly livers and lungs) (Zejc & Gorczyca, 2002) is next example of the natural polysaccharide used as the therapeutic agent. The heparin effects on the all blood clotting phases. Usually is used to treat arterial embolism and thrombosis, heart failure and before surgical operations.

A very important group of the biomedical polymers is macromolecular blood substitutes. They are accountable for the regular osmotic pressure and viscosity, closed to the osmotic pressure and blood viscosity; usually used in the anaphylactic shock, heart failure, intoxication, burns, toxic diarrhoea, embolic-thrombotic complications as well as microcirculation impairment.

A polyvinylpyrrolidone was the first synthetic polymer used as the blood substitutes. Its solutions were mainly used to treat the shock after the burns and, in the case when the blood transfusion was not indicated (Janicki et al., 2002; Florjanczyk & Penczek, 1998; Zejc & Gorczyca, 2002). Likewise, the solutions of polyvinyl alcohol have found the applications as the blood substitutes. However, they were withdrawn from the list of the blood substitutes as result of their undesirable side effects.

The blood susbstitutes with the therapeutic action has also been elaborated as result of incorporation of some therapeutic agents (e.g. penicillin, pelentanic acid, p-aminosalicylic chloride) into polyvinyl alcohol (Janicki et al., 2002; Zejc & Gorczyca, 2002).

Currently, the solutions of polysaccharides (e.g. dextran), modified starch derivatives and modified gelatin products (polygeline, oksopolygelin, liquid gelatin) are commonly used as the blood substititues (Janicki et al., 2002; Zejc & Gorczyca, 2002). The dextran with the average molecular weight ranged from 40 000 to 70 000 Da is used as 6 or 10% solution. This polysaccharide is produced by fermentation of the sucrose solutions in the presence of the *Leuconostoc mesenteroides* bacteria. Obtained glucose is polymerized to dextran in the presence of enzymes.

A hydroksyethyl starch is obtained by hydrolysis of high-amylopectine starch in acidic environment (Zejc & Gorczyca, 2002). The reaction products are neutralized followed by the reaction with ethylene oxide. The starch substituted with hydroxyethyl group is then produced in this reaction.

A polygeline is obtained from the reaction of diisocyanate with the gelatin. As result, linked urea groups are produced, whereas liquid gelatin is produced in the reaction with succinic anhydride (Janicki et al., 2002; Zejc & Gorczyca, 2002).

Macromolecular prodrugs

A prodrug is a modified therapeutic agent, which is metabolized into active precursor in human body (Janicki et al., 2002). Over the recent years, the conception of macromolecular prodrug has appeared as macromolecule that has therapeutic agents in the structure; the released drug becomes pharmacologically active during hydrolytic biodegradation of the polymer (Ouchi & Ohya, 1995). In general, the therapeutic agent could be incorporate into polymer chain, might be end-capped or may form a pendant group of the macromolecular chain (Figure 1).

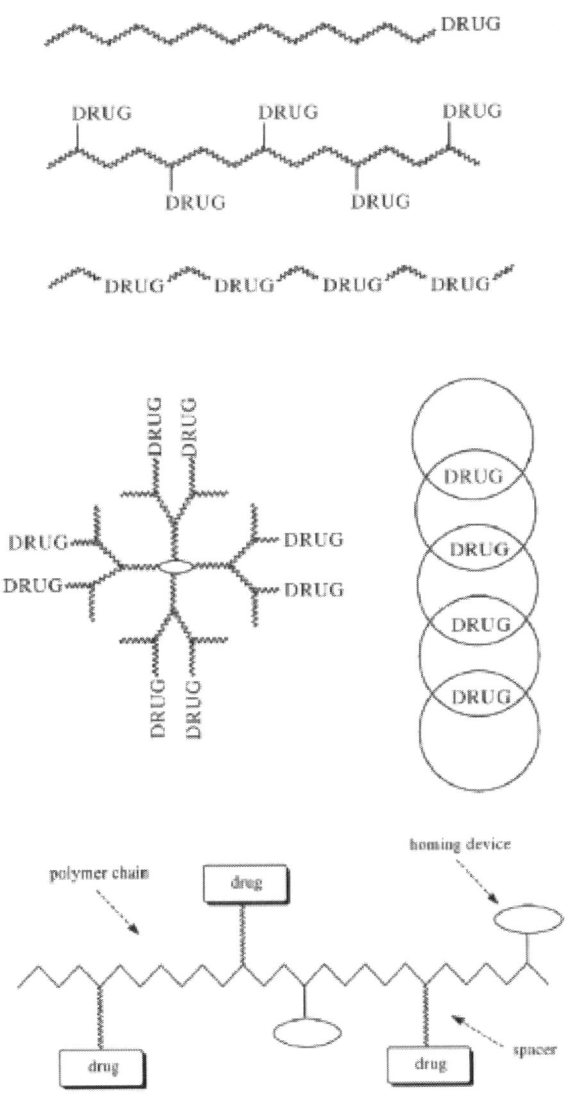

Fig. 1. Structure of the macromolecular prodrugs

Macromolecular prodrugs are mainly used in the cancer therapy. For example, 5- fluorouracil can be applied locally or orally in the therapy of the alimentary tract, urinary bladder and prostate gland cancers. The conjugations of this therapeutic agent as a pendant group to polyethylene glycol (Ouchi et al., 1986, 1992) or to vinyl polymer chain as substituent form examples of its macromolecular prodrugs (Ouchi et al., 1988).

The pharmacokinetics of the macromolecular prodrugs is mainly determined by the structure of the polymer (the rate of hydrolysis under the given conditions and the susceptibility to degradation in the presence of enzymes), its average molecular weight (the ability to the accumulation in blood, lymph, spleen, liver and other organs) or crystallinity (the rate of biodegradation).

The polymers must meet specific criteria to be applied in the synthesis of the macromolecular prodrugs. Namely, macromolecules and their metabolic decomposition products cannot be cumulate in the human body, to be toxic and the most important; the drug should be released from the macromolecule as result of the metabolic processes. The list of the macromolecular prodrugs developed so far is broadly presented in monograph (Ouchi & Ohya, 1995).

Polymer Applications in the Development of Extended-Release Drug Formulations

Macromolecules have also found the application in the technology of prolonged release drug formulations. They are mainly intended to ensure the constant concentration of the therapeutic agent in the certain time (e.g. 8-24 hours), in the patient body. The group of these drugs, therefore, can eliminate the drug multiple dosing during a day and reduce total daily dose of it. The prolonged drug forms are usually applied in the therapy of cardiac and alimentary tract diseases, coronary vessels, diabetics, and psychiatric disorders.

The absorption of the therapeutic agent using prolonged release drug forms can be reduced by coating, incorporation, complexation or bonding on the ionites (Janicki et al., 2002). Polymers applied in this technology, could be generally divided into biodegradable and non-biodegradable. Biodegradable macromolecules are definitely more preferred from the toxicological point of view. In the technology of prolonged release drug formulation, natural polymers and their modified derivative (e.g.: starch, cellulose) as well as synthetic polymers are used e.g.: polyethylene, polypropylene, polyvinyl chloride, polyvinyl alcohol, polyvinyl acetate, polyacrylic acid, polycarbophile, polyacrylamides, polyacrylates, polyethylene glycol, poly (amino amide)s, polyurethanes, siloxanes, homo- or copolymers of lactide and glycolide, poly(ε-caprolactone), polyorthoesters (Cardamone et al., 1997; Ertan et al., 1997; Huang et al., 1994; Lan et al., 1996; Matthews et al., 1996; Merkli et al., 1998; Ouchi & Ohya, 1995; Schierholz 1997; Sintzel et al., 1996; Uhrich et al., 1999; Ulbrich et al., 1996).

The crystals, pellets and granules of the drug might be coated with several polymer layers, according to the expected release rate. The therapeutic agent is gradually released as result of the polymer erosion or diffusion or is rinsing out from the polymer coating (Figure 2) (Uhrich et al., 1999).

Methylcellulose, polyvinylpyrrolidone and polyvinyl alcohol are predominantly applied as the coating substances. The analogous effect can be obtained by coating of thetherapeutic agent with polymeric layers, soluble in different parts of the alimentary tract or under enzymes.

The drug release based on the diffusion takes place when polymers insoluble in the alimentary tract (e.g.: ethyl cellulose, nitrocellulose, cellulose acetate, acrylic and methacrylic ester copolymers) are applied as the coating agents. The coating tablets containing porophors (acrylic and methacrylic ester copolymers, starch, cellulose acetate phthalate or microcrystalline cellulose) are also used. The solubility of these tablets is increased as the effect of porophors dissolution and swelling.

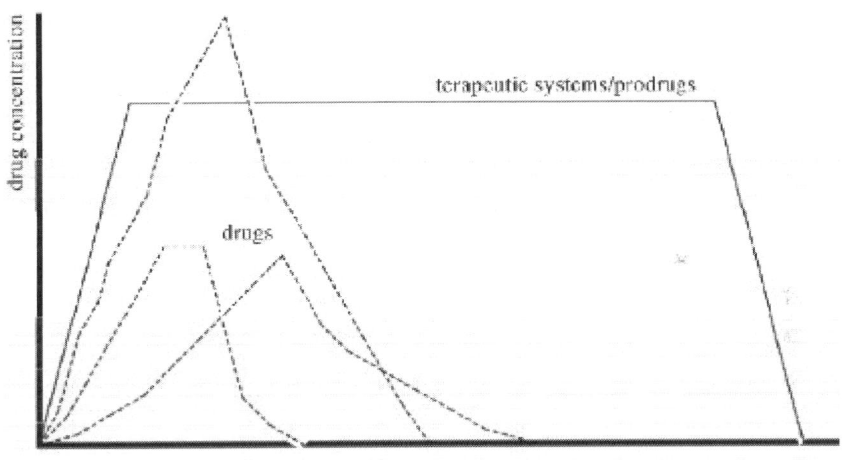

Fig. 2. The mechanism of the controlled release of the therapeutic agent

The incorporation method is relying on the suspension of the therapeutic agent on the prolonged released carrier. Most often as the carriers are used: hydrophobic polymers (e.g.: methylcellulose, acrylic acid polymers) as well as lipopholic polymers and some carriers insoluble in the alimentary tract (e.g.: polyvinyl chloride, polyethylene, cellulose acetate, ethyl cellulose, polystyrene, polyamide, silicone resin and acrylic and metacrylic acids ester copolymers). For instance, when the hydrophilic carrier is used, the tablet is consecutively swelled after passing the alimentary tract followed by creation of high viscous hydogels, which prolonged the drug release. The drug release suspended on the lipophilic carrier is dependant on pH and the presence of enzymes. Matrix tablets contained water-insoluble carriers, however, are stable in the alimentary tract environment. Therefore, the drug is gradually release via the capillaries.

The complexation method involves the creation of poor soluble, therapeutic agent-polymer complexes. The drug is released due to the gradual decomposition of this complex. This technique is also used to produce skin and mucosa antiseptics (iodophors). The iodophorsare the complexes of iodine with water-soluble polymers, which perform a role of carrier. They are high active against bacteria, viruses, fungi and protozoa.

The bonding of the drug on the ionites method is usually applied for acidic or basic drugs. It relies on release of the drug based on ion exchange in the alimentary tract.

Polymers in the therapeutic systems technology

The polymers used in the therapeutic systems are the drug forms that are dosing or releasing drug in the exact time with the controlled rate (Janicki et al., 2002; Müller & Hildebrand, 1998). They are designed to ensure constant concentration of the therapeutic agent in the body (Figure 3).

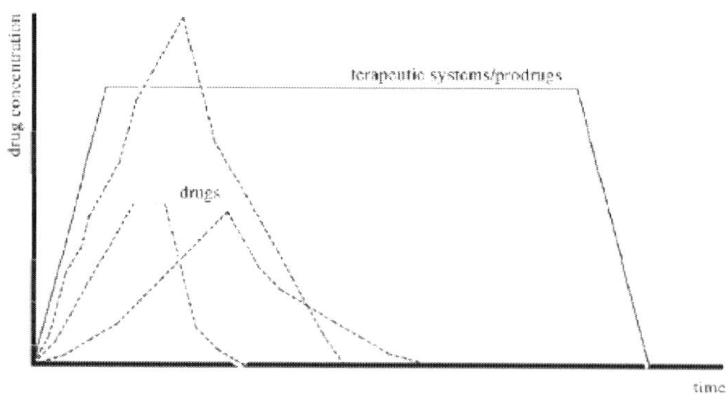

Fig. 3. Drug release profile from the conventional tablets and polymeric therapeutic systems

Therapeutic systems are commonly used in medicine due to their high efficiency in comparison to the conventional drug forms and prolonged release tablets. Considering the way of administration and the location of the drug absorption there are: oral, transdermal, ocular, intra-uterine, implantation and infusion therapeutic systems (Table 1). According to the construction of the element dosing the drug, there are: membrane, matrix and microvessels transdermal therapeutic systems (Knoch & Merkle, 1987; Müller & Hildebrand, 1998), usually used to treat stenocardia, inflammations, motion sickness, chronic hypertensive disease, in the hormonal and anti-nicotinic therapies (De Mey et al., 1989; Fagerström et al., 1993; Hadgraft, 1996; Ho & Chien, 1993; Liedtke et al., 1989; Lin et al., 1993; Man et al., 1993; Monkhouse & Hug, 1988; Sanders, 1996). The novelty comprises the ultrasonic transdermal therapeutic systems and the microelectronics transdermal therapeutic systems, where the drug is released from the polymer carrier under the frequency electric field influence (Prausnitz et al., 1994; Santus & Baker, 1993; Simonin, 1995).

In the ocular therapeutic systems, the drug is released to the lachrymal fluid through the membrane. The intra-uterine therapeutic systems are mostly used in the contraception, whereas implantation therapeutic systems are usually applied under the skin. In their case, the drug release is carried out through the slow diffusion from the polymeric systems to the tissue.

Table 1. The therapeutic systems examples

Therapeutic system	Polymer	Drug
Transdermal therapeutic system	copolymers of acetate vinyl and ethyl, poliacrylate, silicone, polyurethanes, polyolefines, polyethylene glycol	Acetate Noretisterone, Buprenorphine, Clonidyne, Estradiol, Fentanyl, Flurbiprofen, Hyoscine, Isosorbide dinitrate, Nicotyne, Nitroglycerin, Testosterone
Oral therapeutic system	polyvinyl alkohol, polyacetale vinyl, polyamides, polyethylene glycol, poliacrylate, silicone, homo- or copolymers of lactide, glicolide and ε-caprolactone	Acetazolamide, Glipizide, Metoprolol, Nifedipine, Okseprenolol KCl, Li2SO4, FeSO4
Ocular therapeutic system	copolymers of acetate vinyl and ethyl	Pilocarpina
Uterus therapeutic system	silicone	Progesterone
Implantation therapeutic system	copolymers of lactide and glicolide, silicone	Estradiol, Goserelin

UTILIZING NATURAL AND BIOACTIVE COMPOUNDS AS INITIATORS AND CATALYSTS FOR THE PRODUCTION OF BIODEGRADABLE POLYESTERS AND POLYCARBONATES

Biodegradable and bioresorbable polymers such as polyglycolide (PG), polylactide (PLA), poly(ε-caprolactone) (PCL), poly(trimethylene carbonate) (PTMC) and copolymers of glycolide (GL), L-lactides (LA), *rac*-lactide (*rac*-LA), ε-caprolactone (CL), trimethylene carbonate (TMC) or others cyclic esters and carbonates are very often used as polymeric prodrugs, drug delivery or therapeutic systems. Aliphatic polyesters and polycarbonates are degraded *in vivo* by hydrolytic deesterification into glycolic, lactic or other acid monomers. The latter species become involved in the carboxylic acid cycle and are subsequently excreted as carbon dioxide and water. Furthermore, biodegradable and bioresorbable drug forms exhibit unique pharmacokinetics, body distribution and pharmacological efficacy.

There are two methods of the synthesis of aliphatic polyesters or polycarbonates, namely polycondensation of diols, dicarboxylic acids or hydroxycarboxylic acids and ring-opening polymerization (ROP) of cyclic monomers (Platel, 2009; Labet & Thielemans, 2009). The polycondensation is hampered by typical limitations of step polymerization. The polymers obtained in this process are characterized by a high polydispersity. ROP gives polymeric products with the higher molecular weight and lower polydispersity. Therefore, is more preferred route to obtain aliphatic polyesters or polycarbonates (Platel, 2009).

The ROP of cyclic esters, carbonates or ether-esters initiated or catalyzed by the metal complexes or organic compounds yields high molecular weight polymers with the excellent conversion. The metal compounds are used commercially due to their selectivity, rate and lack of side reactions. On the other hand, for some biomedical or pharmaceutical applications, metal residues (Zn, Al, Sn) are undesirable (Albertsson & Varma, 2003; Albertsson & Srivastava, 2008; Varma et al., 2005).

Natural catalysts of ring-opening polymerization of cyclic esters and carbonates

The application of enzymes as catalysts of ROP seems to be a perspective direction in the polymer research. Macromolecules with well-defined structures can be formed by enzyme- catalyzed processes. On the other hand, the use of enzymes has some disadvantages, such as high cost, large quantity of enzymes required for ROP and relatively low molecular weight of the obtained polymers. However, the metal-free method of polymerization and suitable molecular weights of the resulted polymers are desirable conditions for the pharmaceutical applications, especially for the design of new drug delivery systems.

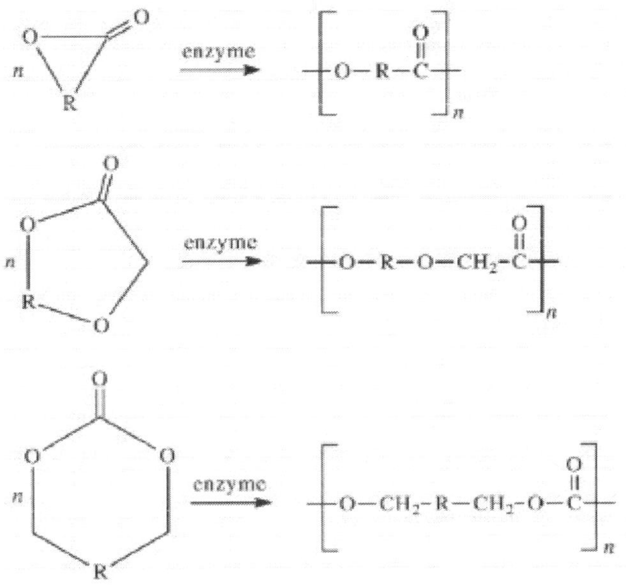

Fig. 4. e-ROP of cyclic esters, ether-esters or carbonates

The primary research on the enzyme ring-opening polymerization (e-ROP) has been carried out for CL. Currently major works concern ROP of six- and seven-membered cyclic esters, cyclic ether-estres or carbonates.

Lipases could also catalyze ROP of cyclic monomers, with different ring sizes as well as monomers containing substituents in the ring: α-methyl-β-propiolactone (1), β–butyrolactone (2), γ-caprolactone (3), α-methyl-valerolactone (4), 1,4-dioxan-2-one (5), δ-caprolactone (6), γ-ethyl-ε-caprolactone (7), ε-heptanolactone (8), δ-decalactone (9), δ-dodecalactone (10), α-methyl-12-dodecanolide (11), α-methyl-15- pentadecanolide (12), L-lactide (13), D-lactide (14),D,L-lactide (15), 1,4-dioxepan-2-one (16), 1,5-dioxepan-2-one (17), 2-methylene-4-oxa-12-dodecanolide (18), 1,3-dioxan-2-one (19), 5-methyl-5-benzyloxycarbonyl-1,3-dioxan-2-one (20),5-benzyloxy-1,3-dioxan-2-one (21), 1-methyl-1,3-dioxan-2-one (22), cyclobis (hexamethylenecarbonate) (23), 2,2'-dimethyl-1,3-dioxan-2-one (24) (Figure 5) (Albertsson & Varma, 2003; Albertsson & Srivastava, 2008; Labet & Thielemans, 2009; Platel, 2009; Varma et al., 2005).

Many families of enzymes were used in ROP of cyclic esters or carbonate: *Aspergillus niger*, *Pseudomonas species*, immobilized *Pseudomonas species*, *Candida rugosa*,\ *Candida antarctica* (Novozyme-435), *Candida cylindracea*, thermophilic *Esterase lipase* CloneZyme ESL-001, cutinase from *Humicola insolens*, immobilized *Pseudomonas species* on celite, *Porcine pancreatic lipase*, immobilized *Porcine pancreatic*,

Lipozyme IM or immobilized lipase from *Thermomyces lanuginose*, *Mucor javanicus*, *Mucor meihei*, *Pseudomonas aeruginosa*, *Pseudomonas cepacia*, *Pseudomonas fluorescens*, *Porcine pancreatic* lipase, *Penicillium roruеforti*, *Tritirachium alkaline proteinase*, *Rhizopus delemer*, *Rhizopus japonicus*, surfactant coated Lipase from *Aspergillus niger*, surfactant coated Lipase from *Pseudomonas species*, surfactant coated Lipase from *Candida rugoza*, surfactant coated Lipase from *Mucor javanicus*, surfactant coated *Pseudomonas species* (Barrera-Rivera et al., 2009; Córdova et al., 1999; Divakar, 2004; Dong, 1998, 1999; Gorke et al., 2007; Henderson et al., 1996; Kobayashi, 2001a, 2001b, 2009; MacDonald et al., 1995; Marcilla et al., 2006; Matsumoto et al., 1999; Mei et al., 2003; Namekawa et al., 1999; Rokicki, 2000; Sivalingam & Madras, 2004; Van Der Mee et al., 2006).

Lipases can accommodate a wide variety of synthetic substrates and still be able to show stereo- and regio-selectivity. They have evolved unusually stable structures that may survive effect of the organic solvents. The lipase-catalyzed hydrolysis in water can be easily reversed in non-aqueous media or bulk into ester synthesis or transesterification (Albertsson& Varma, 2003; Albertsson & Srivastava, 2008; Labet & Thielemans, 2009; Platel, 2009; Varma et al., 2005).

The e-ROP can be carried out in bulk, in organic media and at various interfaces. Enzyme- catalyzed reactions proceed under different reaction conditions (i.e. temperature, pressure, time). As an example, e-ROP of cyclic monomers was performed using lipase as catalyst for 2-720 h. M_n of the resulting polymers was ranged from 1000 to 90 000 Da, when M_w was in the range from 6 000 to 170 000 Da. The yield of the obtained polymers varied from 10 to 100%. The preferred lipase system generally used is a physically immobilized form of *Candida Antarctica*, commercially available as Novozyme-435 (Barrera-Rivera et al., 2009; Córdova et al., 1999; Divakar, 2004; Dong, 1998, 1999; Gorke et al., 2007; Henderson et al., 1996; Kobayashi, 2001a, 2001b, 2009; MacDonald et al., 1995; Marcilla et al., 2006; Matsumoto et al., 1999; Mei et al., 2003; Namekawa et al., 1999; Sivalingam & Madras, 2004; Van Der Mee et al., 2006).

The mechanism of e-ROP of cyclic esters using lipases as catalyst has been proposed by several authors. Monomer activated e-ROP (Scheme 1) involves the activation of the monomer molecules by an enzyme followed by the attack of the activated monomer onto the polymer chain end (Albertsson & Srivastava, 2008; MacDonald et al., 1995; Namekawa et al., 1999).

The ROP of cyclic carbonates catalyzed by enzyme or enzyme derivatives, in which polyesters, poly(ether-ester)s and polycarbonates terminated by hydroxyl groups are obtained, seems very attractive from the pharmaceutical or medical point of view.

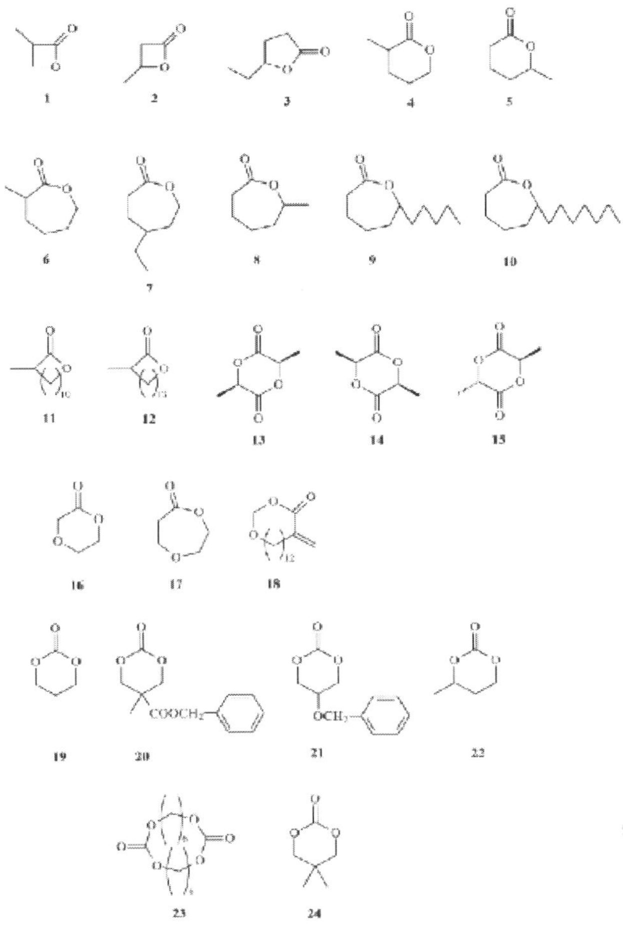

Fig. 5. Representative monomers for e-ROP

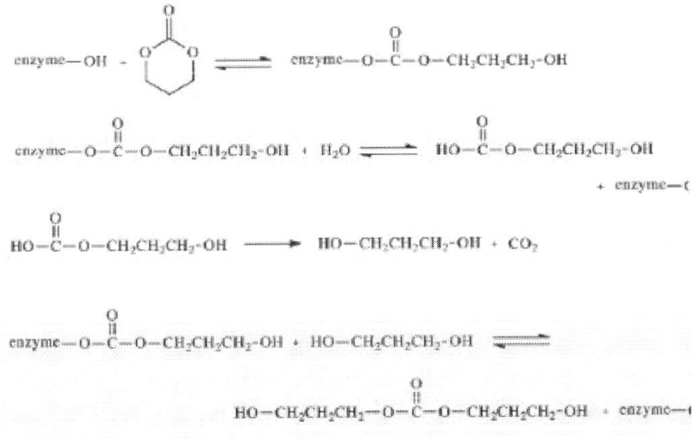

Scheme 1. The mechanism of e-ROP of cyclic esters

Bisht and coworkers proposed a mechanism for chain initiation and propagation for lipase- catalyzed trimethylene carbonate polymerization, based on the symmetrical structure of these products and the end-group structure of high molecular weight chains (Scheme 2) (Bisht et al., 1996).

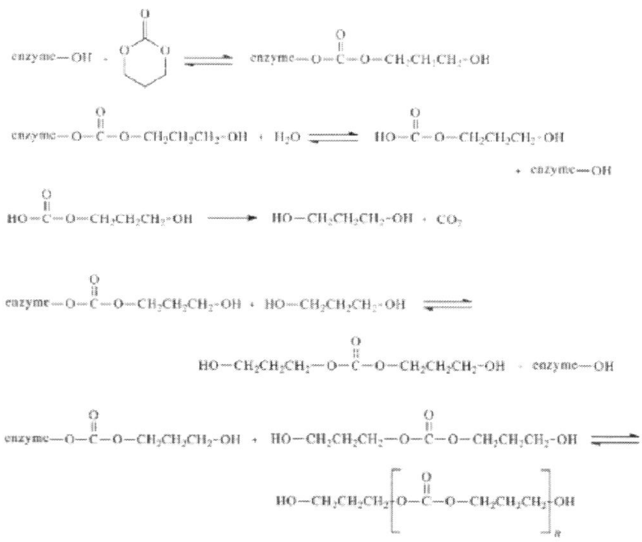

Scheme 2. The mechanism of e-ROP of cyclic carbonates

Natural initiators and organocatalysts of ring-opening polymerization of cyclic esters and carbonates

Recently, many modification approaches on biodegradable and bioresorbable polymers were carried out to meet the requirements of specific medical and pharmaceutical applications. Between them, incorporation of bioactive or biocompatible compounds such as lipids, amino acids into polymer chain or using of natural products as organocatalysts has received considerable attention.

The guanidine is a natural base, existing in human body and some vegetables. Some guanidine derivatives are the components of the therapeutic agents (Kinnel et al., 1998; Ramarao et al., 1993). Application of guanidine derivatives as organocatalysts for the synthesis of biodegradable polymers is an attractive way in the materials science technology. Li and coworkers reported the use of hexabutyl guanidinum acetatae in the living ROP of lactides (LAs) (Li et al., 2004). The polymerization was performed in bulk, producing polylactides (PLAs) with moderate molecular weight and narrow polydispersity. Strong guanidine bases: TBD (1,5,7-Triazabicyclo[4.4.0]dec-5-ene), MTBD (7-Methyl-1,5,7- triazabicyclo[4.4.0] dec-5-ene) and DBU (1,8-Diazabicyclo[5.4.0]undec-7-ene) were applied as effective organocatalysts for ROP of LA, δ-valerolactone (VL) and CL by Lohmeijer`s and coworkers (Lohmeijer et al., 2006). They found that TBD was polymerized LA, VL and CL in the controlled manner while MTBD and DBU polymerized only LA. For VL and CL the addition of thiourea co-catalyst was required. Wang and coworkers were used creatinine as catalysts of ROP of LA, examining the influence of temperature, time and creatinine dosage on the polymerization and properties of the produced biodegradable polymer (Wang et al., 2003). Based on the obtained results they proposed that creatinine is initiated ROP of LA according to the coordination-insertion mechanism. The biogenetic guanidine carboxylates: creatinine acetate (CRA) and creatinine glycolate (CRG) were synthesized and then effectively utilized as single-component initiators of ROP of LAs (Li et al., 2009). The mechanism of ROP was proposed based of the experimental investigation. In our laboratory, other guanidine derivatives: arginine and citrulline were successfully applied as initiators of ROP of LA and CL (Oledzka et al., 2011). The incorporation of α-amino acid molecules into the polymer chain was confirmed using ^{1}H, ^{13}C NMR and FT-IR spectroscopy and MALDI TOF MS spectrometry.

Various carboxylic acids (lactic, tartaric, hexanoic, propionic and citric acids) and natural amino acids (glycine, proline and serine) were engaged as catalysts in living ROP of CL and VL (Casas et al., 2004). The reactions were performed without solvent with the efficient way of recovering of the catalysts. Moreover, the authors found that the order of catalytic efficiency of the organic acid catalysts in ROP was as follows: tartaric acid (pKa=2.98)>citric acid (pKa=3.08)>lactic acid (pKa=3.14)>proline (pKa=1.95).

The fatty acids are found naturally in the human body. They are considered biologically safe and are generally considered suitable candidates for the preparation of biodegradable polymers (Teomim & Domb, 2001).

ROP of CL by organic acids catalyst and oleic acid derivatives initiator systems was investigated by Oledzka and coworker (Oledzka & Narine, 2001). They have found that the polymerizations were efficiently catalyzed by succinic and fumaric acid. The incorporation of fatty acid molecules resulted in less crystallinity and lower melting points of the obtainedpolymer samples. Furthermore, the degradation rate of the received polymers was slower when compared to CL homopolymer, but increase in porosity of the polymers was observed over time.

Natural amino acids are essential components in human nutrition. The studies performed by Liu and coworkers showed that they are also effective as initiators of ROP of CL (Liu & Liu, 2004). Authors found hat the number-average molecular weights of the obtained polymers did not exceed 5700 Da and that used amino acids were incorporated into macromolecule chain (Figure 6). In our laboratory, more detailed studies, involving an extended range of amino-acid initiators applied to ROP of CL and *rac*-LA were previously performed (Sobczak et al., 2008). The polymerization of CL and *rac*-LA was carried out in bulk, at 120–160 ºC. L-alanine,

L- cysteine, L-leucine, L-methionine, L-phenylalanine, L-proline, L-threonine, L-aspartic acid, L- glutamic acid, L-histidine, L-lysine were used as initiators of ROP. Aliphatic polyesters have synthesized with high yield (even ca. 100 % in some cases).

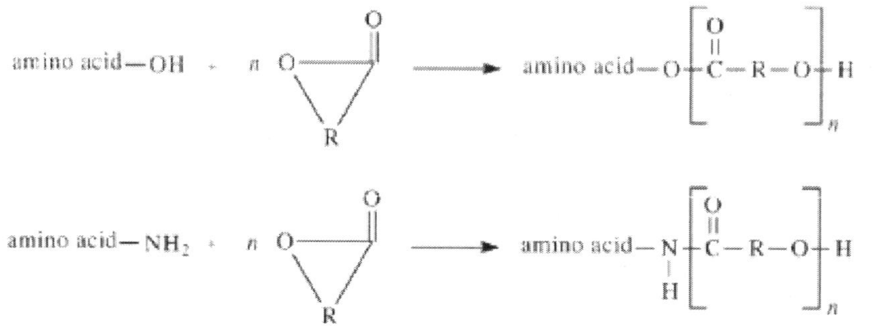

Fig. 6. ROP of cyclic esters using natural amino acids as initiators

ROP of aliphatic cyclic carbonates using natural amino acids was also investigated by Liu and coworkers (Liu et al., 2008). The biodegradable polymers with incorporated amino acids molecules were effectively obtained. The presence of natural amino acids in the polymer chain was proved by nuclear magnetic resonance spectroscopy.

Polyamidoamine (PAMAM) dendrimers are highly hyperbranched synthetic polymers with well-defined structure that allows accurate control of shape, size and functionality of terminal groups (Tomalia et al., 1985). The PAMAM dendrimers have useful applications in pharmaceutical technology e.g. as carriers in drug delivery systems, which can cross cell walls by both paracellular and transcellular pathways (Jevprasesphant et al., 2003). In our laboratory, novel star-shaped biodegradable polyesters were synthesized by ROP of CL using PAMAM dendrimer initiator without any metal catalysts (Oledzka et al., 2011). The nonlinear structure as well as physicochemical properties of the obtained biodegradable polymers were confirmed by nuclear magnetic resonance, gel permeation chromatography, thermal gravimetric analysis and differential scanning calorymetry.

Natural co-initiators of ring-opening polymerization of cyclic esters, ether-esters and carbonates

Metal catalyzed ROP of cyclic esters, ether-esters or carbonates has become the object of intensive studies with respect to their practical applications in the synthesis of biomaterials. Tin(II) 2-ethylhexanoate (SnOct2) is commonly used as a commercial catalyst for the ROP of cyclic monomers. It is effective, relatively cheap, non-toxic, soluble in the most commonly used organic solvents (Labet & Thielemans; 2009). SnOct2 is considered to have a toxicity much lower than other metal compounds, and it is allowed to be used as a food additive in a number of countries.

SnOct2 must be used together with a nucleophilic compound (generally an alcohol) to initiate the reaction if a controlled synthesis of the polymer is to be obtained. The main drawback of SnOct2 is that it requires high temperature, which leads inter- and intra- molecular esterification (Labet & Thielemans; 2009).

According to Kowalski's hypothesis, the first step of the polymerization consists of the production of the active species by reacting the alcohol with the catalyst. The more alcohol is added, the more the equilibrium is displaced towards the right and the more active species are created. With increasing carboxylic acid concentration, the equilibrium shifts to the left and less active species are present in the medium (Scheme 3) (Kowalski et al., 1998). Mechanism of CL and LA polymerization initiated with SnOct2/C4H9NH2 system has also been described (Duda et al., 2005).

$$SnOct_2 + ROH \rightleftharpoons RO{-}SnOct + Oct$$

$$RO{-}SnOct + ROH \rightleftharpoons RO{-}Sn{-}OR + Oct$$

Scheme 3. The formation of active centres in the reaction of Sn(Oct)2 with alcohol (co-initiator)

SnOct2 has also been combined with ureidopyrimidinone-alcohol (UPy) compounds. Using the good soluble alcohols, bearing a 1-ethylpentyl moiety, the ROP was significantly more controlled (Celiz & Scherman, 2008).

Sobczak and Kolodziejski have studied SnOct2/L-carnitine (CA) catalytic system. CA is a hydrophilic amino acid derivative, naturally occurring in human cell. Low-molecular weight PCL, PLA and copolymers of CL and *rac*-LA were obtained by the ROP of cyclic esters in the presence of SnOct2/CA. The molecular mass values averaged over the obtained polyesters were roughly in agreement with the theoretical molecular weights calculated from the feed ratio of the cyclic esters to CA (Sobczak & Kolodziejski, 2009).

Zhang and coworkers used cholesterol (CHL) as an initiator and SnOct2 as a catalyst of ROP of CL (Zhang et al., 2005). The polymerization was carried out under rigorously anhydrous conditions. The optimized ring-opening polymerization conditions have been identified to be 8 h at 140 °C. The molecular weight of CHL-PCL has increased with decreasing cholesterol/CL feed ratio. Incorporation of the cholesteryl moiety into polymer chain has leaded to a slower enzymatic degradation rate. Whereas, Cai and coworkers utilized PAMAM dendrimer as initiator of ROP of LLA (Cai et al., 2003). The star-shaped biodegradable polymers with the average molecular weight about 70000 Da were successfully obtained in that work. The authors also found that the synthesized polymers showed a faster degradation rate than linear homopolymer because of its shortened polymer chains.

AT A GLANCE

The pharmaceutical technology is one of the most important fields of using of polymers. From this review, it is clear that macromolecules have been extremely active research area over the last years. In addition is worth to note, that the progress of modern pharmaceutical technology is not feasible without utilization of natural and synthetic polymers. The discovering of new drug forms, e.g.: new therapeutic systems and macromolecular prodgrugs is simply demanded by the market and industry presently. The elaboration of new medical and pharmaceutical specimens will also require intensive investigations in chemistry and biomedical polymer areas.

As is also evident from this discussion, the spectacular improvement has been achieved with natural compounds applied as initiators, catalysts, organocatalysts or co-initiators of polymerization of cyclic esters, ether-esters and carbonates. The utilized compounds are primarily friendly for environment, safe, non-toxic and irreplaceable for the synthesis of polymers for the pharmaceutical applications. Promising avenues of research have also emerged for the enzymatic approach. Increasing interest has also been dedicated to the polymers containing natural compounds in macromolecules that have been incorporated into though the polymerization process. Clearly, the future development of biodegradable and bioresorbable polymers will be based on discovering macromolecules with not only appropriate chemical, physical and mechanical properties but also suitable biological properties.

REFERENCES

1. Albertsson, A-Ch. & Srivastava, R. K. (2008). Recent developments in enzyme-catalyzed ring-opening polymerization. *Advanced Drug Delivery Reviews*, Vol.60, No.9, (June 2008), pp. 1077–1093, ISSN 0169-409X
2. Albertsson, A-Ch. & Varma, I. V. (2003). Recent developments in ring opening polymerization of lactones for biomedical applications. *Biomacromolecules*, Vol.4, No.6, (November 2003), pp. 1466-1486 ISSN 1525-7797
3. Barrera-Rivera, K.A.; Marcos-Fernández, Á.; Vera-Graziano, R. & Martínez-Richa, A. (2009). Enzymatic ring-opening polymerization of ε-caprolactone by yarrowia lipolytica lipase in ionic liquids. *Journal of Polymer Science, Part A:*

Polymer Chemistry, Vol.47, No.21, (November 2009), pp. 5792-5805, ISSN 0887-624X

4. Bisht, K. S.; Svirkin, Y. Y.; Henderson, L. A.; Gross, R. A.; Kaplan, D. L. & Swift, G. (1997). Lipase-Catalyzed Ring-Opening Polymerization of Trimethylene Carbonate. *Macromolecules,* Vol.30, No.25, (December 1997), pp. 7735–7742, ISSN 0024-9297
5. Cai, Q.; Zhao, Y.; Bei, J.; Xi, F. & Wang, S. (2003). Synthesis and properties of star-shaped polylactide attached to poly(amidoamine) dendrimer. *Biomacromolocules,* Vol.4, No.3, (May 2003), pp. 828-834, ISSN 1525-7797
6. Cardamone, M.; Lofthouse, S. A.; Lucas, J. C.; Lee, R. P.; O'Donoghue, M. & Brandon, M. R. (1997). In vitro testing of a pulsatile delivery system and its in vivo application for immunisation against tetanus toxoid. *Journal of Controlled Release,* Vol.47, No.3, (September 1997), pp. 205-219, ISSN 0168-3659
7. Casas, J.; Persson, P-V.; Iversen, T. & Córdova, A. (2004). Direct organocatalytic ring- opening polymerizations of lactones. *Advanced Synthesis & Catalysis,* Vol.346, No.9- 10, (March 2004), pp. 1087–1089, ISSN 1615-4150
8. Celiz, A. D. & Scherman, O. A. (2008). Controlled Ring-Opening Polymerization Initiated via Self-Complementary Hydrogen-Bonding Units. *Macromolecules,* Vol.41, No.12, (May 2008), pp. 4115–4119, ISSN 0024-9297
9. Córdova, A.; Iversen, T. & Hult, K. (1999). Lipase-catalyzed formation of end-functionalized poly(ε-caprolactone) by initiation and termination reactions. *Polymer,* Vol.40, No.24, (January 1999), pp. 6709-6721, ISSN 0032-3861
10. De Mey, C.; Enterling, D.; Ederhoft, M.;Wesche, H. & Osterwald, H. (1989). Transdermal delivery of mepindolol and propranolol in normal man. 1st communication: Study design, clinical and pharmacological aspects. *Arzneimittel-Forschung/Drug Research,* Vol.39, No.1 A, (March 1989), pp. 1505-1508, ISSN 0004-4172
11. Divakar, S. (2004). Porcine pancreas lipase catalyzed ring-opening polymerization of ε- caprolactone. *Journal of Macromolecular Science: Pure and Applied Chemistry,* Vol. A41, No.5, (February 2007), pp. 537–546, ISSN 1060-1325
12. Dong, H.; Cao, S.-G.; Zheng-Qiang, L. I.; Han, S. I.-P.; You, D. E.-L. & Shen, J.-C. (1999). Study on the enzymatic polymerization mechanism of lactone and the strategy for improving the degree of polymerization. *Journal of Polymer Science, Part A: Polymer Chemistry,* Vol.37, No.9, (January 2000), pp. 1265-1275, ISSN 0887-624X
13. Dong, H.; Wang, H.-D.; Cao, S.-G. & Shen, J.-C. (1998). Lipase-catalyzed polymerization of lactones and linear hydroxyesters. *Biotechnology Letters,* Vol.20, No.10, (October 1998), pp. 905-908, ISSN 0141-5492

14. Duda, A.; Biela, T.; Kowalski, A. & Libiszowski, J. (2005). Amines as (co)initiators of cyclic esters polymerization. *Polimery*, Vol.50, No.7-8, (January 2005), pp. 501-508, ISSN 0032-2725

15. Ertan, G.; Karasulu, E.; Demirtas, D.; Arici, M. & Güneri, T. (1997). Release characteristics of implantable cylindrical polyethylene matrices. *Journal of Pharmacy and Pharmacology*, Vol.49, No.3, (November 1996), pp. 229-235, ISSN 0022-3573

16. Fagerström, K. O.; Säwe, U. & Tonnesen, P. (1993). Therapeutic use of nicotine patches: Efficacy and safety. *Journal of Drug Development*, Vol. 5, No.4, (January 1993), pp. 191-205, ISSN 0952-9500

17. Florjańczyk, Z. & Penczek, S. (1998). *Chemia polimerow*, Wydawnictwo Politechniki Warszawskiej, ISBN 83-86569-35-2, Warsaw, Poland

18. Gorke, J. T.; Okrasa, K.; Louwagie, A.; Kazlauskas, R. J.& Srienc, F. (2007). Enzymatic synthesis of poly(hydroxyalkanoates) in ionic liquids. *Journal of Biotechnology,* Vol.132, No.3, (November 2007), pp. 306–313, ISSN 0168-1656

19. Hadgraft, J. (1996). Pharmaceutical aspects of transdermal nitroglycerin. *International Journal of Pharmaceutics,* Vol.135, No.1-2, (May 1996), pp. 1-11, ISSN 0378-5173

20. Henderson, L. A.; Svirkin, Y. Y.; Gross, R. A.; Kaplan, D. L. & Swift, G. (1996). Enzyme- catalyzed polymerizations of ε-caprolactone: Effects of initiator on product structure, propagation kinetics, and mechanism. *Macromolecules,* Vol.29, No.24, (November 1996), pp. 7759-7766, ISSN 0024-9297

21. Ho, H. & Chien, Y. W. (1993). Kinetic evaluation of transdermal nicotine delivery systems. *Drug Development and Industrial Pharmacy*, Vol.19, No.3, (April 1993), pp. 295-313, ISSN 0363-9045

22. Huang, S. J.; Ho, L-H.; Hong, E. & Kitchen, O. (1994). Hydrophilic-hydrophobic biodegradable polymers: Release characteristics of hydrogen-bonded, ring-containing polymer matrices. *Biomaterials*, Vol.15, No.15, (October 1994), pp. 1243- 1247, ISSN 0142-9612

23. Janicki, S.; Fiebig, A. & Szmitowska, M. (2002). *Farmacja stosowana*, PZWL, ISBN 83-20037-78- 6, Warsaw, Poland

24. Jevprasesphant, R.; Penny, J.; Attwood, D.; McKeown, N. L. B. & D'Emanuele, A. (2003). Engineering of dendrimer surfaces to enhance transepithelial transport and reduce cytotoxicity. *Pharmaceutical Research*, Vol.20, No.10, (June 2003), pp. 1543-1550, ISSN 0724-8741

25. Kinnel, R. B.; Gehrken, H.-P.; Swali, R.; Skoropowski, G. & Scheuer, P. J. (1998). Palau'amine and its congeners: A family of bioactive bisguanidines from the

marine sponge Stylotella aurantium. *Journal of Organic Chemistry*, Vol.63, No.10, (October 1997), pp. 3281–3286, ISSN 0022-3263

26. Knoch, A. & Merkle, H. P. (1987). Polymeric laminates for transdermal delivery, II: In vitro release and release mechanism. *Acta Pharmaceutica Technologica*, Vol.33, No.4, (March 1987), pp. 202-207, ISSN 0340-3157

27. Kobayashi, S. (1999). Enzymatic polymerization: a new method of polymer synthesis. *Journal of Polymer Science, Part A: Polymer Chemistry,* Vol.37, No.16, (January 1999), pp. 3041-3056, ISSN 0887-624X

28. Kobayashi, S. (2009). Recent Developments in Lipase-Catalyzed Synthesis of Polyesters. *Macromolecular Rapid Communication,* Vol.30, No.3-4, (February 2009), pp. 237–266, ISSN 1022-1336

29. Kobayashi, S.; Uyama, H. & Kimura, S. (2001a). Enzymatic Polymerization. *Chemical Reviews,* Vol.101, No.12, (February 2001), pp. 3793-3818, ISSN 990-121l

30. Kobayashi, S.; Uyama, H. & Ohmae, M. (2001b). Enzymatic polymerization for precision polymer synthesis. *Bulletin of the Chemical Society of Japan,* Vol.74, No.4, (September 2002), 613-635, ISSN 0009-2673

31. Kowalski, A.; Duda, A. & Penczek, S. (1998). *Macromolecular Rapid Communication,* Vol.19, No.11, (December 1998), pp. 567–572, ISSN 1022-1336

32. Labet, M. & Thielemans, W. (2009). Synthesis of polycaprolactone: A review. *Chemical Society Reviews,* Vol.38, No.12, (January 2009), pp. 3484-3504, ISSN 0306-0012

33. Lan, P. N.; Corneillie, S.; Schacht, E.; Davies, M. & Shard, A. (1996). Synthesis and characterization of segmented polyurethanes based on amphiphilic polyether diols. *Biomaterials,* Vol.17, No.23, (October 1996), pp. 2273-2280, ISSN 0142-9612

34. Li, H.; Wang, Ch.; Yue, J.; Zhao, X. & Bai, F. (2004). Living ring-opening polymerization of lactides catalyzed by guanidinium acetate. *Journal of Polymer Science Part A: Polymer Chemistry,* Vol.42, No.15, (January 2004), pp. 3775–3781, ISSN 0887-624X

35. Li, H.; Zhang, S.; Jiao, J.; Jiao, Z.; Kong, L.; Xu, J.; Li, J.; Zuo, J. & Zhao, X. (2009). Controlled synthesis of polylactides using biogenic creatinine carboxylate initiators. *Biomacromolecules,* Vol.10, No.5, (February 2009), pp 1311–1314, ISSN 1525-7797

36. Liedtke, R. K.; Chien, L. S.; Mangold, B. & Haase, W. (1989). Clinical pharmacological studies on the transdermal application of mepindolol. Pharmacodynamics and orienting pharmacokinetics. *Arzneimittel-Forschung/ Drug Research*, Vol.39, No.11 A, (August 1989), pp. 1501-1504, ISSN 0004-4172

37. Lin, S. S.; Chien, Y. W.; Huang, W. C.; Li, C. H.; Chueh, C. L.; Chen, R. R. L.; Hsu, T. M.;

38. Jiang, T. S.; Wu, J. L. & Valia, K. H. (1993). Transdermal nicotine delivery systems: Multi-institutional cooperative bioequivalence studies. *Drug Development and Industrial Pharmacy*, Vol.19, No.20, (July 1993), pp. 2765-2793, ISSN 0363-9045

39. Liu, J. & Liu, L. (2004). Ring-opening polymerization of ε-caprolactone initiated by natural amino acids. *Macromolecules*, Vol.37, No.8, (March 2004), pp. 2674-2676, ISSN 0024- 9297

40. Liu, J.; Zhang, Ch. & Liu, L. (2008). Ring opening polymerization of aliphatic cyclic carbonates in the presence of natural amino acids. *Journal of Applied Polymer Science,* Vol.107, No.5, (September 2007), pp. 3275–3279, ISSN 0021-8995

41. Lohmeijer, B. G. G.; Pratt, R. C.; Leibfarth, F.; Logan, J. W.; Long, D. A.; Dove, A. P.; Nederberg, F.; Choi, J.; Wade, Ch.; Waymouth, R. M. & Hedrick, J. L. (2006). Guanidine and Amidine Organocatalysts for Ring-Opening Polymerization of Cyclic Esters. *Macromolecules*, Vol.39, No.25, (September 2006), pp. 8574–8583, ISSN 0024-9297

42. MacDonald, R. T.; Pulapura, S. K.; Svirkin, Y. Y.; Gross, R. A.; Kaplan, D. L.; Akkara, J.; Swift, G. & Wolk, S. (1995). Enzyme-catalyzed ε-caprolactone ring-opening polymerization. *Macromolecules,* Vol.28, No.1, (October 1994), pp. 73-78, ISSN 0024- 9297

43. Man, M.; Chang, C.; Lee, P. H.; Broman, T. & Cleary, G. W. (1993). New improved paddle method for determining the in vitro drug release profiles of transdermal delivery systems, *Journal of Controlled Release*, Vol. 27, No.1, (October 1993), pp. 59-68, ISSN 0168-3659

44. Marcilla, R.; de Geus, M.; Mecerreyes, D.; Duxbury, Ch. J.; Koning, C. E. & Heise, A. (2006). Enzymatic polyester synthesis in ionic liquids. *European Polymer Journal,* Vol.42, No.6, (June 2006), pp. 1215–1221, ISSN 0014-3057

45. Markiewicz, H. & Kwiatkowski, Z. A. (2001). *Bakterie, antybiotyki, lekoopornosc*, PWN, ISBN 83-01-13564-6, Warsaw, Poland

46. Matsumoto, M.; Odach, D. & Kondo, K. (1999). Kinetics of ring-opening polymerization of lactones by lipase. *Biochemical Engineering Journal,* Vol.4, No.1, (September 1999), pp. 73-76, ISSN 1369-703X

47. Matthews, S. E.; Pouton, C. W. & Threadgill, M. D. (1996). Macromolecular systems for chemotherapy and magnetic resonance imaging. *Advanced Drug Delivery Reviews*, Vol.18, No.2, (November 1995), pp. 219-267, ISSN 0169409X

48. Mei, Y.; Kumar, A. & Gross, R. (2003). Kinetics and mechanism of Candida antarctica lipase B catalyzed solution polymerization of ε-caprolactone. *Macromolecules,* Vol.36, No.15, (July 2003), pp. 5530-5536, ISSN 0024-9297

49. Merkli, A.; Tabaabay, C.; Gurny, R. & Heller, J. (1998). Biodegradable polymers for the controlled release of ocular drugs. *Progress in Polymer Science (Oxford),* Vol.23, No.3, (February 1998), pp. 563-580, ISSN 0079-6700

50. Monkhouse, D. C. & Hug, A. S. (1988). Transdermal drug delivery. Problems and promises. *Drug Development and Industrial Pharmacy,* Vol.14, No.2-3, (February 1988), pp. 183- 209, ISSN 0363-9045

51. Müller, R. H. & Hildebrand G. E. (1998). *Technologia nowoczesnych postaci lekow,* PZWL, ISBN 83-20022-06-1;Warsaw, Poland

52. Namekawa, S.; Suda, S.; Uyama, H. & Kobayashi, S. (1999). Lipase-catalyzed ring-opening polymerization of lactones to polyesters and its mechanistic aspects. *International Journal of Biological Macromolecules,* Vol.25, No.1-3, (June 1999), pp. 145–151, ISSN 0141-8130

53. Oledzka, E.; Kaliszewska, D.; Sobczak, M.; Raczak, A.; Nickel, P. & Kołodziejski, W. (2011). Synthesis and properties of a star-shaped poly(ε-caprolactone)-ibuprofen conjugate, *Journal of Biomaterials Science, Polymer Edition,* (March 2011), in press.

54. Oledzka, E. & Narine, S.S. (2011). Organic acids catalyzed polymerization of ε-caprolactone: Synthesis and characterization. *Journal of Applied Polymer Science,* Vol.119, No.4, (May 2010), pp. 1873-1882, ISSN 0021-8995

55. Oledzka, E.; Sokolowski, K.; Sobczak, M.; Kolodziejski W. (2011). α-Amino acids as initiators of ε-caprolactone and L,L-lactide polymerization. *Polymer International,* Vol.60, No.5, (March 2010), pp. 787–793, ISSN 0959-8103

56. Ouchi, T.; Hagihara, Y.; Takahashi, K.; Takano, Y. & Igarashi, I. (1992). Synthesis and antitumor activity of poly(ethylene glycol)s linked to 5-fluorouracil via a urethane or urea bond. *Drug Design Discovery,* Vol.9, No.1, (May 1992), pp. 93-105, ISSN 1055-9612

57. Ouchi, T.; Hagita, K.; Kawashima, M.; Inoi, T. & Tashiro, T. (1988). Synthesis and anti-tumor activity of vinyl polymers containing 5 fluorouracils attached via carbamoyl bonds to organosilicon groups. *Journal of Controlled Release,* Vol.8, No.2, (June 1988), pp. 141-150, ISSN 0168-3659

58. Ouchi, T. & Ohya, Y. (1995). Macromolecular prodrugs. *Progress in Polymer Science (Oxford),* Vol. 20, No.2, (January 1995), pp. 211-257, ISSN 0079-6700

59. Ouchi, T.; Yuyama, H.; Inui, T.; Murakami, H.; Fujie, H. & Vogi, O. (1986). Synthesis of polyether-bound 3-(5-fluorouracil-1-yl)propanoic acid and its hydrolysis reactivity. *European Polymer Journal*, Vol. 22, No.7, (January 1986), pp. 537-541, ISSN 0014-3057

60. Papamatheakis, J. D.; Schultz, R. M.; Chirigos, M. A. & Massicot, J. G. (1978). Cell and tissue distribution of ^{14}C-labeled pyran copolymer. *Cancer Treatment Reports*, Vol. 62, No.11, (February 1978), pp. 1845-1851, ISSN 0361-5960

61. Patrick, L. G. (2003). *Chemia medyczna*, WNT, ISBN 83-204-2833-5, Warsaw, Poland

62. Platel, R. H.; Hodgson, L. M. & Williams, C. K. (2009). Biocompatible initiators for lactide polymerization. *Polymer Reviews*, Vol.48, No.1, (January 2008), pp. 11-63, ISSN 1558- 3724

63. Prausnitz, M. R.; Pliquett, U.; Langer, R. & Weaver, J. C. (1994). Rapid temporal control of transdermal drug delivery by electroporation. *Pharmaceutical Research*, Vol.11, No.12, (May 1994), pp. 1834-1837, ISSN 0724-8741

64. Ramarao, A. V.; Gurjar, M. K. & Islam, A. (1993). Synthesis of a new bicyclic guanidine heterocycle as a potential anti-HIV agent. *Tetrahedron Letters*, Vol.34, No.31, (June 1993), pp. 4993-4996, ISSN 0040-4039

65. Rokicki, R. (2000). Aliphatic cyclic carbonates and spiroorthocarbonates as monomers. *Progress in Polymer Science,* Vol.25, (January 2000), pp. 259–342, ISSN 0079-6700 Sanders, S. W. (1996). Transition from temporal to biological control in the clinical development of controlled drug delivery systems.*Journal of Controlled Release*, Vol.39, No.2-3, (April 1996), pp. 389-397, ISSN 0168-3659

66. Santus, G. C. & Baker, R. W. (1993). Transdermal enhancer patent literature. *Journal of Controlled Release,* Vol.25, No.1-2, (February 1993), pp. 1-20, ISSN 0168-3659

67. Schierholz, J. M. (1997). Physico-chemical properties of a rifampicin-releasing polydimethyl-siloxane shunt. *Biomaterials*, Vol.18, No.8, (April 1997), pp. 635-641, ISSN 0142-9612

68. Simonin, J.-P. (1995). On the mechanisms of in vitro and in vivo phonophoresis. *Journal of Controlled Release*, Vol.33, No.1, (April 1995), pp. 125-141, ISSN 0163-3659

69. Sintzel, M. B.; Bernatchez, S. F.; Tabatabay, C. & Gurny, R. (1996). Biomaterials in ophthalmic drug delivery. *European Journal of Pharmaceutics and Biopharmaceutics*, Vol.42, No.6, (January 1997), pp. 358-374, ISSN 0939-6411

70. Sivalingam, G. & Madras, G. (2004). Modeling of lipase catalyzed ring-opening polymerization of ε-carprolactone. *Biomacromolecules,* Vol.5, No.2, (March 2004), pp. 603-609, ISSN 1525-7797

71. Sobczak, M. & Kolodziejski, W. (2009). Polymerization of cyclic esters initiated by carnitine and tin (II) octoate. *Molecules,* Vol.14, No.2, (January 2009), pp. 621-632, ISSN 1420- 3049

72. Sobczak, M.; Oledzka, E. & Kołodziejski, W. L. (2008). NOTE: Polymerization of cyclic esters using amino acids initiators. *Journal of Macromolecular Science, Part A: Pure and Applied Chemistry*, Vol.45, No.10, (August 2008), pp. 872-877, ISSN 1060-1325

73. Teomim, D. & Domb, A. J. (2001). Nonlinear fatty acid terminated polyanhydrides. *Biomacromolecules*, Vol.2, No.1. (August 2000), pp. 37-44, ISSN 1525-7797

74. Tomalia, D. A.; Baker, H.; Dewald, J.; Hall, M.; Kallos, G.; Martin, S.; Roeck, J.; Ryder, J. & Smith, P. (1985). New class of polymers: starburst-dendritic macromolecules. *Polymer Journal (Tokyo)*, Vol.17, No.1, (August 1984), pp. 117-132, ISSN 0032-3896

75. Tonnesen, H. H. & Karlsen, J. (2002). Alginate in drug delivery systems. *Drug Development and Industrial Pharmacy*, Vol.28, No.6, (August 2001), pp. 621-630, ISSN 0363-9045

76. Uhrich, K. E.; Cannizzaro, S. M.; Langer R. S. & Shakesheff, K. M. (1999). Polymeric systems for controlled drug release. *Chemical Reviews*, Vol.99, No.11, (June 1999), pp. 3181- 3198, ISSN 0009-2665

77. Ulbrich, K.; Strohalm, J.; Šubr, V.; Plocová, D.; Duncan, R. & Říhová, B. (1996). Polymeric conjugates of drugs and antibodies for site-specific drug delivery. *Macromolecular Symposia,* Vol.103, (January 1996), pp. 177-192, ISSN 1022-1360

78. Van Der Mee, L.; Helmich, F.; De Bruijn, R.; Vekemans, J. A. J. M.; Palmans, A. R. A. & Meijer, E. W. (2006). Investigation of lipase-catalyzed ring-opening polymerizations of lactones with various ring sizes: Kinetic evaluation. *Macromolecules,* Vol.39, No.15, (June 2006), pp. 5021-5027, ISSN 0024-9297

79. Varma, I. K.; Albertsson, A.-Ch.; Rajkhowa, R. & Srivastava, R. K. (2005). Enzyme catalyzed synthesis of polyesters. *Progress in Polymer Science,* Vol.30, No.10, (October 2005), pp. 949–998, ISSN 0079-6700

80. Wang, Ch.; Li, H. & Zhao, X. (2004). Ring opening polymerization of l-lactide initiated by creatinine. *Biomaterials*, Vol.25, No.27, (October 2003), pp. 5797-5801, ISSN 0142- 9612

81. Zejc, A. & Gorczyca, M. (2002). *Chemia Lekow*, PZWL, ISBN 83-200-2709-8, Warsaw, Poland Zhang L.; Wang Q. R.; Jiang X. S.; Cheng S. X. & Zhuo R. X. (2005). Studies on functionalization of poly(ε-caprolactone) by cholesteryl moiety. *Journal of Biomaterials Science, Polymer Edition,* Vol.16, No.9, (May 2005), pp. 1095-1108, ISSN 0920-506.

CHAPTER-8

TRANSLATING 2A RESEARCH IN PRACTICAL APPLICATIONS

INTRODUCTION

The process of translating 2A research into practical applications is a pivotal bridge that connects the world of scientific inquiry to real-world solutions. It involves taking the insights and discoveries derived from 2A research, which often delve into cutting-edge scientific or technological realms, and adapting them into tangible, useful, and beneficial practices. This crucial step not only brings academic knowledge to life but also has the potential to revolutionize industries, improve healthcare, and address pressing societal challenges. Whether it's the development of innovative technologies, the implementation of sustainable solutions, or the advancement of medical treatments, translating 2A research into practical applications is a dynamic and transformative journey, driven by the pursuit of progress and the quest for practical impact.

Viruses have evolved a number of unconventional translation strategies to amplify the coding potential of their condensed genetic information. Leaky stop codons may be read- through to produce either the predicted translation product, or at a very low level an extended "read-through" protein. Overlapping (*e.g.* -UAAUG-; -UGAUG-; AUGA-), or highly proximal stop/start codons may give rise to termination accompanied by a low level of re-initiation. There are a number of cases where a single mRNA is translated into more than one protein by recoding, where the rules for decoding are altered through specific sites and signals in the mRNA such as frameshifting and readthrough. Ribosomal "skipping", first identified in the foot-and-mouth disease virus (FMDV), represents yet another translational trick to deliver multiple gene products from limited primary sequence. Briefly, when a ribosome encounters 2A within an open reading frame (ORF), the synthesis of a specific peptide bond is "skipped". The process gives rise to two alternative outcomes: either (i) translation terminates at the end of 2A, or (ii) translation of the downstream sequence occurs. In this manner discrete translation products can be synthesized from a single ORF (for in-depth reviews of recoding see Atkins & Gesteland, 2010).

2A and "2A-like" sequences have been thoroughly studied in the last 25 years. These results, as well as our current understanding of the underlying mechanism, are summarized in the first section of this review. In the next section, important considerations in the design of 2A peptide-linked vectors are discussed. The 2A peptide system has worked in all eukaryotic systems tested and has been used with some spectacular successes in a variety of biotechnology applications. In the final section we provide an overview of the literature highlighting some of these successes.

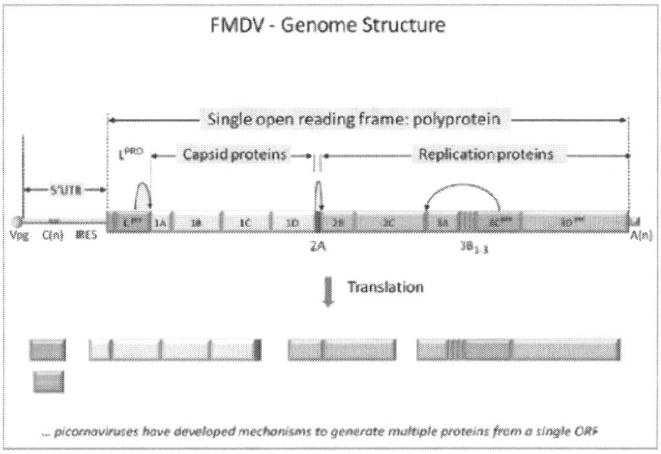

Fig. 1. The FMDV Genome. The FMDV genome is organized like a cellular mRNA: a 5' untranslated region (5'UTR), a single open reading frame (ORF), a 3' untranslated region (3'UTR) and a poly(A) tail. The polyprotein (~2,300aa) undergoes three "primary", co- translational cleavages; L^{pro} cleaves at its own C-terminus, 2A mediates "cleavage" at its own C-terminus and $3C^{pro}$ cleaves between [2BC] and 3A. The 2A oligopeptide is only 18aa long, mediating a "cleavage" by a translational effect "ribosome skipping".

BASIC RESEARCH

The FMDV genome

The FMDV genome organization is similar to that of other picornaviruses, comprising a large single ORF flanked by highly structured 5' and 3' untranslated regions (UTRs) (Fig. 1). The 5' UTR, of approximately 1,300 nucleotides (nt) contains sequence elements controlling the replication of viral RNA, packaging of RNA into capsids, and translation of the viral polyprotein. Preceding the ORF is a type II internal ribosome entry site (IRES), crucial for the cap-independent initiation of translation (for reviews see Jackson et al., 1990; Martĩnez-Salas & Ryan, 2010). The

3’ UTR is about 90 nt long and is thought to contain cis-acting elementsrequired for efficient genome replication (Agol et al., 1999). Moreover, the 3’ end of mRNA has also turned out to be surprisingly important in regulating translation (Wells et al., 1998). The ORF encodes a large protein precursor (polyprotein) which can be divided into three regions, designated P1, P2, and P3. These correspond to the N-terminal capsid protein precursor (P1, containing four capsid proteins 1A-1D), the middle of the polyprotein containing three of the nonstructural proteins (P2, the three proteins 2A-2C), and the most C- terminal segment of the polyprotein containing four non-structural proteins (P3, proteins 3A- 3D) (Palmenberg, 1987). The full-length translation product is never observed within infected cells due to co-translational, intramolecular, cleavages mediated by L^{pro}, 2A and $3C^{pro}$ domains within the polyprotein (for reviews see, Belsham, 2005; Ryan et al., 2004). Besides releasing itself from the polyprotein, L^{pro}, in common with $2A^{pro}$ of the entero- and rhinoviruses, also cleaves the translation initiation factor eIF4G (Glaser & Skern, 2000). This results in the inactivation of cap-dependent translation leading to the shutoff of cellular protein synthesis. The 2A oligopeptide is responsible for the primary cleavage which separates the region comprising the capsid proteins from domains downstream of 2A concerned with the replication of the virus (Ryan et al., 1991; Ryan & Drew, 1994). All picornaviruses encode $3C^{pro}$, which carries out a primary cleavage between 2C and 3A and secondary processing of the [P1-2A], [2BC] and P3 precursors. In FMDV, $3C^{pro}$ also cleaves between 2B and 2C (for review seeMartĩnez-Salas & Ryan, 2010). Aside from the processing sites within the viral polyprotein itself, the enzyme also modifies host cell proteins (Belsham et al., 2000; Li et al., 2001).

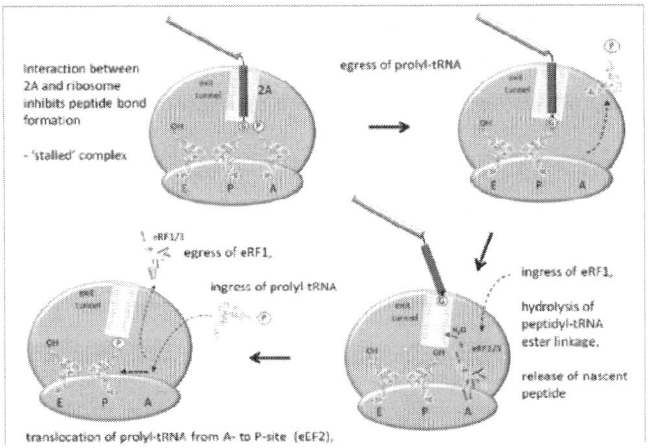

Fig. 2. Analysis of 2A-mediated “cleavage”. Artificial polyprotein cDNA constructs comprising the reporter proteins green fluorescent protein (GFP) and β-glucuronidase (GUS) (left panel). SDS-PAGE of radiolabeled *in vitro* translation products (right panel). The control pGFPGUS construct produces only a single

translation product – the [GFP-GUS] fusion protein. The translation profile from the pGFP2AGUS construct shows 3 major products: uncleaved [GFP2AGUS] and the cleavage products [GFP2A] and [GUS]. The profile from pGUS2AGFP also shows 3 major products: uncleaved [GUS2AGFP] and the cleavage products [GUS-2A] and [GFP]. The cleavage products upstream of 2A are highlighted in red, showing the molar excess over the downstream products shown in blue.

Secondary 3Cpro cleavage of the [1D2A] precursor protein between 1D and 2A shows the FMDV 2A segment is only 18aa long (-LLNFDLLKLAGDVESNPG-) (Belsham, 1993). Analysis of recombinant polyproteins and artificial polyprotein systems in which 2A was inserted between two reporter proteins showed that 2A alone, plus the N-terminal proline of protein 2B, was sufficient to mediate a highly efficient co-translational "cleavage" at the C-terminus of 2A (Ryan & Drew, 1994; Ryan et al., 1991; de Felipe et al., 2003). Translation *in vitro*, together with careful quantification of the products (Fig.2), provided the major finding that a molar excess of protein encoded upstream of 2A accumulated over that downstream – an observation at variance with proteolytic processing (Ryan et al., 1989; Donnelly et al., 2001a).

Extensive protein degradation studies, examining the effects of non-specific premature termination of transcription/translation, have shown that none of these effects account for thisimbalance (Ryan et al., 1999). Addition of puromycin at low concentration to translation reactions programmed with mRNA encoding a 2A containing reporter yields significant product with a size corresponding to the protein up to the 2A site, indicating a pause in translation at this position (Donnelly et al., 2001a). Employing a "toe-printing" approach, Doronina and colleagues confirmed that ribosomes pause at the end of the 2A coding sequence (-NPG↓P-), with glycine and proline in the P- and A- sites, respectively (Doronina et al., 2008b). This front end loading was due to different rates of biosynthesis of each portion of the ORF and constitutes a novel type of recoding (Baranov et al., 2002; Brown & Ryan, 2010).

The Mechanism of Cleavage

The 2A region of the FMDV encodes a sequence that mediates self-processing by a novel translational effect variously referred to as "ribosome skipping" (Ryan et al., 1999), "stop-go" (Atkins et al., 2007) and "stop carry-on" translation (Doronina et al., 2008a). 2A-mediated cleavage occurs between the C-terminal glycine and the proline of the downstream protein 2B (-LLNFDLLKLAGDVESNPG↓P-). The upstream protein contains a short 2A peptide C- terminal fusion, whereas the downstream protein includes a single proline residue on its N- terminus (Ryan and Drew, 1994; Ryan et al., 1991). The translational model of 2A cleavage activity posited is shown in Figure 3. Briefly, the nascent 2A peptide interacts with the exit pore of the ribosome such that the C-terminal portion (-ESNPGP-) is sterically constrained within the peptidyl transferase centre of the ribosome. This inhibits nucleophilic attack of the ester linkage between 2A and $tRNA^{gly}$ by prolyl-tRNA in the A site - effectively stalling, or pausing, translation (Ryan et al., 1999; Donnelly et al., 2001a). It has been shown that this block is relieved by the action of translation release factors eRF1 and eRF3, hydrolysing the ester linkage and releasing the nascent protein (Doronina et al., 2008a & b). Thus two major outcomes are possible; either translation terminates at this point, or, translation effectively 're- initiates' to synthesize the downstream sequences. The latter case would entail; (a) egress of eRF1/3 from the A site, (b) ingress of prolyl-tRNA into the A site, (c) translocation of prolyl- tRNA to the P site and (d) entry of the next aminoacyl-tRNA (for in-depth reviews of the model see Ryan et al., 2002; Martĩnez-Salas & Ryan, 2010; Brown & Ryan, 2010).

The occurrence of 2A and 2A-like sequences

Examining other picornavirus genome sequences showed the DxExNPGP motif to be present in several genera of the *Picornaviridae*: aphtho- cardio-, tescho-, erbo- and certain parechoviruses. Although cardioviruses have much longer natural 2A segments (133 to 143 amino acids) than aphthoviruses, work with *Encephalomyocarditis virus* (EMCV) and *Theiler's murine encephalitis virus* (TMEV) has shown that most of the additional 2A protein is dispensable for primary cleavage activity (Hahn & Palmenberg, 1996; Donnelly et al., 1997). Probing databases for the presence of the motif showed that "2A-like" sequences were also present in a range of non-picornavirus systems. These include a wide range of insect positive-strand RNA viruses belonging to the *Dicistroviridae* and *Tetraviridae* families and the unassigned Iflavirus genus and double-stranded RNA viruses of the *Reoviridae* (insect Cypoviruses and mammalian type C rotaviruses) (Hahn & Palmenberg, 1996; Donnelly et al., 2001b). They are also found in four nonsegmented dsRNA viruses of the *Totiviridae* (Isawa et al., 2011). Analysis of the translation products showed that in all cases these 2As had "cleavage" activity (Luke et al., 2008).

Along with the RNA virus 2As, active 2A-like sequences were also detected in the N-terminal region of the ORFs of non-LTR retrotransposons of *Trypanosoma cruzi* and *T.brucei* – L1Tc and igni, respectively (Donnelly et al., 2001b; Heras et al., 2006). Recently we identified a range of 2As in the purple sea urchin *Stronglocentrotus purpuratus*, then demonstrated their cleavage activities (unpublished data). In this case, 2A-like sequences appear in (i) several copies of non-LTR-retroelements (like trypanosomes) and (ii) the N-terminus of nucleotide binding oligomerization domain (NOD)-like, or CATERPILLER proteins (cited in Brown & Ryan, 2010). It appears, therefore, that this method of controlling protein biogenesis is not confined to viruses or genomic sequences comprising insertion(s) of virus-related sequences (retroelements). 2A and "2A-like" sequences have been shown to function in cells from a wide variety of eukaryotes, ranging from yeast (de Felipe et al., 2003) to plants (Halpin et al., 1999) to insects (Roosien et al., 1990) to mammals (Ryan & Drew, 1994). The only requirement for 2A peptide-based cleavage appears to be translation by 80S ribosomes. The reported proteolysis activity of 1D-2A in *E. coli* cells (Dechamma et al., 2008) was not detected in equivalent constructions in our laboratory showing "cleavage" specificity for eukaryotic systems alone (Donnelly et al., 1997). Although the FMDV 2A sequence (hereafter referred to as "F2A") has been the most widely used, biotechnologists should be aware that many 2A-like sequences have been utilized successfully, including equine rhinitis A virus (ERAV, "E2A"), porcine teschovirus-1 (PTV-1, "P2A") and *Thosea asigna* virus (TaV, "T2A") (Szymczak et al., 2004; Arnold et al., 2004; Osborn et al., 2005; Szymczak & Vignali, 2005; Huang et al., 2006; Scholten et al., 2006; Hart et al., 2008; Sommer et al., 2008; Yang et al., 2008).

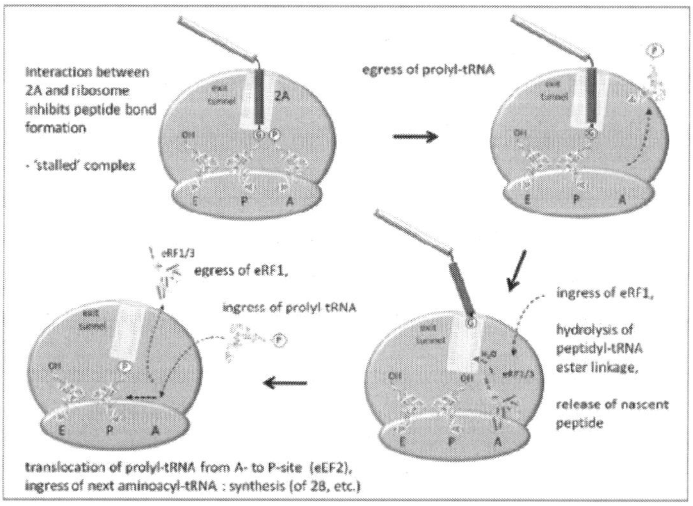

Fig. 3. Schematic representation of the translational model of 2A-mediated "cleavage".

KEY FACTORS WHEN USING 2A PEPTIDE SEQUENCES

Expression of multiple genes

Conventional approaches for the production of multicistronic vectors include the use of IRES elements, multiple promoters, fusion proteins, *etc* (for a review see de Felipe, 2002). Adverse side-effects with multiple promoters on viral vectors include interference between promoters, promoter suppression and rearrangement (Cullen et al., 1984; Emerman & Temin, 1986). IRESes provided the first method of creating eukaryotic polycistronic mRNAs. The internal ribosome entry site serves as a launching pad for internal initiation of translation, allowing expression of two or more genes from a single transcript (for review see Komar & Hatzoglou, 2005). Since genes are under the control of the same promoter and integrated into the same place within the genome, transgenes expressed in this way are co- ordinately regulated. In bicistronic systems, detection of the product encoded by the second cistron is evidence that the first cistron is also being expressed. This approach has been used successfully in gene therapy research in animal systems, and IRESes from different viruses have been tested and shown to function in plant systems (Urwin et al., 2000 & 2002; Dorokhov et al., 2002; Jaag et al., 2003; Bouabe et al., 2008).

On the other hand, there are a couple of limitations using IRES elements. Firstly, the IRES is a relatively large sequence (~500bp) that can cause problems in packaging, especially for size-restricted viral and nonviral vectors. For instance, retro- and lentiviral vectors possess packaging capacities of 8kb and adeno-associated viruses can accommodate <5kb (Thomas et al., 2003). Secondly, expression of the downstream gene can be as much as 10 fold lower than the upstream gene (Mizuguchi et al. 2000; Flasshove et al., 2000; Hasegawa et al., 2007; Ha et al., 2010). In some instances, this can be useful for expressing fluorescent markers or conferring drug resistance during selection (Ngoi et al., 2004). Nevertheless, the obvious advantages of using the 2A sequence vis-á-vis the IRES are its smaller size (~60-70bp) and the stoichiometric production of both upstream and downstream protein products as measured by: i) chloramphenicol acetyltransferase (CAT) and β-glucuronidase (GUS) enzyme activity (Halpin et al., 1999); ii) cell free translation *in vitro* and Western blot (Ryan& Drew, 1994; Donnelly et al., 2001a & b; de Felipe et al., 2003; Torres et al., 2010); iii) GFP/FACS with antibiotic resistance (Lorens et al., 2004); iv) co-fluorescence reporting (de Felipe & Ryan, 2004; Samalova et al., 2006); v) fluorescence resonance energy transfer (FRET) analysis (Szymczak et al., 2004) and vi) protein segregation in genetically engineered animals (Provost et al., 2007; Trichas et al., 2008). Further, if multiple gene expression is required, different members of the 2A peptide family can be selected to disrupt sequence homology to help maintain foreign gene insert stability.

Subcellular targeting of proteins from a 2A polyprotein

A merit of this expression strategy is that individual components of the 2A-polyprotein can be targeted to a range of different sub-cellular sites using both co- and post-translational signal sequences (El Amrani et al., 2004; Lorens et al., 2004; Szymczak et al., 2004). We discovered, however, a major problem with co-expression of some proteins targeted to, or passing through, the mammalian endoplasmic reticulum (ER). When a 2A-based polyprotein comprising an upstream protein bearing an N-terminal signal sequence was followed by a protein lacking any signal sequence, both proteins were translocated into theER (de Felipe & Ryan, 2004). We have identified the source of this problem - the "slipstreaming" effect was due to inhibition of the 2A reaction (formation of fusion protein) by the C-terminal region (immediately upstream of 2A) of some proteins when translocated into the ER – and suggest possible solutions (de Felipe et al., 2010).

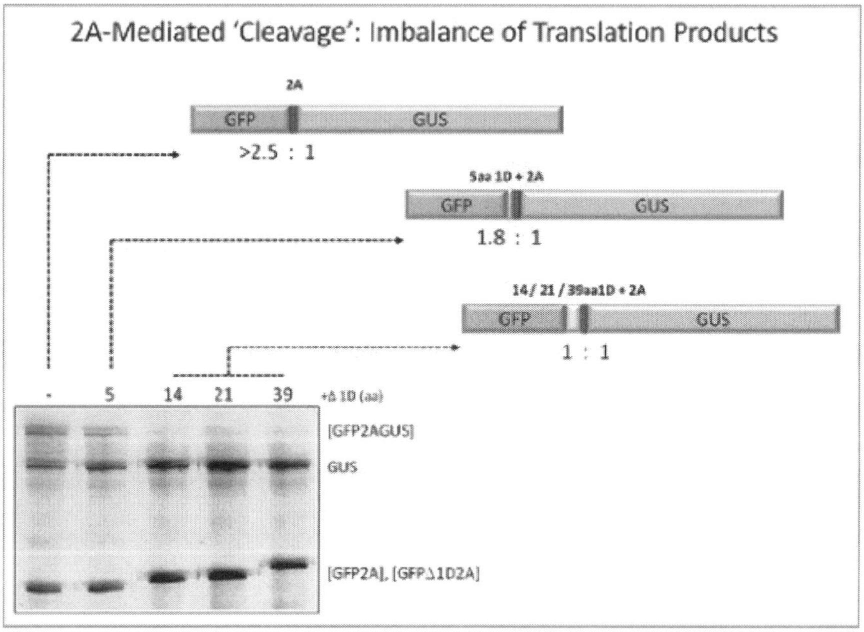

Fig. 4. Translation *in vitro*. Translation products derived from constructs encoding the wild- type 2A sequence are shown together with products derived from constructs encoding N- terminally extended forms of 2A.

The residues that influence cleavage are predicted to reside within the translocon; this length may allow interactions between the nascent peptide and the ribosome that lead to inhibition of the 2A reaction (Ménétret et al., 2000; Beckmann et al., 2001; de Felipe et al., 2010). Solutions to the problem include the use of longer versions of 2A with extra sequences derived from the capsid protein ("1D") (Ryan et al., 1991; Groot Bramel-Verheije et al., 2000; Donnelly et al., 2001b; Klump et al., 2001). Specifically, N-terminal extension of 2A by 5aa of 1D improved "cleavage", but extension by 14aa of 1D or longer (21 and 39aa) produced complete "cleavage" and an equal stoichiometry of the up- and downstream translation products (Donnelly et al., 2001b, see fig. 4). These observations are consistent with our model in which 2A activity is a product of it's interaction with the exit tunnel of the ribosome which is thought to accommodate 30-40aa (Hardesty & Kramer, 2001). Further,the order in which the genes are expressed within the vector needs to be considered. By swapping the order of proteins in several artificial polyproteins the stoichiometry was affected by the gene upstream of 2A (Ma & Mitra, 2002; Lengler et al., 2005; Chinnasamyet al., 2006; Rothwell et al., 2010). Cleavage activity was independent of the immediate downstream sequence (Ryan et al., 1991; Ma & Mitra., 2002). A number of studies show that cleavage efficiency is improved by incorporation of a flexible Gly-Ser-Gly or Ser-Gly- Ser-Gly linker sequence between the upstream protein and the 2A peptide (Lorens et al., 2004; Szymczak et al., 2004; Holst et al., 2006a & b; Provost et al., 2007; Wargo et al., 2009). A noteworthy caveat to attach to this review is that Yan and colleagues argue slipstreaming translocation does not occur in mammalian cells; that is, the second protein downstream of 2A still requires a signal sequence for secretory or membrane-anchored expression (Yan et al., 2010).

The unwanted tags

Cleavage occurs at the end of the 2A peptide sequence, therefore most of the 2A remains attached to the C-terminus of the upstream protein. This may affect the activity of some proteins (*e.g.* if their function is affected by the addition of other tags such as Myc, His, *etc*). In the case of proteins translocated into the ER, a strategy was adopted to include a furin proteinase cleavage site (-RA🡣KR-) between the upstream protein and 2A (Fang et al., 2005). Furin is a cellular endoprotease localized on the trans-Golgi networks of virtually all cell types (Steiner, 1998). Upon entering the lumen of the ER, 2A was trimmed away from the upstream protein (in this case antibody heavy chain), leaving only a 2aa C-terminal extension (-RA). In a follow-on study, the use of alternative furin cleavage sequences consisting of only basic amino acids, which can be efficiently cut by carboxypeptidases (🡣- RRRR-, 🡣-RKRR-, -RRKR-), resulted in the expression of antibodies with no residual amino acids (Fang et al., 2007). Proteins expressed in plants could have their 2A extensions removed by endogenous proteinases acting on similar hybrid linker peptides. A polyprotein precursor consisting of two different marker proteins connected by a linker peptide of Impatiens balsamina (🡣-SNAADEVAT-) followed by F2A was successfully processed in *Arabidopsis thaliana* (François et al., 2002; François et al., 2004).

For biomedical applications using 2A, a concern stems from the addition of 2A-derived sequences to the upstream protein - this protein may act as a carrier to stimulate an anti-2A immune response. Any potential carrier-effects could be abolished by removal of 2A. It should be noted that this "unwanted" tag does confer two advantages. First, antibodies to the 2A-peptide have been generated, allowing detection and/or immunoprecipitation of "upstream" protein products derived from 2A-containing transgenes (de Felipe et al, 2006 & 2010). Second, a shift in protein size is observed in 2A-tagged proteins which can be useful if mutant and endogenous proteins are co-expressed and need to be identified (Szmczak et al., 2004; de Felipe et al., 2010). To our knowledge, the presence of a proline attached to the amino-terminus of the second protein, as a relic of the 2A self-cleaving process, does not interfere significantly with activity and trafficking; it does, however, confer high protein stability (Varshavsky, 1992).

TRANSLATIONAL STUDIES

Gene expression *in planta*

Currently, there are several options available for the introduction of multiple transgenes *in planta*. The different methods include sexual crossing, re-transformation, single-plasmid ormultiple-plasmid co-transformation, and IRES-based transformation. The pros and cons of each have been reviewed previously (François et al., 2002; Halpin, 2005; Luke et al., 2006 & 2010). However, these procedures all suffer from a lack of coordinated expression of the different transgenes. As an alternative, the coding sequences of the genes of interest can be linked *via* 2A in a single transcription unit (Halpin et al., 1999; Ma & Mitra, 2002). The first types of genetically modified organisms created using 2A to co-express multiple proteins were plants, initially as a research tool, but also to improve drought-resistance (Kwon et al., 2004); disease-resistance (François et al., 2004; Geu-Flores et al., 2009) and nutritional qualities (Randall et al., 2004). Plant virus vectors based on potato virus X (PVX), cowpea mosaic virus (CPMV), pepino mosaic virus (PepMV), and bean pod mottle virus (BPMV) have been engineered with F2A and used to produce functional recombinant proteins including vaccines and antibodies (Smolenska et al., 1998; Gopinath et al., 2000; Marconi et al., 2006; Zhang et al., 2010; Sempere et al., 2011).

Metabolic and combinatorial engineering of carotenoid biosynthetic pathways in plants, including those synthesizing important industrial and pharmaceutical products, provide excellent examples of the utility of this approach (Ralley et al., 2004; Ha et al., 2010). Carotenoids have attracted interest not only as a source of pigmentation but also for their beneficial effects on human health. One of the most widely known carotenoids is ?-carotene, which serves as a dietary precursor of vitamin A. In developing countries, where vitamin A deficiency prevails, a promising intervention to existing strategies is to fortify the major staple food, rice, with provitamin A. Golden rice (*Oryza sativa*, GR) is the generic name given to genetically modified rice that produces ?-carotene in the endosperm (Ye et al., 2000, GR1; Paine et al., 2005, GR2). Engineering the provitamin A (?-carotene) biosynthetic pathway into (carotenoid-free) rice endosperm requires two carotenoid biosynthetic genes, phytoene synthase (*psy*) and carotene desaturase (*crtl*) (Lu & Li, 2008). As a step towards the coordinate expression of the two genes, *psy* from *Capsicum* and *crtl* from *Pantoea*, were linked *via* synthetic 2A (*psy-F2A-crtl*) or IRES (*psy-IRES-crtl*) sequences and placed under the control of the rice endosperm-specific globulin promoter (Ha et al., 2010). Collectively, the results demonstrated that the 2A construct performed better than the IRES construct in terms of carotenoid production. In addition, the use of a single promoter (GR1 and 2 require two promoters) reduces the chance of gene silencing and provides more space for transgene stacking.

Optical imaging of gene expression

In order to monitor transgene delivery and expression by optical imaging, the coding region is fused to a fluorescent/luminescent reporter. Another approach is to detect the expressed protein through its activity (conversion of a substrate in a fluorescent product as with γ- galactosidase). We (Halpin et al., 1999; Funston et al., 2008; de Felipe et al., 2010) and others (Samalova et al., 2006 & 2008; Hasegawa et al., 2007; Torres et al., 2010) have successfully used the 2A sequence in a number of *in vitro* and *in vivo* heterologous systems to achieve production of various combinations of fluorescent proteins and proteins requiring discrete co- and post-translational subcellular localization. The zebrafish (*Danio rerio*) has proved to be an excellent vertebrate model system for basic and biomedical science and comparative genomics. The lauded advantage of zebrafish embryos being transparent lends itself remarkably well to the use of fluorescence. To demonstrate the utility of the 2A system inzebrafish, reporter constructs employing eGFP and mCherry separated by the P2A sequence were designed to segregate fluorescent proteins to distinct cellular locations (Provost et al., 2007). Tissue-specific expression of both fluorophores in stably transformed embryos shows this approach could facilitate continuous expression of multiple proteins products at various stages of development in zebrafish. Likewise, Trichas et al (2008) used a bi-cistronic reporter construct containing a single coding sequence for a membrane localized red fluorescent protein (Myr-TdTomato) and a nuclear localized green fluorescent protein (H2B-GFP) separated by the T2A sequence to test 2A function in transgenic mice. Mutually exclusive localization of TdTomato and EGFP to the membrane and nucleus was observed in cultured cells and endogenous vertebrate cells, consistent with complete 2A-mediated processing. For the transgenic mice produced in this study, targeted expression was apparent in all tissues examined throughout development and into adulthood and remained constant across several generations.

In vivo bioluminescent imaging (BLI) allows a low-cost, noninvasive, and real-time analysis of biological processes at the molecular level in living systems. Cao and colleagues used BLI to visualize engraftment, survival, and rejection of transplanted tissues from a transgenic donor mouse that constitutively expresses luciferase (Cao et al., 2005). The donor mouse has a transgene comprised of a hybrid CMV-?-actin promoter, a firefly luciferase gene, a F2A and GFP gene. Isolated haematopoietic stem cells (HSC) from these mice express luciferase at the highest level among different haematopoietic cell types, and all haematopoietic lineages tested (with the exception of erythroblasts and red blood cells) express the reporter gene. As a virtually unlimited source of labelled cells this mouse line represents a valuable resource for stem cell and transplantation studies.

Immunotherapies

Cancer immunotherapy using heat shock protein

In an effort to extend the scope of immunotherapy for the control of advanced ovarian cancer, BLI was used to measure tumour load and distribution in mice vaccinated with irradiated tumour cells secreting heat shock protein 70 (Hsp70) (Chang et al., 2007). Hsps, including Hsp70, are highly effective in potentiating immune responses *via* interaction with several surface receptors on antigen-presenting cells (APCs). Hsp-specific receptors efficiently transport the chaperoned peptide into the major histocompatibility complex (MHC) class 1 cross-presentation pathway leading to recognition and activation of cytotoxic T cells (Udono & Srivastava, 1993; Massa et al., 2004). A retrovirus encoding sHsp70-T2A- GFP was used to introduce the gene for secreted hsp70 directly into mouse ovarian surface epithelial cells (MOSEC) that express luciferase. In summary, the tumour-secreted Hsp70 was capable of generating a potent antigen-specific "cytotoxic" $CD8^+$ T-cell response and CD40 was identified as a likely receptor for Hsp70-mediated cross-presentation.

Immunotherapy using monoclonal antibodies

Advances in recent years delineating the specific components of the immune system that contribute to immune responsiveness point to an important regulatory role for immunomodulators. Monoclonal antibodies (mAbs) are an important class of therapeutic agents for the treatment of cancer, autoimmune disorders, and infectious diseases. Althoughsatisfactory for short-term applications, antibody intravenous infusion is not appropriate in many long-term treatments. Fang et al. (2007) describe a recombinant adeno-associated virus (rAAV) gene delivery system that allows regulated long-term expression of native full- length mAbs *in vivo*. In this study a F2A sequence adjacent to a furin cleavage site (ΔK)RKRR was used to link the antibody heavy and light chain sequences. Notably, the gene expression system included a rapamycin-regulated promoter that can be used to stop mAb production if treatment needs to be terminated. This system potentially offers patients a lifelong mAb therapy that requires only a single administration of an rAAV vector.

Cytotoxic T-lymphocyte-associated antigen (CTLA-4), also known as CD152, is a co- inhibitory molecule that functions to regulate T-cell activation. Antibodies that block the interaction of CTLA-4 with its ligands CD80 (B7-1) and CD86 (B7-2) can enhance immune responses, including anti-tumour immunity (for a brief review, see Chambers et al., 2001). Granulocyte-macrophage colony-stimulating factor (GM-CSF) is a bone marrow growth factor for APCs, which has also been shown to enhance anti-tumour immune responses. Both preclinical animial models and early clinical development indicate synergy between GM-CSF tumour cell vaccination and CTLA-4 blockade (Hurwitz et al., 2000; Hodi et al., 2003; Quezada et al., 2006). To avoid anti-CTLA-4 side effects, tumour cell lines expressing the full-length F2A anti-CTLA-4 mAb in addition to GM-CSF, were administered locally at the immunization site (Simmons et al., 2008). Preliminary results suggest that the delivery of mAbs or proteins locally from immunotherapy cells should prove useful based on promising anti-tumour responses and the reduction of toxicity or adverse immune events associated with systemic exposure.

Cytokines and immunotherapy

The cytokine Interleukin-12 (IL-12) is a growth and maturation factor acting on both the innate and adaptive arms of the immune system. It is produced primarily by APCs and exerts immunoregulatory effects on natural killer (NK) and T cells (Kobayashi et al., 1989; Wolf et al., 1991). The APC-derived IL-12 consists of two subunits, p40 and p35, which are covalently linked (Kobayashi et al., 1989). The expression of this cytokine has been complicated by the observation that p40 homodimers (in excess of the heterodimer) exhibit antagonistic activity (Trinchieri et al., 2003). To ensure the equal expression of both subunits, biologically active IL-12 protein was produced using F2A as a linker between the p40 and p35 subunits (Collins et al., 1998; Kokuho et al., 1999; Chaplin et al., 1999; de Rose et al., 2000; Premraj et al., 2006). Numerous studies have been done which clearly indicate that plasmid expressed F2A IL-12 can modulate and augment the immune response elicited by DNA vaccination against mycobacterial infections (Triccas et al., 2002; Palendira et al., 2002; Martin et al., 2003). Additionally, it has been reported that IL-23 (but not IL-27) increased protection after *M. tuberculosis* challenge (Wozniak et al., 2006). In this study, the genes encoding p19 and p40 chains of IL-23 and EB13 and p28 chains of IL-27 were cloned on either side of the F2A protein.

Enhanced persistence of adoptively transferred tumour-infiltrating lymphocytes has been demonstrated by the administration of growth cytokines such as IL-2 and IL-15 (for reviews see Westwood & Kershaw, 2010; Ngo et al., 2011). However, systemic toxicity and expansion of unwanted cell subsets, such as regulatory T cells limit the use of these cytokines when administered systemically. Transgenic expression of IL-2 and IL-15 has been shown to increaseantigen-specific

T cell expansion *in vivo* and enhance antitumour activity without systemic toxicity in preclinical models (Quintarelli et al., 2007). The 3 genes coexpressed in the cytokine encoding vectors (*iCasp-9, CD34,* and *IL-2* or *IL-15*) were linked using F2A. The truncated form of human CD34 was used as a selectable marker of transduced cells and the inclusion of an iCasp-9 "safety-switch" ensured long-term safety of adoptively transferred lymphocytes.

Gene therapy

In vivo gene therapy

Gene therapy can be defined as the introduction of nucleic acids to somatic cells for a therapeutic purpose (Ylä-Herttuala & Alitalo, 2003). Compared to traditional medicine, gene therapy offers unique possibilities to treat the genetic causes of diseases, such as fatal enzyme deficiencies. Mucopolysaccharidosis type 1 (MPS-1; Hurler syndrome) is a congenital deficiency of ?-L-iduronidase (IDUA), leading to lysosomal storage of glycosaminoglycans. As accumulation and storage continue, tissue and organ damage becomes manifest as loss of function. Patients with MPS-1 present early in life with rapidly progressing disease that usually results in death due to neurological/CNS deterioration and/or cardiovascular/respiratory problems (Neufeld, 1991).

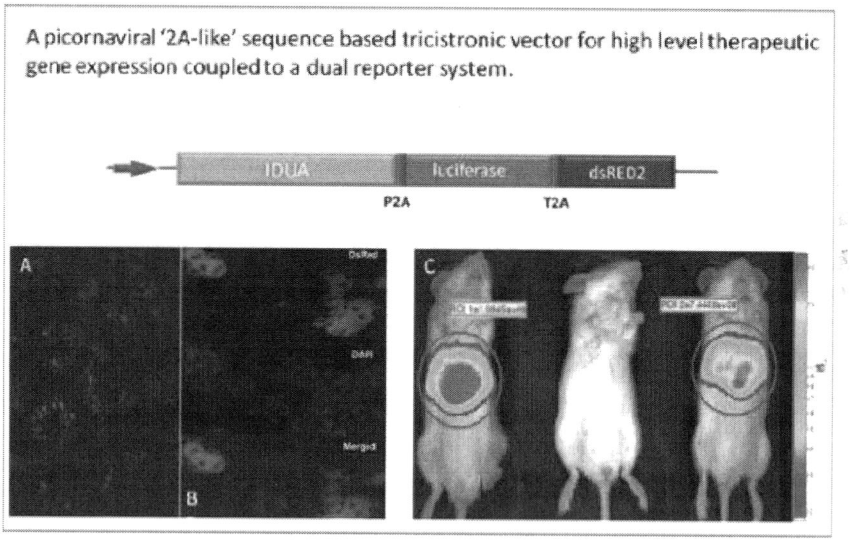

Fig. 5. Schematic of the tricistronic vector construct containing the therapeutic human iduronidase (IDUA) gene along with the firefly luciferase and DsRed2 reporter genes is shown at the top. (A) Whole-organ DsRed2 expression. (B) Cellular DsRed2 expression. (C) Whole-body *in vivo* luciferase imaging. Representative animals of tricistronic plasmid- injected (left), control (*IDUA* injected, middle), and monocistronic luciferase-injected (right) recipients are shown (adapted from Osborn et al., 2005).

Enzyme may be delivered by enzyme replacement therapy (ERT), haematopoietic cell transplantation (HCT) or by gene therapy vectors (Tolar & Orchard, 2008). P2A and T2A were utilized to construct a tricistronic vector bearing the human iduronidase (IDUA) gene along with the firefly luciferase and DsRed2 reporter genes (*IDUA*-P2A-luciferase-T2A- DsRed). Efficient cleavage was observed and all three proteins were functional *in vitro* and *in vivo*, leading to high-level therapeutic gene expression in NOD/*scid* mice that could be tracked by non-invasive whole-body luciferase imaging and at the cellular level using DsRed2 (Fig.5. Osborn et al., 2005).

Ex vivo gene therapy

To improve patient safety and increase the gene transfer efficiency, target cells are taken from the patient, gene-engineered and then adoptively transferred into the patient. Redirecting T cell specificity by T cell receptor (TCR) gene transfer is emerging as an attractive strategy to treat patients suffering from malignant and viral diseasesαβTCR, together with the CD3δε,γε and ζζ signaling subunits, determines the specific $CD4^+$ and $CD8^+$ T cell responses to antigens bound to MHC molecules (Call & Wucherpfennig, 2005; Rudolph et al., 2006). Using the TCR:CD3 complex as a test system, Szymczak and co- workers reported expression of all four proteins that make up CD3 and the two proteins required to make up TCR using just two retroviral vectors (CD3δγεζ-2A and TCRαβ-F2A) (Szymczak et al., 2004; reviewed in Radcliffe & Mitrophanous, 2004). Following the seminal paper of Szymczak et al in 2004, several groups have reported efficient TCR expression using 2A peptide linkers to combine TCRα and β- chain genes (Holst et al., 2006a & b; Scholten et al., 2006; Yang et al., 2008; Leisegang et al., 2008; Wargo et al., 2009). An important consideration in redirecting T cells using TCR genes is the tendency of introduced TCR genes to mispair with endogenous TCR *a*- and β- chains. In this regard, "murinized" receptors improved HLA-A2/LMP2-TCR expression on the surface of human T cells and downregulated expression of endogenous TCRs (Hart et al., 2008).

The feasibility of TCR gene therapy was recently demonstrated in the first bench to bedside experiments with TCR gene-modified T cells in melanoma patients. Johnson et al., (2009) treated metastatic melanoma patients with autologous T cells genetically modified with retroviral vectors to express high-avidity TCRs recognizing tumour-associated antigens MART-1 (MART-1*TCRα* -furinT2A-MART-1*TCRβ*) and gp100 (gp100*TCRα*-IRES- gp100*TCRβ*). Objective cancer regression was observed in 30% - 19% respectively, of patients who received these high affinity TCRs. However, in the study with TCR targeting MART, some patients also experienced toxicity to normal melanocytes in the skin, eye and ear. Another interesting recent study details the first clinical trial involving the adoptive transfer of engineered lymphocytes with optimal TCR complementary determining regions (CDRs) directed against NY-ESO-1, a cancer-testis antigen frequently expressed in melanoma as well as a wide range of non-melanoma epithelial malignancies (Robbins et al., 2011). In contrast to MART-1 and gp100, which are expressed in normal tissues as well as tumours, NY-ESO-1 expression is limited to neoplastic cells and germ line tissue (Chen et al., 1997). The α- and β-chains were expressed in retroviral constructs that contained a furin cleavage site followed by a SGSG spacer and the P2A sequence between the two gene products (Robbins et al., 2008). Response rates of 45% and 67% were observed in patients with melanoma and synovial cell sarcoma, respectively, all of whom had progressive disease after extensive prior treatment.

Induced pluripotent stem cell generation

Embryonic stem (ES) cells have the ability to differentiate into any cell type of the body and to grow indefinitely while maintaining pluripotency. Remarkably, adult somatic cells can be reprogrammed and returned to the naive state of pluripotency seen in embryonic stem cells by ectopic expression of a defined set of transcription factors: Oct 3/4, Sox2, KLF4 and c- Myc (Takahashi & Yamanaka, 2006; Takahashi et al., 2007; for review see Das & Pal, 2010). The delivery of these "Yamanaka factors" to create induced pluripotent stem (iPS) cells has typically required multiple individual viral vectors, carrying the risk of both insertional mutagenesis and viral reactivation (Takahashi and Yamanaka, 2006; Aoi et al., 2008).

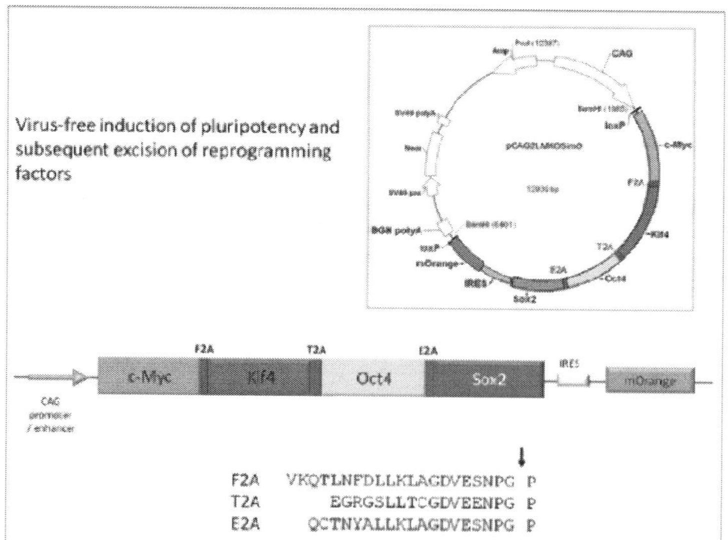

Fig. 6. Schematic diagram of reprogramming cassette. The four reprogramming factors c- *Myc*, *Klf4*, *Oct4* and *Sox2* were fused in-frame *via* 2A sequences and coexpressed as a single ORF and inset: the CAG enhancer/promoter was used to drive the 2A-linked reprogramming cassette and *mOrange* marker, flanked by *loxP* sites (Kaji et al., 2009).

Preliminary work by Okita et al. (2008 & 2010) achieved reprogramming of murine embryonic fibroblasts using repeated transient expressions of two plasmids - one encoding Oct 3/4, Sox2, KLF4 separated by F2A and the other encoding c-Myc. Although the efficiency of iPS cell generation was low, no vector DNA was stably integrated into the iPS cell genome. Sommer et al. (2008) and Carey et al. (2009) reported the derivation of iPS cells from adult skin fibroblasts using polycistronic lentiviral vectors. Sommer's team used a single multicistronic mRNA containing an IRES element separating two fusion cistrons - Oct4 and Sox2 linked *via* F2A and KLF4 and c-Myc linked *via* E2A. In a different way, the Carey group delivered the four factors in a single vector: Oct4, Sox2, KLF4, and c-Myc separated by three different 2A peptides (P2A, T2A and E2A, respectively). Both groups demonstrated reprogramming of fibroblasts to an ES cell-like state, however, in neither case was the polycistronic vector deleted from the genome. Therefore, attempts were made to minimize genome integration by removal of the inserted genes after the reprogramming process was switched-on (Chang et al., 2009; Kaji et al., 2009; Woltjen et al., 2009; Yusa et al., 2009) and more recently by using mRNA/miRNA of the four factors rather than DNA vectors (Warren et al., 2010; Yakubov et al., 2010; Miyoshi et al., 2011). The efficient reprogramming of murine and human embryonic fibroblasts and the traceless removal of factors joined with viral 2A sequences by using the Cre/LoxP or *piggyBac* transposon/transposase systems mark important advances towards achieving clinically acceptable methods of deriving iPS cells (see Fig.6).

SUMMARY

Exciting work of many laboratories in the last few years has clearly established the importance of 2A for co-expression technology. Our increasing knowledge about the cleavage mechanism indicates 2A is not just a novel method of controlling protein biogenesis, but that a crucial aspect of its function is to act as a translational "sensor". Protein synthesis in eukaryotes consumes a high proportion of cellular energy, most of which is used in elongation. During times of energy and/or nutrient deprivation, 2A could act to terminate translation in a stop codon-independent manner – devoting the remainder of the cell's resources into translating only that portion of the ORF upstream of 2A. We envisage that 2A-mediated cleavage could find extra utility in the biomedical and biotechnology fields as a reporter for translational stress.

REFERENCES

1. Agol, V.I., Paul, A.V. & Wimmer, E. (1999). Paradoxes of the replication of picornaviral genomes. *Virus Research*, 62, (August 1999), pp129-147, ISSN: 0168-1702

2. Aoi, T., Yae, K., Nakagawa, M., Ichisaka, T., Okita, K., Takahashi, K., Chiba, T. & Yamanaka, S. (2008). Generation of pluripotent stem cells from adult mouse liver and stomach cells. *Science*, 321(5889), (August 2008), pp699-702, ISSN: 0036-8075

3. Arnold, P.Y., Burton, A.R. & Vignali, D.A.A. (2004). Diabetes Incidence Is Unaltered in Glutamate Decarboxylase 65-Specific TCR Retrogenic Nonobese Diabetic Mice: Generation by Retroviral-Mediated Stem Cell Gene Transfer. *The Journal of Immunology*, 173(5), (September 2004), pp3103-3111.

4. Atkins, J.F., Wills, N.M., Loughran, G., Wu, C-Y., Parsawar, K., Ryan, M.D., Wang, C-H. & Nelson, C.C. (2007). A case for "StopGo": Reprogramming translation to augment codon meaning of GGN by promoting unconventional termination (Stop) after addition of glycine and then allowing continued translation (Go). *RNA*, 13, (June 2007), pp803-810, ISSN: 1355-8382

5. Atkins, J.F. & Gesteland, R.F. (2010). *Recoding: Expansion of Decoding Rules Enriches Gene Expression*, Springer, ISBN: 978-0-387-89281-5, New York.

6. Baranov, P.V., Gesteland, R.F. & Atkins, J.F. (2002). Recoding: translational bifurcations in gene expression. *Gene*, 286, (March 2002), pp187-201, ISSN:0378-1119

7. Beckmann, R., Spahn, C.M.Y., Eswar, N., Helmers, J., Penczek, P.A., Sali, A., Frank, J. & Blobel, G. (2001). Architecture of the protein-conducting channel associated with the translating 80S ribosome. *Cell*, 107, (November 2001), pp361-372, ISSN: 0092-8674

8. Belsham, G.J. (1993). Distinctive features of the foot-and-mouth disease virus, a member of the picornavirus family; aspects of virus protein synthesis, protein processing and structure. *Progress in Biophysics and Molecular Biology*, 60, pp241-260, ISSN: 0079-6107

9. Belsham, G.J., McInerney, G.M. & Ross-Smith, N. (2000). Foot-and-Mouth disease virus 3C protease induces cleavage of translation initiation factors eIF4A and eIF4G within infected cells. *Journal of Virology*, 74(1), (January 2000), pp272-280, ISSN: 0022-538X

10. Belsham, G.J. (2005). Translation and Replication of FMDV RNA. *Current Topics in Microbiology and Immunology*, 288, pp43-70, ISSN: 0070-217X

11. Bouabe, H., Fassler, R. & Heesemann, J. (2008). Improvement of reporter activity by IRES- mediated polycistronic reporter system. *Nucleic Acids Research* 36, (March 2008), pp1-9, ISSN: 0305-1048

12. Brown, J.D. & Ryan, M.D. (2010). Ribosome "Skipping": "Stop-Carry On" or "StopGo" Translation. In: *Recoding: Expansion of Decoding Rules Enriches Gene*

Expression. Eds J.F.Atkins & R.F.Gesteland, pp101-122, Springer, ISBN 978-0-387-89381-5, New York.

13. Call, M.E. & Wucherpfennig, K.W. (2005). The T cell receptor: critical role of the membrane environment in receptor assembly and function. *Annual Review of Immunology*, 23, pp101-125, ISSN: 0732-0582

14. Cao, Y-A., Bachmann, M.H., Beilhack, A., Yang, Y., Tanaka, M., Swijnenburg, R-J., Reeves, R., Taylor-Edwards, C., Schulz, S., Doyle, T.C., Fathman, C.G., Robbins, R.C., Herzenberg, L.A., Negrin, R.S. & Contag, C.H. (2005). Molecular Imaging Using Labeled Donor Tissues Reveals Patterns of Engraftment, Rejection, and Survival in Transplantation. *Transplantation*, 80(1), (July 2005), pp134-139, ISSN: 0041-1337

15. Carey, B.W., Markoulaki, S., Hanna, J., Saha, K., Gao, Q., Mitalipova, M. & Jaenisch, R. (2009). Reprogramming of murine and human somatic cells using a single polycistronic vector. *Proceedings of the National Academy of Sciences of the United States of America,* 106(1), (July 2009), pp157-162, ISSN: 0027-8424

16. Chambers, C.A., Kuhns, M.S., Egen, J.G. & Allison, J.P. (2001). CTLA-4-Mediated Inhibition in Regulation of T Cell Responses: Mechanisms and Manipulation in Tumor Immunotherapy. *Annual Review of Immunology*, 19, pp565-594, ISSN: 0732-0582

17. Chang, C-L., Tsai, Y-C., He, L., Wu, T-C., & Hung, C-F. (2007). Cancer Immunotherapy Using Irradiated Tumor Cells Secreting Heat Shock Protein 70. *Cancer Research*, 67(20), (October 2007), pp10047-10057, ISSN: 0008-5472

18. Chang, C-W., Lai, Y-S, Pawlik, K.M., Liu, K., Sun, C-W., Li, C., Schoeb, T.R. & Townes, T.M. (2009). Polycistronic Lentiviral Vector for "Hit and Run" Reprogramming of Adult Skin Fibroblasts to Induced Pluripotent Stem Cells. *Stem Cells*, 27, pp1042-1049, ISSN: 1066-5099

19. Chaplin, P.J., Camon, E.B., Villarreal-Ramos, B., Flint, M., Ryan, M.D. & Collins, R.A. (1999). Production of Interleukin-12 as a Self-Processing 2A Polypeptide. *Journal of Interferon and Cytokine Research*, 19, (March 1999), pp235-241, ISSN: 1079-9907

20. Chen, Y.T., Scanlan, M.J., Sahin, U., Tureci, O., Gure, A.O., Tsang, S., Williamson, B., Stockert, E., Pfreundschuh, M. & Old, L.J. (1997). A testicular antigen aberrantly expressed in human cancers detected by autologous antibody screening. *Proceedings of the National Academy of Sciences of the United States of America,* 94(5), (March 1997), pp1914-1918, ISSN: 0027-8424

21. Chinnasamy, D., Milsom, M.D., Shaffer, J., Neuenfeldt, J., Shaaban, A.F., Margison, G.P., Fairbairn, L.J. & Chinnasamy, N. (2006). Multicistronic

lentiviral vectors containing the FMDV 2A cleavage factor demonstrate robust expression of encoded genes at limiting MOI. *Virology Journal*, 3, (March 2006), p14, ISSN: 1743-422X

22. Collins, R.A., Camon, E.B., Chaplin, P.J. & Howard, C.J. (1998). Influence of IL-12 on interferon-? production by bovine leucocyte subsets in response to bovine respiratory syncytial virus. *Veterinary Immunology and Immunopathology*, 63, (May 1998), pp69-72, ISSN: 0165-2427
23. Cullen, B.R., Lomedico, P.T. & Ju, G. (1984). Transcriptional interference in avian retroviruses - implications for the promoter insertion model of leukaemogenesis. *Nature*, 307(5948), pp241-245, ISSN: 0028-0836
24. Das, A.K. & Pal, R. (2010). Induced pluripotent stem cells (iPSCs): the emergence of a new champion in stem cell technology-driven biomedical applications. *Journal of Tissue Engineering and Regenerative Medicine*, 4(6), (August 2010), pp413-421, ISSN: 1932-6254
25. Dechamma, H.J., Ashok Kumar, C., Nagarajan, G. & Suryanarayana, V.V.S. (2008). Processing of multimer FMD virus VP1-2A protein expressed in *E. coli* into monomers. *Indian Journal of Experimental Biology*, 46, (November 2008), pp760-763, ISSN: 0019-5189
26. de Felipe, P. (2002). Polycistronic Viral Vectors. *Current Gene Therapy*, 2, (September 2002), pp355-378, ISSN: 1566-5232
27. de Felipe, P., Hughes, L.E., Ryan, M.D. & Brown, J.D. (2003). Co-translational, Intraribosomal Cleavage of Polypeptides by the Foot-and-mouth Disease Virus 2A Peptide. *The Journal of Biological Chemistry*, 13, (March 2003), pp11441-11448, ISSN: 0021-9258
28. de Felipe, P. & Ryan, M.D. (2004). Targeting of Proteins Derived from Self-Processing Polyproteins Containing Multiple Signal Sequences. *Traffic*, 5, (August 2004), pp616-626, ISSN: 1398-9219
29. de Felipe, P., Luke, G.A., Hughes, L.E., Gani, D., Halpin, C. & Ryan, M.D. (2006). *E unum pluribus*: multiple proteins from a self-processing polyprotein. *Trends in Biotechnology*, 24(2), (February 2006), pp68-75, ISSN: 0167-7799
30. de Felipe, P., Luke, G.A., Brown, J.D. & Ryan, M.D. (2010). Inhibition of 2A-mediated 'Cleavage' of Certain Artificial Polyproteins Bearing N-terminal Signal Sequences. *Biotechnology Journal*, 5(2), (February 2010), pp213-223, ISSN: 1860-6768
31. de Rose, R., Scheerlinck, J-P. Y., Casey, G., Wood, P.R., Tennent, J.M. & Chaplin, P.J. (2000). Ovine Interleukin-12: Analysis of Biologic Function and Species Comparison. *Journal of Interferon and Cytokine Research*, 20, (June 2000), pp557-564, ISSN: 1079-9907

32. Donnelly, M.L.L., Gani, D., Flint, M., Monaghan, S. & Ryan, M.D. (1997). The cleavage activities of aphthovirus and cardiovirus 2A proteins. *Journal of General Virology*, 78, (January 1997), pp13-21, ISSN: 0022-1317

33. Donnelly, M.L.L., Luke, G.A., Mehrotra, A., Li, X., Hughes, L.E., Gani, D. & Ryan, M.D. (2001a). Analysis of the aphthovirus 2A/2B polyprotein "cleavage" mechanism indicates not a proteolytic reaction, but a novel translational effect : a putative ribosomal "skip". *Journal General Virology*, 82, (May 2001), pp1013-1025, ISSN: 0022-1317

34. Donnelly, M.L.L., Hughes, L.E., Luke, G.A., Mendoza, H., ten Dam, E., Gani, D. & Ryan, M.D. (2001b). The "cleavage" activities of foot-and-mouth disease virus 2A site- directed mutants and naturally occurring "2A-like" sequences. *Journal General Virology*, 82, (May 2001), pp1027-1041, ISSN: 0022-1317

35. Dorokhov, Y.L., Skulachev, M.V., Ivanov, P.A., Zvereva, Z.D., Tjulkina, L.G., Merits, A., Gleba, Y.Y., Hohn, T. & Atabekov, J.G. (2002). Polypurine (A)-rich sequences promote cross-kingdom conservation of internal ribosome entry. *Proceedings of the National Academy of Sciences of the United States of America*, 99(8), (April 2002), pp5301-5306, ISSN: 0027-8424

36. Doronina, V.A., de Felipe, P., Wu, C., Sharma, P., Sachs, M.S., Ryan, M.D. & Brown, J.D. (2008a). Dissection of a co-translational nascent chain separation event. *Biochemical Society Transaction*, 36(4), pp712-716.

37. Doronina, V.A., Wu, C., de Felipe, P., Sachs, M.S., Ryan, M.D. & Brown, J.D. (2008b). Site- Specific Release of Nascent Chains from Ribosomes at a Sense Codon. *Molecular and Cellular Biology*, 28(13), (July 2008), pp4227-4239, ISSN: 0270-7305

38. El-Amrani, A., Barakate, A., Askari, B.M., Li, X., Roberts, A.G., Ryan, M.D. & Halpin, C. (2004). Coordinate Expression and Independent Subcellular Targeting of Multiple Proteins from a Single Transgene. *Plant Physiology*, 135, (May 2004), pp16-24, ISSN: 0032-0889

39. Emerman, M. & Temin, H.M. (1986). Comparison of promoter suppression in avian and murine retrovirus vectors. *Nucleic Acids Research*, 14(23), (December 1986), pp9381- 9396, ISSN: 0305-1048

40. Fang, J., Qian, J.J., Yi, S., Harding, T.C., Tu, G.H., VanRoey, M. & Jooss, K. (2005). Stable antibody expression at therapeutic levels using the 2A peptide. *Nature Biotechnology*, 23(5), (May 2005), pp584-590, ISSN: 1087-0156

41. Fang, J., Yi, S., Simmons, A., Tu, G., Nguyen, M., Harding, T.C., VanRoey, M. & Jooss, K. (2007). An Antibody Delivery System for Regulated Expression of Therapeutic Levels of Monoclonal Antibodies *In Vivo*. *Molecular Therapy*, 15(6), (June 2007), pp1153-1159, ISSN: 1525-0016

42. Flasshove, M., Bardenheuer, W., Schneider, A., Hirsch, G., Bach, P., Bury, C., Moritz, T., Seeber, S. & Opalka, B. (2000). Type and position of promoter elements in retroviral vectors has substantial effects on the expression level of an enhanced green fluorescent protein reporter gene. *Journal of Cancer Research and Clinical Oncology*, 126, (July 2000), pp391-399, ISSN: 0171-5216

43. François, I.E.J.A., De Bolle, M.F.C., Dwyer, G., Goderis, I.J.W.M., Verhaert, P., Proost, P., Schaaper, W.M.M., Cammue, B.P.A. & Broekaert, W.F. (2002). Transgenic expression in Arabidopsis of a polyprotein construct leading to production of two different antimicrobial proteins. *Plant Physiology*, 128, (April 2002), pp1346-1358, ISSN: 0032-0889

44. François, I.E.J.A., van Hemelrijck, W., Aerts, A.M., Wouters, P.F.J., Proost, P., Broekaert,W.F. & Cammue, B.P.A. (2004). Processing in *Arabidopsis thaliana* of a heterologouspolyprotein resulting in differential targeting of the individual plant defensins. *Plant Science*, 166, (January 2004), pp113-121, ISSN: 0168-9452

45. Funston, G.M., Kallioinen, S.E., de Felipe, P., Ryan, M.D. & Iggo, R.D. (2008). Expression of heterologous genes in oncolytic adenoviruses using picornaviral 2A sequences that trigger ribosome skipping. *Journal of General Virology*, 89, (February 2008), pp389- 396, ISSN: 0022-1317

46. Geu-Flores, F., Olsen, C.E. & Halkier, B.A. (2009). Towards engineering glucosinolates into non-cruciferous plants. *Planta*, 229, (January 2009), pp261-270, ISSN: 0032-0935

47. Glaser, W. & Skern, T. (2000). Extremely efficient cleavage of eIF4G by picornaviral proteinases L and 2A *in vitro*. *FEBS Letters*, 480(2-3), (September 2000), pp151-155, ISSN: 0014-5793

48. Gopinath, K., Wellink, J., Porta, C., Taylor, K.M., Lomonossoff, G.P. & van Kammen, A. (2000). Engineering Cowpea Mosaic Virus RNA-2 into a Vector to Express Heterologous Proteins in Plants. *Virology*, 267, (February 2003), pp159-173, ISSN: 0042-6822

49. Groot Bramel-Verheije, M.H., Rottier, P.J.M. & Meulenberg, J.J.M. (2000). Expression of a Foreign Epitope by Porcine Reproductive and Respiratory Syndrome Virus. *Virology*, 278, (December 2000), pp380-389, ISSN: 0042-6822

50. Ha, S-H., Liang, Y.S., Jung, H., Ahn, M-J., Suh, S-C., Kweon, S-J., Kim, D-H., Kim, Y-M. &Kim, J-K. (2010). Application of two bicistronic systems involving 2A and IRES sequences to the biosynthesis of carotenoids in rice endosperm. *Plant Biotechnology Journal*, 8, (October 2010), pp928-938, ISSN: 1467-7644

51. Hahn, H. & Palmenberg, A.C. (1996). Mutational analysis of the encephalomyocarditis virus primary cleavage. *Journal of Virology*, 70, (Ocotber 1996), pp6870-6875, ISSN: 0022-538X

52. Halpin, C., Cooke, S.E., Barakate, A., El Amrani, A. & Ryan, M.D. (1999). Self-processing 2A-polyproteins – a system for co-ordinate expression of multiple proteins in transgenic plants. *The Plant Journal*, 17(4), pp453-459.

53. Halpin, C. (2005). Gene stacking in transgenic plants – the challenge for 21st century plant biotechnology. *Plant Biotechnology Journal*, 3(2), (March 2005), pp141-155, ISSN: 1467-7644

54. Hardesty, B. & Kramer, G. (2001). Folding of nascent peptide on the ribosome. *Progress in Nucleic Acid Research & Molecular Biology*, 66, pp41-66, ISSN: 0079-6603

55. Hart, D.P., Xue S-A., Thomas, S., Cesco-Gaspere, M., Tranter, A., Willcox, B., Lee, S.P., Steven, N., Morris, E.C. & Stauss, H.J. (2008). Retroviral transfer of a dominant TCR prevents surface expression of a large proportion of the endogenous TCR repertoire in human T cells. *Gene Therapy*, 15(8), (April 2008), pp625-631, ISSN: 0969-7128

56. Hasegawa, K., Cowan, A.B., Nakatsuji, N. & Suemori, H. (2007). Efficient Multicistronic Expression of a Transgene in Human Embryonic Stem Cells. *Stem Cells*, 25, (July 2007), pp1707-1712, ISSN: 1066-5099

57. Heras, S.R., Thomas, M.C., García-Canadas, M., de Felipe, P., García-Perez, J.L., Ryan, M.D. & Lopez, M.C. (2006). L1Tc non-LTR retrotransposons from *Trypanosoma cruzi* contain a functional viral-like self-cleaving 2A sequence in frame with the active proteins they encode. *Cellular and Molecular Life Sciences*, 63, (June 2006), pp1449- 1460, ISSN: 1420-682X

58. Hodi, F.S., Mihm, M.C., Soiffer, R.J., Haluska, F.G., Butler, M., Seiden, M.V., Davis, T., Henry-Spires, R., MacRae, S., Willman, A., Padera, R., Jaklitsch, M.T., Shankar, S., Chen, T.C., Korman, A., Allison, J.P. & Dranoff, G. (2003). Biologic activity ofcytotoxic T lymphocyte-associated antigen 4 antibody blockade in previously vaccinated metastatic melanoma and ovarian carcinoma patients. *Proceedings of the National Academy of Sciences of the United States of America*, 100(8), (April 2005), pp4712-4717, ISSN: 0027-8424

59. Holst, J., Szymczak-Workman, A.L., Vignali, K.M., Burton, A.R., Workman, C.J. & Vignali, DAA. (2006a). Generation of T-cell receptor retrogenic mice. *Nature Protocols*, 1(1), pp406-417 ISSN: 1754-2189

60. Holst, J., Vignali, K.M., Burton, A.R. & Vignali, D.A.A. (2006b). Rapid analysis of T-cell selection *in vivo* using T cell-receptor retrogenic mice. *Nature Methods*, 3(3), March (2006), pp191-197, ISSN: 1548-7091.

61. Huang, X., Wilber, A.C., Bao, L., Tuong, D., Tolar, J., Orchard, P.J., Levine, B.L., June, C.H., McIvor, R.S., Blazar, B.R. & Zhou, X.Z. (2006). Stable gene transfer

and expression in human primary T cells by the Sleeping Beauty transposon system. *Blood*, 107, (January 2006), pp483-491, ISSN: 0006-4971

62. Hurwitz, A.A., Foster, B.A., Kwon, E.D., Truong, T., Choi, E.M., Greenberg, N.M., Burg,M.B. & Allison, J.P. (2000). Combination Immunotherapy of Primary Prostate Cancer in a Transgenic Mouse Using CTLA-4 Blockade. *Cancer Research*, 60, (May 2000), pp2444-2448, ISSN: 0008-5472

63. Isawa, H., Kuwata, R., Hoshino, K., Tsuda, Y., Sakai, K., Watanabe, S., Nishimura, M., Satho, T., Kataoka, M., Nagata, N., Hasegawa, H., Bando, H., Yano, K., Sasaki, T., Kobayashi, M., Mizutani, T. & Sawabe, K. (2011). Identification and molecular characterization of a new nonsegmented double-stranded RNA virus isolated from *Culex* mosquitoes in Japan. *Virus Research*, 155, (January 2011), pp147-155, ISSN: 0168-1702.

64. Jaag, H.M., Kawchuk, L., Rohde, W., Fischer, R., Emans, N. & Pruffer, D. (2003). An unusual internal ribosome entry site of inverted symmetry directs expression of potato leafroll polerovirus replication-associated protein. *Proceedings of the National Academy of Sciences of the United States of America,* 100(15), (July 2003), pp8939-8944, ISSN: 0027-8424.

65. Jackson, R.J., Howell, M.T. & Kaminski, A. (1990). The novel mechanism of initiation of picornavirus RNA translation. *Trends in Biochemical Sciences*, 15, (December 1990), pp477-483, ISSN: 0968-0004.

66. Johnson, L.A., Morgan, R.A., Dudley, M.E., Cassard, L., Yang, J.C., *et al.* (2009). Gene therapy with human and mouse T-cell receptors mediates cancer regression and targets normal tissues expressing cognate antigen. *Blood*, 114(3), (July 2009), pp535-546, ISSN: 0006-4971.

67. Kaji, K., Norrby, K., Paca, A., Mileikovsky, M., Mohseni, P. & Woltjen, K. (2009). Virus-free induction of pluripotency and subsequent excision of reprogramming factors. *Nature*, 458(7239), (April 2009), pp771-775, ISSN: 0028-0836.

68. Klump, H., Schiedlmeier, B., Vogt, B., Ryan, M., Ostertag, W. & Baum, C. (2001). Retroviral vector-mediated expression in HoxB4 in hematopoietic cells using a novel expression strategy. *Gene Therapy,* 8, (May 2001), pp811-817, ISSN: 0969-7128.

69. Kobayashi, M., Fitz, L., Ryan, M., Hewick, R.M., Clark, S.C., Chan, S., Loudon, R., Sherman, F., Perussia, B. & Trinchieri, G. (1989). Identification and purification of natural killer cell stimulatory factor (NKSF), a cytokine with multiple biologic effects on human lymphocytes. *Journal of Experimental Medicine*, 170(3), (September 1989), pp827-845, ISSN: 0022-1007.

70. Kokuho, T., Watanabe, Y., Yokomizo, Y. & Inumaru, S. (1999). Production of biologically active, heterodimeric porcine interleukin-12 using a monocistronic baculoviral expression system. *Veterinary Immunology and Immunopathology*, 72, (December 1999), pp289-302, ISSN: 0165-2427.

71. Komar, A.A. & Hatzoglou, M. (2005). Internal ribosome entry sites in cellular mRNAs: mystery of their existence. *Journal of Biological Chemistry*, 280, (June 2005), pp23425- 23428, ISSN: 0021-9258.

72. Kwon, S-J., Hwang, E-W. & Kwon, H-B. (2004). Genetic Engineering of Drought Resistant Potato Plants by Co-Introduction of Genes Encoding Trehalose-6-Phosphate Synthase and Trehalose-6-Phosphate Phosphatase of *Zygosaccharomyces rouxii*. *Korean Journal of Genetics*, 26(2), (June 2004), pp199-206, ISSN: 0254-5934.

73. Leisegang, M., Engels, B., Meyerhuber, P., Kieback, E., Sommermeyer, D., Xue, S.A., Reu?, S., Stauss, H. & Uckert, W. (2008). Enhanced functionality of T cell receptor- redirected T cells is defined by the transgene cassette. *Journal of Molecular Medicine*, 86, pp573-583.

74. Lengler, J., Holzmuller, H., Salmons, B., Gunzburg, W.H. & Renner, M. (2005). FMDV-2A sequence and protein arrangement contribute to functionality of CYP2B1- reporter fusion protein. *Analytical Biochemistry*, 343, (August 2005), pp116-124, ISSN: 0003-2697.

75. Li, W., Ross-Smith, N., Proud, C.G. & Belsham, G.J. (2001). Cleavage of translation initiation factor 4AI (eIF4AI) but not eIF4AII by foot-and-mouth disease virus 3C protease: determination of the eIF4AI cleavage site. *FEBS Letters*, 507, (October 2001), pp1-5, ISSN: 0014-5793.

76. Lorens, J.B., Pearsall, D.M., Swift, S.E., Peelle, B., Armstrong, R., Demo, S.D., Ferrick, D.A., Hitoshi, Y., Payan, D.G. & Anderson, D. (2004). Stable, stoichiometric delivery of diverse protein functions. *Journal of Biochemical and Biophysical Methods*, 58, (February 2004), pp101-110, ISSN: 0165-022X.

77. Lu, S. & Li, L. (2008). Carotenoid metabolism: Biosynthesis, regulation, and beyond. *Journal of Integrative Plant Biology*, 50, pp778-785.

78. Luke, G.A., de Felipe, P., Cowton, V.M., Hughes, L.E., Halpin, C. & Ryan, M.D. (2006). Self- Processing Polyproteins : A Strategy for Co-expression of Multiple Proteins in Plants. *Biotechnology and Genetic Engineering Reviews*, 23, pp239-252.

79. Luke, G.A., de Felipe, P., Lukashev, A., Kallioinen, S.E., Bruno, E.A. & Ryan, M.D. (2008). The Occurrence, Function, and Evolutionary Origins of "2A-like" Sequences in Virus Genomes. *Journal of General Virology*, 89, (April 2008), pp1036-1042, ISSN: 0022-1317.

80. Luke, G.A., Escuin, H., de Felipe, P. & Ryan, M.D. (2010). 2A to the fore, research, technology and applications. *Biotechnology and Genetic Engineering Reviews*, 26, pp223-260, ISSN: 0264-8725.

81. Ma, C. & Mitra, A. (2002). Expressing multiple genes in a single open reading frame with the 2A region of foot-and-mouth disease virus as a linker. *Molecular Breeding*, 9, pp191- 199, ISSN: 1380-3743.

82. Marconi, G., Albertini, E., Barone, P., DeMarchis, F., Lico, C., Marusic, C., Rutili, D., Veronesi, F. & Porceddu, A. (2006). *In planta* production of two peptides of the Classical Swine Fever Virus (CSFV) E2 glycoprotein fused to the coat protein of potato virus X. *BMC Biotechnology*, 6, (June 2006), p29, ISSN: 1472-6750.

83. Martin, E., Kamath, A.T., Briscoe, H. & Britton, W.J. (2003). The combination of plasmid interleukin-12 with a single DNA vaccine is more effective than *Mycobacterium bovis* (bacilli Calmette-Guèrin) in protecting against systemic *Mycobacterium avium* infection. *Immunology*, 109, pp308-314.

84. Martĩnez-Salas, E. & Ryan, M.D. (2010). Translation and Protein Processing, In: *The Picornaviruses*, eds. E. Ehrenfeld, E. Domingo. & R.P. Roos. pp141-161. ASM Press, Washington, DC.

85. Massa, C., Guiducci, C., Arioli, I., Parenza, M., Colombo, M.P. & Melani, C. (2004). Enhanced Efficacy of Tumor Cell Vaccines Transfected with Secretable hsp 70. *Cancer Research*, 64, (February 2004), pp1502-1508, ISSN: 0008-5472

86. Ménétret, J.F., Neuhof, A., Morgan, D.G., Plath, K., Radermacher, M., Rapoport, T.A. & Akey,C.W. (2000). The structure of ribosome-channel complexes engaged in protein translocation. *Molecular Cell*, 5, (November 2000), pp1219-1232, ISSN: 1097-2765

87. Miyoshi, N., Ishii, H., Nagano, H., Haraguchi, N., Dewi, DL., Kano, Y. *et al.* (2011). Reprogramming of Mouse and Human Cells to Pluripotency Using Mature MicroRNAs. *Cell Stem Cell*, 8(6), pp633-638.

88. Mizuguchi, H., Xu, Z., Ishii-Watabe, A., Uchida, E. & Hayakawa, T. (2000). IRES-dependent second gene expression is significantly lower than cap-dependent first gene expression in a bicistronic vector. *Molecular Therapy*, 1, (April 2000), pp376-382, ISSN: 1525-0016.

89. Neufeld, E.F. (1991). Lysosomal storage diseases. *Annual Review of Biochemistry*, 60, pp57-80, ISSN: 0066-4154.

90. Ngo, M.C., Rooney, C.M., Howard, J.M. & Heslop, H.E. (2011). *Ex vivo* gene transfer for improved adoptive immunotherapy of cancer. *Human Molecular Genetics*, 20, (April 2011), ppR93-R99, ISSN: 0964-6906.

91. Ngoi, S.M., Chien, A.C. & Lee, C.G. (2004). Exploiting internal ribosome entry sites in gene therapy design. *Current Gene Therapy*, 4, (March 2004), pp15-31, ISSN: 1566-5232

92. Okita, K., Nakagawa, M., Hyenjong, H., Ichisaka, T. & Yamanaka, S. (2008). Generation of Mouse Induced Pluripotent Stem Cells Without Viral Vectors. *Science*, 322(5903), (November 2008), pp949-952, ISSN: 0036-8075.

93. Okita, K., Hong, H., Takahashi, K. & Yamanaka, S. (2010). Generation of mouse-induced pluripotent stem cells with plasmid vectors. *Nature Protocols*, 5(3), pp418-428, ISSN: 1754-2189.

94. Osborn, M.J., Panoskaltsis-Mortari, A., McElmurry, R.T., Bell, S.K., Vignali, D.A.A., Ryan, M.D., Wilber, A.C., Scott McIvor, R., Tolar, J. & Blazar, B.R. (2005). A Picornaviral 2A-like Sequence-Based Tricistronic Vector Allowing for High-Level Therapeutic Gene Expression Coupled to a Dual-Reporter System. *Molecular Therapy*, 12, (September 2005), pp569-574, ISSN: 1525-0016.

95. Paine, J.A., Shipton, C.A., Chaggar, S., Howells, R.M., Kennedy, M.J., Vernon, G., Wright, S.Y., Hinchliffe, E., Adams, J.L., Silverstone, A.L. & Drake, R. (2005). Improving the nutritional value of Golden Rice through increased pro-vitamin A content. *Nature Biotechnology*, 23(4), (April 2005), pp482-487, ISSN: 1087-0156.

96. Palendira, U., Kamath, A.T., Feng, C.G., Martin, E., Chaplin, P.J., Triccas, J.A. & Britton, W.J. (2002). Coexpression of Interleukin-12 Chains by a Self-Splicing Vector Increases the Protective Cellular Immune Response of DNA and *Mycobacterium bovis* BCG Vaccines against *Mycobacterium tuberculosis*. *Infection and Immunity*, 70(4), (April 2002), pp1949-1956, ISSN: 0019-9567.

97. Palmenberg, A.C. (1987). Picornaviral processing: some new ideas. *Journal of Cellular Biochemistry*, 33, (March 1987), pp191-198, ISSN: 0730-2312.

98. Premraj, A., Sreekumar, E., Jain, M. & Rasool, T.J. (2006). Buffalo (*Bubalus bubalis*) interleukin- 12: Analysis of expression profiles and functional cross-reactivity with bovine system. *Molecular Immunology*, 43, (March 2006), pp822-829, ISSN: 0161-5890.

99. Provost, E., Rhee, J. & Leach, S.D. (2007). Viral 2A peptides allow expression of multiple proteins from a single ORF in transgenic zebrafish embryos. *Genesis*, 45(10), (October 2007), pp625-629, ISSN: 1526-954X.

100. Quezada, S.A., Peggs, K.S., Curran, M.A. & Allison, J.P. (2006). CTLA-4 blockade and GM- CSF combination immunotherapy alters the intratumor balance of effector and regulatory T cells. *The Journal of Clinical Investigation*, 116(7), pp1935-1945.

101. Quintarelli, C., Vera, J.F., Savoldo, B., Giordano Attianese, G.M.P., Pule, M., Foster, A.E., Heslop, H.E., Rooney, C.M., Brenner, M.K. & Dotti, G. (2007). Co-expression of cytokine and suicide genes to enhance the activity of tumor-specific cytotoxic T lymphocytes. *Blood*, 110, (October 2007), pp2793-2802, ISSN: 0006-4971.

102. Radcliffe, P.A. & Mitrophanous, K.A. (2004). Multiple gene products from a single vector : "self-cleaving" 2A peptides. *Gene Therapy*, 11, (December 2004), pp1673-1674, ISSN: 0969-7128.

103. Ralley, L., Enfissi, E.M.A., Misawa, N., Schuch, W., Bramley, P.M. & Fraser, P.D. (2004). Metabolic engineering of ketocarotenoid formation in higher plants. *The Plant Journal*, 39, pp477-486.

104. Randall, J., Sutton, D., Ghoshroy, S., Bagga, S. & Kemp, J.D. (2004). Co-ordinate expression of α- and β- zeins in transgenic tobacco. *Plant Science*, 167, (August 2004), pp367- 372, ISSN: 0168-9452

105. Robbins, P.F., Li, Y.F., El-Gamil, M., Zhao, Y., Wargo, J.A., *et al.* (2008). Single and Dual Amino Acid Substitutions in TCR CDRs Can Enhance Antigen-Specific T Cell Functions. *The Journal of Immunology*, 180, pp6116-6131.

106. Robbins, P.F., Morgan, R.A., Feldman, S.A., Yang, J.C., Sherry, R.M., *et al.* (2011). Tumor Regression in Patients With Metastatic Synovial Cell Sarcoma and Melanoma Using Genetically Engineered Lymphocytes Reactive with NY-ESO-1. *Journal of Clinical Oncology*, 29, pp917-924.

107. Roosien, J., Belsham, G.J., Ryan, M.D., King, A.M.Q. & Vlak, J.M. (1990). Synthesis of foot- and-mouth disease virus capsid proteins in insect cells using baculovirus expression vectors. *Journal of General Virology*, 71(8), (August 1990), pp1703-1711, ISSN: 0022-1317

108. Rothwell, D.G., Crossley, R., Bridgeman, J.S., Sheard, V., Zhang, Y., Sharp, T.V., Hawkins, R.E., Gilham, D.E. & McKay, T.R. (2010). Functional expression of secreted proteins from a bicistronic retroviral cassette based on FMDV 2A can be position-dependent. *Human Gene Therapy*, 21(11), (November 2010), pp1631-1637, ISSN: 1043-0342

109. Rudolph, M.G., Stanfield, R.L. & Wilson, L.A. (2006). How TCRs bind MHCs, peptides, and coreceptors. *Annual Review of Immunology*, 24, pp419-466, ISSN: 0732-0582

110. Ryan, M.D., Belsham, G.J. & King, A.M.Q. (1989). Specificity of enzyme-substrate interactions in foot-and-mouth disease virus polyprotein processing. *Virology*, 173(1), (November 1989), pp35-45, ISSN: 0042-6822

111. Ryan, M.D., King, A.M.Q. & Thomas, G.P. (1991). Cleavage of foot-and-mouth disease virus polyprotein is mediated by residues located within a 19 amino acid sequence. *Journal of General Virology,* 72, (November 1991), pp2727-2732, ISSN: 0022-1317

112. Ryan, M.D. & Drew, J. (1994). Foot-and-mouth disease virus 2A oligopeptide mediated cleavage of an artificial polyprotein. *The EMBO Journal,* 13, pp928-933.

113. Ryan, M.D., Donnelly, M.L.L., Lewis, A., Mehrotra, A.P., Wilkie, J. & Gani, D. (1999). A model for Nonstoichiometric, Co-translational Protein Scission in Eukaryotic Ribosomes. *Bioorganic Chemistry*, 27, (February 1999), pp55-79, ISSN: 0045-2068

114. Ryan, M.D., Luke, G.A., Hughes, L.E., Cowton, V.M., Ten-Dam, E., Xuejun,L., Donnelly, M.L.L., Mehrotra, A. & Gani, D. (2002). The Aphtho- and Cardiovirus "Primary" 2A/2B Polyprotein "Cleavage". In: *Molecular Biology of Picornaviruses* eds. B.L. Semler & E. Wimmer, pp61-70, ASM Press, ISBN: 1-55581-210-4, Washington

115. Ryan, M.D., Donnelly, M.L.L., Flint, M., Cowton, V.M., Luke, G.A., Hughes, L.E., Knox, C. & de Felipe, P. (2004). Foot-and-Mouth Disease Virus Proteinases. In: *Foot-and-Mouth Disease* eds. F. Sobrino & E. Domingo, pp53-76, Horizon Bioscience, ISBN: 1555812104, Norfolk England

116. Samalova, M., Fricker, M. & Moore, I. (2006). Ratiometric Fluorescence-Imaging Assays of Plant Membrane Traffic Using Polyproteins. *Traffic*, 7, (December 2006), pp1701- 1723, ISSN:1398-9219

117. Samalova, M., Fricker, M. & Moore, I. (2008). Quantitative and Qualitative Analysis of Plant Membrane Traffic Using Fluorescent Proteins. *Methods in Cell Biology*, 85, pp353- 380, ISSN: 0091-679X

118. Scholten, K.B.J., Kramer, D., Kueter, E.W.M., Graf, M., Schoedl, T., Meijer, C.J.L.M., Schreurs, M.W.J. & Hooijberg, E. (2006). Codon modification of T cell receptors allows enhanced functional expression in transgenic human T cells. *Clinical Immunology,* 119, (May 2006), pp135-145, ISSN: 1521-6616

119. Sempere, R.N., Gómez, P., Truniger, V. & Aranda, M.A. (2011). Development of expression vectors based on pepino mosaic virus. *Plant Methods.* 7:6, (March 2011), ISSN: 1746-4811

120. Simmons, A.D., Moskalenko, M., Creson, J., Fang, J., Yi, S., VanRoey, MJ., Allison, J.P. & Jooss, K. (2008). Local secretion of anti-CTLA-4 enhances the therapeutic efficacy of a cancer immunotherapy with reduced evidence of systemic autoimmunity. *Cancer Immunology, Immunotherapy,* 57(8), (August 2008), pp1263-1270, ISSN:0340-7004

121. Smolenska, L., Roberts, I.M., Learmonth, D., Porter, A.J., Harris, WJ., Michael, T., Wilson, A. & Santa Cruz, S. (1998). Production of a functional single chain antibody attached to the surface of a plant virus. *FEBS Letters*, 441, (December 1998), pp379-382, ISSN: 0014-5793

122. Sommer, C.A., Stadfeld, M., Murphy, G.J., Hochedlinger, K., Kotton, D.N. & Mostoslavsky G. (2008). iPS Cell Generation Using a Single Lentiviral Stem Cell Cassette. *Stem Cells*, 27(3), pp543-549, ISSN: 1066-5099

123. Steiner, D.F. (1998). The proprotein convertases. *Current Opinion in Chemical Biology*, 2, (February 1998), pp31-39, ISSN: 1367-5931

124. Szymczak, A.L., Workman, C.J., Wang, Y., Vignali, K.M., Dilioglou, S., Vanin, E.F. & Vignali, D.A. (2004). Correction of multi-gene deficiency *in vivo* using a single "self- cleaving" 2A peptide-based retroviral vector. *Nature biotechnology*, 22(5), (May 2004), pp589-594, ISSN: 1087-0156

125. Szymczak, A.L. & Vignali, D.A.A. (2005). Development of 2A peptide-based strategies in the design of multicistronic vectors. *Expert Opinion on Biological Therapy*, 5, (May 2005), pp627-638, ISSN: 1471-2598

126. Takahashi, K. & Yamanaka, S. (2006). Induction of Pluripotent Stem Cells from Mouse Embryonic and Adult Fibroblast Cultures by Defined Factors. *Cell*, 126(4), (August 2006), pp663-676, ISSN: 0092-8674

127. Takahashi, K., Tanabe,K., Ohnuki, M., Narita, M., Ichisaka, T., Tomoda, K. & Yamanaka, S. (2007). Induction of Pluripotent Stem Cells from Adult Human Fibroblasts by Defined Factors. *Cell*, 131(5), (November 2007), pp861-872, ISSN: 0092-8674

128. Thomas, C.E., Ehrhardt, A. & Kay, M.A. (2003). Progress and problems with the use of viral vectors for gene therapy. *Nature Reviews Genetics*, 4(5), (May 2003), pp346-358, ISSN: 1471-0056

129. Tolar, J. & Orchard, P.J. (2008). ?-L-iduronidase therapy for mucopolysaccharidosis type 1.*Biologics: Targets & Therapy*, 2(4), pp743-751.

130. Torres, V., Barra, L., Garcés, F., Ordenes, K., Leal-Ortiz, S., Garner, C.C., Fernandez, F. & Zamorano, P. (2010). A bicistronic lentiviral vector based on the 1D/2A sequence of foot-and-mouth disease virus expresses proteins stoichiometrically.*Journal of Biotechnology*, 146, (April 2010), pp138-142, ISSN: 0168-1656

131. Triccas, J.A., Sun, L., Palendira, U. & Britton, W.J. (2002). Comparative affects of plasmid- encoded interleukin-12 and interleukin-18 on the protective efficacy of DNA vaccination against Mycobacterium tuberculosis. *Immunology & Cell Biology*, 80(4), (August 2002), pp346-350, ISSN: 0818-9641

132. Trichas, G., Begbie, J. & Srinivas, S. (2008). Use of the viral 2A peptide for bicistronic expression in transgenic mice. *BioMed Central BMC Biology*, 6, (September 2008), 40, ISSN: 1741-7007

133. Trinchieri,G.,Pflanz,S.&Kastelein,R.A.(2003).TheIL-12familyofheterodimeric cytokines: new players in the regulation of T cell responses. *Immunity*, 19(5), (November 2003), pp641-644, ISSN: 1074-7613

134. Udono, H. & Srivastava, P.K. (1993). Heat shock protein 70-associated peptides elicit specific cancer immunity. *The Journal of Experimental Medicine*, 178(1), (October 1993), pp1391-1396, ISSN: 0022-1007

135. Urwin, P.E., Yi, L., Martin, H., Atkinson, H.J. & Gilmartin, P.M. (2000). Functional characterization of the EMCV IRES in plants. *The Plant Journal*, 24, pp583-589.

136. Urwin, P.E., Zubko, E.I. & Atkinson, H.J. (2002). The biotechnological application and limitation of IRES to deliver multiple defence genes to plant pathogens. *Physiological and Molecular Plant Pathology*, P61, (August 2002), pp103-108, ISSN: 0885-5765

137. Varshavsky, A. (1992). The N-End Rule. *Cell*, 69(5), (May 1992), pp725-735, ISSN: 0092-8674

138. Wargo, J.A., Robbins, P.F., Li, Y., Zhao, Y., El-Gamil, M., Caragacianu, D., Zheng, Z., Hong, J.A., Downey, S., Schrump, D.S., Rosenberg, S.A. & Morgan, R.A. (2009). Recognition of NY-ESO-1+ tumor cells by engineered lymphocytes is enhanced by improved vector design and epigenetic modulation of tumor antigen expression. *Cancer Immunology, Immunotherapy*, 58(3), (March 2009), pp383-394, ISSN: 0340-7004

139. Warren, L., Manos, P.D., Ahfeldt, T., Loh, Y-H., Li, H., Lau, F., Ebina, W., Mandal, P.K.,Smith, Z.D., Meissner, A., Daley, G.Q., Brack, A.S., Collins, J.J., Cowan, C., Schlaeger, T.M. & Rossi, D.J. (2010). Highly Efficient Reprogramming to Pluripotency and Directed Differentiation of Human Cells with Synthetic Modified mRNA. *Cell Stem Cell*, 7(5), (November 2010), pp618-630, ISSN: 1934-5909

140. Wells, S.E., Hillner, P.E., Vale, R.D. & Sachs, A.B. (1998). Circularization of mRNA by eukaryotic translation initiation factors. *Molecular Cell*, 2, (July 1998), pp135-140, ISSN: 1097-2765

141. Westwood, J.A. & Kershaw, M.H. (2010). Genetic redirection of T cells for cancer therapy.*Journal of Leukocyte Biology*, 87(5), (May 2010), pp791-803, ISSN:0741-5400

142. Wolf, S.F., Temple, P.A., Kobayashi, M., Young, D., Dicig, M., Lowe, L., Dzialo, R., Fitz, L., Ferenz, C., Hewick, R.M. *et al.* (1991). Cloning of cDNA for natural

killer cell stimulatory factor, a heterodimeric cytokine with multiple biologic effects on T and natural killer cells. *The Journal of Immunology*, 146(9), pp3074-3081.

143. Woltjen, K., Michael, I.P., Mohseni, P., Desai, R., Mileikovsky, M., Hämäläinen, R., Cowling, R., Wang, W., Liu, P., Gertsenstein, M., Kaji, K., Sung, H-K. & Nagy, A. (2009). *piggyBac* transposition reprograms fibroblasts to induced pluripotent stem cells. *Nature,* 458(7239), (April 2009), pp766-770, ISSN: 0028-0836

144. Wozniak., T.M., Ryan, A.A. & Britton, W.J. (2006). Interleukin-23 Restores Immunity to Mycobacterium tuberculosis Infection in IL-12p40-Deficient Mice and Is Not Required for the Development of IL-17 Secreting T Cell Responses. *The Journal of Immunology*, 177, pp8684-8692.

145. Yakubov, E., Rechavi, G., Rozenblatt, S. & Givol, D. (2010). Reprogramming of human fibroblasts to pluripotent stem cells using mRNA of four transcription factors. *Biochemical and Biophysical Research Communications*, 394(1), (March 2010), pp189- 193, ISSN: 0006-291X

146. Yan, J., Wang, H., Xu, Q., Jain, N., Toxavidis, V., Tigges, J., Yang, H., Yue, G. &Gao, W. (2010). Signal sequence is still required in genes downstream of "autocleaving" 2A peptide for secretory or membrane-anchored expression. *Analytical Biochemistry*, 399(1), (April 2010), pp144-146, ISSN: 0003-2697

147. Yang, S., Cohen, C.J., Peng, P.D., Zhao, Y., Cassard, L., Yu, Z., Zheng, Z., Jones, S., Restifo, N.P., Rosenberg, S.A. & Morgan, R.A. (2008). Development of optimal bicistronic lentiviral vectors facilitates high-level TCR gene expression and robust tumor cell recognition. *Gene Therapy*, 15(21), (November 2008), pp1411-1423, ISSN: 0969-7128

148. Ye, X., Al-Babili, S., Klöti, A., Zhang, J., Lucca, P., Beyer, P. & Potrykus, I. (2000). Engineering the Provitamin A (?-carotene) Biosynthetic Pathway into (Carotenoid-Free) Rice Endosperm. *Science*, 287(5451), (January 200), pp303-305, ISSN:0036-8075

149. Ylä-Herttuala, S. & Alitalo, K. (2003). Gene transfer as a tool to induce therapeutic vascular growth. *Nature Medicine*, 9(6), (June 2003), pp694-701, ISSN: 1078-8956

150. Yusa, K., Rad, R., Takeda, J. & Bradley, A. (2009). Generation of transgene-free induced pluripotent mouse stem cells by the *piggyBac* transposon. *Nature Methods*, 6(5), (May 2009), pp363-369, ISSN: 1548-7091

151. Zhang, C., Bradshaw, J.D., Whitham, S.A. & Hill, J.H. (2010). The Development of an Efficient Multipurpose Bean Pod Mottle Virus Viral Vector Set for Foreign Gene Expression and RNA Silencing. *Plant Physiology*, 153(1), (May 2010), pp52-65, ISSN: 0032-0889

CHAPTER-9

REGULATING CELLULAR MOVEMENT USING MICROSCOPIC PATTERNS

INTRODUCTION: AT A GLANCE

- Cell migration is a fundamental process in biology, crucial for various physiological functions and pathological conditions, including tissue development, wound healing, and cancer metastasis. Controlling and studying cell migration is essential for understanding these processes and developing innovative therapeutic strategies. One promising approach to investigate and influence cell migration is through the use of microscopic patterns. These patterns, typically created on substrates, serve as a structured environment that can guide and regulate the movement of individual or groups of cells. This technique has opened up new avenues for research and application in fields such as tissue engineering, regenerative medicine, and drug discovery.
- Long-distance and directional migration of cells is a critical step in development, regeneration, and wound healing. However, physical barriers such as connective tissues and other cells prevent cells from freely migrating towards their destination (Fig. 1a,b). Therefore, cells need to not only mechanically sense the surrounding geometry but they also need to integrate the mechanical information in their migration towards their destination. Technical limitations have meant that the relationship between the surrounding geometry and cell migration has not been well studied.
- Recent advances in soft lithography techniques now allow various designs of micrometre- sized chambers to be easily fabricated on cell-culture vessels. By culturing cells in different micropatterns, the relationship

between geometry and cell response has been studied (Fig. 1c,d). For example, spindle orientation, growth, differentiation, and migration have been shown to be related to micropattern shape. Recently, several groups, including ourselves, have reported that mammalian cells exhibit biased cell movement on asymmetrical periodic micropatterned surfaces. Although it is little wonder that cells migrate asymmetrically in asymmetrical micropatterns, the direction in which they move is not immediately obvious.

- In this chapter, we describe biased cell movement in asymmetrical micropatterns. These studies offer new insights into the migration of cells in response to geometry of their surrounding environment, and we suggest strategies for designing artificial scaffolds that direct cell migration.

BIOLOGICAL SIGNIFICANCE OF CELL MIGRATION

Cell movement: A basic characteristic of life

Movement is a basic characteristic of cells (both unicellular organisms and the various cells of multicellular organisms) that is almost as important as self-renewal (Bray 2001). Some types of cells are extremely motile, while others lack strong motility and are capable of no more than passive movements caused by surrounding forces. Motile cells may sometimes change direction of their own accord in response to changes in their internal state (Oosawa 2001; Nakaoka et al. 2009), but normally they change direction in response to external stimuli, such as chemicals (chemotaxis) and light (phototaxis). For unicellular organisms, the ability to migrate to an environment suited to survival and proliferation is a matter of life and death. Bacteria search for an environment suited to survival by swimming. Social amoebae usually migrate independently of each other, but if the environment deteriorates, they gather to form fruiting bodies (Goldbeter 1996; Gregor et al. 2010). The fact that a great many of the cells of multicellular organisms (particularly animal cells) are capable of migration is also important. In the body plan of multicellular organisms, cells need to be able to do more than just proliferate and differentiate. In the process of development, cells need to migrate to the correct position so that they can adopt their proper shape and properly function (Keller et al. 2008). Cells also need to migrate *en masse* to specific locations to assist in the healing of wounds, to perform immune system functions, and to conduct other aspects of body maintenance (Friedl et al. 2004; Schneider et al. 2010). This suggests that motility is one of the universal characteristics of cells that enable the survival of life forms.

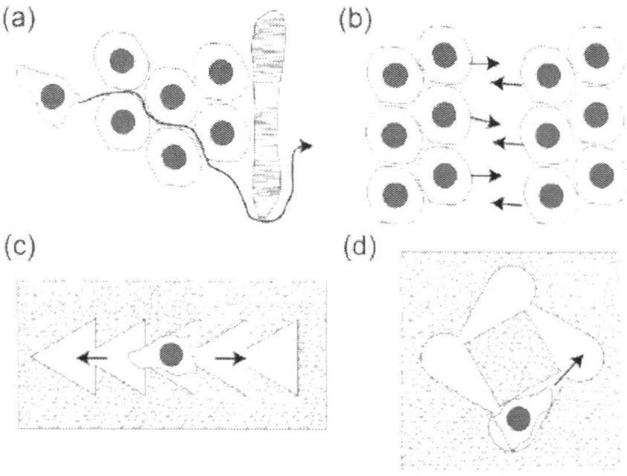

Fig. 1. Schematic illustration of cell migration in a multicellular body (a) and in wound healing (b). *In vitro* cell migration observed in asymmetric micropatterns connected in a linear (c) and circular manner (d).

Migration mechanisms

The way cells move differs markedly according to whether they are non-adherent or adherent cells (Eisenbach et al. 2004). Many non-adherent cells propel themselves using cilia or flagella. The structure of motor-protein complexes, energy balances, and protein response networks related to ciliary and flagellar locomotion have been analysed, and these locomotion mechanisms are becoming increasingly well understood. However, there is much that is still unknown about the movement of adherent cells because they lack specific means of movement such as cilia and flagella. This makes it difficult to clearly separate the different aspects of cell movement, such as deformation, migration, and division.

Fig. 2. Schematic illustration of adherent-cell migration integrating cytoskeleton and focal adhesion molecules complex, called as focal adhesions,

in a fibroblast (A) and a neuronal cell (B) in overhead (a) and cross-sectional view (b–e). STEP 1 (a, b): Polarization of cell. The cell changes from a spherical or hemispherical shape to become anisotropic. STEP 2 (c): Generation of force driving the locomotion. Fibroblasts form filopodia and lamellipodia, and neuronal cells extend neurites. STEP 3 (d): Fixing the leading edge to the substrate. STEP 4 (e): Diving cell body.

However, there are common features in the movements of adherent cells and the basic mechanisms behind such movements are thought to be the same. Adherent cell migration is a result of the integrated dynamics of the cytoskeleton and adhesions molecules. The cytoskeleton runs throughout the cell body and acts as a "skeleton" and "motor" for the cell. The cytoskeleton is a protein complex composed of actin filaments, intermediate filaments, and microtubules. The cells adhere to the extracellular matrix, substrates, and other cells' surfaces with adhesion molecules including integrin to migrate, to survive, and to acquire extracellular information. The cells adhere to the extracellular matrix to transmit signals from extra-cell to intra-cell and vice versa. Therefore, adhesion to substrates through adhesions molecules acts as an "input–output adaptor". During migration, adherent cells kinetically anchor themselves to rearrange their cytoskeleton. Leading fronts, such as lamellipodia in fibroblasts and leading processes in neurons, are repeatedly formed through the extension of the cytoskeleton, which then adhere to the local environment. The posterior side of the cell is then released and retracted (Fig. 2). This process is being investigated by a number of research groups using a diverse range of observation methods because there are still many aspects that are not well understood(Smilenov et al. 1999; Flaherty et al. 2007; Hu et al. 2007).

The Direction of Migration is Determined by Disparities in External Stimuli

The migration direction of cells is determined by a number of different stimuli. The tendency of cells to change direction in response to the direction or gradient (spatial asymmetry) of external stimuli is known as *taxis*. The suffix "-taxis" is attached to prefixes representing specific stimuli to create words such as chemotaxis (movement in response to a chemical stimulus), magnetotaxis (magnetic stimulus), hydrotaxis (water), phototaxis (light), rheotaxis (water current), thermotaxis (temperature), and thigmotaxis (touch). Among the many types of taxes, chemotaxis, which is the tendency to move towards a higher or lower concentration of a specific chemical substance, is well known (Eisenbach et al. 2004). The tendency to move in the direction of a stimulus according to the gradient of the stimulus is known as positive taxis, and movement away from a stimulus is known as negative taxis. It is very important for bacteria, for example, to search for food (sugar) by swimming towards higher concentrations of food, and to avoid poison by swimming towards lower concentrations of poison. When social amoebae form fruiting bodies in response to a deteriorating environment, they are known to move towards higher concentrations of 3',5'-cyclic adenosine monophosphate (cyclic AMP) (Goldbeter 1996; Gregor et al. 2010). Even the cells of multicellular organisms display various taxes. For example, neurons extend neurites towards higher concentrations of netrin, with the cell body following suit and migrating in the same direction (Round and Stein 2007). Neutrophils, which are a type of white blood cell that eliminate invading bacteria by englobing them (phagocytosis), are capable of detecting very slight differences in concentration (1% difference between opposite sides of the neutrophil) of *N*-formyl- methionyl-leucylphenylalanine (FMLP), a protein derived from bacteria, and move towards higher FMLP concentrations (Weiner et al. 1999). Cells can thus change the direction of their movements according to asymmetries in external stimuli, and this ability is extremely important for survival.

Regulating Cellular Migration Through The Utilization Of Microscopic Patterns

Cell migration around spatial obstacles

Much of the research on taxes of adherent cells described in the previous section is based on the results of observation under a microscope of cells adhering to a flat substrate applied to glass. However, in reality cells face a plethora of spatial obstacles (such as surrounding cells, soil and plants in nature, and bone and other connective tissues within the body) that make it difficult for them to migrate freely to their destinations (Fig. 1a). As such, cells need to not only recognize the surrounding geometry mechanically, but also process this geometrical information to determine the direction of their next migration (Ingber 2003). However, research on the relationship between geometry and cell migration has so far been limited. With respect to tactile sensibility (thigmesthesia), some animals are known to display the phenomenon of thigmotaxis, but this refers to the tendency of rats and other animals to hug the edges of walls and so forth when moving, and not to the kind of cell movements with respect to physical obstacles that we are discussing here. Investigation of how cells sense spatial restrictions and respond to them requires the creation of a geometry featuring various shapes on a cellular scale (several micrometres), but while structures on a millimetre scale can be crafted using apparatus such as ordinary lathes and milling machines, creation of structures on a smaller scale is difficult, and this is why research has been limited.

Micro-contact printing

Recent advances in photolithography and other microfabrication techniques have made it possible to create structures that are not toxic to cells and feature all kinds of cell-sized spatial patterns. Of these micropattern techniques, micro-contact printing, a soft lithography technique developed by Whitesides et al. at Harvard University, is particularly well-suited to small-scale research at universities, and is accordingly popular in the field of cellular engineering (Kumar and Whitesides 1993; Kane et al. 1999). Micro-contact printing involves first fabricating a finely patterned master that is then used to produce finely patterned stamps made of the thermosetting silicone elastomer polydimethylsiloxane (PDMS). The stamps are then used to print the patterns associated with cell adherence. Although making masters involves microfabrication techniques such photolithography that requires access to clean rooms and photolithographic equipment, masters do not have to be made in the place where subsequent processes are performed, and so can be made by other research laboratories or companies. Moreover, the subsequent processes can all be performed in a cellular biology laboratory, and enough micro-patterned culture vessels for several experiments can be produced with ease.

Extremely high precision masters can be created by using the silicon wafers that have become synonymous with semiconductor technologies. A technique using SU-8, a UV-curable resin, can be employed to make masters more easily (Ehrfeld et al. 1999). A spin coater is used to coat a silicon wafer or glass slide with a film of SU-8, and films with a thickness of several micrometres to over 100 μm can be created with excellent reproducibility. These thin SU-8 films are cured by irradiating with UV through a patterned mask, after which the uncured parts are washed away, leaving a three-dimensional pattern (Fig. 3a). SU-8 is not so strong, and thus the master can get chipped after repeated casting with PDMS, but because this method enables the production of micrometre-scale masters with high aspect ratios and low cell toxicity using relatively simple apparatus, it is very widely used.

After pouring the PDMS onto masters created in this way and curing at 60°C overnight, the microfabricated stamp can be removed from the master (Fig. 3b,c). Because PDMS keeps its shape very well on thermosetting, it is a superb material for reproducing sub-micrometre structures, and is also known to be non-toxic to cells. Also, PDMS is a pliant material, making it easy to remove stamps from the master and enabling good contact with the surface to be stamped to ensure even printing.

A substance for controlling cell adhesion is applied as "ink" to the PDMS stamp and then stamped onto the culture substrate (Fig. 3d,e). Whitesides et al. utilized self-assembled monolayers (SAMs) by printing with alkanethiol which have a variety of reactive functional groups to anchor cell adhesion related molecules, and created cell adhesion islands. There are two main micropatterning methods—printing with extracellular matrix substances that promote cell adhesion (e.g. collagen, fibronectin, laminin) (Scholl et al. 2000; Kaji et al. 2003; Hou et al. 2009), or printing with substances that impede cell adhesion (Yang et al. 2005; Saravia et al. 2007; Ohnuma et al. 2009). As we explain later, we used the latter method (Fig. 3f). Microfabrication techniques like these have enabled us to create cell adhesion patterns with a variety of geometric shapes and investigate the way that geometric patterns affect cell movement.

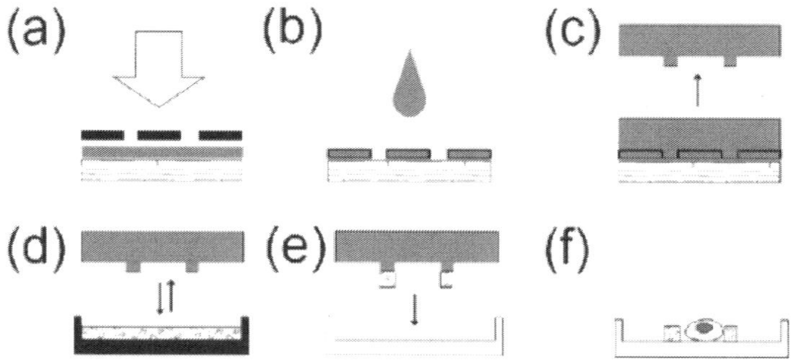

Fig. 3. Schematic illustration of micro-contact printing using a PDMS stamp: (a) fabricating a patterned master from a UV-curable resin by using UV-irradiation through a patterned mask, (b) applying silicone resin, (c) curing the resin to produce a patterned stamp, (d) dipping the patterned surface of the stamp in an ink, (e) printing the ink onto a culture dish, and (f) culturing the cells on the dish.

Selective Motion In Uneven Microscopic Patterns

Asymmetry of cell shape and movement

How does the shape of the space in which a cell can move affect its movement? In the case of chemotaxis, the stimulating substance creates a concentration gradient (spatial asymmetry), and the cell moves in line with that concentration gradient (asymmetrical movement). From this, it is easy to suppose that the asymmetry of the space in which a cell can move imparts a bias on the cell's movement. It is also known that the shape of cells that can move freely on a standard cell culture plate (one that enables cells to adhere and move uniformly) becomes asymmetric. For example, migrating fibroblasts are teardrop-shaped, with their front ends spreading out while the rear ends taper (Fig. 2a); migrating keratocytes, the epithelial cells of the epidermis of fish and frogs are half-moon- shaped (Svitkina et al. 1997; Keren et al. 2009) and neurons put out neurites in the direction in which they are moving (Fig. 2b) (Yamasaki et al. 2010). Adherent cells thus show a strong correlation between the direction of migration and the shape of the cell. Creating a cell adhesion island with an asymmetric geometry is accordingly likely to cause both cell shape and cell migration direction to become asymmetric in line with the asymmetry of the island (Fig. 1c,d). However, just as there are positive and negative taxes, spatial asymmetry needs to be actually measured to determine the direction in which it will bias cell movement.

Cells in a teardrop-shaped micropattern

With respect to the direction of cell movement and geometric pattern of cell adhesion sites, some very interesting research has been carried out on the teardrop shape that fibroblasts assume when moving through an unconfined space. In 2003, Brock et al. reported that cells confined within a polygonal shape tend to form lamellipodia at the corners of the polygon (Brock et al. 2003). Lamellipodia are formed when actin filaments create a mesh and the cell membrane advances, and tend to be formed at the fronts of cells when they are migrating (Fig. 2a). Fibroblasts in an unconfined space tend to advance with the blunt end of the teardrop to the front, but Brock et al.'s observations suggested that cells might also tend to advance towards sharp corners when geometrically confined.

In 2005, Jiang et al. published a very interesting paper on research to elucidate the relationship between cell shape and direction of movement (Jiang et al. 2005). They confined fibroblasts to a teardrop shape, and then used an electric pulse to release them from constraint and observed the direction in which they moved. They found that the fibroblasts moved in the direction of their blunt ends. This result suggested that the shape of the motile cell as a whole was a more important determinant of direction than the ease with which lamellipodia are formed at corners within the geometry. Following this, in 2007, Kumar et al. conducted observations on the direction taken by NIH3T3 fibroblasts in a chain of teardrop-shaped cell adhesion islands (Fig. 1d)(Kumar et al. 2007). They observed the direction of cell movements under various conditions, adjusting the arrangement of the teardrop islands, and the distance separating them, joining them in some cases, and leaving a gap of several micrometres in others, and so forth. They found that the direction of cell movements showed no bias towards either blunt end or sharp end of the teardrop-shaped islands, and that cells chose to move in whichever direction another cell adhesion island lay at the end of the longitudinal axis of each teardrop- shaped island. This suggested that the cell adhesion sites have a major effect on the direction of cell movement. In 2010, in experiments using epithelial cells, Kushiro et al. used the same teardrop-shaped cell adhesion island geometry as Kumar et al. to investigate how modifying the expression of the gene that controls the formation of lamellipodia affects cell movement (Kushiro et al. 2010). Unlike fibroblasts, epithelial cells moving in an unconfined space actively form broad lamellipodia at the front end of thedirection in which they are moving. Kushiro et al.'s findings regarding the direction of cell movement differed markedly from those of Kumar et al. for fibroblasts, but nevertheless showed that the direction of cell movement changes according to the degree of expression of the gene related to lamellipodia formation, and to the arrangement of teardrop-shaped islands and the distance between them.

Cells in ratchet-shaped micropattern

In 2009, Mahmud et al. investigated the same kind of movements using a slightly different geometry, one that used a ratchet-shaped micropattern (chained triangles connected in a linear manner) (Fig. 1c) rather than teardrop shapes as adhesion islands, and also included physical obstacles (Mahmud et al. 2009). They, too, observed bias in the direction of cell movement, and showed that this bias depends upon the type of cell involved. We explain in more detail later, but our group also created a geometrical pattern in which we combined triangles to form a ratchet shape. When we used this geometry to investigate the direction of movement of neuron-like cells, we observed a bias in movement and also found that the location at which the tips of neurites are formed is critical.

Some of these studies were conducted independently during much the same period. Conclusions that can be drawn from the above research using asymmetric geometrical patterns of about the same size as cells are: (1) many different cell types show bias in the direction of their movements; (2) the direction of cell movement changes according to the shape of the geometrical pattern in which cells can move, type of cell, and gene expression; and (3) bias in the direction of cell movement is related to the formation of lamellipodia and neurites, which are thought to be closely involved in cell movement. As such, while we can use geometrical patterns to bias the direction of cell movement, there is still much that we do not know about bias direction and the mechanisms involved in determining it.

Brownian ratchet theory

When discussing the bias direction of cell movement, we have not considered stochastic motion resulting from the spontaneous fluctuation of internal state of cell, which is known to be important in cell migrations (Oosawa 2001; Nakaoka et al. 2009). Here, we consider cell migration in an asymmetric geometrical pattern, taking stochastic motion into account. A Brownian particle, which exhibits stochastic motion due to thermal fluctuations, can be caused move directionally in a spatially asymmetric energy barrier under a non-equilibrium condition, as represented by a flashing ratchet (Fig. 4a)(Reimann 2002). In the case of so called rocking ratchet in which an oscillating force is applied to a Brownian particle (Fig. 4b), directional motion of the particle is also observed. This direction is known to be reversed by changes in the amplitude or period of the oscillating force (Bartussek et al. 1994; Reimann 2002). According to an experiment by Mahmud et al. (Mahmud et al. 2009), the transition probability for a cell in spatially asymmetric micropattern is described by a one- dimensional Brownian model. So, is there a possibility that the cause of the directional motion of a cell is analogous to that of Brownian ratchets? Furthermore, is a reversal of a cell's direction able to be observed when a signal is oscillating in the presence of an asymmetric geometry as is seen in the rocking ratchet (Fig. 4b). It is up to future research to determine whether cell migration is dictated by cell shape and/or by asymmetries in the surrounding space.

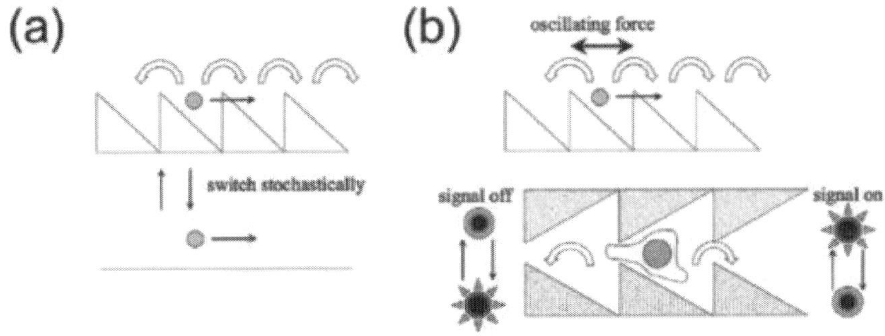

Fig. 4. Illustrated scheme of directional motion caused in a flashing ratchet (a) and in a rocking ratchet (b). In a flashing ratchet, a spatially asymmetric energy barrier, which a Brownian particle is subject to, switches in time stochastically. In a rocking ratchet, an oscillating force is applied to a Brownian particle besides the force exerted by a spatially asymmetric energy barrier.

Extrapolation to cell populations from single-cell migration analysis

Cells in multicellular organisms or wild environments are not independent entities and inevitably interact with the surrounding cells. Therefore, the effects of cell-to-cell interactions through physical contact, chemical signalling, nutrition competition, etc. must be considered to understand the roles of cellular movement and migration in natural contexts. Even if the movements of individual cells are characterized by simple rules, unexpected collective behaviours may emerge at the cell population level due to cell-to- cell interactions.

One of the most understood systems of collective migration is the fruiting-body formation of social amoebae. Upon starvation, thousands of individual cells co-ordinately migrate and aggregate to form fruiting bodies via signalling with cyclic AMP (Goldbeter 1996; Gregor et al. 2010). Fruiting bodies encapsulate spores that can survive severe stress environments for an extended period of time making such collective migration crucial for the survival of the species in harsh environments. Understanding the mechanism of this collective migration requires consideration of the effects of cell-to-cell interactions.

Collective migration also exists in bacteria. Lambert et al. developed a microfluidic device that allows the measurement of the efficiency of chemotactic migration by employing funnel-shaped barriers, and demonstrated that bacteria migrated by chemotaxis beyond the barriers to an area of higher nutrition only when cell density was high (Lambert et al. 2010). Another intriguing study on bacterial migration was reported by Park et al. in which they studied the time-evolution of spatial distributions of bacterial densities in a microfabricated maze (Park et al. 2003a; Park et al. 2003b). Despite the complex topology of the environment, the bacteria migrated and aggregated in a few confined position through chemotaxis towards self-produced signals to create a high cell density. A high cell density is required for the formation of a biofilm, which is resistant to many kinds of stress. A biofilm is a bacterial community-based life-cycle mode that is known to contribute to the virulence of pathogens in bacterial infection (O'Toole et al. 2000; Lewis 2005).

When cells proliferate and divide during migration, another complexity arises. Differences between local environments generally induce different division rates in subpopulations. If subpopulations A and B in different locations have the division rates k_A and k_B ($>k_A$), the difference between the cell densities of the subpopulations grows with $\exp[(k_B - k_A)t]$, where t is time. Therefore, different division rates in a cell population can significantly affect the spatial distribution of cell density when the time-scale of observation is comparable to or longer than the mean doubling time of the cell population. This means that changes in spatial distribution cannot be attributed solely to the effect of cell migration. In phenomena such as embryogenesis or cancer metastasis, or during an immune response, cellular proliferation and migration proceed simultaneously. Cellular proliferation and migration are usually studied separately, but their coupling is an important subject for future research.

The Biotechnological Benefits of Managing Cell Migration with Microscopic Patterns

This kind of cell migration research is of course important from the life science perspective of elucidating the mechanisms behind fundamental cell functions, but it is also important from the perspective of applying the ability to control the direction of cell migration to cellular engineering and medical technology. As mentioned earlier, the migration of individual cells that make up multicellular organisms is a matter of great significance to the development and maintenance of functions of those organisms. The ability to control cell migration could lead to new or improved treatments for developmental disorders, tissue dysfunctions, healing of wounds, cancer metastasis, and so forth. Recently, the relationship between scaffold elasticity, which is closely related to cell migration, and the direction of differentiation has come to be discussed in relation to the induction of stem cell differentiation (Engler et al. 2006). Stem cells are known to be influenced by a huge number of endogenous factors (genes, RNAi, etc.) and exogenous factors such as chemicals and culture environment. The role that cell migration performs in differentiation is likely to attract growing interest.

The control of cell migration also has an important role to play in wound treatment and tissue regeneration (Friedl et al. 2004; Schneider et al. 2010).

The migration not only of fibroblasts mentioned above, but also osteoblasts and osteoclasts in bone, and the cells involved in angiogenesis of blood vessels determine the form of those respective tissues, and are an important factor in the expression of the functions of those tissues.

Contributing to this kind of cellular engineering and medical treatment by equipping tissues with asymmetric spaces capable of controlling cell migration is indispensable to the further development of tailor-made treatment and advanced medical technology. Up to now, the focus of attention has been on cell adhesion substances that ensure that cells remain with and maintain the structure of the tissue to which they belong, but moving forward, development in the area of shaping spatial arrangements of those cells is likely to become an increasingly important endeavour.

CELL MIGRATION CONTROL BY OUR GROUP: CURRENT STUDIES

Introduction: An Overview

To determine the mechanisms of cell migration within cell-sized geometry, we focused on optically tracing two different types of cells that move in completely different manners. We used the rat adrenal pheochromocytoma cell line, PC12, and the fibroblast cell line, NIH3T3. Upon the addition of nerve growth factor, PC12 cells differentiate into sympathetic neuron- like cells with long, extended neurites (Greene et al. 1982; Ohnuma et al. 2006). Although PC12 cells migrate slowly, they are useful for the study of migration via long neurites. NIH3T3 cells are able to move very fast on glass slides and are frequently used as model cells in cell migration studies (Kumar et al. 2007).

To optically trace the migration of these cells in cell-sized geometry over a long period of time and to analyse the dynamics of cell–substrate contact sites (neurite tips for PC12 cells and focal adhesions for fibroblasts), which work both as input sensors for information regarding the local environment and as action sites for locomotion, we employed two different technologies. The first was a micro-contact printing technique using PDMS as the material for both the stamps and also the cell-repellent ink. We enable to keep the cells inside the PDMS micro-chamber for over 18 days (Ohnuma et al. 2009). The second technology was reflection interference contrast microscopy (RICM) (Curtis 1964). Using RICM, the distance between the glass and the cell surface membrane can be visualized as high-contrast images, which are images of the interference between the reflection of light off the glass–medium interface and off the medium–cell interface.

Analysis of the data showed that both PC12 and NIH3T3 exhibited biased migration in asymmetric micropatterns and that migration is likely to proceed from the sharp end of one micropattern unit to the blunt end of the adjacent micropattern unit. The contact sites of each migrating cell on the glass surface, however, expanded in both directions. The mechanism behind biased cell migration has still not been uncovered, but the current experimental setup will give us useful data for the control of cell migration.

Pc12 Migration Control In A Ratchet-Shaped Micropattern

Neuronal cell migration

Long-distance and directional migration of neuronal cells is a critical step in the developing and regenerating nervous system. Some neuronal cells migrate several millimetres to their final destinations. For example, inhibitory neurons originate in the ganglionic eminences, migrate radially to the cortex, and migrate parallel in the cortex surface to their destinations (Marín and Rubenstein 2001). The cellular mechanisms underlying these directional migratory activities have been extensively studied from the basis of chemotaxis (Ayala et al. 2007; Zheng and Poo 2007); however, chemokine gradients decrease with distance and determination of destination by each neuronal cell likely involves chemokine crosstalk. It was found that scaffolds, such as radial glia and blood vessels, provide routes that guide migrating neurons to their destinations (Rakic 1972; Gasser and Hatten 1990; Bovetti et al. 2007). These scaffolds act as "rails" to produce error-free, long-distance migration. However, physical and chemical circumstances, such as connective tissue, prevent cells from freely migrating towards their destination. Cells not only mechanically sense the local geometry, but they also integrate this mechanical information into their migration (Ingber 2003). The relationship between geometry and cell migration has not been well studied.

Experimental set-up

To investigate the relationship between neuronal cells' local geometry and their migration, and, thus, uncover a potential control methodology, we performed a simple *in vitro* experiment (Ohnuma et al. 2009). We focused on periodic structures, which are abundant *in vivo*, by fabricating a ratchet-wheel shaped (gear-type) micropattern, which consists of a series of connected triangles. The micropattern was made by printing a PMDS film onto a collagen-coated culture dish (Fig. 3). Our working hypothesis was that neuronal cells would be able to migrate directionally on a periodic scaffold structure if the periodic unit was asymmetric. We tested the hypothesis using cultured PC12 cells that were attached only to the collagen-coated area between the core and the ratchet-shaped outer frame of the micro- chamber. Because the chamber was designed so that cell migration in the radial direction was restricted and the gap between the teeth and the core was equivalent to the size of a cell body, the cells migrated almost one-dimensionally in the tangential direction. We made both a left (L) and right (R) micropattern (Fig. 5a). The L and R chambers were line- symmetrical to one another and arranged alternately to serve as control chambers for one another. The core diameter, tooth depth, and gap between the teeth and core were approximately 100 μm, 40 μm, and 30 μm, respectively. PC12 cells were plated in neurite outgrowth medium including nerve growth factor in which the cells gradually extend long neurites over approximately 10 days (Greene et al. 1982; Ohnuma et al. 2006). It was previously reported that neuronal cells migrate following neurite extension (Hatten 2002), so it was expected that cell migration in the chambers would differ between cells with short neurites and cells with long neurites (Fig. 5b). Therefore, time-lapse micrographs of the cells in the micropattern were acquired twice: the first acquisition was for 70 hours starting from 1 day after plating, when the cells usually have short neurites; and the second acquisition was for 70 hours starting from 12 days after plating, when the cells usually have long neurites. We defined cell migration in the direction in which the ratchet teeth were tapered as positive migration (Fig. 5c).

Results and discussion

Using the PDMS printed L and R micropatterns, we found the same biased migration in both types of chambers. The PC12 cells in the L and R micropatterns migrated the same distance in a positive direction in both the first and second micrograph acquisition periods. These results suggested that the direction of migration is biased by chamber geometry, and supports our working hypothesis that the periodic nature of the asymmetrical scaffold determines migration direction. Next, we analysed the position of the cell body and the neurite tips because cell migration was strongly related to neurite formation. Although the mean length of the longest neurite was independent of microchamber geometry, the time- course trace of the neurite tips showed that they remained around the tips of the ratchet teeth. We also found that as cells migrated in a positive direction they tended to extend their neurites about one tooth ahead of the cell body and place the neurite tip at the tip of the tooth. The cell body then passed by the neurite tip as it migrated (Fig. 5c). It appears that PC12 cells use the neurite tips as a hook to "climb" the ratchet-shaped geometry. One possible explanation for these results is that since the positively directed neurite bends along the tooth edge while the negatively directed neurite extends in a straight line, the tangential component of the maximal tension of the positive-direction neurite is higher than that of the negative-direction neurite. This results in the probability of continuous forward migration being higher than that of backward migration. Alternatively, the results can be described as thermally fluctuating spring-beads in a rocking ratchet (see 1.3.4).

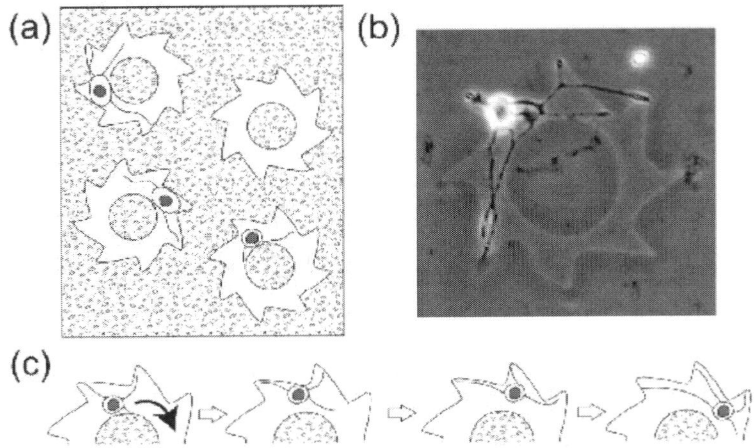

Fig. 5. Schematic illustration (a) and phase contrast microscopy image (b) of a PC12 cell extending its neurites in a gear-type micropattern. (c) The PC12 cell migrates in a positive direction (arrow).

Visualizing Focal Adhesions in Nih3t3 Fibroblasts Within A Teardrop-Designed Microchamber

Focal adhesions in cell migration

As previously mentioned, focal adhesions, which are membrane protein complexes, act as "input–output adaptors". Focal adhesions interact with substrates such as the extracellular matrix to allow the cell to adhere, migrate, and acquire extracellular information. Integrin is a major component of focal adhesions. Integrin is a family of trans-membrane proteins that connect physicochemically between the extracellular matrix and the cytoskeleton proteins inside of the cell to anchor cells and to transmit signals from extra-cell to intra-cell and vice versa. Outside-in signals, which are transmitted from outside to inside the cell, activate the adherent affinity of the complex, especially integrin. Inside-out signals via internal signal transduction proteins also activate integrin. Adherent affinity is regulated by both quantitative (density of the complex and ligand on the substrate) and qualitative properties (attractive interaction of the complex and the substrate). Mature focal complexes that have strongly adhered to a substrate are called focal adhesions.

Cell migration is the integrated dynamics of the cytoskeleton and focal adhesions, but how do cells actually migrate using this dynamics? The following is an accepted mechanism for the movement of adherent cells such as fibroblasts on substrates *in vitro*:

STEP 1: Polarization of cell. The shape of the cell changes from spherical or hemispherical to become anisotropic. Cdc42, PIP3, integrin, and microtubules are important in changing the shape of the cell. The cell membrane at the anterior part of the cell, the leading edge, starts to extend and produce actin filaments called filopodia.

STEP 2: Generation of force driving locomotion. Actin filaments are richly synthesized at the leading edge, and filopodia and lamellipodia are formed in association with the activation of Rac1, generating the force to extend the leading edge. The direction of the filopodia and lamellipodia is determined by the Rho protein family and/or the actin- binding protein family.

STEP 3: Fixing the leading edge to the substrate. At the leading edge, focal complexes are formed by the activation of integrin, which is stimulated by the inside-out and outside-in signals from both the cytoskeleton and the substrate. Focal complexes are also led by Rac1 and Cdc42 activity. The focal complexes mature to become focal adhesions at the leading edge. RhoA, which is activated at the posterior side of cell, is also involved in the formation of focal adhesions, so focal adhesions are also formed at the posterior side of the cell.

STEP 4: Diving cell body by generation of tension. Focal adhesions are the contact sites of the cytoskeleton and substrate outside of the cell. This becomes a point of tension through the cytoskeleton. The focal adhesions formed at the anterior and posterior sides are linked by the cytoskeleton as stress fibres force the cell body to shrink. The tension is regulated by actin motor proteins, myosin, and Rho-kinases (Smith et al. 2008).

STEP 5: Decomposition of the focal adhesions at the posterior side. Although the cell can decompose the focal adhesions at either the anterior or posterior side via RhoA, focal adhesion kinase, Src, or microtubules, the cell selects the focal adhesions at the posterior side for decomposition. This results in the posterior side of the cell shrinking due to the tension created by stress fibres in the cytoskeleton.

On the basis of these mechanisms, the physicochemistry of the migrating cell's surroundings potentially affects the migration direction. This aspect of cell migration should be investigated through the dynamics of the focal adhesions formed at both the anterior and posterior sides of the cell.

Visualization of NIH3T3 Fibroblast Focal Adhesions via Reflective Interference Contrast Microscopy (RICM)

RICM can be used to observe focal adhesions. In the 1970s, electron microscopy revealed that the gap between cells and substrates was less than 30 nm (Abercrombie et al. 1971; Revel and Wolken 1973), however, the cells that were observed were fixed (not living) and the observation chamber was under a vacuum. Therefore, the development of optical microscopy for the observation of the focal adhesions of living cells was considered to be the next step. The principles of RICM were established by Curtis in 1964, (Curtis 1964) who regarded the medium between the glass and cell as a thin film, which allowed the observation of the distance between the glass and the cell surface as a high-contrastimages through the interference of light reflected from glass–medium and medium–cell membrane interfaces. The thickness of the thin medium "layer" is evaluated using the reflective index of the medium, cell membrane, and glass, and the wavenumber and angle of incident light (Bereiter-Hahn et al. 1979; Simson et al. 1998). Izzard and Lochner reported that the nearest distance between cell membranes and substrates is approximately 10 nm, which is shown as dark areas in RICM images. These are focal adhesions (or focal contacts) (Izzard and Lochner 1976). Sackmann et al. and others significantly developed RICM using a model cell membrane that is composed of giant vesicles bearing membrane protein or that has been modified with polymers (Bruinsma et al. 2000; Smith et al. 2008; Limozin and Sengupta 2009; Streicher et al. 2009). RICM has drawn much attention as a non-probing microscopy for observing focal adhesions (Yin et al. 2003; Sengupta et al. 2006; Théry et al. 2006).

Let us briefly summarize the principles of RICM (Fig. 6a). The intensities of the incident light, the light reflected from the interface of the medium and the surface of the glass substrate, and the light reflected from the interface of the medium and the cell membrane are depicted by $I0$, $I01$, and $I12$, respectively. The intensity profile $I(x)$ of the interference between $I01$ and $I12$ is obtained by

$$I(x) = I_{01} + I_{02} + 2(I_{01}I_{02})^{1/2}\left[2kh(x)\cos\theta + \delta\right] \quad (1)$$

where k is the wavenumber of light the phase of which is shifted with δ, and $h(x)$ is the distance of the cell membrane from the glass surface. Using the Fresnel equation with the reflection amplitude coefficients of each interface ($r01$, $r12$), $I01$ and $I12$ are substituted asfollows: $I_{01} = r_{012} I_0$ and $I_{12} = (1-r_{012})r_{122}I_0$.Therefore, the maximum (I_{max}) and minimum intensity (I_{min}) of the interference are obtained from $I_{max} = I_{01}+I_{12}+2(I_{01}I_{12})_{1/2}$ and $I_{min} = I_{01}+I_{12}- 2(I_{01}I_{12})_{1/2}$, which allows the deduction of the following equation (n: refractive index):

$$h(x) = \frac{\lambda}{4\pi n}\left[\arccos\left\{\frac{2I(x) - (I_{max} + I_{min})}{I_{max} - I_{min}}\right\} + \delta\right]$$

When evaluating $h(x)$ of a living cell, the light distribution function should be included. The RICM pattern of the cell gives us the height of the cell membrane and the focal adhesions are the darkest areas with $m = 0$ in the following equation (γ: wavelength of light):

$$2kh(x)\cos\theta + \delta = \lambda m \,. \quad (3)$$

Experimental set-up

Since RICM requires a glass substrate surface, which is weakly cell-adherent, the high cell repellency of the micropattern becomes necessary. In order to obtain images of focal adhesions of cells migrating within micropatterns, we again adopted the micro-contact printing technique to construct a PDMS micropattern consisting of a series of connected teardrop shapes (Fig. 1d) (Kumar et al. 2007). The width of the neck between two teardrop shapes was about 6 μm, which was not notably larger than that of the actual design (5 μm). This convinces us that the current technique for constructing the micropattern worked well.

RICM was conducted with a halogen lamp with a 530 to 550 nm optical band-pass filter as the light source, two polarizing filters, and an objective lens with a quarter-wave plate. The microscope set-up was combined with a culture chamber managed by a temperature and humidity control box, and the cells were kept alive for several days in the chambers (Fig. 6b). This RICM set-up enables observation of the cells and the edges of micropatterns (Fig. 6d).

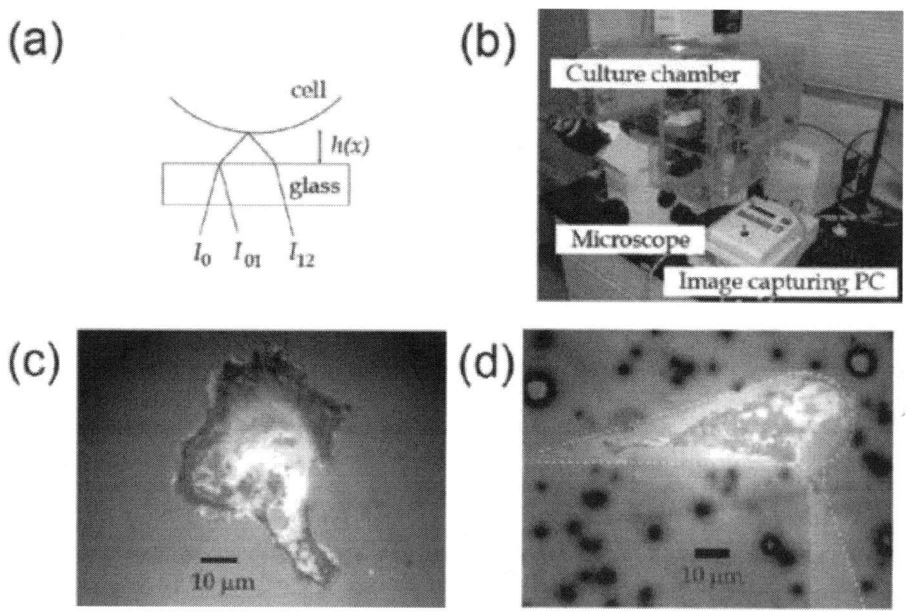

Fig. 6. (a) Schematic illustration of reflection interference contrast microscopy. (b) RICM set- up with a cell culturing system. (c,d) RICM images of NIH3T3 cultured on glass (c) or within the micropatterns (white dashed line) (d). Darkest areas in the images of NIH3T3 cell correspond to areas in most contact with the glass surface, i.e. focal adhesion.

RESULTS AND DISCUSSION

To validate the RICM set-up, we used a latex bead suspension as described (Rädler and Sackmann 1992; Rädler and Sackmann 1993; Kühner and Sackmann 1996; Heinrich et al. 2008). In brief, latex beads were suspended in 200 mM NaCl solution and then observed with the RICM set-up. Using equation (2), the heights of the beads from the glass surface were estimated to be about 10 nm. According to the Derjaguin–Landau–Verwey–Overbeek theory, the height of latex beads in a high ionic-strength suspension is several nanometers. Therefore, we deemed the performance of the RICM set-up to be sufficiently accurate. Immunofluorescence staining revealed that the dark spots in the RICM images were also areas of the cell that contained focal adhesion protein complex (Geiger 1979; Smilenov et al. 1999), suggesting that our RICM set-up was able to accurately visualize focal adhesions.

In RICM images, focal adhesions are dark and lamellipodia at the edge of cells are bright (Fig. 6c). Although the lamellipodia were extended onto the PDMS micropattern, they did not form focal adhesions, suggesting that the cells were restricted to the teardrop-shaped island in the PDMS micropattern. Formation and degradation of focal adhesions at the front and rear of migrating NIH3T3 fibroblasts in the teardrop-type micropattern (Fig. 1d) were clearly seen with our RICM set-up. Anticlockwise biased-migration of NIH3T3 cells in the teardrop-type micropattern was also observed (Figs. 1d, 6d), which was consistent with results previously reported (Kumar et al. 2007). The focal adhesions of each migrating cell in the micropattern, however, expanded both in clockwise and anticlockwise directions. The mechanism behind biased cell migration has still not been uncovered, but the PMDS micropattern and RICM set-up will give us useful data for the control of cell migration.

SUMMARY

- We have described an *in vitro* experimental model of cell migration guided by mechanical information of the local geometry. The fact that not only fibroblasts NIH3T3 but also neuronal PC12 cells, robustly exhibited biased movement within the micropatterns is indeed a surprise. Biased movement from the blunt end to the sharp end of the micropatterns resembles colloidal motion in a ratchet pattern. Further progress in the RICM imaging of focal adhesions will no doubt reveal the precise mechanism of cell migration and control within micropatterns.
- Results from our group's study are expected to contribute the science of cell migration and the understanding of multi-cellular organisms. For example, the question of whether cell migration results from probabilistic (Brownian) or deterministic components of factors internal or external to cells will be solved when the current set-up is combined with fluorescent microscopy and protein-specific probes. The resultant knowledge on cell migration may also stimulate the research field of soft micromachines

which can involve sensory motor coupling (Borckmans et al. 2009; Toyota et al. 2009; Masubuchi et al. 2011). At present, there is no evidence that scaffolds with asymmetrical surface structures exist *in vivo*. However, both repetitive structures, including the somite and the cortical layer, and asymmetric protein distributions are abundant *in vivo*. These studies offer new insights into the migration of cells controlled by mechanical stimulation and suggest strategies for designing artificial scaffolds that direct cell migration.

REFERENCES

1. Abercrombie, M., Heaysman, J. E. & Pegrum, S. M. (1971). *The locomotion of fibroblasts in culture. IV. Electron microscopy of the leading lamella*, Exp Cell Res, Vol. 67, No. 2, pp. 359-367
2. Ayala, R., Shu, T. & Tsai, L. H. (2007). *Trekking across the brain: the journey of neuronal migration*, Cell, Vol. 128, No. 1, pp. 29-43
3. Bartussek, R., Hanggi, P. & Kissner, J. G. (1994). *Periodically Rocked Thermal Ratchets*, Europhys Lett, Vol. 28, No. 7, pp. 459-464
4. Bereiter-Hahn, J., Fox, C. H. & Thorell, B. O. (1979). *Quantitative reflection contrast microscopy of living cells*, J Cell Biol, Vol. 82, No. 3, pp. 767-779
5. Borckmans, P., De Kepper, P. & Khokhlov, A. R. (2009). Chemomechanical Instabilities in Responsive Materials, I SBN 9048129923, Springer Verlag
6. Bovetti, S., Hsieh, Y. C., Bovolin, P., Perroteau, I., Kazunori, T. & Puche, A. C. (2007). *Blood vessels form a scaffold for neuroblast migration in the adult olfactory bulb*, J Neurosci, Vol. 27, No. 22, pp. 5976-5980
7. Bray, D. (2001). Cell Movements: From Molecules to Motility, I SBN 0815332823, Garland Publishing
8. Brock, A., Chang, E., Ho, C. C., LeDuc, P., Jiang, X., Whitesides, G. M. & Ingber, D. E. (2003). *Geometric determinants of directional cell motility revealed using microcontact printing*, Langmuir, Vol. 19, No. 5, pp. 1611-1617
9. Bruinsma, R., Behrisch, A. & Sackmann, E. (2000). *Adhesive switching of membranes: experiment and theory*, Phys Rev E, Vol. 61, No. 4, pp. 4253-4267
10. Curtis, A. S. (1964). *The Mechanism of Adhesion of Cells to Glass. A Study by Interference Reflection Microscopy*, J Cell Biol, Vol. 20, pp. 199-215
11. Ehrfeld, W., Hessel, V., L we, H., Schulz, C. & Weber, L. (1999). *Materials of LIGA technology*, Microsystem technologies, Vol. 5, No. 3, pp. 105-112
12. Eisenbach, M., Lengeler, J. W. & Varon, M. (2004). Chemotaxis: A New Approach to Global Strategy and Leadership, I SBN 1860944132, Imperial College Pr

13. Engler, A. J., Sen, S., Sweeney, H. L. & Discher, D. E. (2006). *Matrix elasticity directs stem cell lineage specification*, Cell, Vol. 126, No. 4, pp. 677-689

14. Flaherty, B., McGarry, J. P. & McHugh, P. E. (2007). *Mathematical models of cell motility*, Cell Biochem Biophys, Vol. 49, No. 1, pp. 14-28

15. Friedl, P., Hegerfeldt, Y. & Tusch, M. (2004). *Collective cell migration in morphogenesis and cancer*, Int J Dev Biol, Vol. 48, pp. 441-450

16. Gasser, U. E. & Hatten, M. E. (1990). *Central nervous system neurons migrate on astroglial fibers from heterotypic brain regions in vitro*, Proc Natl Acad Sci U S A, Vol. 87, No. 12, pp. 4543-4547

17. Geiger, B. (1979). *A 130K protein from chicken gizzard: its localization at the termini of microfilament bundles in cultured chicken cells*, Cell, Vol. 18, No. 1, pp. 193-205

18. Goldbeter, A. (1996). Biochemical oscillations and cellular rhythms: the molecular bases of periodic and chaotic behaviour, I, Cambridge University. Cambridge. GB

19. Greene, L. A., Burstein, D. E. & Black, M. M. (1982). *The role of transcription-dependent priming in nerve growth factor promoted neurite outgrowth*, Dev Biol, Vol. 91, No. 2, pp. 305-316

20. Gregor, T., Fujimoto, K., Masaki, N. & Sawai, S. (2010). *The onset of collective behavior in social amoebae*, Science, Vol. 328, No. 5981, pp. 1021-1025

21. Hatten, M. E. (2002). *New directions in neuronal migration*, Science, Vol. 297, No. 5587, pp. 1660-1663

22. Heinrich, V., Wong, W. P., Halvorsen, K. & Evans, E. (2008). *Imaging biomolecular interactions by fast three-dimensional tracking of laser-confined carrier particles*, Langmuir, Vol. 24, No. 4, pp. 1194-1203

23. Hou, S., Li, X. X., Li, X. Y., Feng, X. Z., Guan, L., Yang, Y. L. & Wang, C. (2009). *Patterning of 293T fibroblasts on a mica surface*, Anal Bioanal Chem, Vol. 394, No. 8, pp. 2111- 2117

24. Hu, K., Ji, L., Applegate, K. T., Danuser, G. & Waterman-Storer, C. M. (2007). *Differential transmission of actin motion within focal adhesions*, Science, Vol. 315, No. 5808, pp. 111-115

25. Ingber, D. E. (2003). *Tensegrity I. Cell structure and hierarchical systems biology*, J Cell Sci, Vol. 116, No. Pt 7, pp. 1157-1173

26. Izzard, C. S. & Lochner, L. R. (1976). *Cell-to-substrate contacts in living fibroblasts: an interference reflexion study with an evaluation of the technique*, J Cell Sci, Vol. 21, No. 1, pp. 129-159

27. Jiang, X., Bruzewicz, D. A., Wong, A. P., Piel, M. & Whitesides, G. M. (2005). *Directing cell migration with asymmetric micropatterns*, Proc Natl Acad Sci

U S A, Vol. 102, No. 4, pp. 975-978

28. Kühner, M. & Sackmann, E. (1996). *Ultrathin hydrated dextran films grafted on glass: preparation and characterization of structural, viscous, and elastic properties by quantitative microinterferometry*, Langmuir, Vol. 12, No. 20, pp. 4866-4876
29. Kaji, H., Takoh, K., Nishizawa, M. & Matsue, T. (2003). *Intracellular Ca^{2+} imaging for micropatterned cardiac myocytes*, Biotech Bioeng, Vol. 81, No. 6, pp. 748-751
30. Kane, R. S., Takayama, S., Ostuni, E., Ingber, D. E. & Whitesides, G. M. (1999). *Patterning proteins and cells using soft lithography*, Biomaterials, Vol. 20, No. 23-24, pp. 2363- 2376
31. Keller, P. J., Schmidt, A. D., Wittbrodt, J. & Stelzer, E. H. (2008). *Reconstruction of zebrafish early embryonic development by scanned light sheet microscopy*, Science, Vol. 322, No. 5904, pp. 1065-1069
32. Keren, K., Yam, P. T., Kinkhabwala, A., Mogilner, A. & Theriot, J. A. (2009). *Intracellular fluid flow in rapidly moving cells*, Nat Cell Biol, Vol. 11, No. 10, pp. 1219-1224
33. Kumar, A. & Whitesides, G. M. (1993). *Features of gold having micrometer to centimeter dimensions can be formed through a combination of stamping with an elastomeric stamp and an alkanethiol "ink"followed by chemical etching*, App Phys Lett, Vol. 63, No. 14, pp. 2002-2004
34. Kumar, G., Ho, C. C. & Co, C. C. (2007). *Guiding cell migration using one-way micropattern arrays*, Adv Mater, Vol. 19, No. 8, pp. 1084-1090
35. Kushiro, K., Chang, S. & Asthagiri, A. R. (2010). *Reprogramming directional cell motility by tuning micropattern features and cellular signals*, Adv Mater, Vol. 22, No. 40, pp. 4516- 4519
36. Lambert, G., Liao, D. & Austin, R. H. (2010). *Collective Escape of Chemotactic Swimmers through Microscopic Ratchets*, Phys Rev Lett, Vol. 104, No. 16, pp. 168102
37. Lewis, K. (2005). *Persister cells and the riddle of biofilm survival*, Biochemistry (Moscow), Vol. 70, No. 2, pp. 267-274
38. Limozin, L. & Sengupta, K. (2009). *Quantitative reflection interference contrast microscopy (RICM) in soft matter and cell adhesion*, ChemPhysChem, Vol. 10, No. 16, pp. 2752- 2768
39. Mahmud, G., Campbell, C. J., Bishop, K. J. M., Komarova, Y. A., Chaga, O., Soh, S., Huda, S., Kandere-Grzybowska, K. & Grzybowski, B. A. (2009). *Directing cell motions on micropatterned ratchets*, Nature Phys, Vol.5, 606-612
40. Marín, O. & Rubenstein, J. L. (2001). *A long, remarkable journey: tangential migration in the telencephalon*, Nat Rev Neurosci, Vol. 2, No. 11, pp. 780-790

41. Masubuchi, M., Toyota, T., Yamada, M. & Seki, M. (2011). *Fluidic shear-assisted formation of actuating multilamellar lipid tubes using microfabricated nozzle array device*, Chem Comm, Vol. 47, 8433-8435

42. Nakaoka, Y., Imaji, T., Hara, M. & Hashimoto, N. (2009). *Spontaneous fluctuation of the resting membrane potential in Paramecium: amplification caused by intracellular Ca^{2+}*, J Exp Biol, Vol. 212, No. 2, pp. 270-276

43. O'Toole, G., Kaplan, H. B. & Kolter, R. (2000). *Biofilm formation as microbial development*, Ann Rev Microbiol, Vol. 54, No. 1, pp. 49-79

44. Ohnuma, K., Hayashi, Y., Furue, M., Kaneko, K. & Asashima, M. (2006). *Serum-free culture conditions for serial subculture of undifferentiated PC12 cells*, J Neurosci Methods, Vol. 151, No. 2, pp. 250-261

45. Ohnuma, K., Toyota, T., Ariizumi, T., Sugawara, T. & Asashima, M. (2009). *Directional migration of neuronal PC12 cells in a ratchet wheel shaped microchamber*, J Biosci Bioeng, Vol. 108, No. 1, pp. 76-83

46. Oosawa, F. (2001). *Spontaneous signal generation in living cells*, Bull Math Biol, Vol. 63, No. 4, pp. 643-654

47. Park, S., Wolanin, P. M., Yuzbashyan, E. A., Lin, H., Darnton, N. C., Stock, J. B., Silberzan, P. & Austin, R. (2003a). *Influence of topology on bacterial social interaction*, Proc Natl Acad Sci U S A, Vol. 100, No. 24, pp. 13910-13915

48. Park, S., Wolanin, P. M., Yuzbashyan, E. A., Silberzan, P., Stock, J. B. & Austin, R. H. (2003b). *Motion to form a quorum*, Science, Vol. 301, No. 5630, pp. 188

49. Rädler, J. & Sackmann, E. (1992). *On the measurement of weak repulsive and frictional colloidal forces by reflection interference contrast microscopy*, Langmuir, Vol. 8, No. 3, pp. 848- 853

50. Rädler, J. & Sackmann, E. (1993). *Imaging optical thicknesses and separation distances of phospholipid vesicles at solid surfaces*, Journal de Physique II, Vol. 3, No. 5, pp. 727- 748

51. Rakic, P. (1972). *Mode of cell migration to the superficial layers of fetal monkey neocortex*, J Comp Neurol, Vol. 145, No. 1, pp. 61-83

52. Reimann, P. (2002). *Brownian motors: noisy transport far from equilibrium*, Phys Reports, Vol. 361, No. 2-4, pp. 57-265

53. Revel, J. P. & Wolken, K. (1973). *Electronmicroscope investigations of the underside of cells in culture*, Exp. Cell Res, Vol. 78, pp. 1-14

54. Round, J. & Stein, E. (2007). *Netrin signaling leading to directed growth cone steering*, Current opinion in neurobiology, Vol. 17, No. 1, pp. 15-21

55. Saravia, V., Kupcu, S., Nolte, M., Huber, C., Pum, D., Fery, A., Sleytr, U. B. & Toca-Herrera, J. L. (2007). *Bacterial protein patterning by micro-contact printing of*

PLL-g-PEG, J Biotech, Vol. 130, No. 3, pp. 247-252

56. Schneider, L., Cammer, M., Lehman, J., Nielsen, S. K., Guerra, C. F., Veland, I. R., Stock, C., Hoffmann, E. K., Yoder, B. K. & Schwab, A. (2010). *Directional cell migration and chemotaxis in wound healing response to PDGF-AA are coordinated by the primary cilium in fibroblasts*, Cell Physiol Biochem, Vol. 25, No. 2-3, pp. 279-292
57. Scholl, M., Sprossler, C., Denyer, M., Krause, M., Nakajima, K., Maelicke, A., Knoll, W. & Offenhausser, A. (2000). *Ordered networks of rat hippocampal neurons attached to silicon oxide surfaces*, J Neurosci Methods, Vol. 104, No. 1, pp. 65-75
58. Sengupta, K., Aranda-Espinoza, H., Smith, L., Janmey, P. & Hammer, D. (2006). *Spreading of neutrophils: from activation to migration*, Biophys J, Vol. 91, No. 12, pp. 4638-4648
59. Simson, R., Wallraff, E., Faix, J., Niew hner, J., Gerisch, G. & Sackmann, E. (1998). *Membrane bending modulus and adhesion energy of wild-type and mutant cells of Dictyostelium lacking talin or cortexillins*, Biophys J, Vol. 74, No. 1, pp. 514-522
60. Smilenov, L. B., Mikhailov, A., Pelham, R. J., Marcantonio, E. E. & Gundersen, G. G. (1999). *Focal adhesion motility revealed in stationary fibroblasts*, Science, Vol. 286, No. 5442, pp. 1172-1174
61. Smith, A. S., Sengupta, K., Goennenwein, S., Seifert, U. & Sackmann, E. (2008). *Force-induced growth of adhesion domains is controlled by receptor mobility*, Proc Natl Acad Sci U S A, Vol. 105, No. 19, pp. 6906-6911
62. Streicher, P., Nassoy, P., Barmann, M., Dif, A., Marchi-Artzner, V., Brochard-Wyart, F., Spatz, J. & Bassereau, P. (2009). *Integrin reconstituted in GUVs: A biomimetic system to study initial steps of cell spreading*, Biochim Biophys Acta (BBA)-Biomembranes, Vol. 1788, No. 10, pp. 2291-2300
63. Svitkina, T. M., Verkhovsky, A. B., McQuade, K. M. & Borisy, G. G. (1997). *Analysis of the actin-myosin II system in fish epidermal keratocytes: mechanism of cell body translocation*, J Cell Biol, Vol. 139, No. 2, pp. 397-415
64. Théry, M., P pin, A., Dressaire, E., Chen, Y. & Bornens, M. (2006). *Cell distribution of stress fibres in response to the geometry of the adhesive environment*, Cell motility and the cytoskeleton, Vol. 63, No. 6, pp. 341-355
65. Toyota, T., Maru, N., Hanczyc, M. M., Ikegami, T. & Sugawara, T. (2009). *Self-propelled oil droplets consuming "fuel" surfactant*, J Am Chem Soc, Vol. 131, No. 14, pp. 5012- 5013
66. Weiner, O. D., Servant, G., Welch, M. D., Mitchison, T. J., Sedat, J. W. & Bourne, H. R. (1999). *Spatial control of actin polymerization during neutrophil chemotaxis*,

Nature cell biology, Vol. 1, No. 2, pp. 75-81

67. Yamasaki, E., Tanaka, D. H., Yanagawa, Y. & Murakami, F. (2010). *Cortical GABAergic interneurons transiently assume a Sea Urchin-like nonpolarized shape before axon initiation*, J Neurosci, Vol. 30, No. 45, pp. 15221-15227
68. Yang, I. H., Co, C. C. & Ho, C. C. (2005). *Alteration of human neuroblastoma cell morphology and neurite extension with micropatterns*, Biomaterials, Vol. 26, No. 33, pp. 6599-6609
69. Yin, C., Liao, K., Mao, H. Q., Leong, K. W., Zhuo, R. X. & Chan, V. (2003). *Adhesion contact dynamics of HepG2 cells on galactose-immobilized substrates*, Biomaterials, Vol. 24, No. 5, pp. 837-850
70. Zheng, J. Q. & Poo, M. M. (2007). *Calcium signaling in neuronal motility*, Annu Rev Cell Dev Biol, Vol. 23, pp. 375-404

PART-III
MICROBIAL BIOTECHNOLOGY

CHAPTER-10

MICROBIAL EXPRESSION PLATFORMS AND PRODUCTION FROM A MARKET AND ECONOMIC VIEWPOINT

INTRODUCTION: AN OVERVIEW

Examining microbial expression platforms and production processes from a market and economic perspective is essential in today's biotechnology landscape. These platforms, including bacterial and yeast systems, are not only instrumental in producing a wide array of biotherapeutics, enzymes, and industrial chemicals but also have a significant impact on the overall cost-effectiveness of these processes. By assessing market trends, demand dynamics, and economic considerations, researchers and businesses can make informed decisions to optimize production strategies, enhance product yield, and ensure competitiveness in the biopharmaceutical and industrial sectors. This approach not only fosters innovation but also contributes to the sustainable growth of the biomanufacturing industry.

Biotechnology generates global sales in the order of well over 200 billion US$ in all markets and has thus become an important economic factor in manufacturing. The buzz word 'biotechnology' carries expectations that it can provide sustainable solutions for greenhouse gas reduction in manufacturing industries, trigger a 'clean tech' boom and create new jobs. It is no wonder that biotechnology has gained significant attention even in high level politics as it can give a 'green' touch to administrations. Many consumers are not even aware of the surprising array of products and services which biotechnology can or could provide today; these range from high-tech pharmaceutical applications to snow making. Table 1 lists some new or unusual applications of biotechnology products which include, for example, skin protection compounds from the oceans or biopolymers for drag reduction in transport pipelines.

- Microbial secondary metabolites for the bio-control of invasive mussels in water pipes
- Microbial products for rust removal and anticorrosion
- Proteins for plant protection by induction of the plant's natural defence systems
- Glycoproteins radically affecting the palate and sensorial perception
- Compounds from deep-sea microorganisms for skin protection
- Nanoscaffolds (also functionalized) based on biomaterial for e.g. tissue replacement and repair
- Biopolymers and biosurfactants for drag reduction in transport pipelines
- Biopolymers and enzymes for plywood production
- Channel proteins for H2O desalination and other purposes
- Enzymes as toxic gas antidotes for military applications
- Microbially derived innovative lubricants
- Biologicial production of solar cells

Table 1. A few examples of new and unusual applications of biotechnology and its products in different fields.

These high expectations are merited due for 4 reasons:

- The unmatched precision in the production and assembly of small and large molecules. This precision of the natural biosynthetic machinery cannot be reached using chemical approaches.
- The fantastic speed, at which these production systems can reproduce themselves. The reason for this is that bacteria have by far the largest surface-to-volume ratio in the living world, leading to maximal metabolic rates. A single bacterium, weighing about 10^{-12} grams, grows so fast that its biomass would theoretically reach the mass of the earth in only a few days!
- The inherent safety of biological systems as metabolic heat makes run-away reactions impossible, when compared to organic chemistry.
- The biocatalyst and biomass are fully recyclable.

Consequently, biotechnology will have an especially high impact in the production of complex chemicals used for pharmaceuticals, fine chemicals and specialities (Meyer, 2011). Other promising areas are biopolymers and protein-based novel biomaterials for consumer goods, car parts, medical devices or as support for the 2D and 3D cultivation of tissue and organ replacements.

It is industrial or white biotechnology which is of growing academic and private interest, as it represents an equal or even bigger business potential than red biotechnology in the long term. But how can application fields and markets of biotechnology be classified? One way to describe the different markets of biotechnology is the colour code (red, white, green, blue and grey).

THE MARKETS

A useful way to classify the applications of biotechnology is the colour code of biotechnology shown in Table 2.

Estimates and definitions may vary, but there is one common denominator in various assessments, namely that the proportion of products manufactured using biotechnology will increase significantly. While the development and the market introduction of new biopharmaceuticals such as monoclonal antibodies will continue at its present rate, it is especially industrial biotechnology which is expected to realise high growth rates.

The different "biotechnologies" do overlap and especially the boundaries especially between red and white biotechnologies for pharmaceutical applications can be confusing. There is one important additional difference between the red biotechnology of therapeutic proteins and monoclonal antibodies and the white biotechnology pharmaceuticals which includes a large variety of products: red biotechnology is characterised more by its products whereas white biotechnology is defined more by its technology platform.

The chemical market

The sales of global chemical markets are expected to grow from 2292 billion Euros in 2007 to 3235 billion Euros in 2015 and to 4012 billion Euros in 2020 (Perlitz, 2008). It is estimated that only about 3-4% of all chemical sales have been generated with some help from biotechnology (Nieuwenhuizen et al., 2009), but this figure is anticipated to grow faster than the average market rate. It is speculated that at least 20% of the global chemicals will be derived usingindustrial biotechnology in 2020, which translates into almost 1000 billion Euros. This means that the sales generated by industrial biotechnology will increase by an order of magnitude as the recent estimates of the global sales of industrial biotechnology products vary between 50 and 150 billon Euros, depending on whether biofuels are included or not. There is a consensus that biotechnology will play a much greater role in future manufacturing as it can deliver complex products using economically and ecologically sustainable processes.

Table 2. Classification of the applications of biotechnology with respect to markets; for each class sales volumes, compound annual growth rate and number of companies globally active in the field are noted; there is of course an overlap between industrial and pharmaceuticals biotechnology, some sectors of industrial biotechnology are dependent on cheap and reliable sources from agro biotechnology (Clive, 2010). The market size of>11 billion US$ for agro biotechnology refers to "seed biotech". The global value of marketed harvested goods resulting from these seeds would be much larger, by one or two orders of magnitude.

	Industrial Biotechnology	Pharma Biotechnology	Environmental Biotechnology	Agro Biotechnology	Marine Biotechnology
Markets served	Many different markets such as small molecule pharma & fine chemicals, flavour & fragrance, bulk chemicals etc.	Monoclonal antibodies and other therapeutic proteins	Environmental biotechnology, services & solution for bioremediation and waste treatment	Transgenic or genetically modified (GM) seeds and plants	Products and lead substances from the marine environment
Color code	White	Red	Grey	Green	Blue
Market size US$	> 50 bn without biofuels	> 100 bn	n.a.	> 11 bn	n.a.
CAGR	15%	> 20%	n.a.	15-20%	n.a.
Companies	4'000	6'000	n.a.	> 50	n.a.

The products potentially produced by biotechnology range from commodities (e.g. succinic acid), biopolymers (e.g. polyhydroxybutyric acid), flavour & fragrance products (e.g. vanillin), agroproducts (*Bacillus thuringiensis*) to small molecule pharmaceuticals and more.

Unfortunately there are no shortcuts in biotechnology, and in order to meet the anticipated 1000 billion Euros derived from white or industrial biotechnology and to keep red biotechnology humming we need to further develop appropriate tools while keeping in mind that it took 200 years to complete today's chemical toolbox (Ghisalba et al., 2010).

The feed, food and dietary supplement markets

Biotechnology plays an important role in the kitchen through the food and beverage industry. Enzymes are used in large scale food production of glucose from starch (hydrolytic enzymes), high fructose syrup (isomerase), conversion of lactose to galactose and glucose (hydrolase), cheese production (proteases), meat processing (proteases) andmany more. Phytase (phosphohydrolase) is an example of an enzyme used in the feed industry. However, only about 25% to 30% of all industrial enzymes are used for food and feed purposes. It is also worthwhile to note that over 90% of industrial enzymes sales come from less than 30 enzymes, most of them hydrolytic enzymes. As these enzymes are used as an auxiliary material in the food manufacturing process, most of them can be produced using recombinant technology. Thus microbial expression systems play a crucial role for enzyme production.

The market of functional foods more than doubled between 2001 and 2010 and is estimated at 5 billion € (Welck & Ohlig, 2011). Other important biotechnology products are amino acids, vitamins or PUFAs (polyunsaturated fatty acids).

Flavours and fragrances are also biotechnology-relevant markets with a volume of over 22 billion US$ (Leffingwell & Associates, 2011). More than 10% of the supply is derived from bioprocesses, with more than 100 commercial aroma chemicals derived via different biotechnological methods (Berger, 2009). In order to relieve the pressure on natural resources, companies are increasingly turning towards novel biotechnological sources and methods including genetic engineering approaches for the production of these raw materials.

This is also true for a part of the pharma market described below. To supply the anticancer secondary metabolite taxol (paxclitaxel) from its original natural source (the bark of the Pacific yew tree *Taxus brevifolia*) would be impossible. Taxol is produced in minute amounts (0.4 g/kg of bark) and synthetic alternatives including biotechnological steps had to be developed for this blockbuster drug with sales of about 3.5 billion US$. It is difficult to estimate the global market value of botanical or plant derived drugs (BCC Research, 2009), but they seem to grow at a CAGR of 11% and alternative commercially viable expression and production systems will be needed for their production to replace the extraction from endangered or slow growing plants.

The pharmaceutical market

According to the IMS Health Forecasts, the global pharmaceutical market is expected to grow at an overall rate of 5-7% per annum, to reach $880 billion US$ in 2011 (Gatyas, 2011). Of this total pharmaceutical market, monoclonal antibodies alone represented 43 billion US$ and therapeutic proteins 66 billion US$ (2009 figures, Market Research News, 2011). The numbers for small molecule pharmaceuticals, which represent a very fragmented market, are even more difficult to estimate. Franssen et al. estimated the market segment in 2010 for pharmaceutical ingredients using industrial biotechnology at about 20 bn US$ (Franssen et al, 2010). This could be conservative, as the number of small molecule APIs (Active Pharmaceutical Ingredients) using biotechnology in their (chemical) synthesis is rapidly increasing. Blockbuster such as Merck's Sitagliptin or Pfizer's Atorvastatin now include biotechnological steps in their manufacturing. Atorvastatin (Lipitor®), a cholesterol- lowering small molecule drug is the largest selling drug in history, which peaked at 12.7 billion US$ annual sales. The chemical synthesis includes a biocatalysis step using a ketoreductase. The same is true for MERCK's Sitagliptin, the small molecule compound used in the antidiabetes drug Januvia®, which includes a transaminase step in its chemical synthesis. Taking such products into consideration, the number will increase well above 30 billion US$ in 2011. Many classical biotechnological products like antibiotics (market of 42 billion US$ in 2009) or steroids are even not included in these 30 billion US$. What isimportant is that the small molecule market remains by far the most important driver for innovation in industrial biotechnology (Meyer et al., 2009).

It is especially red biotechnology with therapeutic proteins and monoclonal antibodies which is thriving and driving growth with established technologies. Remicade was the top selling mAbs brand followed by Avastin®, Rituxan®, Humira® and Herceptin®. The top 5 brands had sales of over $ 5 billion each. However, besides that they often serve narrower disease phenotypes due to their specificity. The downsides of therapeutic proteins and monoclonal antibodies are that

- they are not consumable as pills
- logistics are more complicated due to their instability
- they are more expensive than small molecules.

Although the market value of small molecule drugs which use one or more biocatalysis steps in their synthesis amounts to well over 30 billion US$ and is growing, the share of biotechnology is well below what it could be. The pharmaceutical industry is under great pressure due to exploding R&D costs, politically due to the health care cost burden and from the market side due to an expected reduction of growth from 7% to 3%.

Both large and small molecules could face a technology gap and manufacturing bottlenecks to meet these production challenges. In both cases we need innovative, sustainable and cost- effective production methods.

The vaccine market is a special subset of the pharmaceutical market, in which innovative solutions are sought for cancer, infectious diseases (e.g. malaria), pandemics (influenza) or bioterrorism. The vaccine market of 20 billion US$/year is expected to grow to 35 billion US$ in five years. Unusual new solutions such as production in transgenic plants may be required to achieve the scale and price targets (Langer, 2011).

The number of market introductions of small molecule pharmaceuticals or NCEs (New Chemical Entities) has been steadily decreasing since the late 1980s, whereas new therapeutic proteins and monoclonal antibodies have increased. To make things worse, small molecule drugs are positioned in markets that are becoming increasingly generic, thereby adding further pressure. The only small molecule drugs with an annual growth of over 10% were high potency drugs and to a lesser extent peptides. However, keeping healthcare costs under control will require efficient and affordable drugs, which are generally smaller entities without complicated and folded structures such as large proteins which need expensive production and logistics and can only be administered by injection.

However, solutions of the manufacturing challenges in the pharmaceutical industry will also provide solutions for other markets.

We need new cost-effective production for large and small molecule drugs (Meyer & Turner, 2009) with more chirality, more complex functionalities, and composed of various chemical structures. Examples are the glycosylation of proteins, aryl- or alkyl-organics drugs for modification of their biological efficacy and the functionalisation of novel biomaterials for medical devices and scaffolds for tissue generation using stem cells. The current biotechnological and chemical toolbox is reaching its technical limits and needs expansion to meet the economic and ecological manufacturing standards of the future (Meyer & Werbitzky, 2011).

Chinese hamster ovary cells, baby hamster kidney cells

Table 3. This table gives an overview of all biological methods and frequently used genus which can be used for the production of small and large molecules. The microbial pro- and eukaryotes are further tabulated in Table 6.

Organism	Genus, species	Products	Comment
Prokaryotes	*Bacillus* *Actinomycetes*	*Small and large molecules*	*Generalists, well established technology, cost efficient, remains the standard method*
Gram+	*Escherichia coli* *Pseudomonas* *Gluconobacter*	Large molecules and enzymes Secondary metabolites	Host for secreted proteins
Gram-	*Myxobacterium*	Small and large molecules	The standard microbial workhorse
		Secondary metabolites	Source of biologically active compounds but tricky to manufacture with

Organism	Genus, species	Products	Comment
Eukaryotes	*Saccharomyces Pichia*	*Large and small molecules*	*Mostly specialist, niche applications with increasing importance*
Fungi		Proteins and enzymes	
	Penicillium Aspergillus		Will lose its importance
Algae		Secondary metobolites	Uniform human like N-linked glycans, host for industrial enzymes
Plants	*Chlamidomonas*	Citric acid, enzymes, proteins Mainly small molecules	
Protozoa	*Zea mais Nicotiana tabacum Pisum sativum*	Small and large molecules Bovine trypsin	Nutritional flexibility, efficient secreters, industrially widely used
Insect cells			
	Tetrahymena Trochoplusia	Different products	PUFAs, pigments, in discussion for biofuels
Mammalian cells		Large molecules	Potential host for human antibodies, expression in 3 genomes
	Spodoptera Trochoplusia	Large molecules	
	CHO , BHK1 C127, NSO	Cells, tissues, organs	Cost advantage, vaccines, "plantibodies" (antibodies)
	PerC6 Stem cells		
	hESC, iPS		Expression system for ciliates for the production of therapeutic proteins
			Veterinary products, easier cultivation than mammalian cells
			Several pharma blockbusters, established technology
			Mouse cell lines Human cell line
			Tissue and organ repair, mass production in development
			Human embryonic stem cells, induced pluripotent cells

THE BIOTECHNOLOGY TOOLBOX

Table 3 gives an overview of the production systems available today for the production of small and large molecules. Old technologies still dominate the industry, especially for pharmaceuticals. Other than pharmaceuticals still extracted from plants, animals (mostly now forbidden), only a few biotechnological methods are used. For example of the 130 recombinant protein products in the US and European markets, 48 are expressed in microbes (34% *E. coli* specifically, 1% other bacteria, yeast 13%, see also Figure 6). Another 43% of products are produced in mammalian cells, primarily in CHO cells. *E. coli* and CHO cells have the longest history of use since the commercialisation of the first recombinant proteins in the 1980s. CHO and *E. coli* account for 64% of the expression systems used in manufacturing of currently marketed recombinant therapeutics (US and Europe). Novel expression systems in evaluation for biopharmaceuticals include e.g. *Pseudomonas fluorescens, Staphylococcus carnosus, Bacillus subtilis, Caulobacter crescentus, Chrysosporium lucknowense, Arxula sp.* as we will see later.

Cultivation options

While Table 3 lists the organism which can be used, it does not include the cultivation methods which are listed in Table 4 below.

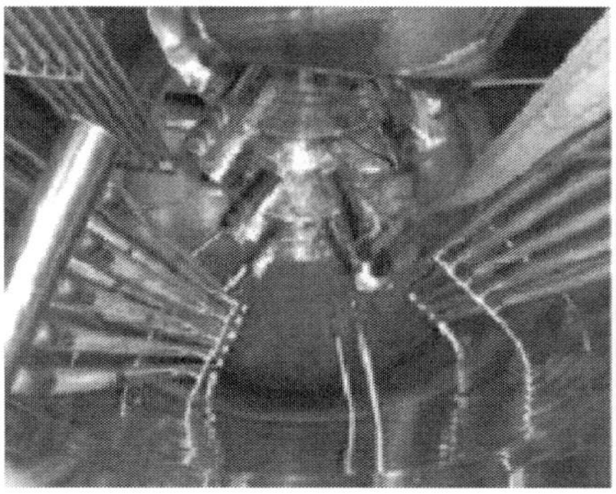

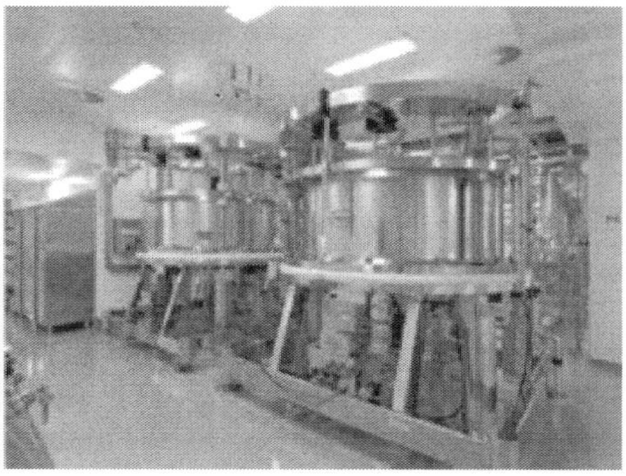

Table 4. Overview of the principal biotechnological production methods. Fermentation in sterile containments is by far the most important and suitable for bacteria, fungi, protozoa, algae, plants cells, insect cells, mammalian cells. Stem cells are preferentially produced adherent in biofilms, but the mass production for therapeutic purposes will probably also use suspension culture.

The standard manufacturing procedure in Table 4 is the submersed production of organisms in sterile containments as shown in Figure 1. This method allows the controlled growth of one organisms in a closed tank (fermenter). In over 15 years Armin Fiechter of the Swiss Federal Institute of Technology in Zürich developed a standardised biological test system and systematically tested many different bioreactor designs at the m^3 scale. Although many new and interesting bioreactor designs exist, nothing has really changed since the 1980s and the numbers prove that the classic stirred tank with Rushton impellers remains the most versatile fermenter design (Meyer, 1987). It can be used for unicellular and filamentous cells, and is even used in large scale fermentation for mammalian and plant cell cultures. Moreover, the design gives good performance over a wide range of viscosities. It is a truly multipurpose equipment, with decades of depreciation times for the invested capital for the tanks themselves. The average cost for a large scale sterile fermentation plant (without down stream processing) ranges between 130000 €/m^3 for an ISO plant to 300000 €/m^3 nominal volume for a cGMP plant for injectable products.

The technology transfer and engineering challenges of fermentation, which include control of the physicochemical environment, mass and gas transfer and other items are discussed elsewhere (Sharma et al., 2011; Meyer & Klein, 2006; Meyer & Rohner, 1995; Meyer & Birch, 1999; Hoeks et al., 1997). Table 4 is a very general overview and many variants of suspension cultures cannot be discussed here. For example the immobilisation of cells by filtration in combination with continuous culture is a powerful tool to reach outstanding productivities with mammalian and microbial cells if sterile operation can be achieved and the cell type allows extended number of cell divisions and generations Hoeks et al., 1992).

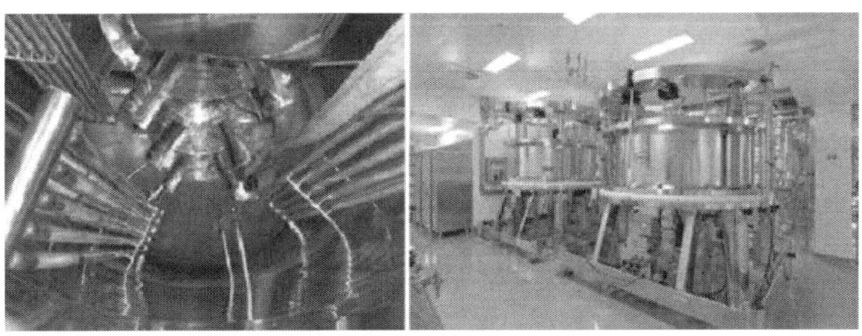

Fig. 1. Left side: photograph of a 15m^3 state of the art sterile containment for high cell density fermentations.

LONZA has built and operates such high performance equipment up to the 75m^3 for a range of organisms including methylotrophic yeasts such as *Pichia pastoris.* The flow breakers are designed as cooling elements needed to remove metabolic heat. The stirrers are concave Rushton impellers which, in combination with specially designed dip pipes (not shown), allow optimal mixing. In most cases the stirrer configuration is completed by the use of a top radially downward pumping marine impeller. Right: Two chromatography columns for the purification of injectable therapeutic proteins with a diameter of 2 meters to demonstrate size and complexity of down stream processing (Copyright@Lonza).

The production organisms

The production strain is not everything, but everything is nothing without a good production strain. Irrespective of the organism used in suspension culture for biotechnological manufacturing, there are numerous common problems, but also some typical differences, which will be discussed. Strain development is, however, the key issue for a commercially viable bioprocess. We have seen above, that companies have invested in expensive multipurpose fermentation equipment. Thus, the biology must be adapted to existing equipment and notthe other way round. Table 5 summarises our experience with regard to the different factors influencing manufacturing costs with mammalian and bacterial cell cultures.

Generally, the strain and its growth characteristics define medium composition, cycle time and final product concentrations. Consequently, the number of steps needed in downstream processing (DSP) and volumetric sterile productivity (the two key cost drivers) are directly related to the choice of strain. The ideal strain is genetically stable, has a high specific (qp) and volumetric productivity (Qp), forms no by-products, and uses a well-defined medium resulting in a DSP with a limited number of steps. Fermentation is where value is created, downstream processing needs to conserve that created value.

A key issue in achieving a high specific and volumetric production rate is the choice of a highly efficient expression system and finally choosing the right recombinant strain for production.

Table 5. Effect of strain, process and plant on the overall process outcome. A highly productive strain is the most important factor in a bioprocess. I & D = Interest & depreciation. Y = Yield.

	Process Flexibility	Effect on O_2	Effect on sterility	Effect on cost
Strain	key	key	cycle time	key
Process				
Medium	limited	limited	can be important	medium
Parameters	limited	very limited	usually limited	small
Plant				
Fermentation	very limited	moderate	key	I&D
DSP	yes	small	product dependent	I&D&Y

Prokaryotes and lower eukaryotes

a, yeast and fungi

Prokaryotes such as bacteria and lower eukaryotes such as yeasts or fungi are by far the most productive organisms in biotechnology. We will call them collectively microorganisms in this paper. Because of their small size (Figure 2), microorganisms have by far the largest surface to volume ratio in the living world which allows them to maximize their metabolic rates because of a high rate of exchange of molecules through their surface. With the right cultivation conditions, microorganisms grow exponentially according to the equation

$$X_t = X_o * e^{(\mu * t)}$$

X_o is the biomass concentration at time zero, or the start of cultivation. X_t is the biomass concentration at the time of harvest. μ is a strain specific growth rate. Some of the fastest growing bacteria weighing maybe 10^{-12} g are theoretically able to duplicate and grow so fast that their biomass would reach the mass of the earth ($9 * 10^{54}$ tons) in less than a week. This means, that if a bacterial strain produces a protein or another product which can be industrially applied, large amounts can theoretically be produced economically.

Not only are microorganisms able to increase biomass at breathtaking rates, many are also able to grow under different conditions and on a great variety of substrates. This metabolic flexibility requires the ability to produce thousands of different enzymes and other proteins for all sorts of reactions and purposes. With the advent of genetic engineering, these enzymes and proteins can be overproduced in great quantities. Today entire pathways are modified and completed with heterologous genes, which allows the expression and production of molecules foreign to the species.

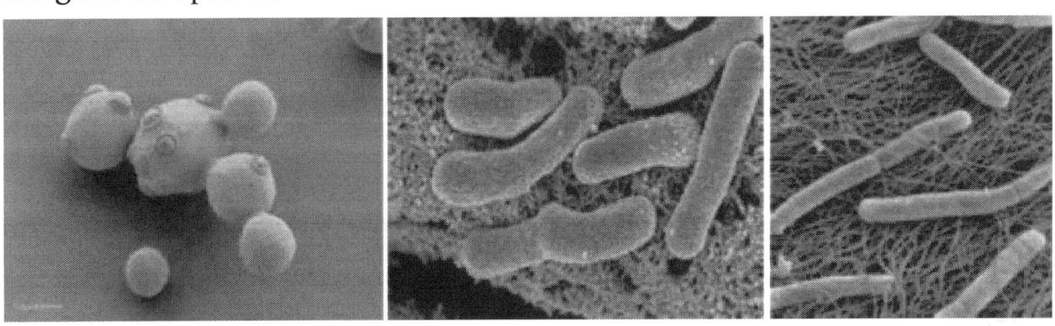

Fig. 2. Left a budding cell of the yeast *Saccharomyces cerevisiae*. During the reproductive phase a cell multiplies by forming buds. After the buds enlarge, nuclear division occurs and a cross wall is formed between the two cells. *S. cerevisiae* is easy to cultivate at large scale and it serves as host in many biotransformation processes, mainly after genetic recombination. The photograph in the center shows a few cells of the workhorse par excellence in biotechnology: *Escherichia coli*. The diameter of the *E. coli* cells varies between 0.5 and 1.5 micrometer. The bacterium to the right belongs to the genus *Rhizobium*, of which the strain HK1349 is used at large scale for the production of L-carnitine (Copyright@Lonza).

However, it does not end with the ability of microorganisms to produce large amounts of biomass and a great variety of different enzymes and proteins in a short tisme. There is an even more important reason why one wants to use microorganisms, namely the chemo-, regio- and enantio-selectivity of the microbial toolbox, and this is true for all living matter. No chemical manufacturing technology can match the precision of natural systems for the production of chiral and complex small and large molecules.

While this article focuses on the production of recombinant proteins and small molecules with *Escherichia coli*, it is important to understand the advantages of the different expression and production systems and compare them with the work-horse *Escherichia coli*.

There are now such a multitude of microbial host and expression systems theoretically available (Meyer et al., 2008) that it becomes difficult to choose the right combination. Emerging alternative production systems including those for glycosylated biopharmaceutical proteins for therapeutic use also are numerous (Jostock, 2007). *Escherichia coli* is currently the almost exclusively used prokaryotic production system but alternatives in discussion and developed on top of the commercially available expression systems are listed in Table 6. Examples are *Caulobacter crescentus*, *Proteus mirabilis*, *Pseudomonas* strains, *Staphylococcus carnosus*, *Streptomyces*, as well as fungi and yeasts such as *Pichia*, *Hansenula*, *Arxula*, *Yarrowia*, *Aspergillus* or *Trichoderma*.

Microbial Expression Systems and Manufacturing from a Market and Economic Perspective

Table 6. Commercialised microbial expression platforms and hosts; B, bacterial; F, fungal; Y, yeast.

Name	Microorganism	Commercial Source	Type
Bacillus megaterium	Bacillus megaterium	Technical University Carolo Wilhelmina	B
XS TechnologiesTM, Bacillus	Bacillus subtilis	Lonza	B
Bacillus subtilis (superoxidizing strains)	Bacillus subtilis	University of Groningen	B
PurePro Caulobacter expression system	Caulobacter	Research Corporation Technologies Inc. (RCT)	B
Clostridium Expression System	Clostridium botulinum	Wisconsin Alumni Research Foundation	B
Corynex	Corynebacterium glutamicum	Ajinomoto	B
Clean genome, stripped down E. coli	Escherichia coli	Scarab Genomics LLC	B
XS TechnologiesTM, E.coli	Escherichia coli	Lonza	B
PaveWay ProTM	Escherichia coli	Avecia	B
ESETECTM	Escherichia coli	Wacker Biotech	B

Name	Microorganism	Commercial Source	Type
Glycovaxin Bioconjugates	Escherichia coli	Glycovaxin	B
pET System	Escherichia coli	Novagen	B
Flavobacterium heparinum	Flavobacterium heparinum	BioMarin Pharmaceutical Inc	B
Lactococcus lactis	Lactococcus lactis	GTP Technology S.A., NIZO B.v., Bioneer Corp.	B
Methylobacterium extorquens	Methylobacterium extorquens	National Research Council of Canada	B
Pseudoalteromonas haloplanktis	Pseudoalteromonas haloplanktis	University of Naples	B
PfenexTM	Pseudomonas fluorescens	Dow Chemical Co.	B
Ralstonia eutropha	Ralstonia eutropha	Dartmouth College	B
Staphylococcus carnosus	Staphylococcus carnosus	Boehringer Ingelheim Pharma KG	B
Streptomyces	Streptomyces	Plant Bioscience Ltd, Tsukuba Industrial LCRC	B
Cangenus, Streptomyces	Streptomyces lividans	Cangene Corp.	B
C1 Expression	Chrysosporium lucknowense	Dyadic International Inc.	F
NeuBIOS Expresison	Neurospora crassa	Neugenesis Corp.	F
Neurospora Expression	Neurospora crassa	Genencor International	F
Arxula adeninivorans Expression	Arxula adeninivorans	Pharmed Artis GmbH	Y
Hansenula Expression	Hansenula polymorpha	Rhein Biotech Ltd, Artes Biotechnology GmbH	Y
Hansenula polymorpha Expression	Hansenula polymorpha	Artes Biotechnology GmbH	Y
Kluyveromyces lactis	Kluyveromyces lactis	DSM Biologics, New England Biolabs	Y
XS TechnologiesTM	Pichia pastoris	Lonza	Y
Pichia pastoris	Pichia pastoris	Research Corporation Technologies Inc. (RCT)	Y
Pichia pastoris	Pichia pastoris	Alder Biopharmaceuticals	Y
GlycoFi Technology	Pichia pastoris	GlyocoFi, Merck	Y
Saccharomyces cerevisiae expression	Saccharomyces cerevisiae	Delta Biotechnology Ltd. (Novozymes Delta)	Y

Name	Microorganism	Commercial Source	Type
ApoLife Yeast Expression	Saccharomyces cerevisiae	ApoLife	Y
EASYEAST	Saccharomyces cerevisiae	Biomedal, S.L.	Y
Zygosaccharomyces bailii	Zygosaccharomyces bailii	University of Milano Bicocca	Y

Fig. 3. The fungus *Aspergillus niger* is a "generally recognised as safe" (GRAS) organism used industrially for the production of many substances such as citric acid or gluconic acid, enzymes (e.g. glucose oxidase, glucoamylase, alpha-galactosidase) and other compounds. It is also a versatile host for the heterologius expression of proteins (Copyright@Lonza).

Aspergilli (Figure 3) are ideally suited for recombinant protein expression as they have an enormous nutritional flexibility combined with a particularly efficient secretion system and secretion capacity (Fleissner & Dersch, 2010). They are amongst preferred organisms for the production of commercial food enzymes. Genetic engineering of different *Aspergillus* host strains has also allowed the synthesis of industrially relevant amounts of various heterologous proteins (such as human lactoferrin, calf chymosin or the plant-derived sweeteners thaumatin or neoculin peptide sweeteners, Nakajiama et al., 2008). Proteins are also efficiently glycosylated in *Aspergilli*, while undesired hyperglycosylation is usually not observed. Whole genome sequences of several *Aspergillus* species are now available.

Algae

Recently, *Chlamydomonas reinhardtii*, a unicellular eukaryotic green algae has been proposed as a host to produce several forms of a human IgA antibody directed against herpes simplex virus (Franklin & Mayfield, 2005; Specht et al., 2010). The main reason for turning to algae is the claimed cost advantage and the absence of viral or prion contaminations that can harm humans. One can frequently read that microalgae grow faster (by an order or two of magnitude!) than terrestrial plants and biomass titers of 600-1000 mg/l dry weight are reached. However, one has to be careful. Microalgae can grow very fast when grown heterotrophically on glucose for example. In this case the performance of high cell density fermentations are almost as productive as those reported for bacteria and yeasts (Xiong et al., 2008). But things look different when microalgae are mass-produced phototrophically! For a number of technical and biological reasons not discussed here, doubling time of around 10 hours are probably a realistic assumption, leading to growth rates of μ= 0.07 per hour. This is very low when compared to the bacterium *Escherichia coli* where growth rates of 1 and higher are quite common in large scale. But growth rates are perceptibly higher than with all other eukaryotes except the yeasts and fungi mentioned above. The advantageover terrestrial plants is the smaller amount of soil used which leads to much better productivities per hectare for algae.

Algae might also have some other distinctive features, which could give them an advantage over other expression system for selected products. For example all three genomes (chloroplast, mitochondrial, nuclear) have been sequenced and can be transformed and each has distinct transcriptional, translational and post-translational properties. Proteins can accumulate at particularly high levels in the chloroplasts because of the absence of the silencing mechanisms. However, proteins in the chloroplasts are not glycosylated. Another feature of algae is, that they can be grown using sunlight or heterotrophically, or using a combination of both.

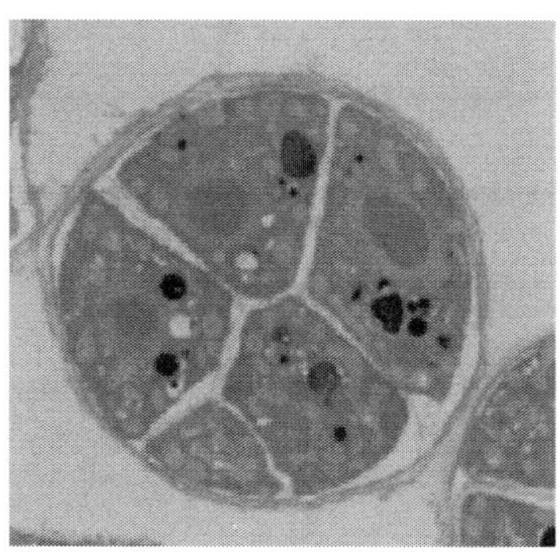

Fig. 4. Polyunsaturated fatty acids (PUFAs) such as docosohexanoic (DHA) acid are important building blocks of human brain tissue or the retina of the eye. The molecule can be produced by fermentation of marine algae. The picture shows the algae *Ulkenia sp.* which is used for the industrial production of DHA. They grow to form "footballs" consisting of single cells – 5 in this case with the product accumulated in the cells as oil drops. The size of the "football" is about 2 µm. Picture by Stefan Geimer, University of Bayreuth, Germany for Lonza (Copyright@Lonza).

The reality is that algae are routinely used for the production of polyunsaturated fatty acids only, and that other small molecule products such as carotenoids (Fernández-Sevilla et al., 2010) are merely in discussion. Although their growth is slow when compared to microbes, algae grow faster than terrestrial plants especially when grown heterotrophically with sugar as carbon and energy source. However, large-scale mass cultivation of algae using sunlight is far from being solved, and the calculations are sobering. Algae using CO_2 as a carbon source are a theoretically ideal solution but are a long way from being cost-competitive in practice (Van Beilen, 2010).

Whether algae will ever play a role for recombinant protein production and especially for recombinant monoclonal antibodies is very doubtful for two reasons. Firstly, the industry is very risk averse, conservative and one does not want to change well established production systems. Secondly, there are microbial systems being successfully developed which are also able to provide human-like post translational modifications as we will see later.

Higher Eukaryontsants

Whole plants. Pharming or molecular farming describes the use of transgenic plants (potato, tobacco, banana, tomato, maize, rice, lettuce) and animals for the production of recombinant therapeutic proteins or other recombinant drugs. As with algae, the driver to use plants for recombinant proteins and monoclonal antibodies is the lower unit cost of agricultural production combined with easy scalability while post-translational modifications are possible (Ahmad et al., 2010).

Starting in the late 1980s human interferon, mouse immunoglobulins and human serum albumin were the first recombinant proteins to be tested for transgenic plant production. The first molecular farming project was in 1999 to trigger an immune response in humans with a safe and cost-effective edible vaccine (Langer, 2011). By 2007, about 370 plant made pharmaceuticals (PMPs) were undergoing field trials and about 16 of them were reported to be in clinical trials (Spök, 2007). Recombinant plants are an interesting form of production, and we see important differences to algae.

- If the product is in an edible form (for example in seeds or fruits) formulation may not be necessary
- the product in plant material will be easily and stably stored
- and
- logistics can be easier as cooling may not be needed for storage and transport.

Because of the cost advantage, it is claimed that insulin could be a candidate for transgenic plant production as diabetes becomes a globally widespread disease, where affordable insulin is an absolute necessity. Vaccines in particular can be interesting products for plant made therapeutic proteins and other pharmaceuticals. MEDICAGO has developed a plant- based manufacturing platform (Proficia vaccine and antibody production system) with PHILIP MORRIS that produces (proficia) vaccine doses (H5N1 flu) in *Nicotiana benthamiana* leaf cells, a wild Australian relative of cultivated tobacco (Vécina et al., 2011). It is a transient expression system based on *Agrobacterium tumefaciens*, infecting plant cells and transferring genetic information to leaf cells. 600 million US$ have been invested in NOVARTIS plant R&D facilities with 167 acres of land. SIGMA ALDRICH uses molecular farming to make avidin, aprotinin, lysozyme, lactoferrin and trypsin at small scale for use as chemicals. TrypZean™ of SIGMA ALDRICH FINE CHEMICALS is a commercial recombinant bovine trypsin from transgenic corn.

Plant seeds are also being investigated as bioreactors for recombinant protein production (Lau & Sun, 2009), because they naturally contain large amounts of proteins but have low protease activities and low water content. Antibodies, vaccine antigens and other recombinant proteins have been shown to accumulate at high levels in seeds and remain stable and functional for years at ambient temperatures. As seeds can be eaten they allow oral delivery of vaccine antigens or pharmaceutical proteins for immunisation where oral delivery is an option. Unfortunately, most protein based therapeutics are usually not biologically active after oral consumption.

However, recombinant production of, for example, proteins in genetically modified plants has also several drawbacks.

- The time to create a stable transgenic host
- low protein titers
- extraction and purification from plant organs (if needed)
- non mammalian type glycosylation.

Plant transformation using physical methods is rather inefficient (gene gun). Transient expression is less time consuming but limited in scale. Recombinant *Agrobacterium tumefaciens* or plant viruses like the tobacco mosaic virus are used for transient expression.

Let us consider growth rates and use rapeseed as an example as it is one of the industrially used genetically modified crops. At the end of August one hectare is inoculated with 3-4 kg of rapeseeds. At the end of July in the following year (the plant needs a cold period to thrive – the so called vernalisation) the same hectare yields on average 4000 kg of seeds, from which 1700 l of rapeseed oil can be extracted. The biomass has increased by a factor of 1000 in 11 months. This represents a growth rate of $\mu = 0,05\ h^{-1}$ when calculated for the seeds growing on one hectare. This is relatively high, but it should be considered that one uses 1 hectare during one full year to produce 1'700 l of oil. Using heterotrophic algal fermentation probably 2000 times more oil could probably be produced on the same hectare. But then again, one still needs hectares to produce the necessary carbon and energy source for the fermentation, and we are thus back to square one with algae.

Coming back to transgenic plants and recombinant proteins: because of strong negative pressure from non-governmental environmental organisations and consumer organisations, transgenic plants in general will be slow to be used, also for the manufacturing of therapeutic proteins, unless they offer some of the advantages mentioned, such as edible and cheap recombinant drugs and vaccines for example.

Plant cell culture. An alternative to grow whole plants is to grow plant cells in suspension cultures. GREENVAX produces influenza vaccines by growing tobacco using XCELLEREX XDR single use bioreactors. GREENOVATION (Greenovation 2011) proposes bryotechnology (bryophytes = mosses) for recombinant protein production. PROTALIX's BIOTHERAPEUTICS Inc produces the enzyme replacement drug taliglucerase alfa for the treatment of Gauchers disease in disposable 800 litre bioreactors with carrot cells. Despite the fact that large companies such as PFIZER also stake a claim in plant cell-made biopharmaceuticals (taligurase alfa in carrot cells against Gaucher's disease of PROTALIX, Ratner, 2010), we conclude that commercial heterologous protein production in transgenic plants will not be economically relevant in the foreseeable future.

Whole plants and plant cell suspension culture, both will not play a relevant role in the production of therapeutic proteins and monoclonal antibodies. However, as with whole plants the situation is a different one for the recombinant production of other products and molecules such as secondary metabolites. The shift of wealth from west to east and north to south will fortunately eliminate economic inequalities. But an increasing global prosperity also increases pressure on naturally sourced raw materials especially from plants. Plant cell culture, recombinant or not, is a great alternative and will help to relieve pressure on partly even endangered species.

Protozoa

Heterologous expression of proteins or protein fragments is also possible in protozoa. *Dictyostelium discoideum* (Han et al., 2004), *Leishmania tarentolae*, *Perkinsus marinus* and especially *Tetrahymena thermophila* are organisms used, for which recombinant protozoa techniques exist. The ciliate-based expression system (CIPEX) of CILIAN AG (Cilian, 2011) is a proven tool which makes protozoa a potential competitor of mammalian cells such as CHO cells. For example viral influenza haemaglutinin, parasite surface proteins and a human intestinal alkaline phosphatase and a human DNase I were expressed and secreted in the unicellular non-pathogenic protozoan ciliate *Tetrahymena thermophila* (Hartmann et al., 2010; Aldag et al., 2011). The genome of *Tetrahymena thermophila* is entirely sequenced, and it is one of the best-characterized unicellular eukaryotes as it has served for long time as a laboratory model in biology.

Protozoa are naturally mostly feeding on bacteria. However, they can also easily be grown by pinocytosis in a culture medium containing only soluble components. Growth rates of protozoa in non-optimized media and culture conditions reach values of $\mu = 0.02\ h^{-1}$ and viable cell densities of $1 \times 10^7\ ml^{-1}$. Protozoa can also be cultivated in normal continuously stirred tank bioreactors and possess the sub-cellular machinery to perform eukaryotic post- translational protein modifications. However, protozoan-based expression systems have not yet made the transition from a laboratory model to an established recombinant protein platform at large scale. The main advantage of protozoa is that they are free of endogenous infectious agents as their genetics and phylogenetic distance to higher animals make viral infection unlikely. As with other expression systems high gene doses allow relatively high volumetric productivities [Qp] and scale-up has been proven up to the 1.5 m^3 scale. Protozoa have a consistent oligo-mannose N-glycosylation albeit not of mammalian nature.

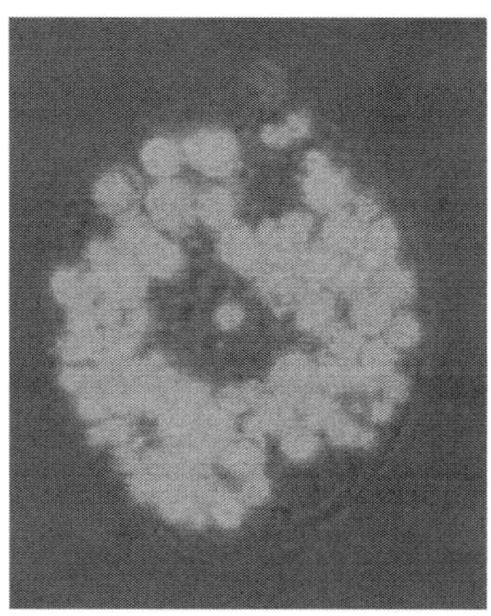

Fig. 5. Picture of *Tetrahymena thermopile*, one of the best characterized unicellular non- pathogenic protozoa, expressing a recombinant green fluorescent protein in phagosomes; protozoa can also secrete recombinant products. The ciliates, distributed on the cell surface, cannot been seen because the photograph was taken with a light microscope. The cell measures about 50 µm in diameter (Copyright@Cilian).

In the case of *Tetrahymena thermophila* (Figure 5) and with optimized media and culture conditions, cell densities of over 2 x 10^7 cells per milliliter and dry biomass titers of 8 g/l canbe reached in high cell density protozoa fermentations and 50 g/l in continuous fermentations with cell retention. Generation times become very short and values between1.4 and 3 hours can be obtained.

In conclusion, protozoa such as *Tetrahymena thermophila* have a certain potential for the recombinant production of selected recombinant proteins. To some extent they combine the advantages of the microbial and the mammalian world as they can grow rapidly to relatively high cell densities and have a posttranslational modification apparatus. If the necessary tools for the manufacturing with stable expression, conserved sequences and target glycosylation can be established, we may see cultivation of protozoa in large scale fermenters in the future for human enzymes, monoclonal antibodies and, in particular protein vaccines. Because of their faster growth, the first vaccine targets in preclinical phase are already in development using protozoan recombinant expression.

Insect Cells

Insect cells culture. A few recombinant proteins made using insect cell lines have already been approved for veterinary use. Only one vaccine, Cervarix® of GLAXO SMITH KLINE, has been approved for human use within the EU (in 2007) and in the US in 2009. Production with insect cells has the following advantages:

- easy to culture and faster than many mammalian cell lines
- high tolerance of osmolality
- advantageous and low ratio by-product vs expressed product

The *Spodoptera frugiperda* cell lines (Sf-9 and Sf-21, Figure 6) and *Trichoplusia ni* are frequently used host cell lines for recombinant protein expression and production via infection with a genetically modified baculovirus expression vector system, BEVS (Invitrogen, 2011; Quiagen, 2011; Tiwary et al., 2010). As for other cell lines used for production, insect cell lines need to be "immortalized" or rendered permanent.

Fig. 6. Cells from the ovarium of this moth *Spodopera frugiperda* are used in combination with baculoviruses for the recombinant production of proteins in suspension culture in fermenters. The method was first introduced in 1982 and in 1985 the first recombinant protein interleukin-2 was expressed in moth cells (Copyright@Canadian Biodiversity Information Facility, 2011).

Taticek et al. (2001) compared the growth and recombinant protein expression in suspension culture and attached cells culture of *Spodoptera frugiperda* and *Trichoplusia ni* (Taticek et al., 2001) expressing *Escherichia coli* beta-galactosidase or human secreted alkaline phosphatase. The production of both enzymes varied as a function of inoculum size, media, culture conditions etc. Cell densities of 5 x 10^6 viable cells/ml and doubling of 20 hours were reached, which corresponds to a growth rate of about $\mu = 0.04\ h^{-1}$. It is not clear whether a typical large scale fed batch cycle for insect cells would also be around 4 to 5 days which would about 2 to 3 times faster than a batch with mammalian cell culture. However, generally insect cells have two major disadvantages:

- the baculovirus system results in cell death and lysis of the host insect cells and the release of cell proteins, which offsets again the high productivity of the cells
- N-glycosylation of insect cells is different from mammalian cells.

Vermasvori et al. compared the production of a model protein (Negative factor or Nef) in *Escherichia coli*, *Pichia pastoris* and the *Drosophila* S2 cell line (Vermasvuori et al., 2009). When studying the systems purely economically the microbial system had significantly lower manufacturing costs than the insect cell lines. The most significant difference between the manufacturing costs of the two microbial systems was due to the much longer strain construction time with the *Pichia pastoris*. The manufacturing costs for the production of 100 mg Nef protein were

1.	Escherichia coli	6456 €
2.	*Pichia pastoris*	13382 €
3.	*Drosophila* S2	21111 €

Omitting the strain construction costs, the microbial systems were cost-wise fairly comparable. These numbers from small scale experiments insinuate that insect cells could be theoretically a threat for mammalian cell culture. However, there are no compelling data showing an advantage by switching to insect cells for large scale production.

Transgenic insects. The production of recombinant proteins in live animals is an option, as production of therapeutic proteins and monoclonal antibodies is an option in living plants. The most frequently proposed method is that of producing and extracting recombinant proteins from the milk of transgenic livestock. However, insects such as silkworms (Fraser & Jarvis, 2010), can theoretically also be used for recombinant protein production. However, we believe that transgenic insects will not play a role in the production of pharmaceuticals rather than in the control of insect populations and prevention and control of parasites and disease transmission in man, animals and plants.

lls

Mammalian cells in culture. The clinical and commercial success of mainly monoclonal antibodies has led to the rapid development of mammalian cell culture within a short time after the landmark findings of Cesar Milstein and Georges Köhler in the mid 70s. Mammalian cell culture is a well established production method for the production of therapeutic proteins and monoclonal antibodies. One of the most frequent mammalian hosts used is the CHO cell line (Chinese hamster ovary) in combination with the GS expression system or the DHFR (dihydroolate reductase) expression system (Birch & Racher, 2006). The mouse NS0 cells cell line is another option which is less frequently chosen today.

Continuously stirred tank fermenters are operated (Varley & Birch, 1999) with volumes of up to 20 m^3. An average batch lasts about two weeks with specific productivities up to 100 pg/cell x day. Product titers for antibodies of several grams per litre are almost routine today, making product recovery and purification the greater challenge from a manufacturing perspective. Defined media are available for high growth rates, stable expression, low lactate and byproduct formation to facilitate fermentation, isolation and purification. Lonza (Lonza, 2011) recently introduced a novel medium and feeding platform for its GS-expression system™, which leads to product titers of up to 10 g per litre!

Rapid and efficient development of stable production cell lines is a critical step and transfection alternatives for the quick preparation of stable mammalian production cell lines for recombinant proteins are needed (Wurm, 2004; Birch et al., 2005). The PiggyBac transposon is an example of a novel delivery vehicle for rapid and efficient recombinant cell line generation (Matasci et al., 2011).

Unfortunately, the generation of a stable production cell line remains a time consuming endeavour. However, for early feasibility tests with a protein, transient gene expression and protein production may be sufficient (Ye et al., 2009). EXCELLGENE SA (Excellgene, 2011) has successfully established transient gene expression for fast protein delivery. Production and recovery of several grams of purified protein can be done within 3-4 weeks. However, transient recombination is limited by two factors:

- the need for large amounts of plasmid DNA for transfection
- the requirement for cell culture medium exchange before transfection

Nevertheless, transient technology allows a reduction of cost and time for biopharmaceuticals with mammalian cells. It is not yet a valid method for regular production or even material for clinical material testing of e.g. drug candidates. Cell lines may not be stable but at least production can start within a matter of weeks. The combination of transient expression with disposable reactors, which are now available up to the m^3 scale (Eibl et al., 2010), makes the production of several 100 g of new proteins possible in a short time.

Transgenic animals: Clones and transgenics of livestock and companion animals have become a reality in animal farming, meat production and even pharmaceutical production since the first transgenic animals were created in the laboratory. Mice were the first mammals to be genetically modified in the early 1980s and in 1987 the breeding of a tPA (human tissue plasminogen activator) producing transgenic mouse was reported by Simons *et al.* (1987). As with any other organism mentioned above, it is also possible to introduce or delete existing genes in an animal cell and have the modification passed on to the next generation resulting in a new phenotype of the living animal.

In 2004 about a dozen companies had products in development in transgenic animals (Keefer, 2004) but many of them have been abandoned. Human antithrombin III (hAT) of GTC BIOTHERAPETICS, a glycoprotein controlling blood clotting, was the first approved biopharmaceutical from transgenic animals. NEXIA BIOPHARMACEUTICALS published a patent in 2002 for the production of spider silk biofilaments in transgenic animals using milk or urine-specific promoters (Karazas & Turcotte, 2003) but the process is not competitive with microbial production. The Dutch company GEN PHARMING was a pioneer in the field of production of human lactoferrin in ruminants. Alpha 1-antitrypsin, fibrinogen,tissue plasminogen activator, vaccines, human monoclonal and polyclonal antibodies are other example of products.

This technology has, however, been the subject of controversial discussions (for example "cloned meat") as these biotechnology applications have been judged with evident hierarchies of acceptability (Einsiedel, 2005). Transgenic animals are particularly sensitive but they nevertheless offer options to produce a therapeutic protein in a fluid of the animal (milk, blood, urine) of mice, goats or cows. Coupling the target protein to a signal one can direct the expression into the mammary glands and milk - the optimal choice.

Nonmammalian animals such as birds or insects (see above) can also be used for the production of glycosylated therapeutic proteins and monoclonal antibodies, especially in the egg white or egg yolk of transgenic chicken.

Patel et al., 2007 claim, that animals are more cost effective bioreactors, with 16 therapeutic proteins in development in transgenic animals (sheep 5, pig 2, goat 4, cow 5) plus several monoclonal antibodies in cows, goats and chicken. Productivities in terms of milk containing recombinant protein given are ~8000 litres containing a total 40 – 80 kg of recombinant protein. However we disagree, and think that transgenic animals will only be used in a very few and exceptional cases for the following reasons:

- low success rate of gene transfer (e.g. 0,1% with cows)
- cost and time to produce transgenic animals
- possible infectious agents from mammals
- low consumer acceptance in view of better alternatives
- many alternatives exist

Take productivity as one example, a single cow will produce ~60 kg of a therapeutic protein in one full year which has still to be isolated and purified. In the case of a goat or a sheep that number drops to 4 kg and 2.5 kg per year, respectively. This should be compared with the productivity of a $20m^3$ bioreactor with mammalian cells which produces on average 125 kg in a single month! That means one bioreactor replaces 25 cows, 375 goats or 600 sheep.

How do depreciations between 25 cows and a $20m^3$ plant differ? The cost of one transgenic animal is high, 500000 US$ for one calf according to Keefer (2004). This is due to inefficiencies in this technique as over 1000 bovine, 300 sheep and 200 goat oocytes must be injected. Goats and sheep with their shorter generation interval are less costly, and the figures may look somewhat better today. On the other hand and based on numbers given by Patel et al (2007), we calculated an average of about 50000 US$ in value created per animal and per year. It now all depends how fast a productive herd can be created by cloning and how long the productive period of a cloned transgenic animal is? Over how many productive years do we have to depreciate the animals? One important disadvantage of a capital investment in a transgenic cow is that a cow is a dedicated “plant” while a bioreactor is a multipurpose and multiproduct installation.

We do not believe that transgenic animals are an attractive alternative except for a few very special exceptions. However, the one argument for transgenic animals (as for transgenic plants and plant seeds) is the case in which the product can be consumed directly with the milk. It would make therapy affordable, and one can imagine that an oral protein-based vaccine would be an ideal candidate for a transgenic animal.

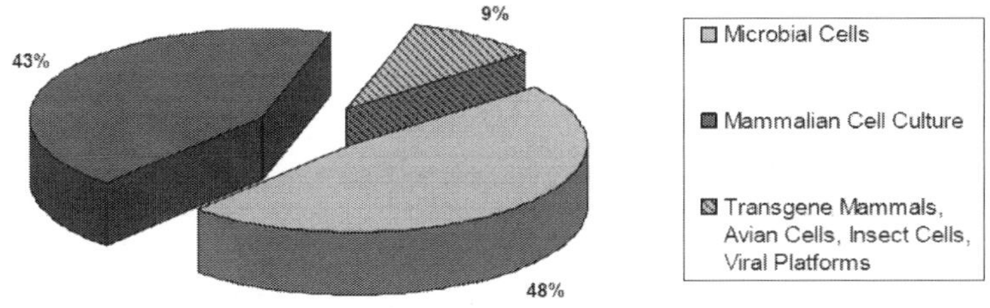

Fig. 7. Microbial and mammalian cell culture are used in 93% of all cases for the production of therapeutic proteins. See also Figure 8 with the spread of the individual expression systems.

In summary, microbial fermentation and mammalian cell culture will continue to carry the main burden for the production of recombinant proteins as it is already the case today (Figure 7). Other expression systems, especially plant-based and algae, will have potential for recombinant protein niche applications. The situation is different for small molecule pharmaceuticals, neutraceuticals and fine chemicals, where a more varied host-expression system combination will be needed. However, even in the latter case one will first fall back on proven methods. We will now describe in more details the beacon in recombinant microbial expression – *Escherichia coli.*

ESCHERICHIA COLI AS WORK HORSE

In the year 1885 the German paediatrician Theodor Escherich (1857-1911) described a bacterium, which he called "*Bacterium coli comunale*". At that time nobody could anticipate that this bacterium, which later on was named after him *Escherichia coli*, would become world famous as a model organism in the field of molecular biology and as "the" minifactory for recombinant protein manufacturing (Piechocki, 1989).

This is best demonstrated by statistical figures related to expression platforms in use (Figure 8). In the reported year 34% of all recombinant therapeutic proteins registered in the US and EU were produced by means of *Escherichia coli* based expression technology. The second and third most successful expression platforms were Chinese Hamster Ovary cells with a 30% and yeast systems, mostly *Sacharomyces cerevisiae*, with a 12% shares respectively (Rader, 2008).

Why is *Escherichia coli* such a popular expression host?

Although there is no gold standard platform in microbial expression, expression systems based on *Escherichia coli* have dominated microbial expression for more than 30 years. One can only speculate on the reasons for this domination. *Escherichia coli* and its phages were early objects and models for studying molecular biology topics, especially aspects related to the understanding of gene functions and regulation. More than 10 scientists received the Nobel prize for exciting discoveries connected to research on *Escherichia coli* (Piechocki, 1989). Worth mentioning is the isolation and purification of a restriction enzyme for the first time by Werner Arber in 1968. These enzymes are enabling tools in the area of rDNA technology. The rapid pace in the development of expression technology and of genetic engineering tools is best reflected by the quite early launch of a first biopharma product, expressed in *Escherichia coli*, recombinant human insulin in 1982 (Humulin®, licensed by GENENTECH to ELI LILLY). This is even more remarkable if one considers the lengthy approval procedure for therapeutics. *Escherichia coli* based biotechnology profited directly from the multitude of fundamental discoveries made on this model organism, giving this species a timely technical advantage in use as expression host.

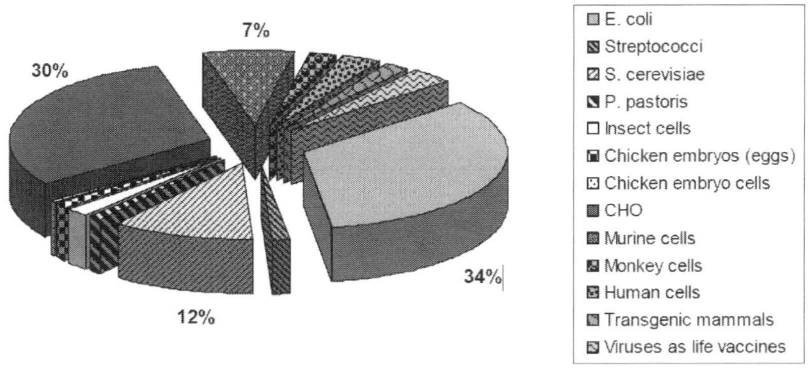

Fig. 8. Percentage of expression platforms used for the manufacture of biotherapeutics in the US and the EU. The figure is based on numbers published by Rader (2008).

Other explanations for the success of this microorganism are low genome complexity and the extra-chromosomal genetic elements, plasmids, which ease both (a) in-vitro manipulation of genetic elements and (b) insertion of homologous and foreign genes into the organism.

Besides low safety concerns and high regulatory acceptance, ease of use and familiarity with the organism was in favour of *Escherichia coli*. There is hardly a student in biology who has not run at least one cloning experiment in one of the *Escherichia coli* expression systems used in academia. Since its first industrial applications, *Escherichia coli* expression technology has been continuously improved with the aims of gaining control of the quality of the recombinant products and increasing the product titre in fermentation, the latter obviously being crucial to process economy.

Characteristics of an industrial *Escherichia coli* expression platform

Incremental improvements led to the development of *Escherichia coli* based expression platforms that are suitable for industrial use. More precisely these systems allow for robust, reliable and scalable processes and economical manufacturing. High performance expression technology is characterized by two properties: (a) high volumetric productivity Qp, preferentially due to a high specific product production rate qp and (b) high control on product quality, meaning that no or only a negligible amount of product variants are produced.

Industrial expression systems distinguish themselves from academic systems by an optimized combination of the various components of which an expression system is made. Basically, a bacterial expression system is composed of a host and a vector which contains the product coding DNA, a selection marker and various regulatory elements. Regulatory elements are promoters, signal sequences, ribosome binding sites, transcription terminators and vector replication or integration regions.

Host. The host organism provides specific features to an expression system as a result of its genetic background; these features include:

- growth characteristics such as specific growth rate ?
- maximum achievable cell densities
- nutritional needs
- robustness at cellular and genetic level
- control of product degradation
- secretion capacity preferentially into the medium
- amount of endotoxins produced
- post-translational modifications

High cell densities are most desirable since a positive correlation exists between the amount of biomass (X) and the product production rate (r_P). The corresponding equation is $r_P = q_P * X$ (product production rate = specific product production rate x biomass).

The relationship above should not be confused with growth rate dependency on the product production rate, which can be optimal at high or low growth rates. It is possible that in the worst case maximal specific production rates r_P correlate with very low growth rate close to maintenance (Meyer & Fiechter,1985). In that case production requires two separate phases, growth and production phase.

Commonly used *Escherichia coli* host strains are listed in Table 7. BL21 is the most frequently used *Escherichia coli host*. BL21 popularity is based on

- *lon* and *ompT* protease deficiencies
- beneficial growth and metabolic characteristics
- Insensitivity to high glucose concentration.

The organism is not sensitive to high glucose concentration due to its active glyoxylate shunt, gluconeogenesis and anaplerotic pathways and a more active TCA cycle, which leads to better glucose utilisation and lower acetate production (Phue et al., 2008). However, when used in combination with the T7 expression system and when exposed to stress, this host is at risk of bacteriophage DE3 excision. For this reason laboratories started to promote the use of BLR, a $recA^-$ mutant of BL21. In our experience an increased use of W3110 is taking place in the industry. This can be attributed to the excellent production capabilities of this host. Orgami strains may allow for better formation of disulfide bonds in the cytoplasm due to lower reducing power in the cytoplasm (Novagen, 2011). The endA^{-} and recA^{-} hosts DH5? and JM109 are the organisms of choice for the manufacture of pDNA. The lack of endonuclease 1 which degrades double stranded DNA positively affects stability of pDNA (Phue et al., 2008). In conclusion, product nature and product characteristics determine the selection of the most optimal host.

Table 7. Frequently used *Escherichia coli* host strains and related specific characteristics.

Escherichia coli Host Strains	Strain Characteristics
BL21 HMS174 BLR	B strain, *lon* and *ompT* protease deficiencies K-12 strain, recA^{-}
Orgami strains Rosetta strains	B strain, recA^{-} mutant of BL21 with decreased likelihood of excision of DE3 K-12 or B strains with mutations in *trxB* and *gor*
W3110 MG1655 RV308	K-12 or B strains which supply tRNAs for codons that are rare in E. coli K-12 strain
DH5α	K-12 strain K-12 strain
	K-12 strain, recA^{-}, endA^{-}, often used for pDNA manufacture

Promoters. Promoters control the expression to the extent of how much and at which point in time mRNa is synthesized. As a consequence they control production of product. A large number of promoters that allow modulation of the mode of induction in a desired way are used in the industry. Lactose or lactose-analogue IPTG induced T7 promoter- based expression systems currently dominate the market. Apart from T5, araB and phoA, other classical promoters such as lambda, lac, trp, P_L, P_R, tetA and trc/tac are rather seldom used.

Novel promoters are under development and continuously make their way into industrial applications. New disaccharide inducible promoters, which induce protein production during the stationary growth phase, have recently been successfully applied in *Escherichia coli* based biopharmaceutical processes. Some of these are part of Lonza's XS Technologies^{TM} *Escherichia coli* platform, which has been chosen as an example to discuss performance of current leading industrial *Escherichia coli* expression platforms (Lonza). Depending on the promoter the induction signal is of a chemical or physical nature. Some of the above mentioned *Escherichia coli* promoters have been successfully used in other bacterial systems such as *Bacillus subtilis* (Alexander et al., 2007).

State of the art industrial expression platforms allow for product specific modulation of the rate of protein synthesis. Proteins of high complexity, having disulphide bonds are typically best produced at a lower production rate. In contrast proteins of low complexity are often produced at a high production rate, thus achieving high concentrations after a short time of fermentation. Productivity is often affected by interaction between specific promoters and recombinant target proteins. Therefore, in general, it makes sense to screen for the performance of different promoters.

Signal Sequences. Signal sequences determine whether a product is directed through the cellular membrane and out of the cytoplasma; the signal sequence is cleaved during the secretion step. Secretion is desirable in many cases, since a large proportion of target proteins do not fold correctly in the reducing cytoplasmic environment. Folding requires oxidative conditions which are provided outside the cytoplasm. Secretion sequencesfrequently used in *Escherichia coli* are MalE, OmpA and PelB. Yeast organisms such as *Saccharomyces*, *Pichia*, *Hansenula*, *Yarrowia* and Gram-positive bacteria such as *Bacillus* and *Corynebacterium* secrete proteins which carry a secretion signal into the medium, whereas Gram-negative genera such as *Escherichia*, *Pseudomonas* and *Ralstonia* direct the product through the inner membrane into the periplasmic space. This is what the theory says. According to the authors' experience, the *Escherichia coli* outer membrane is leaky for a large proportion of secreted proteins which are supposed to accumulate in the periplasmic space. The observed partitioning of the secreted protein between fermentation medium and periplasmic space can be influenced to some extent by modifying the fermentation conditions. The latter behaviour is product dependent and for the time being not predictable.

Selection markers. Selection markers are necessary for the cloning process and crucial for controlling plasmid stability. Typical microbial selection markers are antibiotic resistance genes. However, the prevalence of β-lactam allergies strongly suggests avoidance of the use of ampicillin and other β-lactam derivatives for the purpose of selective pressure in the manufacture of clinical products. Optional stabilization systems used in *Escherichia coli* are based on antidote and poison gene systems with the poison gene being integrated into the bacterial chromosome and the antidote gene located on the plasmid, respectively (Peubez et al., 2010). Constitutive expression of the antidote gene stabilizes plasmid-containing cells. A system based on the mode of action described above is marketed by DELPHI GENETICS Inc (Delphigenetics, 2011).

Besides the above mentioned regulatory aspect, Rozkov et al. (2004) note another one that should be taken into consideration when selecting the plasmid stabilizing system. According to these authors, the presence of an antibiotic selection marker imposes a huge metabolic burden on an expression system. They found that the product of the selection marker gene accounted for up to 18% of the cell protein. A negative effect on the recombinant expression of the genes of interest is highly likely. Due to constitutive expression this is the case even in the absence of the corresponding antibiotic in the medium. One way to circumvent this problem is to use complementation markers, i.e. marker genes that complement an auxotrophic chromosomal mutation.

A majority of successful technologies, genetic elements and related know-how, are subject to patent protection or trade secrets, as shown also in Table 6. In particular, multiple license requirements for the use of a specific production technology can lead to an unfavourable economic situation. On the other hand, off-patent expression systems and elements thereof are usually not state of the art. Since process economy depends to a large extent on productive and robust strains, outsourcing strain development to a specialised laboratory is often justified, given that licensing cost remain reasonable. The resulting economic benefits on the process side typically offset the costs related to accessing a productive and robust state of the art industrial strain platform.

A more critical view on *Escherichia coli* expression platforms

Despite their dominant position within microbial expression *Escherichia coli* based expression platforms also exhibit weaknesses which should not be ignored. These drawbacks are shared with other commercialised Gram-negative expression platforms as *Pseudomonas* and *Ralstonia*. Among these disadvantages are

- sswq of expressing complex, multimeric proteins with a high number of disulfide bondsare often referred to as disadvantages. These apparent drawbacks can, however, be turned to advantages depending on the target protein's specifics.

Table 8 compares the suitability of the 3 leading expression platforms related to characteristics of the expression candidate protein. Apart from the two characteristics (a) requirement for human-like glycosylation, which includes monoclonal antibodies whose efficacy depends on Fc effector functions and (b) peptide nature of the recombinant target, most of the aspects captured in the table, do not give a clear indication regarding choice of the ideal expression platform. There is a large grey zone which typically needs to be explored empirically.

Active enzymes up to a size of 220 kDa and 250 kDa recombinant spider silk protein have been successfully expressed in *Escherichia coli* at high concentrations, questioning the dogma that bacterial systems are not suitable for the expression of large proteins. This thesis is further supported by successful expression of complex heterodimers, such as aglycosylated functional antibodies, in bacterial systems. For an in-depth analyis of expression of complex heterodimers in *Escherichia coli* we recommend the paper of Jeong et al. (2011). We also question the criticism towards inclusion body formation that often is cited as a disadvantage. Rather than a drawback we consider this as a capability that adds flexibility to the use of *Escherichia coli* based platforms. Industrial expression platforms allow for inclusion body concentrations as high as 10 g/l culture broth and above. This consideration combined with an efficient refolding process provides high potential for a competitive process from a cost point of view.

Some therapeutic protein candidates are not glycosylated, such as a non-glycosylated version of an antibody. In particular, recombinant proteins produced by yeast expression systems may carry undesired O-glycans. In these cases a lack of glycosylation capability can be considered as advantage rather than a system weakness. Intracellular expression in Escherichia coli may lead to product variants (a) with N-terminal formyl-methionine and (b) without formyl-methionine at the N-terminus. Methionine cleavage by the methionyl- aminopeptidase depends on the characteristics of the adjacent amino acid, which consequently determines the ratio of the 2 product fractions.

Table 8. Criteria that drive selection of an expression platform. Legend: –, not suitable; • low, •• medium, ••• high suitability; 1) mostly cited as limiting criterion, nevertheless, up to 220 kDa proteins have been expressed in Lonza's *E. coli* XS Technologies™ platform with a very high titre, 2) Unigene and Lonza developed *E. coli* based peptide platforms, 3) secretion required for most recombinant proteins, 4) proteins exhibiting low solubility or a high aggregation propensity are often expressed at high titres as inclusion bodies, 5) yeast type glycosylation, mainly mannose comprising oligosaccharides, is highly immunogenic, 6) N-terminal product variants are frequently observed with *Pichia pastoris* and *Hansenula polymorpha* as a result of incomplete N-terminal processing.

Protein Characteristics	Bacterial (Gram-) Systems		Yeast Systems		Mammalian Systems
size: small to mid size	•••		•••		••
size: large proteins	•	1)	••		•••
peptides	•••	2)	α		–
monomers	•••		•••		•••
homo-multimers	•••		•••		•••
hetero- multimers	•		••		•••
disulphide bonds (folding)	••	3)	•••		•••
hydrophilic proteins (soluble)	•••		••		••
hydrophobic proteins (low solubility)	••	4)	•		•
human (like) glycosylated	–		•	5)	•••
not-glycosylated	•••		–		••
Protein prone to proteolytic digest	•••		•		•••
(N-terminal product variants)				6)	

Earlier on endotoxin formation and low control of secretion into the medium were mentioned as problematic aspects for expression systems which are based on Gram-negative bacteria such as *Escherichia*, *Pseudomonas* and *Ralstonia*. On the other hand Table 8 alsoindicates some weaknesses of yeast platforms. On the one hand yeast N- and O- glycosylation capability can negatively impact product quality so that adverse immunogenic reactions in the clinic are the result. Another problem often observed with *Pichia* and *Hansenula* are product variants produced through incomplete N-terminal processing and proteolytic degradation (Meyer et al., 2008). This, together with an on average lower observed productivity, negatively affects broad usage of yeast systems, despite their advantageous secretion capability.

Table 9 compares bacterial Gram-negative and yeast platforms to selected bacterial Gram- positiveexpressionplatforms,i.e.to ***Bacillus*** and ***Corynebacterium*** platforms (White, 2011). The comparison suggests that the disadvantages of the existing bacterial Gram-negative platforms and yeast platforms can be overcome by moving into bacterial Gram-positive platforms. Gram-positive bacteria, in contrast to Gram-negative bacteria do not produce endotoxins and they naturally secrete proteins. Comparing them to yeast, they do not Glycosylate proteins and there are no N-terminal processing problems. Both ***Bacillus*** and ***Corynebacterium*** hosts need to be engineered to resolve the problematic aspects of the corresponding wildtype strains such as low plasmid stability and secretion of undesired proteases.

Table 9. Suitability of yeast, Gram-negative and Gram-positive expression platforms related to classical microbial platform weaknesses.

Problematic Characteristics	Yeast Platforms *Pichia Hansenula*	Gram+ Platforms *Bacillus Corynebacterium*	Gram- Platforms *Escherichia Pseudomonas*
Endotoxins	suitable	suitable	not suitable
Control of secretion	suitable	suitable	not suitable
N-terminal product variants	not suitable	suitable	suitable
Undesired glycosylation	not suitable	suitable	suitable

Production performance of relevant industrial *Escherichia coli* expression platforms

In contrast to the expression of antibodies in CHO cells, expression success cannot be predicted in microbial expression systems. What is good for a specific recombinant protein A does not necessarily work for protein B, even if B is a protein variant of A. An integral part of the various strain platforms are generic high cell density fermentations. When considering industrialisation, strains and fermentation procedures should be looked at as single entities rather than separate process aspects. This is the main reason for the difficulty in judging the performance of expression platforms in general. Data from one single product are not sufficient, since the performance of one expression platform can differ greatly from product to product for as yet unknown reasons. One platform typically shows exceptional productivity only for a small number of products and rather low productivity for the majority of desired expression targets.

Expression titres of commercial products are typically handled as trade secrets. The authors have access to an informative set of expression titre data of leading *Escherichia coli* expression systems which are part of Lonza's XS Technologies^{TM} platform (Figure 9). This platform is a broad one which in itself encompasses various *Escherichia coli, Pichia pastoris* and *Bacillus subtilis* platforms. In our experience, heterogeneity of the recombinant protein pipeline demands access to a variety of powerful expression tools in order to cope with specific expression challenges. On a few occasions the platform performance could be directly benchmarked against competitive CMO and other commercialised platforms based on *Escherichia coli* and *Pseudomonas*. On these occasions XS Technologies^{TM} showed superior or equal performance. Therefore we consider the performance data shown in Figure 10 as representative for leading bacterial Gram-negative expression platforms.

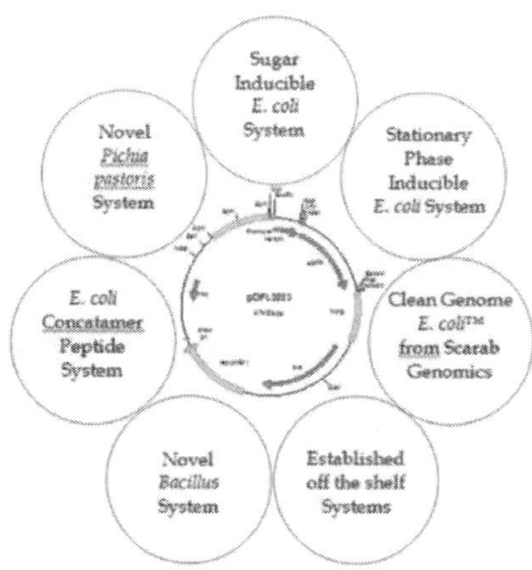

Fig. 9. Example of an industrial expression platform, XS Technologies^{TM} (Lonza). The platform comprises a number of powerful expression technologies for expressing recombinant proteins in *Escherichia, Pichia* and *Bacillus* in order to cope with the expression challenges related to the heterogeneity of the recombinant proteins pipeline, including recombinant peptides and pDNA.

With Gram-negative organisms such as *Escherichia coli* and *Pseudomonas,* the recombinant product can be localized in different spaces, either intracellular (cytoplasmic) or extracellular. We define the latter as proteins expressed with a secretion sequence, and thus directed through the inner membrane, which means that the recombinant protein can be localized either in the periplasm or in the cell free medium. As a second aspect to consider, product is formed in either a soluble form or as insoluble aggregates. Apart from intentional inclusion body formation, production in a soluble, functional form is preferred. Therefore 4 effective expression modes are to be distinguished. Recombinant protein can be localised (C1) in the cytoplasm, insoluble as inclusion bodies, (C2) in the cytoplasm in a soluble form, (C3) in the cell-free medium in a soluble form and (C4) in the periplasm in a soluble form. Periplasmic insoluble material is typically not accessed and therefore ignored in the productivity figures.

Figure 10 shows expression levels of 24 recombinant proteins, mostly biopharmaceuticals that are expressed in ***Escherichia coli*** platforms. Induction is platform-dependent either by the addition of a corresponding sugar or by entering the stationary phase.

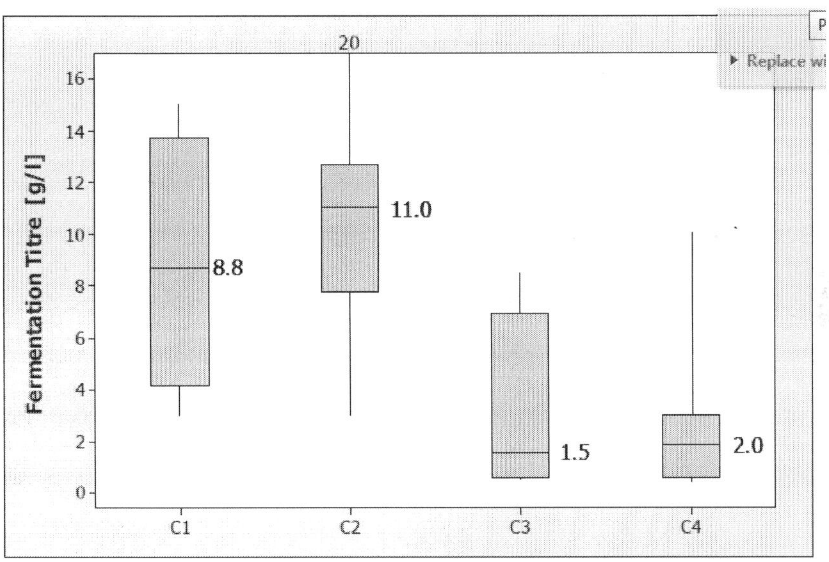

Fig. 10. Expression titres obtained for 24 different recombinant proteins mostly biopharmaceuticals. The proteins were expressed in either one of the sugar inducible or one of the stationary phase inducible *Escherichia coli* systems belonging to Lonza XS Technologies™ platform. Among the 24 recombinant products were fragment antibodies, Fab-fusions, single- chain antibodies, virus-like particles, novel non-antibody type binders, growth factors, recombinant enzymes, amphipathic proteins, recombinant vaccines, peptides (hormones and others), affinity ligands and monomers of biopolymers; size of the proteins varied between 2 and 220 kDa. Legend: C1, insoluble as inclusion bodies in cytoplasm; C2, soluble in cytoplasm; C3, soluble in cell-free medium; C4, soluble in periplasm.

Cytoplasmic expression (categories C1 and C2). Among the products expressed in the cytoplasm, either soluble or insoluble as inclusion bodies, were highly soluble recombinant proteins as well as proteins prone to high aggregation propensity belonging to product classes such as recombinant vaccines, novel non-antibody based binders, recombinant therapeutic and non-therapeutic enzymes, virus like particles (VLPs), peptides (hormones and others), monomers of biopolymers, affinity ligands and others. The proteins were mostly monomeric with the size ranging from 2 to 40 kDa. Highest expression titres are obtained in the case of cytoplasmic soluble expression (C2 in Figure10) with a median titre of 11 g/l culture broth and a range of 3 – 20 g/l. Intentional intracellular expression of recombinant protein in an insoluble state as inclusion bodies (C1 in Figure 10) resulted in a median titre of about 9 g/l, with a range of 3 - 15 g/l dependent on the target protein.

Extracellular expression, periplasmic and into cell free medium (categories C3 and C4). Products that were expressed with a signal sequence were fragment antibodies (Fab), Fab fusion proteins, single chain antibodies (scFv), growth factors, enzymes and various formats of amphipathic proteins. The size of the corresponding products varied between 20 and 220kDa. Among them were both soluble and fairly soluble monomers and multimers, homo- and heteromers. Extracellular product (C3 in Figure 10) reached concentrations in the range of 0.5 to 8.5 g/l in the cell free medium with a median of 1.5 g/l. Proteins which accumulated in the periplasm (C4 in Figure 10) reached titres of functional product between0.5 and 10 g/l with a median titre of 2.0 g/l. Dependent on the product-specific aggregation propensity sometimes significant amounts of precipitated recombinant protein were observed in the periplasm. This fraction has been ignored, since it does not contribute to functional product. The extent of product precipitation can be influenced by the choice of the promoter system, the related induction mode and fermentation conditions. Similarly, the distribution of product between the periplasm and the cell-free medium can be partly controlled by changes in physical and chemical environmental conditions. However, ideal conditions need to be identified empirically. The above mentioned product titres have been typically obtained within 36 to 72 hours of fermentation.

Posttranslational modification in *Escherichia coli*

Proteins often require posttranslational modification in order to gain full biological activity. Therefore, missing posttranslational protein modification capabilities such as glycosylation, formation of pyroglutamic acid at the N-terminus, N-terminal acylation and C-terminal amidation are frequently cited as a disadvantage of bacterial expression.

However, over the last decade big advances have been made in understanding glycosylation mechanisms and in glycoengineering of microbial organisms. Gerngross and coworkers (Choi et al., 2003) and Contreras and coworkers (Vervecken et al., 2004) were among the first to succeed in glycoengineering yeast more precisely, *Pichia pastoris*, towards the formation of defined glycoforms. The yeast related work culminated in successful expression of human- like glycosylated antibody in a *Pichia pastoris* host, that enables specific human N- glycosylation with high fidelity (Potgieter et al., 2009).

In parallel it became evident that protein glycosylation is also abundant in prokaryotes. Whereas N-linked protein glycosylation is the most abundant posttranslational modification in eukaryotes, within prokaryotes it seems to be restricted to the domain of the Archea where S-layer proteins show N-linked glycosylation. Already in 2002 Aebi and coworkers (Wacker et al., 2002) demonstrated successful transfer of the *Campylobacter jejuni* protein N-glycosylation machinery into *Escherichia coli*. This opened up an exciting opportunity to produce N-glycoproteins within bacterial expression platforms. Nevertheless, two features were inhibitory to a broad application of the new system. (a) The *Campylobacter jejuni* glycan is immunogenic for humans. (b) The glycan is linked to asparagine through an unusual deoxysugar, bacillosamin. Recently the system has been further developed towards formation of the required N-acetylglucosamin-asparagine linkage that is commonly found in glycoproteins of eukaryotic origine (Schwarz et al., 2010). The same paper proposes a semi-synthetic approach towards human glycosylation based on the new developed technology.

The goal of any microbial glycoengineered system must be to overcome the weaknesses of the existing mammalian platforms that are, (a) mammalian glycosylation is characterized bynaturally occurring heterogeneity in the glycan structure and (b) by limited possibilities to tailor glycosylation towards improved therapeutic performance. Consequently microbial glycoengineered expression platforms should allow for tailored, homogenous and human- like glycosylation. However the challenges on the way to the development of a well performing microbial glycoengineered platform are manifold. The following lists the technical obstacles that need to be addressed:

- glyoform homogeneity, ideally one glycoform should be formed
- tailoring, access to a number of specific glycoforms through defined host backgrounds
- productivity, volumetric productivity should not be below the productivity of existing mammalian systems

- O-glycosylation, existing yeast O-glycosylation causes immunogenic reactions with humans
- glycosylation efficiency, the whole of the target protein is expected to be glycosylated
- secretion efficiency, needs to be high, since glycosylation is connected to secretion
- expression of complex proteins such as antibodies, capability to produce hetero- multimers (disulphide bridges)
- plug and play, access to stable glycoengineered hosts, such that only the target gene needs to be inserted
- proteases, deletion of all undesirable proteolytic activity
- good growth characteristics, system viability is affected by the amount of genetic changes
- N-terminal variability, often seen in yeast systems, needs to be under control

As mentioned before, existing mammalian expression technology is not fulfilling all of the desirable requirements and there is an even longer way to go for the existing yeast systems in order to compete with mammalian systems. Not all of the above mentioned technical challenges have been successfully addressed in yeast. Even further away from technical maturity are bacterial glycoengineered systems. Nevertheless technical advances are achieved at high pace. The authors would not be surprised if bacterial expression technology would one day be a viable solution for large scale manufacturing of glycoproteins.

Cost considerations

From a commercial point of view, bacterial and yeast systems share many advantages over mammalian systems such as high growth rate, the potential to reach high biomass concentration, structural and segregational robustness and a higher product production rate rP, resulting in significantly shorter fermentation times. While mammalian cells such as CHO cells are characterized by a high specific product production rate qP, volumetric productivity QP is typically negatively affected by a relatively low growth rate and more importantly by the lower achievable biomass concentration as compared to *Pichia pastoris* (Kunert et al., 2008). The same is true, when comparing CHO cells to *Escherichia coli*. Other aspects such as time required for the development of a stable CHO cell line and media costs should be considered as well. All these aspects add to the attractiveness of microbial and yeast systems when the manufacture of aglycosylated non-antibody type of recombinant proteins is considered. Table 10 shows cost drivers in fermentation of the current key biopharmaceuticals production platforms.

Table 10. Comparison of bacterial, yeast and mammalian system characteristics which drive cost of goods in fermentation; 1) the figures have been modelled based on typical production key figures and assuming an equal product titre of 5 g/l at the end of the fermentation; USP, upstream processing.

Characteristics driving USP cost	Bacteria	Yeast	Mammalian Cells
Growth rate ? [1/h]	0.7	0.2	0.02
Final dry biomass concentration [g/l]	60-70	80-100	3-8
Typical duration of fermentation [days]	2-3	4-5	15-20
Specific product production rate qP [g/gh] 1)	0.002	0.001	0.005
Volumetric productivity QP [g/lh] 1)	0.10	0.05	0.01
Medium cost	low	low	high
Strain development cost and duration	low	medium	high
Equipment standard	steel	steel	steel, disposable

Methylotrophic yeast fermentation can be very demanding on equipment performance as a result of the high oxygen demand, high cooling requirements and explosion-proof design because of methanol feeding. Corresponding bioreactor layout requirements are described by Hoeks et al. (2005).

Figure 7 shows that about 9% of all recombinant DNA products are supposedly manufactured with transgenic animals, avian cells, insect cells and viral platforms. On top of these, there are early projects of recombinant expression in plants, filamentous fungi, plants and protozoa. The decision to opt for one of these systems is mostly driven by specific product aspects, cost or IP reasons in order to gain freedom to operate. A cost advantage through higher productivity or lower depreciation compared to more conventional systems is not obvious. Cost allocated to fermentation is typically in the range of 30% to 50% of overall manufacturing costs. Irrespective of the recombinant biosynthesis method used, the DSP costs remain. Therefore the sometimes cited 10X overall cost improvement through the use of one specific expression system and the related USP production platform is difficult to understand if not unrealistic.

The cost of downstream processing (DSP) is more or less independent of the chosen system, if we assume product localization in the cell-free medium. When using Gram-negative expression technology special attention needs to be paid to endotoxin removal. On the other hand a mammalian system makes viral clearance mandatory.

Intracellular production obviously requires cell disruption or product release from the cells followed by a usually more complex biomass removal step. The latter is more or less standardized for conventional expression technologies. Other operations such as inclusion body isolation and purification followed by protein refolding typically drive DSP costs up. Theses higher costs for DSP can only be justified through higher upstream productivity as shown in Figure 10 or a lack of production alternatives. It is also obvious that no significant cost advantage is to be expected on the DSP side, if product needs to be extracted andpurified out of whole plants. However, in the latter case a significant cost advantage arises if for example, a therapeutic or a vaccine is administered through oral consumption of the whole plant or a non-purified low-cost plant extract.

Please note that other costs for so called secondary manufacturing (e.g fill and finish, formulation) accrue for the finished product, which we can not discuss here.

SUMMARY

The industry has become very conservative, risk averse and reluctant to change established and successful manufacturing platforms because of a very strict interpretation of regulatory guidelines. This is also the main reason why the authors think that the main load of biotechnological manufacturing production has remained with the already industrially established microbial (*E. coli*, yeast) and mammalian production systems and will continue to do so. Nevertheless, regulatory government bodies do welcome novel manufacturing methods for the production of affordable pharmaceuticals because of ever increasing health care costs. Indeed, it cannot be denied that cost pressure and novel applications will help to disturb the established situation. We consider two alternative expression systems to have some potential.

- Transgenic plants have the possibility to combine therapeutic with nutrition needs. The production of edible vaccines for human or veterinary applications for example appear to be an attractive option especially as the active crop can be phototrophically and cheaply grown locally.
- Due to their short doubling times and easier cultivation, protozoa offer themselves as a possibility between microbial and mammalian cell culture. Insect cell culture seem to be not as attractive as protozoa as they do not grow as fast and the frequently used BEVS results in more complex isolation and purification procedures.

These two options, however, will again be hampered by another expected or even partly realised breakthrough: the successful targeted humanised glycosylation in yeast and later in bacteria. On top of that, we will sooner or later experience the realisation of extensive pathway engineering and synthetic biology principles, where production organisms will be designed using engineering principles as in the automotive or aerospace industry. It is even harder to imagine how and where alternatives such as plants or protozoa can beat such advanced microbial or mammalian options.

REFERENCES

1. Ahmad, A., Pereira, E.O., Conley, A.J., Richman, A.S., Menassa, R. (2010). Green Biofactories: Recombinant Protein Production in Plants. *Recent Patents on Biotechnology,* 4, pp. 242-259.
2. Aldag, I., Bockau, U., Rossdorf, J., Laarmann, S., Raaben, W., Herrmann, L., Weide, T., Hartmann, M. W. W. (2011). Expression, secretion and surface display of a human alkaline phospatase by the ciliate *Tetrahymena thermopile. BMC Biotechnolology,* 11, pp. 1-11. Download: http://www.biomedcentral.com/1472-6750/11/11
3. Alexander, P., Rudolph, D.B., Underwood, S.A., Desai, S.G., Liu, X.M. (2007). Optimizing Microbial Fermentation and Mammalian Cell Culture: An Overview. Biopharm International, Supplement, (May 2007), pp. 16-24
4. BCC Research LLC (2009). Botanical and Plant-Derived Drugs: Global Markets, Press Release. Download: http://www.bccresearch.com/pressroom/report/code/BIO022E
5. Berger, R.G. (2009). Biotechnology of flavours-the next generation. *Biotechnoogy Letters,* 31(11), pp. 1651-1659
6. Birch, J.R., Racher, A.J. (2006). Antibody production. *Advanced Drug Delivery Reviews,* 58(5- 6), pp. 671-685.
7. Birch, J.R., Mainwaring, D.O., Racher, A.J. (2005). Use of Glutamine Synthetase (GS) Expression System fort the Rapid Development of Highly Productive Mammalian Cell Processes, In: *Modern Biopharmaceuticals*, Knäblein, J., Müller, R.H., pp. 1-24, Wiley-VCH Verlag GmbH & Co. KGaA, ISBN 3-527-311834-X, Weinheim
8. Canadian Biodiversity Information Facility, http://www.cbif.gc.ca
9. Choi, B.-K., Bobrowicz, P., Davidson, R., Hamilton, S.R., Kung, D.H., Li, H., Miele, R.G., Nett, J.H., Wildt, S., Gerngross, T.U. (2003). Use of combinatorial genetic libraries to humanize N-linked glycosylation in the yeast *Pichia pastoris. PNAS,* 100, 9, (April 29, 2003), pp. 5022-5027
10. Cilian AG, http://www.cilian.com
11. Clive, J. (2010). Global Status of Commercialized Biotech/GM Crops: 2010. ISAAA Brief No. 42. ISAAA ISBN: 978-1-892456-49-4, Ithaca, NY
12. Delphi Genetics, http://www.delphigeneitcs.com

13. Eibl, R., Kaiser, S., Lombriser, R., Eibl, D. (2010). Disposable bioreactors : the current state- of-the-art and recommended applications in biotechnology, *Applied Microbiology Biotechnology,* 86, pp. 41-49.
14. Einsiedel, E. F. (2005). Public perception of transgenic animals, *Rev. sci. tech. Off. int. Epiz.,* 24(1), pp. 149-157.
15. Excellgene, http://excellgene.com
16. Fernández-Sevilla, J. M., Acién Fernández, F.G.A., Molina Grima, E. (2010). Biotechnological production of lutein and its applications, *Applied Microbiology Biotechnology,* 86, pp. 27-40.
17. Fleissner, A., Dersch, P. (2010). Expression and export: recombinant protein production systems for Aspergillus. *Applied Microbiology and Biotechnology,* 87, pp. 1255-1270.
18. Franklin, S.E., Mayfield, S.P. (2005). Recent developments in the production of human therapeutic proteins in eukaryotic algae, *Expert Opinion on Biological Therapy,* 5(2), pp. 1-11.
19. Franssen C.R., Kircher, M., Wohlgemuth, R. (2010). Industrial Biotechnology in the Chemical and Pharmaceutical Industries, In: *Industrial Biotechnology. Sustainable Growth and Economic Success.* Vandamme, E. pp. 323-351, ISBN 978-3-527-31442-3, Weinheim.
20. Fraser, M.J, Jarvis, D., 2010. Production of human gylcosylated proteins in transgenic insects, *United States Patent Application Publication,* US 20100186099 A1 20100722.
21. Gatyas, G. (2011). IMS Health Forecasts Global Pharmaceutical Market Growth of 5-7 Percent in 2011, Reaching $ 880 Billion, IMS Health Incorporated Press Release. Download: http:// www.imshealth.com/portal/site/imshealth/menuitem.a46c6d4df3db4b3 8f611019418c22a /?vgnextoid=119717f27128b210VgnVCM100000ed152ca2RCR D&vgnextchannel=b5e57900b55 a5110VgnVCM10000071812ca2RCRD&vgnextfmt=default
22. Ghisalba, O., Meyer, H.-P., Wohlgemuth, R. (2010). Industrial Biotransfromation, In: *Encyclopedia of Industrial Biotechnology,* M.C. Flickinger, Editor, pp.1-18, John Wiley & Sons Inc., ISBN 978-0471799306, Hoboken, NJ, United States.
23. Han, S.I., Firehs, K., Flaschel, E. (2004). Improvement of a synthetic medium for *Dictyostelium discoideum*. *Process Biochemistry,* 39, pp. 925-930.
24. Hartmann, M., Sachse, C., Apelt, J., Bockau, U. (2010). Viral protein recombinant expression in ciliates and vaccine uses, Brit. *UK Pat. Appl. GB 2471093 A 20101222.*
25. Hoeks, F.W.J.M.M., Kulla, H., Meyer, H.-P. (1992). Continuous cell-recycle process for L- carnitine production: performance, engineering and downstream processing aspects compared with discontinuous processes, *J. Biotechnol.* 22, pp. 117-128.

26. Hoeks, F.W.J.M.M., Ven Wees-Tangerman C., Gasser, K., Mommers, H.M., Schmid, S., Lyuben, Ch.A.M. (1997). Stirring as Foam Disruption (SAFD) Technique in Fermentation Processes, *Can. J. Chem. Eng.* 75, pp. 1018-1029.

27. Hoeks, F.W.J.M.M., Schmidhalter, D.R., Gloeckler R., Herwig C., Theriault, K., Frie, S., van den Broek, W., Laukel, F. (2005). PBMSS – Lonza's Biopharmaceutical Small Scale Plant Started cGMP Manufacturing in September 2004, *Chimia*, 59 (2005), 1(2), pp. 31-33, ISSN 0009-4293

28. Invitrogen (2002). Guide to Baculovirus Expression Vector Systems (BEVS) and Insect Cell Culture Techniques. *Invitrogen life technologies instruction manual*, http://www.invitrogen.com

29. Jeong, K.J., Jang, S.H., Velmurugan, N. (2011). Recombinant antibodies: Engineering and production in yeast and bacterial hosts. *Biotechnology Journal*, 6, pp. 16-27

30. Jostock, T., 2007. Emerging Alternative Production Systems. In: *Handbook of Therapeutic Antibodies*, Dübel, S. (Ed), pp. 445- 466. WILEY-VCH Verlag GmbH & Co KGaA, ISBN 978-3-527-31453-9, Weinheim.

31. Karatzas, C.N., Turcotte, C. (2003). Methods of producing silk polypeptides and products thereof. *PCT International Application*, WO 2003057727 A1.

32. Keefer, C.L. (2004). Production of bioproducts through the use of transgenic animal models. *Animal Reproduction* Science, 82-82, pp. 5-12.

33. Kunert, R., Gach, J., Katinger, H. (2008). Expression of a Fab Fragment in CHO and *Pichia pastoris*, a Comparative Case Study. *Bioprocess International*, Supplement, (June 2008), pp. 34-40

34. Langer, E. (2011). New Plant Expression Systems Drive Vaccine Innovation and Opportunity. *BioProcess International*, (April 2011), pp. 16-22.

35. Lau, O.S., Sun, S.S.M. (2009). Plant seeds as bioreactors for recombinant protein production. *Biotechnology Advances*, 27, 1015-1022.

36. Leffingwell & Associates (2011). 2006-2010 Flavour & Fragrance Industry Leaders. Download: http://www.leffingwell.com/top_10.htm Lonza, http://www.lonza.com

37. Market Research News (2011). Biologic Therapeutic Drugs: Technologies and Global Markets. Download: http://www.salisonline.org/market-research/biologic-therapeutic-drugs- technologies-and- global-markets/

38. Matasci, M., Baldi, L., Hacker, D.L., Wurm, F.M. (2011). The PiggyBac transposon enhances the frequency of CHO stable cell lione generation and yields recombinant lines with superior productivity and stability. *Biotechnology Bioengineering*, Epub April 4, PMID: 21495018

39. Meyer, H.-P., Fiechter, A. (1985). Production of cloned human Leucocyte interferon by *Bacillus subtilis*: optimal production is conntected with restrained growth, *Appl. Env. Microbiology*, 50 (2), pp. 503-507

40. Meyer, H.-P. (1987). Reference Fermentations. In: *Physical Aspects of Bioreactor Performance,* Crueger, W., pp. 144-157, Dechema, ISBN 3-921567-89-0, Frankfurt am Main.

41. Meyer, H.-P., Rohner, M. (1995). Applications of modelling for bioprocess design and control in industrial production, *Bioprocess Engineering,* 13, pp. 69-79.

42. Meyer, H.-P., Birch, J. (1999). Production with Bacterial and Mammalian Cells – Some Experiences. *Chimia*, 53(11), pp. 562-565.

43. Meyer, H.-P., Klein, J. (2006). About concrete, stainless steel and microbes, *PharmaChem*. March, pp. 6-8.

44. Meyer, H.-P., Brass, J., Jungo, C., Klein, J., Wenger, J., Mommers, R. (2008). An Emerging star for Therapeutic and Catalytic Protein Production. *BioProcess International Supplement,* June 2008, pp.10-22.

45. Meyer, H.-P., Ghisalba, O., Leresche, J.E. (2009). Biotransformation and the Pharma Industry, In: *Handbook of Green Chemistry*, Editor R. H. Crabtree, Editor, pp. 171- 212, WILEY-VCH Verlag GmbH & Co. KGaA, ISBN 978-3-527-32498-9, Weinheim

46. Meyer, H.-P., Turner, N. (2009). Biotechnological Manufacturing Options for Organic Chemistry, *Mini-Reviews Organic Chemistry.* 6(4), pp. 300-306.

47. Meyer, H.-P. (2011). Sustainability and Biotechnology. *Organic Process Research & Development*, 15(1), pp. 180-188.

48. Meyer, H.-P., Werbitzky, O. (2011). How Green Can the Industry Become with Biotechnology. In: *Biocatalysis for Green Chemistry and Chemical Process Development.* A. Tao and R. Kaslauskas, Editors, pp. 23-44, John Wiley & Sons Inc.; ISBN 9780470437780. Hoboken, NJ, United States, in print

49. Mücke, M., Ostendorp, R., Leonhartsberger, S. (2009). *E. coli* Secretion Technologies Enable Production of High Yields of Active Human Antibody Fragments. *BioProcess International*, (September 2009), pp. 40-47

50. Nakajima, K.-I., Asakura, T., Maruyama, J.-I., Morita, Y., Hideaki, S. Shimizu, I., Ibuka, A., Misaka, T., Sorimachi, H.-R., Arai, S., Kitamoto, K., (2008). Extracellular production of neoculin, a sweet-tasting heterodimeric protein with taste- modifying activity, by Aspergillus oryzae *Applied and Environmental Microbiology,* 72(5), pp. 3716-3723.

51. Nieuwenhuizen, P., Lyon, D., Laukkonen, J., Hartley, M. (2009). A rose in the bud? Anticipating opportunities in industrial biotechnology. *A.D. Little Prism* 2, pp. 39- 55.

52. Novagen, http://www.novagen.com

53. Patel, R.P., Patel, M.M., Patel, N.A. (2007). Animal Pharming for the Production of Pharmaceutical Proteins. *Drug Delivery*, 7, pp. 47-54.

54. Perlitz, U. (2008). *Chemieweltmarkt: Asiatische Länder auf dem Vormarsch*, Just, T., Editor, Deutsche Bank Research; Frankfurt am Main, Germany, http://www.dbresearch.de

55. Peubez, I., Chaudet N., Mignon, C., Hild, G., Husson, S., Courtois, V., De Luca, K., Speck, D., Dodoyer, R. (2010). Antibiotic-free selection in E. coli: new considerations for optimal design and improved production, *Microbial Cell Factories*, 9, 65, doi: 10.1186/1475-2859-9-65

56. Phue, J.-N., Lee, S.J., Trinh, L., Shiloach, J. (2008). Modified *Escherichia coli* B (BL21), a Superior Producer of Plasmid DNA Compared with *Escherichia coli* K (DH5?), *Biotechnology and Bioengineering*, 101, 4(Nov 2008), pp. 831-836

57. Piechocki, R. (1989). *Das berühmteste Bakterium*, Aulis Verlag Deubner und CO KG, ISBN 3- 7614-1258-4, Köln

58. Potgieter, T.I., Cukan, M., Drummond, J.E., Houston-Cummings, N.R., Jiang, Y., Li, F., Lynaugh, H., Mallem, M., McKelvey, T.W., Mitchell, T., Nylen, A., Rittenhour, A., Stadheim, T.A., Zha, D., d'Anjou, M. (2009). Production of monoclonal antibodies by glycoengineering *Pichia pastoris*. *Journal of Biotechnology*, 139, 2009, pp. 318-325

59. Quiagen, http://www.quiagen.com

60. Rader. R. A. (2008). *Biopharmaceutical Expression Systems and Genetic Engineering Technologies: Current and Future Manufacturing Platforms*, Bioplan Associates Inc., ISBN 1-934106-14-3, Rockville

61. Rozkov, A., Avignone-Rossa, C. A., Ertl, P.F., Jones, P., O'Kennedy, R.D., Smith, J.J., Dale, J.W., Bushell, M.E. (2004). Characterization of the Metabolic Burden on Escherichia coli DH1 Cells Imposed by the Presence of a Plasmid Containing a Gene Therapy Sequence, *Biotechnology and Bioengineering*, 88, 7, (Dec 2004), pp. 909-915

62. Ratner, M. (2010). Pfizers stakes a claim in plant cell-made bipharmaceuticals, *Nature Biotechnology*, 28(2), pp. 107-108.

63. Schwarz, F., Huang, W., Li, C., Schulz, B.L., Lizak, C., Palumbo, A., Numao, S., Neri, D., Aebi, M., Wang, L.-X. (2010). A combined method for producing homogeneous glycoproteins with eukaryotic N-glycosylation. *Nature Chemical Biology*, 6, (4), pp. 264-266

64. Sharma, S., Whalley., A., McLaughlin, J., Brello, F., Bishop, B., Benerjee, A. (2011). Fermentation process technology transfer for production of a recombinant vaccine component. *BioPharm International* July, pp. 30-39.

65. Simons, J.P., McClenaghan, M., Clark, A.J. (1987). Alteration of the quality of milk by expression of sheep beta-lactoglobulin in transgenic mice, *Nature* 328, pp. 530-532

66. Specht, E., Miyake-Stoner, S., Mayfield, S. (2010). Micro-algae come of age as platform for recombinant protein production, *Biotechnology Letters,* 32, pp. 1373-1383.

67. Spök, A. (2007). Molecular Farming on the Rise: GMO Regulators Still Walking a Tightrope. *Trends in Biotechnology,* 25(2), pp. 74-82.

68. Taticek, R.A., Choi, C., Phan, S.E., 2001. Palomares LA, Shuler ML. (2001). Comparison of growth and recombinant protein expression in two different insect cell lines in attached and suspension culture, *Biotechnology Progress,* 17(4), pp. 676-84.

69. Tiwary, S., Saini, S., Upmanyu, S., Benjamin, B., Tandon, S., Saini K.S., Sahdev, S. (2010). Enhanced expression of recombinant proteins utilizing a modified baculovirus expression vector, *Molecular Biotechnology*, 46, pp. 80-89.

70. Van Beilen, J.B. (2010). Why microalgal biofuels won't save the internal combustion machine, *Biofuels, Bioproducts & Biorefining (Biofpr)*, 4(1), pp. 41-52.

71. Vécina, L.-P., D'Aoust, M.-A., Landry, N., Couture, M.M.J., Charland, N., Ors, F., Barbeau, B., Sheldon, A.J. (2011). Plants as an Innivative and Accelerated Vaccine- Manufacturing Solution, *Supplement to BioPharm International*, May, pp. s27-s30.

72. Varley, J., Birch, J. (1999). Reactor design for large scale suspension animal cell culture, *Cytotechnology,* 29, pp. 177-205.

73. Vermasvuori, R., Koskinen, J., Salonen, K., Sirén, N., Weegar, J., Dahlbacka, J., Kalkkinen, N., van Weymarn, N. (2009). Production of Recombinant HIV-1 Nef Protein Using Different Expression Host Systems: A Techno-Economical Comparison, *Biotechnology Progress,* 25(1), pp. 95-102

74. Vervecken, W., Kaigorodov, V., Callewaert, N., Geysens, S., De Vusser, K., Contreras, R. (2004). In vivo synthesis of Mammalian-Like, Hybrid-Type N-Glycans in *Pichia pastoris. Applied and Environmental Microbiology*, 70, (5), pp. 2639-2646

75. Wacker, M., Linton, D., Hitchen, P.G., Nita-Lazar, M., Haslam, Stuart, M., North, S.J., Panico, M., Morris, H.R., Dell, A., Wren, B.W., Aebi, M. (2002). N-linked Glycosylation in Campylobacter jejuni and its functional transfer into *E. coli. Science*, Washington, DC, United States, 2002, 298, (5599), pp. 1790-1793.

76. Welck, H., Ohlig, L. (2011). Netzwerk Bioaktive Pflanzliche Lebensmittel *GIT Laborzeitschrift*, 6, pp. 408-409.

77. White, J. (2011). Protein Expression in *Corynebacterium glutamicum. Bioprocessing Journal*, Vol. 9, (2, Winter 2010/2011), pp. 53-55, ISSN 1538-8786

78. Wurm, F. (2004). Production or recombinant protein therapeutics in cultivated mammalian cells. *Nature Biotechnology,* 11, pp. 1393-1398.

79. Xiong, W., Li, X.F., Xian, J.Y., Wu, Q.Y. (2008). High-density fermentation of microalga *Chlorella protothecoides* in bioreactors for microbio-diesel production. *Applied Microbiology Biotechnology*, 78, pp. 29-36.

80. Ye, J., Kober, V., Tellers, M., Zubia, N., Salmon, P., Markusen, F.F. (2009). High-Level Protein Expression in Scalable CHO Transient Transfection, *Biotechnology Bioengineering*, 103(3), pp. 542-551.

CHAPTER-11

UTILIZING EXTERNAL CATALASE GENE EXPRESSION TO AMPLIFY METABOLIC PERFORMANCE AND BIOMATERIAL PRODUCTION IN HOST MICROBES

INTRODUCTION: AN OVERVIEW

The exogenous expression of catalase genes serves as a powerful tool for boosting the metabolic activity and the production of biomaterials within host microorganisms. Catalase enzymes play a crucial role in breaking down harmful reactive oxygen species (ROS) and effectively managing cellular oxidative stress. When foreign catalase genes are introduced into host microorganisms, they enhance the cell's capacity to mitigate oxidative damage and maintain metabolic homeostasis. This, in turn, fosters an environment conducive to increased biomaterial production, whether it involves biofuels, pharmaceuticals, or other valuable compounds. Harnessing exogenous catalase gene expression offers a promising avenue for optimizing microbial biomanufacturing processes and improving overall yields, thereby advancing biotechnology and industrial applications.

Heterologous gene expression is a widely used and vital biotechnology in basic and applied biology research fields. In particular, this technology (bioengineering) is emerging as a useful tool in fields of applied biology, such as medical, pharmaceutical, and agricultural sciences including food science. With this technology, animals, fishes, plants, and eukaryotic and prokaryotic microorganisms can be used as host organisms for transformation, and various types of vectors have been developed and have become commercially available. It is only about 40 years since a heterologous cloned DNA was expressed in *Escherichia coli* cells (Annie et al., 1974; Old & Primrose, 1989). Development of various types of gene transfer systems has assisted the widespread use of gene engineering technologies in all research fields.

The production of high-value compounds is a requisite purpose of gene engineering for all the fields of applied biology. Such production can occur in two ways: host cells are used to produce a product that is new to them by expressing a foreign gene(s); or host cells are enhanced to produce higher levels of the target product(s), which can be also inherently synthesized at normal levels by host organisms. The examples of the former are the production of insulin (Goeddel et al., 1979) or α-fetprotein (Nishi et al., 1988) by *E. coli* cells that were transformed with genes encoding these proteins. The literature on the latter means of production has accumulated rapidly. In such cases, two or more kinds of foreign gene carried in one or more vectors was used to transform the host organisms, by which a high-value compound(s) can be generated as the new product in the transformed host organisms or can enhance the metabolic activities of the host organisms.

Yumoto et al. (1998) isolated a bacterium with a remarkably high catalase activity from a waste pool at a fish-processing factory in Hokkaido, Japan. This bacterium was identified as *Vibrio rumoiensis* strain S-1^{T} (strain S-1 hereafter; Yumoto et al., 1999). Details of this bacterium and its catalase protein (VktA) and gene (*vktA*) are described in the following section **(Section 2, VktA catalase and its gene)**. The VktA catalase had a significantly high specific activity after being purified (Yumoto et al., 2000). The *vktA* gene encoding VktA was cloned and expressed in various strains of *E. coli* (Ichise et al., 1999; Orikasa, 2002). Cell-free extracts prepared from *vktA*-transformed *E. coli* cells exhibited almost the same specific activity of catalase as those prepared from the parent strain S-1. Biochemical and molecular studies on VktA prepared from the parent and *vktA*-transformed *E. coli* transformants showed that strain S-1 could accumulate VktA protein at a level as high as a concentration of 2% of total soluble proteins and that the high catalytic activity of purified VktA enzyme is 4 times greater than that of bovine liver catalase (Yumoto et al., 2000).

In this chapter, we describe two biotechnological uses of the *vktA* gene as a foreign gene. First it was used to enhance the nitrogen-fixing activity in a root nodule bacterium **(Section 3, Enhancement of nitrogen fixation by *vktA* in root nodule bacteria)**, which has its own catalase. Second, it was used to enhance eicosapentaenoic acid (EPA) biosynthesis in *E. coli* that had already been transformed with the EPA biosynthesis gene cluster (*pfa* genes) cloned from a marine bacterium **(Section 4, Enhancement of eicosapentaenoic acid production in**

***E. coli* through expression of *vktA*)**. The host *E. coli* cell has its own catalase but no ability to synthesize EPA. In both cases, the catalase activity of *vktA*-transformed host cells increased remarkably. Furthermore, the nitrogen-fixing ability in *R. leguminosarum* was definitely enhanced, and the EPA contents in *E. coli* transformed with *pfa* and *vktA* became greater than the EPA content in *E. coli* transformed with *pfa* genes only. The physiological and molecular roles of the increased catalase activity in catalase gene (*vktA*)-engineered bacterial cells are discussed in each section. **Section 5** contains concluding remarks and discusses possibilities for further use of VktA. Catalase comparable with VktA or those with much higher performance than VktA are described as a tool for producing biomaterials by biotechnologies in various research fields.

VKTA CATALASE AND ITS GENE

The VktA catalase is the sole catalase protein detected in strain S-1 (Yumoto et al., 1999). This enzyme is characterized by its significantly high catalytic activity when compared with other known catalases. General information on catalases and detailed characteristics of VktA and its gene (*vktA)* are provided in this section.

Catalases and their genes

Aerobic organisms metabolize oxygen through respiration for production of energy to sustain their life. During aerobic respiration, organisms generate toxic reactive oxygen species (ROSs), such as superoxide anion radicals ($O_2•^-$), hydrogen peroxide (H_2O_2), and hydroxyl radicals (OH•), as by-products. The presence of H_2O_2 in cells has a possibility to generate a more toxic ROS, OH•, by a Fenton-type reaction (Halliwell & Gutteridge, 1999). Excessive amounts of H_2O_2 and OH• are harmful to cell components. Therefore, aerobic organisms eliminate H_2O_2 with scavenging enzymes, such as catalase, peroxidase, and glutathione peroxidase. On the other hand, elimination of extracellular H_2O_2 by catalase is also important for either aerobic or anaerobic organisms to distribute their niche among several microorganisms. For example, parasitic and symbiotic microorganisms are attacked by ROSs which are produced by the external defense system of their host organisms (Katsuwon & Anderson, 1992; Rocha et al., 1996; Visick & Ruby, 1998).

The dismutation reaction of H_2O_2 has evolved into three unrelated groups in the category of catalase (EC 1.11.1.6). The first group consists of so-called typical catalase. These catalases consist of four identical subunits each equipped with protoheme IX (heme *b*) or heme *d* in the active site as the prosthetic group, and their subunit molecular masses are 55-84 kDa. These catalases can be subdivided into small subunits (subunit molecular mass: 55-69 kDa) possessing heme *b* as the prosthetic group and large subunits (75-84 kDa) possessing mainly heme *d* as the prosthetic group. This group of catalases is the most widespread in nature and exhibits efficient catalytic reactions. These enzymes have a broad pH optimum range, are specifically inhibited by 3-amino-1,2,4-triazole, which reacts with a catalytic intermediate state, compound I (see below), and are resistant to reduction by dithionite (Kim et al., 1994; Nadler et al., 1986). The second group of catalases is catalase-peroxidase. Catalase- peroxidases exhibit a bifunctional character: catalase and peroxidase activities. The maximal catalatic activities of catalase-peroxidases are two or three orders of magnitude lower than those of typical catalases. These enzymes typically have a dimeric or tetrameric structure with a subunit with a molecular mass of approximately 80 kDa, containing only one or two hemes *b* per molecule. Catalase-peroxidases have been detected in Bacteria, Archaea and Eukarya domains (although only in fungi in Eukarya). In addition, these enzymes have a sharp pH optimum, are not inhibited by 3-amino-1,2,4-triazole, and are readily reduced by dithionite (Hochman & Shemesh, 1987; Kengen et al., 2001). The deduced primary structures of these enzymes are closely related to each other and their three-dimentional structures are similar to those of plant peroxidase. The third group of catalases consists of manganese catalase (Mn-catalase). Mn-catalases, in contrast with the other two catalase groups, are not equipped with heme as the prosthetic group; rather these enzymes use manganese ions. Therefore, activities of these enzymes are not inhibited by cyanide or azide, which are inhibitors of catalases in the other groups. Mn-catalases are mostly hexameric and the molecular size of their subunit ranges 28 kDa to 35 kDa (Kono & Fridovich, 1983; Allgood & Perry, 1986). These catalases, which are sometimes referred to as pseudocatalases, are distributed in lactic acid bacteria and thermophilic bacteria.

All catalases possessing heme as the prosthetic group commonly exhibit a two-step mechanism for the degradation of H2O2. In the first step, one H2O2 molecule oxidizes theenzyme in the resting state (Fe^{3+} Por) to ferryl porphyrin with a porphyrin π-cation radical(Fe^{4+}=O $Por^{+\bullet}$), so-called compound I. In the second step, compound I oxidizes a secondH2O2 molecule to molecular oxygen and water (eqs. 1 and 2; Deisseroth & Dounce, 1970; Schonbaum & Chance, 1976).

Enzyme (Fe^{3+} Por) + H2O2 → Compound I (Fe^{4+}=O $Por^{+\bullet}$) + H2O (Reaction 1)
Compound I (Fe^{4+}=O $Por^{+\bullet}$) + H2O2 → Enzyme (Fe^{3+} Por) + H2O + O2 (Reaction 2)
Phylogenic analysis based on the amino-acid sequence deduced from the gene sequence oftypical catalases has revealed their subdivision into three distinct clades (Klotz et al., 1997). Clade 1 catalases are small-subunit catalases and are mainly of plant origin, but also includes one algal representative and a subgroup of bacterial origin. Clade 2 catalases are all large-subunit catalases of bacterial, archaeal, and fungal origins. The one archaeal enzymebelonging to clade 2 catalase is postulated to have arisen in a horizontal transfer event from *Bacillus* species. This clade of catalases exhibits a strong resistance to denaturation by heat and proteolysis. Clade 3 catalases are small-subunit catalases and their origins are bacteria, archaea, fungi, and other eukaryotes. There are no pronounced functional difference between clade 3 and clade 1 catalase. Bacteria harboring clade 3 catalase as the single catalase isozyme are distributed to a restricted environment.

Most aerobic bacteria contain one or more catalases, which are produced in response to oxidative stress or depending on the growth phase. *Escherichia coli* possess two types of catalase gene, *katG* and *katE*. These encode periplasmic HPI (catalase-peroxidase) and cytoplasmic catalase HPII (typical catalase, clade 2), respectively. *KatG* is induced by H_2O_2, while *katE* is induced by the entry into the stationary phase of growth (Storz & Zheng, 2000). Both of the genes are regulated by the alternative sigma factor, σ^S, which is produced by the *rpoS* gene. The expression of *katG* is regulated by OxyR, a transcriptional regulator that senses H_2O_2 (Ivanova et al., 1994; Storz & Zheng, 2000). OxyR can switch rapidly between reduced and oxidized states, and only the oxidized form acts as a transcriptional activator for target genes (Aslund et al., 1999; Christman et al., 1985).

CHARACTERISTICS FEATURES OF VKTA CATALASE AND ITS GENE

Even though there have been many reports of bacterial oxidative stress responses, there had been few reports on the microorganisms that are able to survive in highly oxidative environments. Therefore, studies were conducted in order to understand how a bacterium adapts to an oxidative environment and what kind of H_2O_2 eliminating system it possesses. A facultatively psychrophilic bacterium exhibiting high catalase activity was isolated from a drain pool of fish egg processing factory that uses H_2O_2 as a bleaching agent (Yumoto et al., 1998, 1999). The isolate, strain S-1, was identified as a new species, *Vibrio rumoiensis*. The catalase activity in cell extract of strain S-1 was 2 orders higher than those of *E. coli* and *Bacillus subtilis*. Although S-1 cells exhibit high catalase activity, individual cells do not exhibit strong resistance to H_2O_2. It is probably due to the fragility of the cell structure (Ichise et al., 1999).

Catalase (VktA) from strain S-1 has been purified and characterized (Yumoto et al., 2000). Molecular mass of the subunit of the catalase is 57.3 kDa and the enzyme consists of four identical subunits. The enzyme was not apparently reduced by dithionite. The activity showed a broad optimum pH range (pH 6–10) and was inhibited by 3-amino-1,2,4-triazole.

Therefore, the enzyme belongs to the typical small subunit catalase. The catalase activity of VktA was 1.5 and 4.3 times faster than *Micrococcus luteus* and bovine catalases, respectively, under 30 mM H_2O_2 in 50 mM potassium phosphate buffer (pH 7) at 20°C. Therefore, VktA is considered to be a catalase with very high activity and one with the highest turnover numbers in all known catalases. The thermoinstability of VktA was significantly higher than that of *M. luteus* and bovine catalases. It is suggested that the unique properties of VktA may reflect protective strategies for the survival of strain S-1 under oxidative environmental conditions, where this bacterium was inhabited.

The gene of VktA, *vktA*, has been isolated and sequenced. The *vktA* consisted of an open reading frame of 1530 bp encoding a 508 amino-acid protein with a calculated molecular mass of 57657.79 Da (Ichise et al., 2000). A putative ribosome binding site (AGGAGA) wasfound 5 bases upstream of the start codon (Fig. 1). In further upstream, putative promoter sequences, (TATAAT) and (TTGGCT), corresponding to -10 and -35, respectively were also found. The promoter sequences were probably recognized by RNA polymerase carrying the housekeeping sigma subunit σ^{70} (Hawley & McClure, 1983). Another putative promoter binding site for OxyR was in further upstream (Fig. 1).

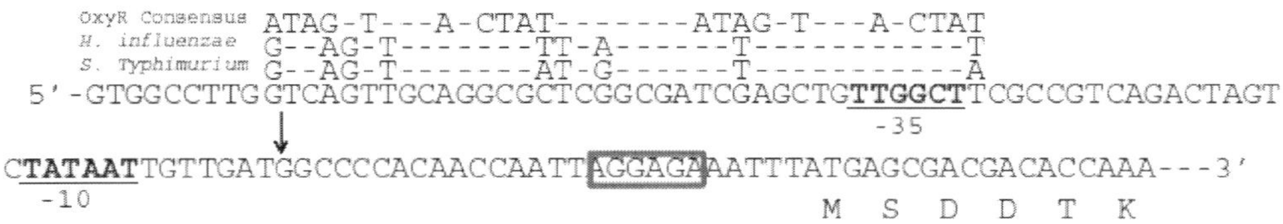

Fig. 1. Nucleotide sequence of the *vktA* promoter region. Nucleotides in the putative ribosome binding site, Shine-Dalgarno ribosome-binding site (SD) is in the boxed. Arrow indicates the putative transcription initiation site. Potential promoter sites of -10 and -35 are indicated as underlined bold letters. An alignment with OxyR consensus sequences of OxyR consensus (Tartaglia et al., 1989; Toledano et al., 1994), *Haemophilus influenza* (Bishai et al., 1994); and *Salmonella typhimurium* (Tartaglia et al., 1989) is shown upstream of the putative -35 region.

The deduced amino acid sequence of VktA shows high similarity with that of typical small subunit catalases belonging to clade 3 (Fig. 2). Many calatases possessing as the sole isozyme in parasitic or symbiotic microorganisms belong to clade 3. Therefore, there is a possibility that strain S-1 inherently be a symbiont of fish, which is supported by the fact that the strain was originally obtained from the fish egg processing factory. It is also pointed that bacteria possessing clade 3 catalases as the sole isozyme like strain S-1 was naturally selected in the environment that is frequently exposed to H_2O_2.

Localization of the VktA catalase in the cytoplasmic and periplasmic spaces has been demonstrated by the enzymatic activity of fractionated cells and immunological detection methods (Ichise et al., 1999; Yumoto et al., 2000). In addition, it has been shown that cell density and release of VktA from disrupted cells play an important role in the survival of strain S-1 cells when they reacted with H_2O_2 (Ichise et al., 1999). However, almost no growth hindrance is observed when 100 mM H_2O_2 is introduced into the culture. Accounts of strain S-1's strong tolerance to H_2O_2 in high concentrations of H_2O_2 during cultivation, concomitant with very rapid elimination of H_2O_2 due to the strong catalase activity of the cells themselves, have remained unresolved. The contribution of catalase to H2O2 tolerance and the presence of cell surface catalase have been demonstrated by H2O2 tolerance of catalase-deficient *E. coli* strain UM2 carrying *vktA* and immunoelectron microscopic observation on strain S-1 and catalase-deficient mutant, strain S-4, derived from strain S-1 using an antibody for the intracellular catalase of strain S-1, respectively (Ichise et al., 2008). Cell surface catalase is considered to contribute to the elimination of extracellular H2O2.

To further characterize the VktA catalase protein, it was analyzed by polyacrylamide gel electrophoresis (PAGE) using partially purified bovine liver catalase (Sigma, St Louis, USA), two types of *Aspergillus niger* catalase (one from Sigma and the other from NAGASE Co., Ltd, Tokyo, Japan), and *Corynebacterium glutamicum* catalase (Sigma) as references. All samples except for strain S-4 exhibited one activity band detected by the method by Uriel (1958) in native gel (data not shown). Fig. 3 shows protein band profiles and indications of the activityband position of all the samples in SDS-PAGE under different conditions. In the presence of SDS with neither 2-mercaptoethanol (2-ME) nor heating, all samples except for strain S-4 showed one activity band, whereas no activity band was observed in the presence of SDS, 2- ME and heating. The bands with activity are considered to be proteins with a dimeric form. It is interesting that two *A. niger* and *C. glutamicum* catalases were stained positive for activity in the presence of SDS, 2-ME, but no heating. The dimeric form of these catalases was resistant to reduction by 2-ME, at room temperature. By contrast, VktA was sensitive to 2-ME treatment at room temperature. A very faint oxygen bubble formation was observed for the position of the dimeric form of bovine liver catalase in the presence of SDS, suggesting that the tetrameric form of bovine liver catalase is apt to be easily dissociated into monomers in the presence of SDS only. These results suggest that VktA has an intermediate characteristic of bovine liver catalase and other microbial catalases against treatment with 2-ME at room temperature. The VktA catalase, which is tetrameric in its native form (Yumoto et al. 2000), can be dissociated into active dimers in the presence of SDS; however, the structure and activity of the dimers are sensitive to treatment with 2-ME at room temperature, suggesting that VktA could be a structurally and actively flexible enzyme.

Fig. 2. Phylogenic tree of catalases. The tree was constructed by the CLUSTAL W program (Thompson et al., 1994) with multiple alignment using neighbor-joining method (Saitou & Nei, 1987). Numbers at the branches are bootstrap percentages based on 1000 replicates. Bar, 0.1 changes per amino acid position.

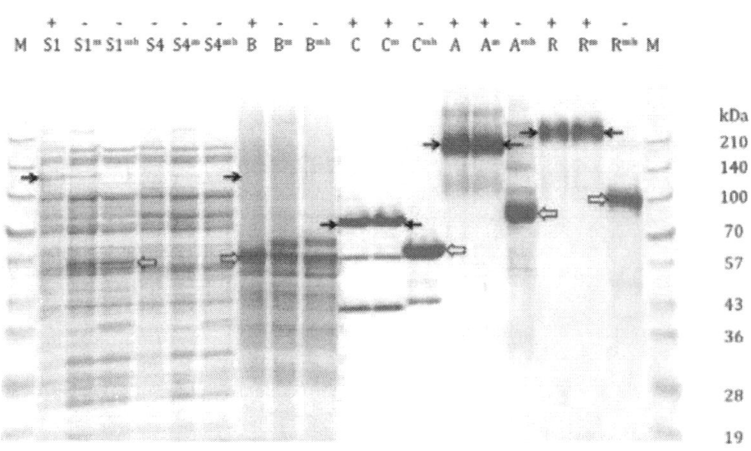

Fig. 3. SDS-PAGE profiles of cell-free extracts from *Vibrio rumoiensis* strain S-1 (S1) and its catalase-deficient mutant strain S-4 (S4), and commercially available catalases: B from bovine liver; C from *Corynebacterium glutamicum;* A

from *Aspergillus niger* (Sigma); R from *A. niger* (Ryonet F Plus; Nagase Co., Ltd). +, catalase activity positive (oxygen bubble formation) at the band shown by solid arrows; –, no catalase activity band; superscripts, m and mh, SDS-PAGE conditions in the presence of 2-mercaptoethanol (2-ME) at room temperature for 5 min and in the presence of 2-ME with heating at 100ºC for 5 min, respectively. No superscript means SDS- PAGE conditions with SDS but no 2-ME. The open arrows indicate the position of monomeric subunits of catalases. VktA, bovine liver (Sigma) and two *A. niger* catalases from Sigma and Nagase have a tetrameric native structure with a molecular mass of 230 kDa, 240 kDa, 250 kDa and approximately 350 kDa, respectively. Information on the native form of the *C. glutamicum* catalase is not available. Gels (e-Pagel, 10%; Atto, Tokyo) were stained with Coomassie Brilliant Blue. Lane M, molecular marker standard (kDa).

ENHANCEMENT OF NITROGEN FIXATION BY *VKTA* IN ROOT NODULE BACTERIA

Root nodule bacteria are characterized by their ability to fix dinitrogen in root nodules, where microaerobic conditions are maintained. However, the nitrogen-fixing process requires a large amount of ATP, suggesting that molecular oxygen is requisite for oxidative phosphorylation. This section describes the use of *vktA* to modulate oxygen metabolism in root nodule bacteria. The *vktA* gene was introduced into *Rhizobium leguminosarum* cells and the strain with a remarkably high catalase activity was constructed. Results show that the increase of catalase activity in rhizobia could be a valuable way to improve the nodulation and nitrogen-fixing ability of nodules (Orikasa et al., 2010).

Nitrogen Fixation and Catalases in Rhizobia: An Overview

Rhizobia infect the roots of leguminous plants and form nodules with elaborate control by exchanging molecular signals between two partners (Wei et al., 2008, 2010). The rhizobia are present as bacteroids such as microsymbiotic organelles in the nodules and convert atmospheric dinitrogen into biologically available ammonia (nitrogen fixation). Nitrogen fixation is an energy-requiring process and needs large amounts of ATP, which is supplied by oxidative phosphorylation. This process is extremely oxygen-sensitive and the partial pressure of oxygen inside the nodules is maintained at very low levels, resulting in strongly reduced conditions with the production of ROSs such as H_2O_2(Tjepkema & Yocum, 1974). Leghemoglobins present in nodules play a role in the effective diffusion of oxygen and their autoxidation results in the production of $O_2{\bullet}^-$ and H_2O_2 (Appleby, 1984; Puppo et al., 1981). However, as the host plant can control the nodulation in such a way that the infection during symbiotic interaction is aborted due to ahypersensitive reaction, it has been suggested that the release of ROS such as H_2O_2, termed the oxidative burst, may occur in the early stage of the infection thread formation (Vasse et al., 1993). It was reported that a striking release of ROS occurred under the conditions of plant defenses against pathogens (Mehdy, 1994). Since it is believed that ROSs such as H_2O_2, $O_2^{\bullet-}$ and OH• are also generated naturally during the metabolism of cells growing aerobically and that these ROS could damage the protein, lipids, and DNA components in organisms (Sangpen et al., 1995), these reports suggest that the response of rhizobia to ROSs such as H_2O_2 would exert a serious influence on both the nodulation process and nitrogen-fixing abilities.

Catalase in rhizobia cells was not previously considered to play a crucial role in the nitrogen-fixing abilities because *Sinorhizobium meliloti* has three kinds of catalase (KatA, KatB, KatC) and the *katA*-minus mutant did not impair both nodulation and nitrogen fixing abilities (Herouart et al., 1996). Rhizobia are symbiotic microorganisms. They were previously considered to be able to rely on their host plant for defense against toxic forms of oxygen such as H_2O_2; that is, plant-derived catalases, as well as defense systems against H_2O_2 toxicity such as the ascorbate-glutathione reaction system, were considered to be able to contribute to the removal of H_2O_2 toxicity. It was reported that the production of oxidative protection enzymes by plant cells in nodules was positively correlated with the increase in nitrogenase activity and leghemoglobin content (Dalton et al., 1986). In addition, parasitic bacteria were reported not to need their own catalase because they could depend upon their host catalase (Steiner et al., 1984). However, subsequent experiments showed that the double-catalase mutants (Δ*katA*Δ*katC* or Δ*katB*Δ*katC*) had considerably lower capability for both nodulation and nitrogen fixation (Jamet et al., 2003; Sigaud et al., 1999), and their activities decreased during nodule senescence (Hernandez- Jimenez et al., 2002). These results showed that the catalase in rhizobia could be important for the efficient function of both nodulation and nitrogen fixation. Another rhizobium strain, *Bradyrhizobium*, showed the important role of catalase for H_2O_2 decomposition, and the *katG*-minus mutant resulted in the loss of both catalase activity and exogenous H_2O_2 consumption (Loewen et al., 1985b; Panek & Obrian, 2004).

Unexpectedly, rhizobia had a tendency for less catalase activity than other genera of aerobic and facultative anaerobic bacteria, resulting in higher susceptibility to H_2O_2 (Ohwada et al., 1999). We compared catalase and peroxidase activities in rhizobia (11 strains) with those in other genera of bacteria (six strains). Catalase activities (units per mg protein) of the rhizobia strains tested were in the range of 0.9-5.8, although the addition of H_2O_2 increased activities in all strains to the range of 2.5-11.3. By contrast, the activities in the other genera strains tested ranged 12.3 to 893.3, and, in the presence of H_2O_2, the levels of all strains were increased to the range of 16.1-1,460.2, indicating that the catalase activity of the tested rhizobia tended to be considerably lower than those of the other genera tested (Fig. 4). Results of Southern analyses imply that rhizobia have no DNA region similar to *katE* of *E. coli*, which is inducible tenfold during the stationary phase of growth (Loewen et al., 1985a), and the kinetics of catalase induction in *R. leguminosarum* bv. *phaseoli* was different from that in *E. coli* (Crockford et al., 1995; Hassan & Fridovich, 1978). Crockford et al. (1995) mentioned that, during growth, unidentified compounds accumulate in the cells and repress catalase activity, although the significance of growth-phase-dependent regulation of catalase activity remains obscure.

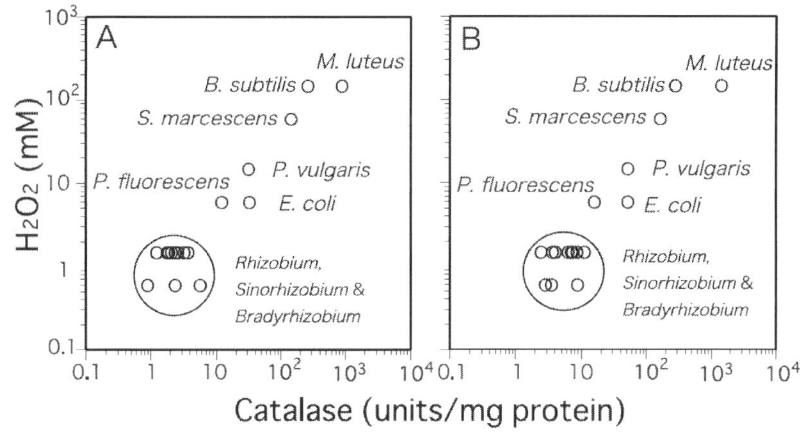

Fig. 4. Correlation between catalase activities and H_2O_2 tolerance. Cells were incubated with H_2O_2 (0.6, 1.5, 5.9, 14.7, 29.4, 58.8, and 147 mM) and the maximum concentration of H_2O_2 (mM) in which the cells could grow within 24 h (96 h for *Bradyrhizobium japonicum*) was plotted as ordinate. Catalase activities in the cells in the early stationary phase with (panelB) or without (panel A) H_2O_2 (0.6 mM) were plotted as the abscissa. *Rhizobium, Sinorhizobium* and *Bradyrhizobium* strains tested are enclosed within a circle. Data points are means of results from at least three trials. *Rhizobium, Sinorhizobium* & *Bradyrhizobium* strains used: *Rhizobium leguminosarum* bv. *phaseoli* USDA2667, 2676; *R. leguminosarum* bv. *trifolii* USDA2053, 2145; *R. leguminosarum* bv. *viciae* USDA2370, 2443; *Sinorhizobium fredii* USDA191, 206; *S. Meliloti* USDA1021, 1025; *Bradyrhizobium japonicum* S32. Other strains used: *Bacillus subtilis* AHU1390; *Escherichia coli* JM109; *Micrococcus luteus* AHU1427; *Proteus vulgaris* AHU1144; *Pseudomonas fluorescens* AHU1719; *Serratia marcescens* AHU1488. Adapted with permission from Ohwada et al. (1999).

The relationships between catalase activities and H_2O_2 tolerance in all bacteria tested are shown in Figure 4. The results indicate that a tendency for a positive and mutual correlation between them for all strains tested. Particularly, both the catalase activities and the H_2O_2 tolerance of the rhizobia tested were lower than those of the others. On the other hand, peroxidase activities in cells with different electron donors (NADH, *o*-dianisidine, and *p*- phenylenediamine) were considerably lower than the catalase activities and there was no significant difference between rhizobia and the others. In addition, the growth repression caused by the addition of catalase inhibitor in the presence of H_2O_2 was observed in rhizobia such as *R. leguminosarum* bv. *phaseoli*, bv. *trifolii* and *S. meliloti*. These results indicate that catalase could be mainly responsible for the defense mechanism against H_2O_2 toxicity. For *E. coli*, it was reported that catalase-deficient and catalase-overproductive mutants were more sensitive and resistant to H_2O_2, respectively (Greenberg & Demple, 1988; Loewen, 1984). The peroxisome-targeting signal (SKL sequence) of catalase (KatA) in *S. meliloti*, which is supposed to be connected with the export into the periplasmic region, suggests that the *S. meliloti* catalase is located in this region (Herouart et al, 1996). The location of protecting enzymes such as catalase against H_2O_2 seems to be important because periplasmic enzymes could be advantageous to the defense against exogenous H_2O_2. Therefore, the location of catalase in rhizobia would have an effect on the mutual correlation between catalase activities and H_2O_2 tolerance.

Since rhizobia are symbiotic microorganisms, they might be able to rely on their host plant in nodules to some extent for defense against toxic forms of oxygen such as H_2O_2. However, the catalase activity of *Bradyrhizobium* strain from effective nodules was reported to be higher than that in a strain from ineffective nodules (Francis & Alexander, 1972). These results prompted the construction of rhizobia with higher catalase activity to improve the nitrogen-fixing ability of nodules.

Improving Nitrogen Fixation via vktA Expression in Rhizobium leguminosarum

The DNA fragment (4.9 kbp) including the catalase gene (*vktA*), which is controlled by its own promoter activity, was ligated into pBluescriptII SK+ to construct pBSsa1 (Ichise et al., 2000). The *Bam*HI-*Xho*I fragment including a coding region of *vktA* (4.9 kbp) was isolated from the pBSsa1 and ligated into a broad host range vector, pBBR1MCS-2 (Kovach et al., 1995) at the site of *Bam*HI-*Xho*I. Then, the recombinant plasmid was introduced into *R. leguminosarum* bv. *phaseoli* USDA2676 cells to construct the *vktA*-transformant by triparental mating (Simon, 1984). The result showed that the *vktA*-transformed *R. leguminosarum* exhibited a remarkably high catalase activity of up to around 10,000 units per mg protein. This activity was three orders of magnitude greater than that of the parent strain and comparable to that of strain S-1 (Yumoto et al., 1999). To confirm the production of VktA catalase, activity staining of the catalase and immunoblot analysis with anti-VktA antiserum were conducted. The activity staining was carried out according to Clare *et al.* (1984). Results showed an obvious band of VktA catalase in both logarithmic and stationary phases.

Although the parent strain of *R. leguminosarum* originally had two bands of different mobility from the VktA catalase, their intensities were considerably weak compared with that of the VktA catalase and were not major in the *vktA*-transformant. Additionally, the immunoblot analysis was conducted using polyclonal anti-VktA and a goat anti-rabbit IgG- horseradish peroxidase conjugate as primary and secondary antibodies, respectively. The results indicated that a positive antigen-antibody reaction occurred as a single band between the anti-VktA antiserum and the VktA catalase. These results clearly showed that the efficient production of VktA catalase was responsible for the high catalase activity in the *vktA*-transformant. As described above, since catalase could be mainly responsible for the H_2O_2 tolerance in bacteria and the catalase activities in the rhizobia tested were considerablylower compared with the other genera tested, the growth of almost all of rhizobia tested was severely repressed even in the presence of 1.5 mM H_2O_2. However, the *vktA*-transformant showed almost the same cell density as the parent in the presence of 10 mM H_2O_2; even with 50 mM H_2O_2, the *vktA*-transformant could grow and the cell density reached levels half that of the parent at 24 h, then almost the same as the parent at 30 h after incubation. These results indicate that the *vktA*-transformant acquired resistance against H_2O_2 through the enhancement of catalase production in the cells.

Table 1. Number, weight and acetylene reduction activity (ARA) of nodules formed in the combination of *Phaseolus vulgaris* (L.) cv. Yukitebou and *vktA*-transformed *R. leguminosarum*. Values were obtained from 20 determinants of at least two independent experiments. The values given are the means ± S. D. of 20 different tests. Adapted with permission from Orikasa et al. (2010).

Cultivation condition	Strain	Nodule number (per plant)	Nodule weight (per dry wt per plant)	ARA[1)] (per mg dry wt of nodule)	ARA[1)] (per plant)
Seed bag	Parent	129 ± 8.2	34.1 ± 1.6	0.7 ± 0.1	25.6 ± 2.0
	vktA-transformant	111 ± 8.2	31.6 ± 1.8	1.2 ± 0.1 [2)]	36.5 ± 2.7 [2)]
Pot	Parent	75 ± 11.4	16.6 ± 2.8	0.7 ± 0.1	11.3 ± 2.7
	vktA-transformant	89 ± 12.2	21.9 ± 1.8 [2)]	1.6 ± 0.2 [2)]	34.3 ± 4.5 [2)]

- Acetylene reduction activity (μmol of C_2H_2/h).
- Significant differences were evaluated by Student's *t* test ($P < 0.01$).

The host plant, *Phaseolus vulgaris* (L.), was inoculated with *vktA*-transformed *R. leguminosarum* cells (10^6 cells per seed) and, after cultivation in a seed bag with Norris and Date medium (Dye, 1980) or in a pot filled with vermiculite, the number, weight, and nitrogenase activity (acetylene reduction activity, ARA) of the nodules were measured (Table 1). For the seed bag, the number and weight of nodules did not show a significant difference between *vktA*- transformant and the parent cells. However, the acetylene reduction activity (ARA) of nodules formed with *vktA*-

transformed cells was significantly higher than that formed with the parent cells, and around 1.7 times as many nodules were formed as with the parent cells (around 1.4 times per plant). For the pot, the number and weight of the nodules formed with *vktA*- transformant were larger than those of the parent cells, with around 1.2 and 1.3 times those of the parent, respectively, although these levels tended to be lower than those for seed bag cultivation. Higher levels of ARA in the nodules formed with *vktA*-transformant were also observed and the levels reached around 2.3 times those of the parent (around 3.0 times per plant). Another set of experiments with the combination of *vktA*-transformed *S. fredii* and *Glycine max* (L.) also showed that the production of VktA significantly increased the ARA per nodule or plant weight. These results indicate that enhancing the catalase activity in *Rhizobium* cells significantly increased the nodules' nitrogen-fixing ability.

Next, catalase production in bacteroids of *vktA*-transformed *R. leguminosarum* was measured. Bacteroids were separated immediately after the nodules were detached from the plant roots (Kouchi & Fukai, 1989). The result showed that the *vktA*-transformant maintained an even higher catalase activity compared with the parent (around 150 units per mg protein). Results of western blot analysis using the anti-VktA antiserum showed a single band for VktA catalase, indicating that higher production of VktA catalase resultedin a high catalase activity even in bacteroids. However, the catalase activity in bacteroids was considerably low as compared with free-living cells. Given that a decrease in the relative amounts of DNA, as well as the dynamic conversion of cellular metabolism such as the repression of sugar degradation, was reported during the differentiation process of bacteroids (Bergersen & Turner, 1967; Verma et al., 1986; Vierny & Laccarino, 1989), the loss of a certain number of the *vktA*-recombinant plasmids and/or the repressive production of VktA catalase might occur through the differentiation to bacteroids in the absence of antibiotics. The localization of the VktA catalase in free-living cells and bacteroids of *vktA*-transformant was studied by immunoelectron microscopy using the polyclonal antiserum against VktA with a secondary anti-rabbit antibody, which was coupled with gold particles. The number of gold particles at the periphery of the free- living cells including periplasm accounted for about 57.4 % of the sum total. For bacteroids, a relatively large number of gold particles (about 52.3% of the sum total), were observed at the periphery of the bacteroids including the symbiosome. These results indicate that the VktA catalase was preferentially distributed at the peripheral part of the cells for both free-living cells and bacteroids. H_2O_2 and leghemoglobin contents in the nodule formed with *vktA*-transformant were also measured. Nodules were detached 35 days after planting and H_2O_2 was extracted by grinding in 1 M $HClO_4$ (Ohwada & Sagisaka, 1987). The H_2O_2 content in the extracts was measured by Quantitative Hydrogen Peroxide Assay (OXIS International, Portland, USA).

The extraction and quantification of leghemoglobin components using capillary electrophoresis were carried out according to Sato et al. (1998). The results showed that the H_2O_2 content (nmol/g fresh wt of nodule) in the nodules formed with the parent cells was around 21.0, but this level was decreased to around 15.4 by the production of VktA catalase in the cells. By contrast, the VktA production increased the content of the leghemoglobins (Lba and Lbb) and the levels in the nodules formed with *vktA*-transformant were around 1.2 (Lba) and 2.1 (Lbb) times higher than those with the parent cells.

Considering that ROSs such as H_2O_2 are released from the plant root not only under pathogenic conditions but also during the infection process (Mehdy, 1994; Vasse et al., 1993), it is possible that *Rhizobium* cells with a higher catalase activity are advantageous to the infection process because they decrease the amounts of H_2O_2 around them. This supports the possibility that the VktA catalase is preferentially located near the surface area of the cells, suggesting that they could be effective in decomposing H_2O_2. The peripheral distribution of VktA was also observed in strain S-1 (Ichise et al., 2000). In nodules, lack of the ability to remove H_2O_2 caused the reduction of both nodulation and nitrogen-fixing ability (Bergersen et al., 1973). Given that electron microscopic observation did not seem to reveal any difference in the density of bacteroids inside the nodules between *vktA*-transformant and the parent, it is thought that the enhancement of the ability to decrease H_2O_2 by higher catalase activity is responsible for the increased levels of nitrogen-fixing activity. On the other hand, it was reported that leghemoglobins accumulated in the infected plant cells before nitrogen fixation in order to decrease the partial pressure of oxygen inside the nodule and protect nitrogenase from inactivation by oxygen (Appleby, 1984). Adding leghemoglobin to bacteroid suspensions enhanced the nitrogenase-mediated reactions, and the nitrogenase activity of bacteroids was dependent on the concentration of leghemoglobin (Bergersen et al., 1973). Furthermore, the deficiency of leghemoglobin synthesis in nodules of *Lotus japonicus* using RNAi led to theabsence of symbiotic nitrogen fixation (Ott et al., 2005). Therefore, it is considered that the increase of leghemoglobin content also contributed toward the improvement of nitrogen- fixing ability, although the accelerated mechanism of the leghemoglobin production is still under investigation. It was reported that the effective nodules of white clover and soybean contained higher activity of catalase compared with the ineffective nodules (Francis & Alexander, 1972). It seems that catalase is disadvantageous to protect nitrogenase from the cytotoxic effect of H_2O_2 because oxygen, which represses nitrogenase activity, is generated through the decomposition of H_2O_2. However, considering that a large amount of ATP, which could be supplied by bacteroidal oxidative phosphorylation, is required for the nitrogen-fixing reaction and that the leghemoglobins maintain a high oxygen flux for respiration through the facilitated oxygen diffusion (Ott et al., 2005; Tajima et al., 1986; Wittenberg et al., 1975), it might be possible that the oxygen generated by the catalase reaction could also be useful for energy production. The results here show that an increase in catalase activity reduced H2O2 levels in the nodules concomitantly with the enhancement of leghemoglobin contents, followed by improvement of the nitrogen-fixing ability in the nodules. The enhanced nitrogen fixation from the expression of *vktA* in rhizobia would lead to the growth of the host plant with reduced use of chemical nitrogen fertilizer.

BOOSTING EICOSAPENTAENOIC ACID SYNTHESIS IN E. COLI VIA VKTA EXPRESSION

Eicosapentaenoic acid (EPA) is an essential nutrient for humans and animals. Its derivatives, such as eicosanoids, are known as signal compounds in blood and nervous systems. Therefore, the ethyl ester of EPA is used a medicine. Fish oils, which have been the most widely used source of EPA to date, have been recognized as unsuitable because of their low EPA content and their unavoidable contamination with heavy metals from seawater; therefore new sources of EPA have been sought. Bacteria or fungi, which inherently produce EPA, constitute one of such possible source. Another possibility is the heterologous expression of EPA biosynthesis genes or chain elongation/desaturase genes of fatty acids in various types of host organism. This section describes the EPA biosynthesis in *E. coli* transformed only with EPA biosynthesis genes and the enhancement of EPA biosynthesis by coexpression of the *vktA* gene.

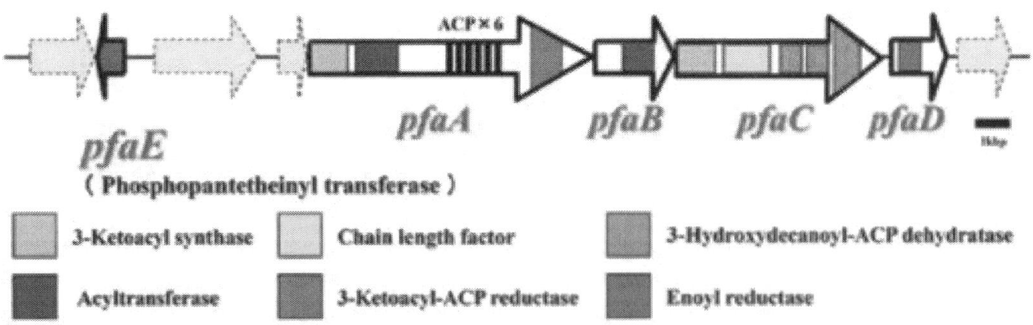

Fig. 5. Domain structure of *pfa* genes responsible for the biosynthesis of EPA from *Shewanella pneumatophori* SCRC-2738.

Bacterial biosynthesis of EPA

Bacterial species belonging to *Shewanella*, *Vibrio*, *Flexibacter*, and *Halomonas* (Salunkhe et al., 2011) are known to produce EPA as a major long-chain polyunsaturated fatty acid. EPA is synthesized de novo in a polyketide biosynthesis mode by the enzyme complex consisting of PfaA, PfaB, PfaC, PfaD, and PfaE, which are encoded by *pfaA*, *pfaB*, *pfaC*, *pfaD*, and *pfaE*, respectively. These five genes (designated as an EPA biosynthesis gene cluster) generally locate in proximity on the chromosome (Fig. 5). PfaA and PfaC are multifunctional proteins and have some functional domains (Fig. 5). Only one functional domain for each of acyltransferase, enoyl reductase, and phosphopantetheinyl transferase is found, in PfaB, PfaD, and PfaE, respectively. Since the EPA gene cluster was first cloned from *Shewanella* sp. SCRC 2738 (*S. pneumatophori* SCRC-2738; Hirota et al., 2005) in 1996 (Yazawa, 1996), much attention has been paid to increasing the content of EPA in *E. coli* host cells and to its heterologous expression of these genes in various organisms, such as bacteria, yeast, and plants (Yazawa, 1996). The EPA gene clusters have been successfully expressed in various types of *E. coli* strains (Orikasa et al., 2004). Furthermore, numerous attempts have been made to express bacterial EPA biosynthesis genes in bacteria other than *E. coli* and in eukaryotic cells. However, to our knowledge, the report by Yu et. al. (2000) is the only one, in which a marine cyanobacterium is used as a host organism to express the EPA gene cluster.

Improved EPA Synthesis by vktA Expression in E. coli with pfa Genes

The enhanced production of EPA was observed in recombinant systems of *E. coli* that carried both EPA biosynthesis genes (*pfa*) and a *vktA* catalase gene. Although no molecular mechanism has been determined for this enhanced production of EPA, this technique may become another useful method to increase the productivity of EPA using recombinant systems. Docosahexaenoic acid (DHA) can be synthesized also in bacteria using DHA biosynthesis *pfa* genes, because the two *pfa* genes have a very similar structure (Okuyama et al., 2007).

- *coli* DH5α transformants carrying pEPAΔ1 that included *pfaA-E* genes to the host cell led to the production of EPA (approximately 3% of total fatty acids; Table 2). The production of EPA in host organisms carrying pEPAΔ1 was increased to 12% of total fatty acids by the introduction of a *vktA* insert in pGBM3 [strain DH5α (pEPAΔ1) (pGBM3::*vktA*)]. The empty pGBM3 had no effect on EPA production. In strain DH5α carrying (pEPAΔ1) and partially deleted *vktA* in pGBM3(pGBM3::Δ*vktA*), EPA made up 6% of total fatty acids. The increase in EPA production in strain DH5α (pEPAΔ1)(pGBM3::*vktA*) was accompanied by a decrease in the proportions of palmitoleic acid [16:1(9)] (Table 2). When pGBM3 and pGBM3::*vktA* were replaced in the *E. coli* transformants with pKT230 and pKT230::*vktA*, respectively, similar trends were observed (data not shown). The yield of EPA per culture was approximately 1.5 μg/ml for DH5α(pEPAΔ1) and DH5α(pEPAΔ1)(pGBM3). It increased to 7.3 μg/ml for DH5α(pEPAΔ1) (pGBM3::*vktA*). The yield of EPA from DH5α (pEPAΔ1) pGBM3::Δ*vktA*) was 3.3 μg/ml (Table 2). *E. coli* DH5α has an inherent catalase activity of 2–3 U/mg protein (Nishida et al., 2006).

The plasmid pEPAΔ1 had no effect on the catalase activity of the host cells. Catalase activity was increased to 535 U/mg protein for DH5α(pEPAΔ1)(pGBM3::*vktA*). However, there was no enhancement of catalase activity in DH5α(pEPAΔ1)(pGBM3::Δ*vktA*)

(Table 2). Figure 6 shows the profiles of proteins prepared from various *E. coli* DH5α transformants using SDS-PAGE. A significant amount of protein in the VktA band of 57 kDa, was detected only for DH5α(pEPAΔ1)(pGBM3::*vktA*). No notable novel band was observed in DH5α (pEPAΔ1)(pGBM3::Δ*vktA*) or in any of the other transformants.

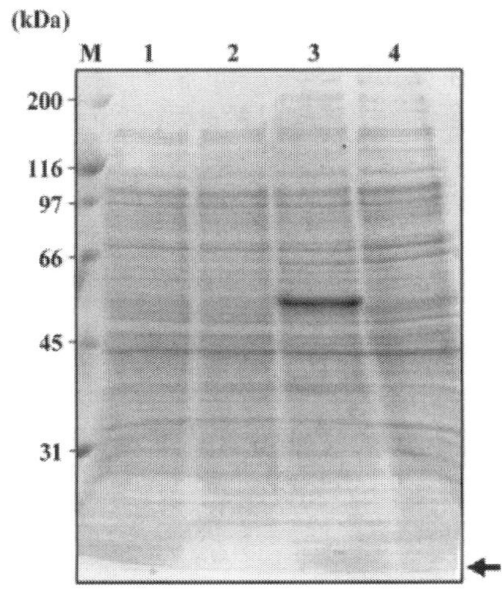

Fig. 6. SDS-PAGE profiles of cell-free extracts from various *Escherichia coli* DH5α transformants. Lane 1, *E. coli* DH5α carrying pEPAΔ1; lane 2, *E. coli* DH5α carrying pEPAΔ1 plus empty pGBM3; lane 3, *E. coli* DH5α carrying pEPAΔ1 plus pGBM3::*vktA*; lane 4, *E. coli* DH5α carrying pEPAΔ1 plus pGBM3::Δ*vktA*. Lane M, molecular marker standard (kDa). Arrow indicates the position of running dye. Adapted with permission from Orikasa et al. (2007).

It is evident that bacterial EPA (and DHA) is synthesized by the polyketide biosynthesis pathway, and that this process operates independently of the de novo biosynthesis of fatty acids up to C16 or C18 (Metz et al., 2001; Morita et al., 2000). However, it is likely that acetyl-CoA would be commonly used as a priming substrate in both processes, as specific inhibition of the de novo synthesis of fatty acids up to C18 by cerulenin enhanced the production of EPA and DHA in bacteria and probably also in *Schizochytrium* (Hauvermale et al., 2006). This is analogous to the situation in the unsaturated fatty acid auxotroph *E. coli fabB*$^-$ that was transformed with bacterial *pfa* genes, where EPA accounted for more than 30% of total fatty acids (Metz et al., 2001; R.C. Valentine & D. L Valentine, 2004). All of these findings suggest that the metabolic regulation of host organisms carrying *pfa* genes responsible for EPA biosynthesis could potentially be usedcommercially to enhance the production of EPA. *V. rumoiensis* S-1 accumulates high levels of VktA protein, the amount of which is calculated approximately 2% of total soluble proteins (Yumoto et al., 2000). A significant accumulation of VktA was observed in DH5α(pEPAΔ1)(pGBM3::*vktA*) (Fig. 6). However, the fact that a slight increase in EPA production was also observed in DH5α(pEPAΔ1)(pGBM3::Δ*vktA*) excludes the possibility that the catalytic activity of VktA protein per se was involved in this increased EPA production.

Table 2. Fatty acid composition of *E. coli* DH5α and its various transformants and recovered amount of EPA from cultures. Adapted with permission from Orikasa et al. (2007).

Strains [1]	Fatty acid[2] (% total)					Content of EPA
	16:0	16:1(9)	18:1(11)	EPA	Others[3]	(μg/ml)
E. coli DH5α	36.0 ± 1.0	29.6 ± 0.7	22.0 ± 0.6	0	12.5 ± 1.4	0
E. coli DH5α (pEPAΔ1)	35.6 ± 0.9	26.9 ± 1.5	21.8 ± 0.9	2.5 ± 0.2	13.2 ± 2.7	1.7 ± 0.1
E. coli DH5α (pEPAΔ1)(pGEM3)	38.6 ± 1.8	28.2 ± 0.6	20.8 ± 0.3	3.2 ± 1.7	9.2 ± 1.1	1.5 ± 1.3
E. coli DH5α (pEPAΔ1)(pGEM3::*vktA*)	35.9 ± 3.1	18.5 ± 0.4	22.9 ± 1.9	12.3 ± 0.7	10.3 ± 0.8	7.3 ± 1.2
E. coli DH5α (pEPAΔ1)(pGEM3::Δ*vktA*)	34.0 ± 0.7	26.7 ± 0.2	24.1 ± 1.2	5.9 ± 0.2	9.2 ± 1.7	3.3 ± 0.2

- The cells were grown at 20°C until the culture had an OD660 of 1.0
- Fatty acids are denoted as number of carbon atoms:number of double bond. The Δ-position of double bond is presented in parenthesis
- Others incude 12:0, 14:0, 18:0 and 3-hydroxyl 14:0.

At present, the mechanism for the enhanced production of EPA in *E. coli* recombinant systems carrying DH5α(pEPAΔ1)(pGBM3::*vktA*) is unknown. One possibility is that the increase in production of EPA is a response against intracellular stress. DH5α(pEPAΔ1)(pGBM3::*vktA*) accumulated a large amount of VktA protein, which may have increased the stress for the host cells. This would have delayed their growth. Nishida et al. (2006) provided evidence that cellular EPA has an antioxidative function against extracellular H_2O_2 in bacterial recombinant systems expressing EPA biosynthesis (*pfa*) genes. Interestingly, levels of protein carbonyls were much lower in *E. coli* carrying *pfa* genes (with EPA) than in *E. coli* carrying no vector (without EPA), even if they had not been treated with H_2O_2. That is, cellular EPA may exert an antioxidative effect on ROS produced intracellularly (Nishida et al., 2006). A variety of stressful conditions, such as heat shock, osmotic shock, nutrient deprivation, and oxidative stress, are known to induce the synthesis of specific proteins. In *E. coli*, the induction of a protein was elicited in response to the overexpression of foreign proteins (Arora et al., 1995). However, to our knowledge, instances where the expression of one foreign gene (DNA) induces the expression of another foreign gene(s) have not been reported.

Clarification of the mechanism of increased EPA (and probably DHA) biosynthesis and the combined use of this technique with the others described above would create the possibility of greater production of these useful polyunsaturated fatty acids.

CONCISE SUMMARY

The VktA catalase is characterized by its high specific activity (Yumoto et al., 1998; 1999; 2000). However, the molecular mechanism of this notable future has not been clarified by itsprimary structure of protein. VktA accumulate predominantly in the periplasmic space at a level of approximately 2% of total soluble proteins of strain S-1 cells (Yumoto et al., 2000) and part of it is localized at the surface of cells. Such specific distribution of VktA may protect it from attack by protein-degrading enzymes. We are not able to conclude whether the high specific activity of VktA and/or VktA accumulation in the cell are involved in enhancing the nitrogen-fixing activity in *R. leguminosarum* and the increased production of EPA (and probably DHA) in *E. coli* cells. If the high accumulation of catalase with a significantly high specific activity is essential for metabolic modifications (discussed above) in *vktA*-transformed host cells, it is desirable to use other kinds of catalase that accumulate in the cells and have a high specific activity. Such catalases are the *Exiguobacterium oxidotolerans* catalase (EKTA catalase; Hara et al., 2007) and the *Psychrobacter piscatorii catalase* (Kimoto et al., 2008), whose specific activity is comparable to that of VktA, indicating that these could be used instead of VktA.

REFERENCES

1. Annie, C.; Chang, A. C. Y. & Cohen, S. N. (1974). Genome Construction Between Bacterial Species *In Vitro:* Replication and Expression of *Staphylococcus* Plasmid Genes in *Escherichia coli. Proceedings of National Academy of Science USA*, Vol. 71, No. 4, pp. 1030–1034.
2. Allgood, G. S. & Perry U. J. (1986). Characterization of a Manganese-Containing Catalase from the Obligate Thermophile *Thermoleophilum album. Journal of Bacteriology*, Vol. 168, No. 2, 563-567.
3. Appleby, C. A. (1984). Leghemoglobin and *Rhizobium* Respiration. *Annual Review of Plant Physiology*, Vol. 35, pp. 443-478.
4. Arora, K. K. & Pedersen P. L. (1995). Glucokinase of *Escherichia coli*: Induction in Response to the Stress of Overexpressing Foreign Proteins. *Archives of Biochemistry and Biophysics,* Vol. 319, No. 2, pp. 574–578.
5. Aslund, F.; Zheng M.; Beckwith, J. & Storz, G. (1999). Regulation of the OxyR Transcription Factor by Hydrogen Peroxide and the Cellular Thiol-Disulfide Status. *Proceedings of National Academy of Science USA*, Vol. 96, No.11, pp. 6161-6165.

6. Bergersen, F. J. & Turner, G. L. (1967). Nitrogen Fixation by Bacteroid Fraction of Breis of Soybean Root Nodules. *Biochimica et Biophysica Acta*, Vol. 141, No.3, pp. 507-515.

7. Bergersen, F. J.; Turner, G. L. & Appleby, C. A. (1973). Studies of the Physiological Role of Leghaemoglobin in Soybean Root Nodules. *Biochimica et Biophysica Acta*, Vol. 292, No. 1, pp. 271-282.

8. Bishai, W. R.; Smith, H. O. & Barcak G. J. (1994). A Peroxide/Ascorbate-Inducible Catalase from *Haemophilus influenza* is Homologous to the *Escherichia coli* KatE Gene Product. *Journal of Bacteriology*, Vol. 176, No. 10, pp. 2914-2921.

9. Christman, M. F.; Morgan, R. W.; Jacobson, F. S. & Ames, B. N. (1985). Positive Control of Regulon for Defenses Against Oxidative Stress and Some Heat-Shock Proteins in *Salmonella typhimurium*. *Cell*, Vol. 41, No. 3, pp. 753-762.

10. Clare, D. A.; Duong, M. N.; Darr, D.; Archibald, F. & Fridovich, I. (1984). Effects of Molecular Oxygen on Detection of Superoxide Radical with Nitroblue Tetrazolium and on Activity Stains for Catalase. *Analytical Biochemistry*, Vol. 140, No. 2, pp. 532-537.

11. Crockford, A. J.; Georgina, A. D. & Williams, H. D. (1995). Evidence for Cell-Density- Dependent Regulation of Catalase Activity in *Rhizobium leguminosarum* bv. *phaseoli*. *Microbiology*, Vol. 141, No. 4, pp. 843-851.

12. Dalton, D. A.; Russell, S. A.; Hanus, F. J.; Pascoe, G. A. & Evans, H. J. (1986). Enzymatic Reactions of Ascorbate and Glutathione that Prevent Peroxide Damage in Soybean Root Nodules. *Proceedings of National Academy of Science USA*, Vol. 83, No. 11, pp. 3811-3815.

13. Deisseroth, A. & Dounce A. L. (1970). Catalase: Physical and Chemical Properties, Mechanism of Catalysis, and Physiological Role. *Physiological Reviews*, Vol. 50, No. 3, pp. 319-375.

14. Dye, M. (1980). Function and Maintenance of a *Rhizobium* Collection, pp. 435-471. In Rao, N. S. S. (Ed.), *Recent advance in biological nitrogen fixation*, Holmes and Meier Publishers Inc., New York.

15. Francis, A. J. & Alexander, M. (1972). Catalase Activity and Nitrogen Fixation in Legume Root Nodules. *Canadian Journal of Microbiology*, Vol. 18, No. 6, pp. 861-864.

16. Goeddel, D. V.; Kleid, D. G.; Bolivar, F.; Heyneker, H. L.; Yansura, D. G.; Crea, R.; Hirose, T.; Kraszewski, A.; Itakura, K. & Riggs A. D. (1979). Expression in *Escherichia coli* of Chemically Synthesized Genes for Human Insulin. *Proceedings of National Academy of Science USA*, Vol. 76, No. 1, pp. 106–110.

17. Greenberg, J. T. & Demple, B. (1988). Overproduction of Peroxide-Scavenging Enzymes in *Escherichia coli* Suppresses Spontaneous Mutagenesis and

Sensitivity to Redox- Cycling Agents in *oxyR*- Mutants. *The EMBO Journal*, Vol. 7, No. 8, pp. 2611-2617.

18. Halliwell, B. & Gutteridge, J. M. C. (1999). *Free Radical in Biology and Medicine* (3rd ed.). Clarendon Press, Oxford, United Kigdom.
19. Hara, I.; Ichise, N.; Kojima, K.; Kondo, H.; Ohgiya, S.; Matsuyama, H. & Yumoto I. (2007). Relationship Between Size of Bottleneck 15 Å Away from Iron in Main Channel and Reactivity of Catalase Corresponding to Molecular Size of Substrates. *Biochemistry*, Vol. 46, No. 1, pp. 11-22.
20. Hassan, H. M. & Fridovich, I. (1978). Regulation of the Synthesis of Catalase and Peroxidase in *Escherichia coli. The Journal of Biological Chemistry*, Vol. 253, No. 18, pp. 6445-6450.
21. Hauvermale, A.; Kuner, J.; Rosenzweig, B.; Guerra, D.; Diltz, S. & Metz, J. G. (2006). Fatty Acid Production in *Schizochytrium* sp.: Involvement of a Polyunsaturated Fatty Acid Synthase and a Type I Fatty Acid Synthase. *Lipids,* Vol. 41, No. 8, pp. 739– 747.
22. Hawley, D. K. & McClure W.R. (1983). Compilation and Analysis of *Escherichia coli* Promoter DNA Sequences. *Nucleic Acids Research*, Vol. 11, No. 8, pp. 2237-2255.
23. Hochman, A. & Shemesh, A. (1987). Purification and Characterization of Catalase- Peroxidase from the Photosynthetic Bacterium *Rhodopseudomonas capsulate. The Journal of Biological Chemistry*, Vol. 262, No.14, pp. 6871-6876.
24. Hernandez-Jimenez, M. J.; Mercedes, L. M. & Rosario de Felipe, M. (2002). Antioxidant Defense and Damage in Senescing Lupine Nodules. *Plant Physioogy and Biochemistry*, Vol. 40, No. 6-7, pp. 645-657.
25. Herouart, D.; Sigaud, S.; Moreau, S.; Frendo, P.; Touati, D. & Puppo, A. (1996). Cloning and Characterization of the *katA* Gene *of Rhizobium meliloti* Encoding a Hydrogen Peroxide-Inducible Catalase. *Journal of Bacteriology*, Vol. 178, No. 23, pp. 6802- 6809.
26. Hirota, K.; Nodasaka, Y.; Orisaka, Y.; Okuyama, H. & Yumoto, I. (2005). *Shewanella pneumatophori* sp. nov., Eicosapentaenoic-Acid-Producing Marine Bacterium Isolated from the Intestines of Pacific Mackerel (*Pneumatophorus japonicus*). *International Journal of Systematic and Evolutionary Microbiology*, Vol. 55, No. 6, pp. 2355-2359.
27. Ichise, N.; Hirota, K.; Ichihashi, D.; Nodasaka, Y.; Morita, N.; Okuyama, H. & Yumoto, I. (2008). H2O2 Tolerance of *Vibrio rumoiensis* S-1^{T} is Attributable to Cellular Catalase Activity. *Journal of Bioscience and Bioengineering*, Vol. 106, No. 1, pp. 39- 45.

28. Ichise, N.; Morita, N.; Hoshino, T.; Kawasaki, K.; Yumoto, I. & Okuyama, H. (1999). A Mechanism of Resistance to Hydrogen Peroxide in *Vibrio rumoiensis* S-1. *Applied and Environmental Microbiology*, Vol. 65, No. 1, pp. 73-79.

29. Ichise, N.; Morita, N.; Kawasaki, K.; Yumoto, I. & Okuyama, H. (2000). Gene Cloning and Expression of the Catalase from the Hydrogen Peroxide-Resistant Bacterium *Vibrio rumoiensis* S-1 and its Subcellular Localization. *Journal of Bioscience and Bioengineering*, Vol. 90, No. 5, pp. 530-534.

30. Ivanova, A.; Miller, C.; Gilinsky, G. & Eisenstark, A. (1994). Role of *rpoS* (*katF*) in *oxyR*- Independent Regulation of Hydroperoxidase I in *Escherichia coli*. *Molecular Microbiology*, Vol. 12, No. 4, pp. 571-578.

31. Jamet, A.; Sigaud, S.; Van de Sype, G.; Puppo, A. & Herouart, D. (2003). Expression of the Bacterial Catalase Genes During *Sinorhizobium meliloti-Medicago sativa* Symbiosis and Their Crucial Role During the Infection Process. *Molecular Plant-Microbe Interaction*, Vol. 16, No.3, pp. 217-225.

32. Katsuwon, J. & Anderson A. J. (1992). Characterization of Catalase Activities in Root Colonizing Isolates of *Pseudomonas putida*. *Canadian Journal of Microbiology*, Vol. 38, No. 10, pp. 1026-1032.

33. Kengen, S.W.M.; Bikker, F. J.; Hagen, W.R.; Vos, W. M. & van der Oost, J. (2001). Characterization of Catalase-Peroxidase from the Hyperthermophilic Archaeon *Archaeoglobus fulgidus*. *Extremophiles*, Vol. 5, No. 5, pp. 323-332.

34. Kim, H.; Lee, J. S.; Hah, Y. C. & Roe, J. H. (1994). Characterization of the Major Catalase from *Streptmyces coelicolor* ATCC 10147. *Microbiology*, Vol. 140, No. 12, pp. 3391- 3397.

35. Kimoto, H.; Matsuyama, H.; Yumoto, I. & Yoshimume, K. (2008). Heme Content of Recombinant Catalase from *Psychrobacter* sp. T-3 Alters by Host *Escherichia coli* Cell Growth Conditions. *Protein Expression and Purification*, Vol. 59, No. 2, pp. 357-359.

36. Klotz, M. G.; Klassen, G. R. & Loewen, P. C. (1997). Phylogenetic Relationships Among Prokaryotic and Eukaryotic Catalases. *Molecular Biology and Evolution*, Vol. 14, No. 9, pp. 951-958.

37. Kono, Y. & Fridovich, I. (1983). Isolation and Characterization of the Pseudocatalase of *Lactobacillus plantarum*. *The Journal of Biological Chemistry*, Vol. 258, No. 10, pp. 6015-6019.

38. Kouchi, H. & Fukai, K. (1989). Rapid Isolation of Bacteroids from Soybean Root Nodules by Percoll Discontinuous Gradient Centrifugation. *Soil Science and Plant Nutrition*, Vol. 35, No. 2, pp. 301-306.

39. Kovach, M. E.; Elzer, P. H.; Hill, D. S.; Robertson, G. T.; Farris, M. A.; RoopII, R. M. & Peterson, K. M. (1995). Four New Derivatives of the Broad-Host-

Range Cloning Vector pBBR1MCS, Carrying Different Antibiotic-Resistance Cassettes. *GENE*, Vol. 166, No. 1, pp. 175-176.

40. Loewen, P. C. (1984). Isolation of Catalase-Deficient *Escherichia coli* Mutants and Genetic Mapping of *katE*, a Locus that Affects Catalase Activity. *Journal of Bacteriology*, Vol. 157, No. 2, pp. 622-626.
41. Loewen, P. C.; Switala, J. & Triggs-Raine, B. L. (1985a). Catalases HPI and HPII in *Escherichia coli* are Induced Independently. *Archives of Biochemistry and Biophysics*, Vol. 243, No. 1, pp. 144-149.
42. Loewen, P. C.; Triggs, B. L.; George, C. S. & Hrabarchuk, B. E. (1985b). Genetic Mapping of *kat*G, a Locus that Affects Synthesis of the Bifunctional Catalase-Peroxidase Hydroperoxidase I in *Escherichia coli*. *Journal of Bacteriology*, Vol. 162, No. 2, pp. 661-667.
43. Mehdy, M. C. (1994). Active Oxygen Species in Plant Defense Against Pathogens. *Plant Physiology*, Vol. 105, No. 2, pp. 467-472.
44. Metz, J. G.; Roessler, P.; Facciotti. D.; Levering, C.; Dittrich, F.; Lassner, N.; Valentine, R.; Lardizabal, K.; Domergue, F.; Yamada, A.; Yazawa, K.; Knauf, V. & Browse, J. (2001). Production of Polyunsaturated Fatty Acids by Polyketide Synthases in Both Prokaryotes and Eukaryotes. *Science*, Vol. 293, No. 5528, pp. 290–293.
45. Morita, N.; Tanaka, N. & Okuyama, H. (2000). Biosynthesis of Fatty Acids in the Docosahexaenoic Acid-Producing Bacterium *Moritella marina* Strain MP-1. *Biochemical Society Transactions*, Vol. 28, No. 6, pp. 943–945.
46. Nadler, V.; Goldberg, I. & Hochman, A. (1986). A Comparative Study of Bacterial Catalases. *Biochimica et Biophysica Acta, Vol.* 882, No. 2, pp. 234-241.
47. Nishi S.; Koyama Y.; Sakamoto T.; Soda M. & Kairiyama C. B. (1988). Expression of Rat α-Fetoprotein cDNA in *Escherichia coli* and in Yeast. *The Journal of Biochemistry*, Vol. 104, No. 6, pp. 968-972.
48. Nishida, T.; Orikasa, Y.; Ito, Y.; Yu, R.; Yamada, A.; Watanabe, K. & Okuyama, H. (2006). *Escherichia coli* Engineered to Produce Eicosapentaenoic Acid Becomes Resistant Against Oxidative Damages. *FEBS Letters*, Vol. 580, No. 11, pp. 2731–2735.
49. Ohwada, T. & Sagisaka, S. (1987). An Immediate and Steep Increase in ATP Concentration in Response to Reduced Turgor Pressure in *Escherichia coli* B. *Archives of Biochemistry and Biophysics*, Vol. 259, No. 1, pp. 157-163.
50. Ohwada, T.; Shirakawa, Y.; Kusumoto, M.; Masuda, H. & Sato, T. (1999). Susceptibility to Hydrogen Peroxide and Catalase Activity of Root Nodule Bacteria. *Bioscience, Biotechnology and Biochemistry*, Vol. 63, No. 3, pp. 457-462.

51. Old, R. W. & Primrose, S. B. (1989). Principles of Gene Manipulation: An Introduction to Genetic Engineering. Blackwell Scientific Publications, Oxford, Boston.

52. Orikasa, Y. (2002). *Construction of Root Nodule Bacteria Expressing High Catalase Activity and Their Ability of Nitrogen Fixation.* Unpublished master's thesis, Obihiro University of Agriculture and Veterinary Medicine, Japan.

53. Orikasa, Y.; Ito, Y.; Nishida, T.; Watanabe, K.; Morita, N.; Ohwada, T.; Yumoto, I. & Okuyama, H. (2007). Enhanced Heterologous Production of Eicosapentaenoic Acid in *Escherichia coli* Cells that Co-Express Eicosapentaenoic Acid Biosynthesis *pfa* Genes and Foreign DNA Fragments Including a High-Performance Catalase Gene, *vktA*. *Biotechonology Letters*, Vol. 29, No. 5, pp. 811-812 (Erratum).

54. Orikasa, Y.; Nodasaka, Y.; Ohyama, T.; Okuyama, H.; Ichise, N.; Yumoto, I.; Morita, N.; Wei, M. & Ohwada, T. (2010). Enhancement of the Nitrogen Fixation Efficiency of Genetically-Engineered *Rhizobium* with High Catalase Activity. *Journal of Bioscience and Bioengineering*, Vol. 110, No. 4, pp. 397-402.

55. Orikasa, Y.; Yamada, A.; Yu, R.; Ito, Y.; Nishida, T.; Yumoto, I.; Watanabe, K. & Okuyama, H. (2004). Characterization of the Eicosapentaenoic Acid Biosynthesis Gene Cluster from *Shewanella* sp. Strain SCRC-2738. *Cellular and Molecular Biology,* Vol. 50, No. 5, pp. 625–630.

56. Ott, T.; Dongen, J. T.; Günther, C.; Krusell, L.; Desbrosses, G.; Vigeolas, H.; Bock, V.; Czechowski, T.; Geigenberger, P. & Udvardi, M.K. (2005). Symbiotic Leghemoglobins Are Crucial for Nitrogen Fixation in Legume Root Nodules But Not for General Plant Growth and Development. *Current Biology*, Vol. 15, No. 6, pp. 531-535.

57. Okuyama, H.; Orikasa, Y.; Nishida, T.; Watanabe, K. & Morita, N. (2007). Bacterial Genes Responsible for the Biosynthesis of Eicosapentaenoic and Docosahexaenoic Acids and Their Heterologous Expression. *Applied and Environmental Microbiology,* Vol. 73 No. 3, pp. 665-670.

58. Panek, H. R. & Obrian, M. R. (2004). KatG is the Primary Detoxifier of Hydrogen Peroxide Produced by Aerobic Metabolism in *Bradyrhizobium japonicum*. *Journal of Bacteriology*, Vol. 186, No. 23, pp. 7874-7880.

59. Puppo, A.; Rigaud, J. & Job, D. (1981). Role of Superoxide Anion in Leghemoglobin Autoxidation. *Plant Science Letters*, Vol. 22, No. 4, pp. 353-360.

60. Rocha, E. R.; Selby, T.; Coleman, J. P. & Smith, C. J. (1996). Oxidative Stress Response in an Anaerobe, *Bacteroides fragilis*: a Role for Catalase in Protection Against Hydrogen Peroxide. *Journal of Bacteriology*, Vol. 178, No. 23, pp. 6895-6903.

61. Saitou, N. & Nei, M. (1987). The Neighbor-Joining Method: a New Method for Reconstructing Phylogenetic Trees. *Molecular Biology Evolution*, Vol. 4, No. 4, 406- 425.

62. Salunkhe D.; Tiwari N.; Walujkar S. & Bhadekar R. (2011). *Halomonas* sp. nov., an EPA- Producing Mesophilic Marine Isolate from the Indian Ocean. *Polish Journal of Microbiology,* Vol. 60, No. 1, pp. 73-78.

63. Sangpen, C.; Skorn, M.; Paiboon, V. & Mayuree, F. (1995). Usual Growth Phase and Oxidative Stress Protection Enzymes, Catalase and Superoxide Dismutase, in the Phytopathogen *Xanthomonas oryzae* bv. *oryzae*. *Applied and Environmental Microbiology*, Vol. 61, No. 1, pp. 393-396.

64. Sato, T.; Yashima, H.; Ohtake, N.; Sueyoshi, K.; Akao, S.; Harper, J. E. & Ohyama, T. (1998). Determination of Leghemoglobin Components and Xylem Sap Composition by Capillary Electrophoresis in Hypernodulation Soybean Mutants Cultivated in the Field. *Soil Science and Plant Nutrition*, Vol. 44, No. 4, pp. 635-645. Schonbaum, G. R. & Chance, B. (1976). Catalase, In: *The Enzymes*. Boyer, P. D., (Ed), p. 363- 408, Academic Press, New York, USA.

65. Sigaud, S.; Becquest, V.; Frendo P.; Puppo, A. & Herouart, D. (1999). Differential Regulation of Two Divergent *Sinorhizobium meliloti* Genes for HPII-like Catalases During Free-Living Growth and Protective Role of Both Catalases During Symbiosis. *Journal of Bacteriology*, Vol. 181, No. 8, pp. 2634-2639.

66. Simon, R. (1984). High Frequency Mobilization of Gram-Negative Bacterial Replicons by the *In Vitro* Constructed Tn5-Mob Transposon. *Molecular and General Genetics*, Vol. 196, No. 3, pp. 413-420.

67. Steiner, B.; Wong, G. H. W. & Graves, S. (1984). Susceptibility of *Treponema pallidum* to the Toxic Products of Oxygen Reduction and the Non-Treponemal Nature of Its Catalase. *The British Journal of Venereal Diseases,* Vol. 60, pp. 14-22.

68. Storz, G. & Zheng, M. (2000). Oxidative Stress. In Storz, G & Hegge-Aronis, R. (Eds.), *Bacterial Stress Response*, pp. 47-59, American Society for Microbiology Press, Washington, DC, USA.

69. Tjepkema, J. D. & Yocum, C. S. (1974). Measurement of Oxygen Partial Pressure within Soybean Nodules by Oxygen Microelectrodes. *Planta*, Vol. 119, No. 4, pp. 351-360.

70. Tajima, S.; Kimura, I. & Sasahara, H. (1986). Succinate Metabolism of Isolated Soybean Nodule Bacteroids at Low Oxygen Concentration. *Agricultural and Biological Chemistry*, Vol. 50, No. 4, pp. 1009-1014.

71. Tartaglia, L. A.; Storz, G. & Ames, B. N. (1989). Identification and Molecular Analysis of oxyR-Regulated Promoters Important for the Bacterial Adaptation to Oxidative Stress. *Journal of Molecular Biology*, Vol. 210, No. 4, pp. 709-719.

72. Thompson, J. D.; Higgins, D. G. & Gibson, T. J. (1994). CLUSTAL W: Improving the Sensitivity of Progressive Multiple Sequence Alignment Through Sequence Weighting, Position-Specific Gap Penalties and Weight Matrix Choice. *Nucleic Acid Research*, Vol. 22, No. 22, pp. 4673-4680.

73. Toledano, M. B.; Kullik, I.; Trinh, F.; Baird, P. T.; Schneider, T.D. & Storz, G. (1994). Redox-Dependent Shift of OxyR-DNA Contacts Along an Extended DNA- Binding Site: A Mechanism for Differential Promoter Selection. *Cell*, Vol.78, No. 5, pp. 897-909.

74. Uriel, J. (1958). Detection des Activites Catalasiques et Peroxydasiques de L'hemoglobine Apres Electrophorese en Gelose. *Bulletin de la Société de chimie biologique*, Vol. 40, pp. 277-280.

75. Valentine, R. C. & Valentine, D. L. (2004). Omega-3 Fatty Acids in Cellular Membranes: a Unified Concept. *Progress in Lipid Research*, Vol. 43, No. 5, pp. 383–402.

76. Vasse, J.; Billy, F. D. & Truchet, G. (1993). Abortion of Infection During a *Rhizobium meliloti*-Alfalfa Symbiotic Interaction is Accompanied by a Hypersensitive Reaction. *The Plant Journal*, Vol. 4, No. 3, pp. 555-566.

77. Verma, D. P. S.; Fortin, M. G.; Stanley, J.; Mauro, V. P.; Purohit, S. & Morrison, N. (1986). Nodulins and Nodulin Genes of *Glycine max*. *Plant Molecular Biology*, Vol. 7, No. 1, pp. 51-61.

78. Vierny, C. & Laccarino, M. (1989). Comparative Study of the Symbiotic Plasmid DNA in Free Living Bacteria and Bacteroids of *Rhizobium leguminosarum*. *FEMS Microbiology Letters*, Vol. 60, No. 1, pp. 15-20.

79. Visick, K. L. & Ruby, E. G. (1998). The Periplasmic, Group III Catalase *Vibrio fisheri* is Required for Normal Symbiotic Competence and is Induced Both by Oxidative Stress and by Approach to Stationary Phase. *Journal of Bacteriology*, Vol. 180, No. 8, pp. 2087-2092.

80. Wei, M.; Takeshima, K.; Yokoyama, T.; Minamisawa, K.; Mitsui, H.; Itakura, M.; Kaneko, T.; Tabata, S.; Saeki, K.; Omori, H.; Tajima, S.; Uchiumi, T.; Abe, M.; Ishii, S. & Ohwada, T. (2010). Temperature-Dependent Expression of Type III Secretion System Genes and Its Regulation in *Bradyrhizobium japonicum*. *Molecular Plant- Microbe Interaction*, Vol. 23, No. 5, pp. 628-637.

81. Wei, M.; Yokoyama, T.; Minamisawa, K.; Mitsui, H.; Itakura, M.; Kaneko, T.; Tabata, S.; Saeki, K.; Omori, H.; Tajima, S.; Uchiumi, T.; Abe, M. & Ohwada, T. (2008). Soybean Seed Extracts Preferentially Express Genomic Loci of *Bradyrhizobium japonicum* in the Initial Intraction with Soybean, *Glycine max* (L.) Merr. *DNA Research*, Vol. 15, No. 4, pp. 201-214.

82. Wittenberg, J. B.; Appleby, C. A.; Bergersen, F. J. & Turner, G. L. (1975). Leghemoglobin: The Role of Hemoglobin in the Nitrogen-Fixing Legume Root Nodule. *Annals of the New York Academy of Sciences*, Vol. 244, No. 1, pp. 28-34.

83. Yazawa, K. (1996). Production of Eicosapentaenoic Acid from Marine Bacteria. *Lipids* (Supplement), Vol. 31, No. 1, pp.S297-S300.

84. Yu R.; Yamada A.; Watanabe K.; Yazawa K.; Takeyama H.; Matsunaga T. & Kurane R. (2000). Production of Eicosapentaenoic Acid by a Recombinant Marine Cyanobacterium, *Synechococcus* sp. *Lipids*, Vol. 35, No. 10, pp. 1061-1064.

85. Yumoto, I.; Ichihashi, D.; Iwata, H.; Istokovics, A.; Ichise, N.; Matsuyama, H.; Okuyama, H. & Kawasaki, K. (2000). Purification and Characterization of Catalase from the Facultatively Psychrophilic Bacterium Vibrio rumoiensis S-1T Exhibiting High Catalase Activity. Journal of Bacteriology, Vol. 182, No. 7, pp. 1903-1909.

86. Yumoto, I.; Iwata, H.; Sawabe, T.; Ueno, K.; Ichise, N.; Matsuyama, H.; Okuyama, H. & Kawasaki, K. (1999). Characterization of a Facultatively Psychrophilic Bacterium, Vibrio rumoiensis sp. nov., that Exhibits High Catalase Activity. Applied Environmental Microbiology, Vol. 65, No. 1, pp. 67-72.

87. Yumoto, I.; Yamazaki, K.; Kawasaki, K.; Ichise, N.; Morita, N.; Hoshino, T. & Okuyama, H. (1998). Isolation of Vibrio sp. S-1 Exhibiting Extraordinarily High Catalase Activity. Journal of Fermentation and Bioengineering, Vol. 85, No. 1, 113-116.

CHAPTER-12

ACUPUNCTURE IN THE MANAGEMENT OF COMMON OBESITY: FUNDAMENTALS AND CLINICAL CONSIDERATIONS

INTRODUCTION: AN OVERVIEW

Acupuncture is gaining recognition as a potential therapeutic approach for addressing simple obesity, both from fundamental and clinical perspectives. Simple obesity, characterized by an excess accumulation of body fat, has become a global health concern due to its association with various chronic diseases. Basic research in acupuncture's mechanisms suggests that it may influence appetite regulation, metabolism, and the distribution of body fat. Clinical studies are exploring its efficacy as a complementary therapy for weight management. By delving into both the fundamental underpinnings and practical applications of acupuncture in the context of simple obesity, we aim to shed light on its potential as a valuable tool in the fight against this prevalent health issue.

Obesity is a serious, prevalent, and refractory health problem. Individuals who are overweight or obese are at greater risk for a variety of medical conditions including diabetes, hypertension, dyslipidemia, fatty liver, cardiovascular disease, and polycystic ovary syndrome.

Typical therapy for obesity includes: diet restriction, regulation of physical activity, behavior treatment, pharmacotherapy, operation, or the use of any of these methods in combination. The cost of treatment of obesity and obesity-related diseases is significant in general health expenditures of various countries. Moreover, pharmacotherapy and surgical operations have side effects and may be unsafe for some people. Even the behavioral treatments, including diet restriction and regulation of physical activity, seem to produce unfavorable psychological changes. The introduction of new therapies for obesity is in demand.

Acupuncture, practiced for several thousand years in China as monotherapy or complementary therapy that is safe and inexpensive, is increasingly used worldwide in the treatment of a wide spectrum of clinic symptoms and diseases. Acupuncture has been found effective in weight control since 1980s'. An increasing body of evidence demonstrates that acupuncture has good effects for weight loss without adverse reactions. Although relatively new, acupuncture therapy for obesity is increasingly accepted by more and more people around the world. In addition, acupuncture can help with the treatment of obesity-related diseases. Extensive research on acupuncture weight loss has been conducted in both basic and clinic areas in recent years. In this chapter we try to provide a comprehensive review of the most recent basic and clinical advances relating to acupuncture in the treatment of obesity.

THE TRADITIONAL CHINESE MEDICINE PERSPECTIVE ON EXCESS WEIGHT AND OBESITY

Obesity is a medical condition in which excess body fat has accumulated to the extent that it may have an adverse effect on health. The major symptoms of obesity are excessive weight gain and the presence of large amounts of fatty tissue. People are considered obese if they weigh 20 percent or more above average for their height and build. A widely applied crude population measure of obesity is the body mass index (BMI), a person's weight (in kilograms) divided by the square of his or her height (in metres). A person with a BMI of 30 or more is generally considered obese. As Asian populations develop negative health consequences at a lower BMI than Caucasians, some nations have redefined obesity; the Japanese have defined obesity as any BMI greater than 25 (Kanazawa et al., 2002) while China uses a BMI of greater than 28 (Beifan et al., 2002).

In the view of Western medicine, obesity is primarily caused by a combination of excessive food energy intake, lack of physical activity, and genetic susceptibility. Although excess food (calories) in any form can be converted into fat and stored, the amount of fat in a person's diet may have a greater impact on weight than the number of calories it contains. A sedentary lifestyle can contribute to positive energy balance and the spared calories are stored as fat (adipose) tissue. Eating habits and patterns of physical activity also play a significant role in the amount of weight a person gains. Genetic factors can influence how the body regulates the appetite and the rate at which it turns food into energy (metabolic rate). Psychological factors such as depression and low self-esteem may, in some cases, also play a role in weight gain.

Not surprisingly, the molecular biology of obesity is only partially understood. This is likely due to the heterogeneity of "garden variety" obesity and the fact that it is caused, like other complex diseases, not by a single genetic mutation but by multiple allelic defects, which determine susceptibility to environmental factors. Individuals who carry only one or some of these alleles may still not develop the disease, because they either lack another allele (gene-gene interaction) or are not exposed to the precipitating environment (gene- environment interaction). Furthermore, there is controversial evidence for a direct association between genotypes and lifestyle or anatomical phenotypes of obesity.

Obesity can be classified in several different ways: for example, by BMI intervals, by anatomic phenotypes or by the stage of life a person becomes obese. Most common form of obesity classification is done using BMI values. According to the World Health Organization (WHO), obesity is classified into 3 classes as listed in table 1.

Table 1. Obesity classification using BMI values

BMI	Classification	BMI(for Asian)
< 18.5	underweight	< 18.5
18.5–24.9	normal weight	18.5–22.9
25.0–29.9	overweight	23.0–24.9
30.0–34.9	class I obesity	25.0–29.9
35.0–39.9	class II obesity	≥ 30.0
≥ 40.0	class III obesity	

The most common anatomical characterization refers to a prevalently visceral or a prevalently subcutaneous deposition of fat. The ratio of waist circumference to hip circumference (WHR) has served the purpose of defining the degree of central (ie visceral) vs. peripheral (ie subcutaneous) obesity. It is known that visceral adiposity is a major risk factor for metabolic complications of obesity, while subcutaneous fat seems to be much more benign, and in some cases even protective against the development of metabolic complications.

At what stage of life a person becomes obese can affect his or her ability to lose weight. In childhood, excess calories are converted into new fat cells (hyperplastic obesity), whileexcess calories consumed in adulthood only serve to expand existing fat cells (hypertrophic obesity). Since dieting and exercise can only reduce the size of fat cells, not eliminate them, persons who were obese as children can have great difficulty losing weight, since they may have up to five times as many fat cells as someone who became overweight as an adult.

Obesity Causes as Viewed Through the Lens of Traditional Chinese Medicine

In Traditional Chinese Medicine (TCM), good health consists of the body's systems acting in harmony according to the individual's constitution. If all is working well, there will not be any weight problem. TCM takes a holistic approach to obesity by focussing on the underlying changes in the body. According to TCM principles, development of obesity is due to the following pathological changes (Integrated Chinese Medicine Holdings LTD. [ICMHL], Shen-Nong Info. a).

Stomach Heat and Impaired Spleen Function as Outcomes of Dyspepsia from a Traditional Chinese Medicine Perspective

Over consumption of heavy, greasy and spicy foods or alcohol facilitate production of heat evils in the stomach. Meanwhile, inadequate exercise after eating these types of foods damages the spleen function. The over-heated stomach will ripen an excessive amount of food.Therefore, the stomach will digest food easier and make an individual feel hungry, but the spleen cannot handle an excessive food load causing it to under function and be unable to carry out its transformation and transportation functions properly. As a result, the spare metabolic products turn into turbid fluid and phlegm which intermix with blood and qi (vital energy) filling up the organs, bones and muscles.

Exogenous evils giving rise to obesity

Invasion of exogenous evils or over consumption of greasy foods leads to poor transformation and transportation functions of the spleen. Dampness evils then begin to accumulate in the middle burner, which is part of the triple burner (the passage through which water, food and fluid are transported). When dampness and turbid fat enter these passages, they are further distributed by the lungs, allowing penetration into all the organs internally. Additionally, exogenous evils can also penetrate the skin, subcutaneous tissue and muscles through the body's surface giving rise to obesity.

Qi (vital energy) stagnation causes turbid phlegm accumulation

For those who are emotionally disturbed, experience trauma, have menstruation problems or are elderly, the liver can fail to regulate qi flow which in turn affects digestion and blood flow. The resulting sluggish qi and blood flow tend to block the meridians. Therefore, in these people, dampness is likely to endure in the body. Over time, this will congeal into phlegm and result in obesity.

Kidney essence exhaustion leads to disharmony

Lifestyles, which consume kidney essence, such as being sexually over active can lead to the excitation of the internal ministerial fire. The excessive ministerial fire is a kind of "evil fire" which makes the body produce an over abundance of heat. This "evil fire" affects the middleburner, leading to a malfunction of the stomach and spleen. When this persists over a long period, the vaporization processes in the bladder and triple burner are impaired causing more evils to accumulate and worsen the obesity condition.

From TCM experience, the above causes of obesity can appear together or separately. In short, the fundamental causes of obesity are spleen and kidney deficiencies, which manifest as an overflow of body fluids, accumulation of dampness and phlegm evils and stagnation in blood flow. Sometimes stomach heat and qi stagnation are associated. Moreover, improper vaporization of body fluids by the triple burner may also appear. All of these factors play an important role in the development of obesity.

Types of obesity from a TCM perspective

Syndrome identification is the premise and foundation of TCM treatment. Currently there are still no standardized obesity patterns, ranging from 3 to 12 patterns by clinical reports. Most often, simple obesity was classified into deficiency syndrome and excess syndrome by syndrome differentiation of TCM, which further classified into four types (ICMHL, Shen-Nong Info.b).

An overabundance of internal mucus and moisture resulting from a weakened spleen

Phlegm is an important concept in TCM. TCM holds that fat or adipose tissue is mostly due to phlegm and dampness evils. The spleen is regarded to be at the root of all phlegm production. When the spleen becomes damaged, such as eating too many sweet foods and getting too little exercise, it will fail in its duty to move and transform waste fluids and foods. Instead these metabolic wastes will gather, collect and transform into evil dampness. If dampness evils endure, over time they will congeal into phlegm, and become fat tissue. The excessive internal phlegm manifests itself as excess weight, accompanied by tiredness, body heaviness, chest and/or stomach distension, and in some cases poor appetite. The tongue has a slimy covering of fur, while there is a rolling, taut pulse. This type of obesity is generally due to an eating disorder, or secondarily by some other illness.

Treatment revolves around sweeping away phlegm and removing stagnation. Once the phlegm is swept away, the qi can move smoothly and easily. This promotes the movement of phlegm and reduction of fat with the ultimate result of decreasing obesity.

Stagnation of qi and blood

Patients exhibiting this condition may suffer from irritability or low motivation, chest or breast fullness, insomnia, a dreamy state, menstrual disorder or amenorrhea (absence of menstruation), and infertility. Some patients may complain of headaches. There may be dizziness and numbness of the four extremities; and the tongue is dark red with a white thin fur or a thin and greasy fur. The pulse is thready and rolling. This is because the movement of blood is not smooth or easily flowing. Stasis obstructs the vessels and inhibits the qi mechanism. Therefore, fat and dampness collect and accumulate within the vessels, making the blood more viscous. If this continues over time, obesity and arteriosclerosis (thickening and hardening of the arteries) will result.

The principle treatment is to speed up the blood flow and remove stagnation.

Yang deficiency of spleen and kidney

In the case of yang deficiency of spleen and kidney, there is not enough qi to transform or melt the phlegm. People in this category often feel exhausted or fatigued. They may experience lower back and knee weakness, shortness of breath, impotence or low libido. The pulse is deep and fine, the tongue is pale and without any fur covering. Genetic factors can play a part in this condition. It may also be the result of other illnesses, stress or an unhealthy lifestyle.

Treatment involves fortifying the spleen and rectifying the kidney deficiency.

Liver stagnation

Liver stagnation caused by prolonged strong emotions or depression leads to disharmony between the spleen and the liver and gives rise to fluid retention. Due to the liver being depressed, the gall bladder is also depressed and exhausted; the ebb and flow of these organs become unbalanced, and the qi mechanism does not flow freely. Hence fat turbidity is difficult to be transformed and over time it leads to obesity.

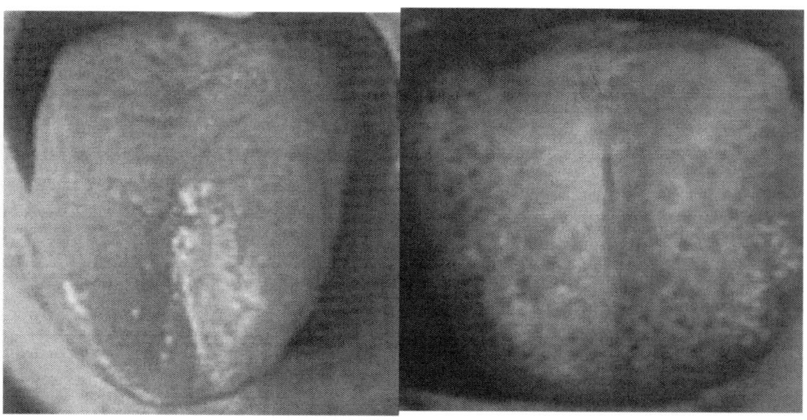

Fig. 1. Changes of tongue coating in different types of fat person(ICMHL, Shen-Nong Info.b)

People in this category tend to have excessive fatty material deposited in the abdomen. The physique is bloated and individuals feel drained of energy. Individuals may also experience excess sputum secretion, dizziness, vertigo, retching, a dry mouth, lack of desire for food or drink and discomfort in chest and abdomen. A white glossy or greasy coating usually covers their tongues. The pulse is rolling. Treatment involves improving liver functioning, unblocking the gallbladder and moving stagnation.

INDIVIDUALIZED ACUPUNCTURE FOR THE TREATMENT OF OBESITY: EFFECTS AND METHODS

In terms of more healthy and holistic methods of weight loss in Chinese medicine, the fundamental prescriptive methodology is to "bian zheng lun zhi"—base treatment on the patient's personal pattern discrimination. TCM doctors diagnose the name of a disease,followed by the differentiation diagnosis of syndromes, for prescribing a treatment. Acupuncture treatment based on a patient's syndrome differentiation is both safe and effective because it addresses that person's own metabolic reasons for being overweight or obese. Using this method, each obese patient can receive his or her own individually tailored acupuncture treatment plan.

Very basically, acupuncture is the insertion of stainless steel filiform needles into precisely specified acupoints on the body's surface , in order to influence physiological functioning of the body.

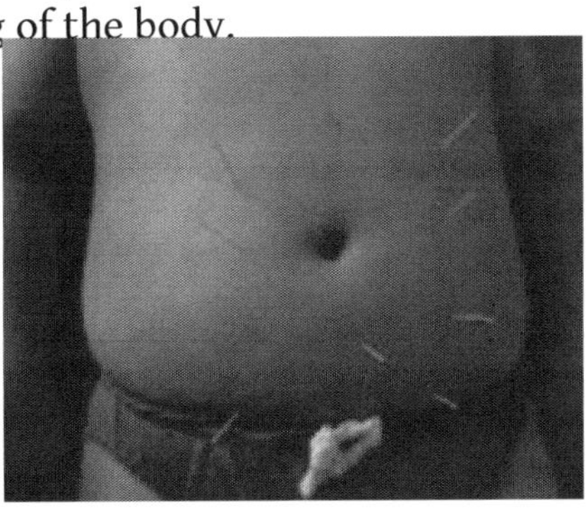

Fig. 2. Acupuncture being applied to the abdomen

According to TCM, the meridian system provides the transportation channel for the fundamental substances of **qi**, **blood**, and **body fluids**, and along the fourteen **main meridians** there are a total of 365 acupoints have been identified, each point belongs to a particular meridian and connects to a corresponding organ that make it exert particular therapeutic properties.

TCM holds that obesity is caused by anomaly transportation and transformation of the body fluid, accumulation of water-dampness and phlegm turbidity, which are the result of disorders of zang-fu organs, stagnation of qi and blood, disharmony of the Thoroughfare and Conception vessels. Therefore the weight loss can be achieved by needling meridian points to balance yin-yang, regulate zang-fu organs, promote flow of qi and blood of the meridians, and eliminate the inner pathogenic factors by means of dredging meridian and collateral. Accordingly, different set of points would be used, depending on which organ(s) needed to be energized or inhibited. Furthermore, acupuncture needles can be twirled, electrically stimulated, penetrated to different depths and left in place for variable lengths of time.

Individualized acupuoints selection

Acupuncture for weight loss refers to the therapeutic approach applying acupuncture or moxibustion on some special points under the guidance of meridian theory of the TCM. For point selection, the acupoints zhong-wan, xia-wan, liang-men and tai-yi were offen used to regulate stomach qi, remove dampness to restore normal function of the spleen. Acupoints tian-shu and da-heng were selected to promote qi circulation and remove obstruction in the collaterals. Acupoints qi-hai and guan-yuan were used to reinforce the kidney. Acupoints wai- ling, shui-dao, qu-chi, zhi-gou and nei-ting were selected to eliminate the dampness and heat, induce diuresis to alleviate edema, and promote qi flow to relax the bowels.

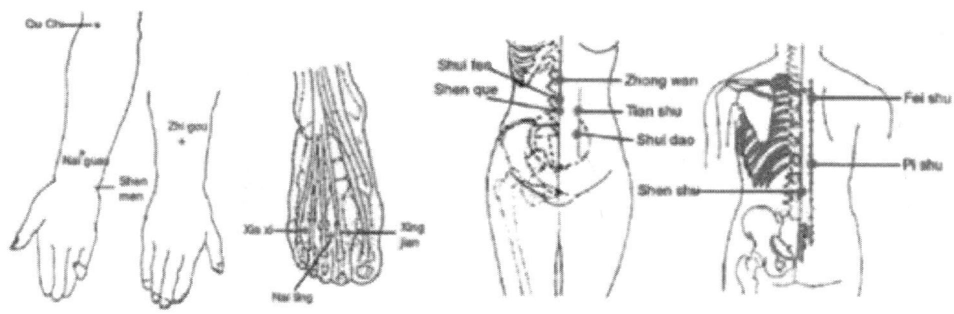

Fig. 3. Some examples of acupoints used for promoting weight loss (ICMHL, Shen-Nong Info. c)

Besides these routine points (mainly the points of Conception Vessel, Spleen, Stomach, Kidney, and Bladder Meridians) which were often selected as chief acupoints in the acupuncture treatment of obesity, different adjunct acupoints were added depending on the types of disharmony pattern.

Liu et al. (2004) used the following therapeutic principle and acupoints selection in their clinical practice:

- Pattern of excessive heat in the stomach and intestines:
- The treatment was designed to clear away heat from the stomach and intestines. The auricular points selected were external nose, small intestine and large intestine, and the body acupoints selected were nei-ting, shang-ju-xu, tian-shu, and qu-chi.
- Pattern of liver qi stagnation:

- The treatment was designed to soothe the liver, regulate qi, activate blood and disperse blood stasis. The auricular points selected were liver, heart, pancreas and gall bladder, and the body acupoints selected were gan-shu, ge-shu, tai-chong, and qu-quan
- Pattem of damp accumulation by spleen deficiency:
- The treatment was designed to clear away heat, remove dampness, dry up dampness and strengthen the spleen. To clear away heat, the auricular points selected were san- jiao, spleen and lung, and the body acupoints selected were shui-fen, qi-hai, yin-ling- quan, zu-lin-qi. To dry up dampness and strengthen the spleen, the auricular points selected were spleen, kidney and san-jiao, and the body acupoints selected were pi-shu, zhong-wan, zu-san-li and tai-bai.
- Pattem of deficiency in both heart and spleen:
- The treatment was designed to reinforce the heart and spleen. The auricular points selected were heart, spleen and endocrine, and the body acupoints selected were xin- shu, pi-shu, nei-guan and zu-san-li.
- Pattem of deficiency in both the spleen and kidney:
- The treatment was designed to tonify the kidney,strengthen the spleen and benefit qi.The auricular points selected were spleen, kidney and endocrine, and the body acupoints selected were shen-shu, pi-shu, tai-xi and zu-san-li.
- Pattem of yin deficiency in the liver and kidney:
- The treatment was designed to nourish the liver and kidney. The auricular points selected were liver, kidney and endocrine, and the body acupoints selected were gan- shu, shen-shu, guan-yuan and san-yin-jiao.
- Pattem of deficiency in both lung and spleen:
- The treatment was designed to tonify and benefit the lung and spleen. The auricular points selected were lung, spleen and san-jiao, and the body acupoints selected were fei-shu, pi-shu, zu-san-li and lie-que.
- Pattern of qi deficiency in the heart and lung:

The treatment was designed to tonify and benefit the heart and lung. The auricular points selected were heart, lung and ear shen-men, and the body acupoints selected were xin-shu, pi-shu, fei-shu, nei-guan and dan-zhong.

In case of heart palpitations and shortness of breath add shen-men and nei-guan; for scanty urine add shui-fen and yin-ling-quan; for qi depression, nei-guan and tai-chong need to be added; for yin deficiency and heat, tai-xi and zhao-hai need to be added; for menopausal obesity, qi-hai, guan-yuan, pi-shu, shen-shu, tai-xi, or ming-men need to be added; for complications of high blood sugar (e.g., deficiency of both qi and yin), yang-chi, wan-gu, ran-gu, san-yin-jiao, yi-shu and shen-shu need to be added; for excessive appetite (e.g., excessive heat of spleen and stomach), liang-men, liang-qiu, nei-ting, gong-sun, fei-shu and wei-shu need to be added.

Selection of acupuncture weight loss methods

Currently there are a variety of acupuncture weight loss methods being used in clinic practice, including body acupuncture, electric acupuncture, magnetic acupuncture, laser acupuncture, warming acupuncture, acupressure, auricular acupuncture, integrative acupuncture and herbal prescription, integrative acupuncture and tuina, integrative warming acupuncture and herbal moxibustion, elongated-needle therapy, catgut embedment, gua-sha(scraping), etc.

Classical body acupuncture

The filiform needles with diameter of 0.25-0.30 mm and length of 40-75 mm were typically selected based on the case's obesity degree. In applying simple body acupuncture, needling maneuvers were stressed. The reducing method with twirling-rotating techniques was used for excessive syndrome, and the reinforcing method with twirling-rotating techniques was used for deficient syndrome.

For assessment of therapeutic effects of simple obesity, most acupuncturists and Traditional Chinese Medicine practitioners in China use the criteria stipulated at 3rd national conference on obesity by integrated Chinese and western medicine in 1991(Wei, 1992).

Recent clinical cure: Body weight decreases to the standard level; BMI is near to 23.0 kg/m^2. Remarkable effect: Body weight decreases>5.0 kg, and BMI decreases ≥ 4.0 kg/m^2.

Effect: Body weight decreases>2.0 kg but<5.0 kg, and BMI decreases ≥ 2.0 kg/m^2 but<4.0 kg/m^2.

Failure: Body weight decreases<2.0 kg, and BMI decreases<2.0 kg/m^2.

Table 2. Needle techniques in acupuncture treatment (ICMHL, Shen-Nong Info. d)

Needle Techniques	Reinforcement	Reduction
Lifting and Thrusting	After the needle is inserted into a given depth and the needling sensation appears, reinforcement is obtained by thrusting the needle heavily and then lifting it gently. This is repeated in a slow and delicate manner from shallow to deep.	After the needle is inserted into a given depth and the needling sensation appears, reduction is obtained by thrusting the needle gently and then lifting it heavily. This is repeated in a quick manner from deep to shallow.

Needle Techniques	Reinforcement	Reduction
Twirling and Rotating	After the needle is inserted into a given depth and the needling sensation appears, reinforcement is obtained by twirling in small amplitude with a gentle and slow pace, and only for a short duration.	After the needle is inserted into a given depth and the needling sensation appears, reduction is obtained by twirling in large amplitude with a fast, heavy and quick pace; manipulation should be of long duration.
Insertion and Withdrawing	Insert slowly, twirl the needle slightly, and then withdraw it quickly.	Insert quickly, twirl the needle vigorously, and then withdraw it slowly.
Keeping the hold open or close	Press the hold after the needle is withdrew.	Shake and enlarge the hold while withdrawing the needle.
Means of respiration	Insert the needle when the patient breathes out and withdraw the needle when the patient breathes in.	Insert the needle when the patient breathes in and withdraw the needle when the patient breathes out.

Zhang (2008) reported a 72 cases of obesity treated by body acupuncture which were divided into two groups: heng-gu, da-he, qi-xue, si-man, zhong-zhu and zhi-gou were used in the first group, and da-chang-shu, guan-yuan-shu, xiao-chang-shu, pang-guang-shu, bai- huan-shu and tai-xi were selected in the second group, among which bilateral points were alternately used. A lifting-thrusting and twirling reduction method was applied in the first two weeks, which was followed by a lifting-thrusting and twirling uniform reinforcing- reducing method with the intensity tolerable to patients. Needles were retained for 30 minutes, during which needles were manipulated twice. The treatment was given 5 times weekly in the first two weeks and then followed by 3 times a week, and 3 months of treatments constituted a therapeutic course. Short-term results showed that 16 cases were clinically cured, 18 cases markedly effective, 34 cases effective and 4 cases failed.

Sun (2008) treated 31 cases of abdominal obesity and 52 cases of symmetrical obesity with the same acupuncture methods for 3 months: zhong-wan, cheng-man, tian-shu, shui-dao, qi- hai, zu-san-li and san-yin-jiao were used as main points. A reinforcing maneuver was used for qi-hai, zu-san-li and san-yin-jiao, and an even maneuver for others. Liang-men, dai-mai, feng-long, ji-men, yin-bao, nao-hui and zhi-gou were selected as subordinate points and punctured with an even maneuver. The symptomatic points: liang-qiu and nei-ting for stomach heat, reducing; tai-chong for liver depression, reducing; yin-ling-quan for deficiency of spleen, reinforcing; and guan-yuan for deficiency of qi, reinforcing. Needleswere remained for 20 min, the treatment was given 3 times weekly, and one month constituted a course. In abdominal obesity group, body mass index (BMI), waist circumference

(WC) and skin fat thickness (SFT) in the upper limbs, trunk and abdomen were very significantly reduced ($P<0.01$) after the 1st and the 2nd course of treatment respectively, but no significant difference was found in all indices after the 3rd course of treatment ($P>0.05$). In symmetrical obesity group, all the indices of BMI, WC and SFT were reduced in the successive 3 months of treatment ($P<0.01$).

A majority of obesity patients have substantial accumulation of fat in abdomen and waist, and the more severe the accumulation, the higher relative risk of obesity-related diseases. Mu et al. (Mu &Yuan, 2008) applied abdomen acupuncture therapy to 30 obese patients with elongated needles (75 mm in length). After routine disinfection in the acupoints areas, needles were inserted into the abdominal points of zhong-wan, xia-wan, qi-hai and guan- yuan to reinforce the spleen and kidney; of bilateral hua-rou-men and wai-ling to regulate qi and blood; bilateral da-heng to reinforce spleen and dispel dampness; bilateral zhi-gan to regulate the body's qi movement; and of shui-dao to clear heat in triple energizer and downbear the urine and stool. After insertion, the needles were perpendicularly punctured into the earth level (deep level) and retained for 3-5 min to await qi, then the needles were manipulated with twirling-rotating techniques to produce needling sensation. The needles were retained for 30 mm and manipulated once every 5 min to strengthen the needling sensation. The treatment was given once every day in the first 5 times, and once every other day in the latter l0 times. Fifteen treatments constituted a therapeutic course. Another 30 comparable obese patients receiving the same acupuncture treatment at the points of zhong- wan, tian-shu, da-heng, shui-dao, qu-chi, zhi-gou, yin-ling-quan, shang-ju-xu, feng-long and nei-ting were taken as body acupuncture control group in this randomized controlled trial. Before treatment, after l treatment course and after 2 treatment courses were selected as observation time point. The results showed that there was no significant difference in BMI between the two groups after 1-course treatment, but the BMI of abdomen acupuncturegroup was significant lower than body acupuncture group after 2-course treatment ($P < 0.05$). The cure and total effective rates in abdomen acupuncture group were also higher than those in body acupuncture group after 2-course treatment (see table 3), indicating thecurative effect is better in the abdomen acupuncture group.

Abdomen acupuncture divides the insertion depth into three levels of heaven, earth and human. Heaven level (shallow needling) is for those with shorter duration or the pathogen in exterior. Human level (middle needling) is for those have long duration but the pathogen doesn't affect zang-fu organs or pathogen in interstices. For those have long duration and the pathogen involve zang-fu organs or the pathogen in interior, earth level (deep needling) is used.

De qi is viewed as essential to acupuncture's therapeutic effectiveness. The therapeutic depth is the depth to which a needle can be manipulated to achieve the characteristic de qi reaction. Although many studies have used computed tomography (CT) to measure the safe depths (the distance from the surface of the skin of the acupoint to the transverse fascia of the abdominal cavity) of acupoints, few studies have reported on the relative ratio between the therapeutic depth and the safe depth. This ratio may be of clinical importance because it may have an impact on the safety and the therapeutic effectiveness of acupuncture. Chen et al. (2009) analyzed the ratio between the therapeutic depth and safe depth of 12 abdominalacupoints with factors sex, body weight, age, and waist girths by one-way analysis of variance and multiple linear regression analysis to show that the therapeutic depth of abdominal acupoints is closer to the safe depth in overweight and in older children aged 7 to 15 years old, ranged from 0.67 to 0.88 and increased significantly with body weight, age, and waist circumference, but there was no significant difference between genders.

Table 3. Comparison of the therapeutic effects between two groups (Cases)

Group	N	Time point	BMI ($\bar{x}\pm s$)	Recent clinical cure	Remarkable effect	Effect	Failure	Total effective rate(%)
Body acupuncture	30	Before treatment	28.36±3.14					
		After 1-course treatment	27.47±2.99	0	0	19	11	63.3
		After 2-course treatment	26.79±2.86	0	11	14	5	83.3
Abdomen acupuncture	30	Before treatment	28.31±2.99					
		After 1-course treatment	26.97±2.89	0	7	20	3	90.0
		After 2-course treatment	25.95±2.85	7	18	5	0	100.0

Classical auricular acupuncture

From a TCM viewpoint, the ears are an important pivot point for the meridians to communicate with each other. When the organs are in disharmony, it will be reflected on the auricle of the ear. Some of the common auricle acupoints selected for needling in weight loss are: large intestine,small intestine, lung, triple burner, endocrine, subcortex, hunger center, thirst center, constipation center, sympathetic, stomach, esophagus, mouth, adrenal gland and spleen (See Fig.4). Obese patient with spleen and kidney disharmonies, for instance, can be treated by stimulating corresponding acupoints on the ears to regulate these organs' functions.

Special stainless steel thumbtack form ear needles were used for auricular acupuncture. After sterilizing the acupoints with 75% alcohol, the ear needles were inserted into the auricular acupoints using forceps or fingers. In each treatment, three to five acupoints were needled to induce soreness, numbness, or heat sensation. The bilateral points can be simultaneously used and the needles retained for 30 minutes, once a day, or be alternately used and the needles be embedded with 3M ventilation tape and kept on the ear for several days. Moreover, Cowherb seeds instead of ear needles can be fixated on the auricular acupoints (acupoint embedding) and the points were pressed several times a day or press when hungry by the patients themselves to cause pain sensation.

Although both body acupuncture and auricular acupuncture were effective for weight reduction in obese subjects, combining the application of both body acupuncture and otopuncture has a better result in reducing body weight with its reliable short-term and stable long-term effect.

Fig. 4. Selected aural acupoints in weight loss (ICMHL, Shen-Nong Info. e)

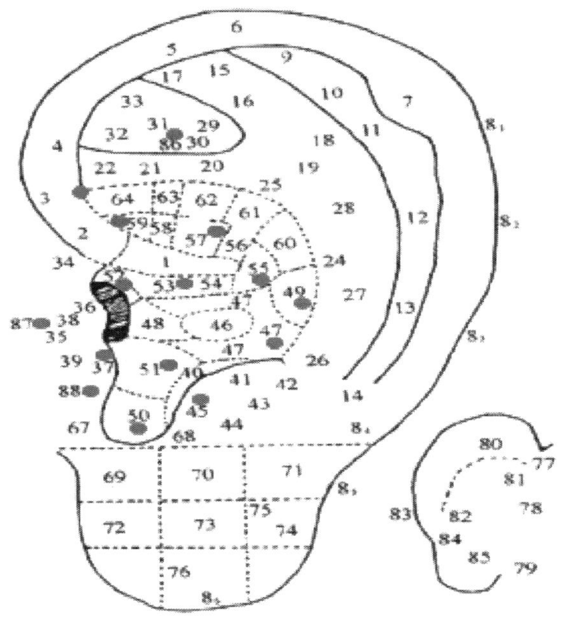

37	adrenal gland
45	sub-cortex
47	lung
49	spleen
50	endocrine
51	triple burners
52	mouth
53	esophagus
55	stomach
57	small intestine
59	large intestine
64	sympathetic
86	constipation center
87	thirst center
88	hunger center

Electroacupuncture

Meng et al. (Zhang, 2008, as cited in Meng et al, 2002) treated 180 cases of female simple obesity by using electroacupuncture (EA) and 60 cases by manual acupuncture as control. Zhong-wan, tian-shu, guan-yuan and zu-san-li were selected as main points in both groups. In the EA treatment group, bilateral tian-shu were stimulated by a G6805 electric apparatus with disperse and dense wave and the intensity tolerable to patients. Needles were retained for 40 minutes, and the treatment was given 5 times a week followed by a 2- day-interval in both groups, and 20 sessions made up of a therapeutic course. The total effective rate of 97.8% and 88.0% was achieved in the EA treatment and control group respectively. Yin (Zhang, 2008, as cited in Yin, 2000) selected zhong-wan, da-heng, guan- yuan and san-yin-jiao as main points, and added secondary points according to differentiation of symptoms and signs. After the arrival of qi by lifting and thrusting for reinforcing and reducing, a G6805 electric apparatus was applied to the main points with continuous waves and 20/ sec in frequency and intensity tolerable to patients. The treatment was given once every other day, and 10 treatments constituted a therapeutic course with an interval of 3 days between two courses. The total effective rate after two courses of treatments was 87.5%.

As the parameters of the EA can be precisely characterized and the results are more or less reproducible, an attempt was made by Han Jisheng's research team to clarify whether EA of strictly identified parameters is effective to suppress the simple obesity induced by high energy diet in a rat model. In the diet-induced obese rats, EA was applied at the hind leg acupoints three times per week for 4 weeks with high energy diet and water provided adlibitum. A significant reduction of the body weight accompanied by a reduction in food intake was observed. 2 Hz EA was more effective than 100 Hz EA (Tian et al., 2005).

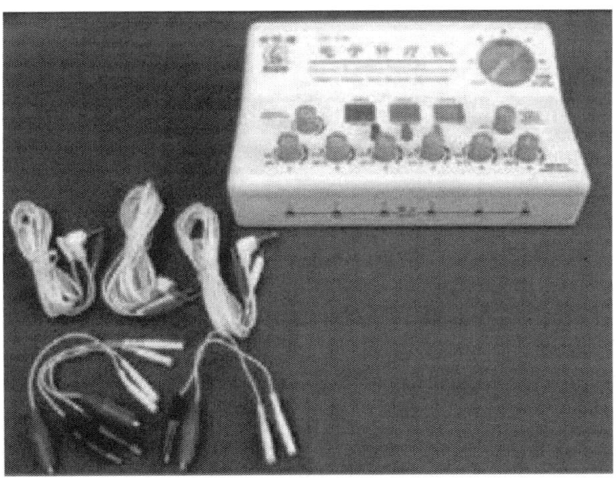

Fig. 5. An electric apparatus for electroacupuncture stimulation

As was known to all, diet-induced obese rats showed an increased level of plasma cholesterol and triglyceride. EA stimulation produced a reduction of plasma level of total cholesterol and triglyceride. In this respect, 100 Hz EA was more effective than 2 Hz EA. If it is verified that 2 Hz EA is more effective in body weight loss and 100 Hz EA more effective in decreaing plasma lipid content, it may be worthwhile to try the 2/100 Hz alternative mode of stimulation to cover both sides of the disorder.

Laser acupuncture

One of the latest developments in acupuncture stimulation methods are laser needles, which are applied to the surface of the skin but are not inserted into the skin. This non-invasive, painless laser stimulation can induce reproducible peripheral and specific cerebral changes that can be measured in different ways, for example, cerebral blood flow velocity. Several studies show that the cerebral effects induced by laser needles are similar to those evoked by manual needle acupuncture. The "low-power" segment of the beam was postulated to be responsible for the clinical therapeutic effects. Laser devices were manufactured in which power densities and energy densities of laser were lowered to a point where no photothermal effects occurred; but the photoosmotic, photoionic, and photoenzymatic effects were still operative. The latest new laser devices are designed at infrared wavelength combined with highfrequency pulses that allow the photons to penetrate deep into tissue without heat effect.

It has been observed that laser acupuncture application to obese people increases excitability of the satiety center in the ventromedial nuclei of the hypothalamus, thus suppressing appetite. John et al (2008) made a randomized controlled pilot study on the effects of laser acupuncture on body weight with subjects divided into control and experimental groups. The experimental group was treated with an activated laser and received 16 J of laser energy output to the he-gu and qu-chi. The control group was given a sham low-level laser therapytreatment with no power output. During the treatment period, each subject received 2 treatments per week for 12 weeks with 4 minutes of active laser or sham treatment to the acupoints in each treatment. Perhaps due to the subjects recruited were not overt obese and the limited laser acupoint stimulation time in this study, no significant weight reduction was observed after the laser acupoint treatment. More studies are needed to investigate the effect of laser acupuncture therapy on body weight.

For the first time, laser needle acupuncture allows simultaneous optical stimulation of individual point combinations. Systematic, double-blind studies of acupuncture can also be performed using optical stimulation because the patient does not notice the activation or deactivation of red or infrared laser needles.

Warming acupuncture

Yang (Zhang, 2008, as cited in Yang, 2002) used moxibustion with warming needle to treat 32 cases of simple obesity of deficiency type by selecting qi-hai, guan-yuan, zu-san-li, tian- shu, yin-ling-quan, and san-yin-jiao as the main points and secondary points according to differentiation of symptoms and signs. Following the arrival of qi, 1-2 lighted moxa sticks about 2 cm in length were consecutively put on the handles of the needles of the 2-3 main points, and the other needles were retained as usual. The treatment was given 6 times weekly, and 30 sessions constituted a therapeutic course. A total effective rate of 90.6% was achieved after one course of treatments.

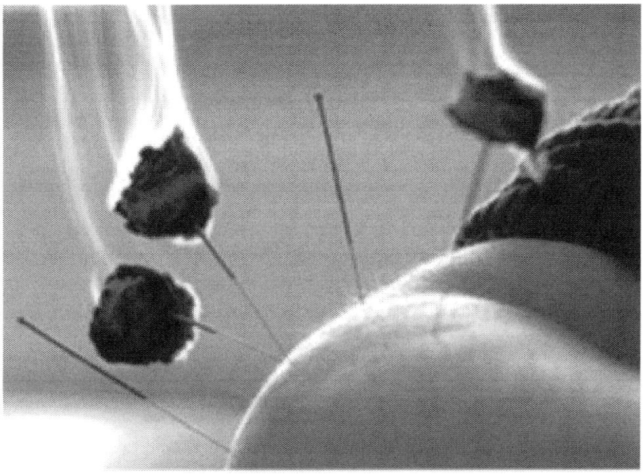

Fig. 6. A pattern of warming needle moxibusiton

Shi et al. (2008) investigated the clinical effect of acupuncture-moxibustion therapy on simple obesity due to spleen deficiency. Sixty-eight cases of simple obesity of deficiency syndrome types, including internal dampness due to spleen deficiency, qi deficiency of lung-spleen, yang deficiency of spleen-kidney, were randomly allocated into two groups, treatment group (36 cases) and control group (32 cases). The former group was treated with warm needling moxibustion method, and the latter was treated with electroacupuncture. Zhong-wan, shui-fen, qi-hai, zhong-ji and bilateral tian-shu, shui-dao, nei-guan, he-gu, xue- hai, zu-san-li, feng-long, san-yin-jiao were selected as main points in both groups, Biliteralda-heng, fu-jie, yin-ling-quan, gong-sun, pi-shu, wei-shu and qi-hai-shu were added as adjunct acupoints for pattern of internal dampness due to spleen deficiency; dan-zhong and biliteral chi-ze, lie-que, yin-ling-quan, fei-shu, pi-shu, gao-huang-shu were added for pattern of qi deficiency of lung-spleen; guan-yuan, ming-men and bilateral gui-lai, shou- san-li, tai-xi, fu-liu, pi-shu, shen-shu were added for pattern of yang deficiency of spleen- kidney. After the needling sensation was obtained by routine acupuncture, the main acupoints were connected to the electroacupuncture apparatus in the control group, with continuous wave and at frequency of 2 Hz, by the intensity of stimulation within the patients's tolerance. For the treatment group, warming needle moxibustion was done on 3-4 pairs acupoints for each pattern [qi-hai and biliteral shui-dao, yin-ling-quan, san-yin-jiao were selected for pattern of internal dampness due to spleen deficiency; shui-fen and biliteral chi-ze, zu-san-li, san-yin-jiao for pattern of deficiency of lung-spleen qi; shui-fen, guan-yuan and biliteral tai-xi, zu-san-li for pattern of yang deficiency of spleen-kidney].

Two cones of moxa roll with length of 1.5-2.0 cm were inserted into the needle handle and light it. The needles were retained for 30 min. The treatment was done every other day and 15 times made up of a course. After one course of treatment, the therapeutic efficacy was analyzed and indicated that the weight losing value of treatment group was obviously higher than that of control group. It indicated that for treating simple obesity due to spleen deficiency, warming needle moxibustion method has more advantage than electroacupuncture method. Some cases in two groups were followed up to 6 months after the treatment ended, no obvious rebound phenomenon of body weight were found in two groups, moreover, for some cases treated by warming needle moxibusiton, their body weight continued to decrease in different degrees.

Warm acupuncture is the combination of neediing and moxibustion. Acupuncture has better effect in dredging the channels and collaterals, qi and blood; moxibustion has the double function of warm-dredge and warm-tonify. The combination of acupuncture and moxibustion can give both reinforcing and reducing, address both the symptoms and root causes.

Usually acupuncture weight loss consists of such three phases as fast, stable, and slow weigh-loss phases (Xu et al., 2004). Clinically most people showed marked effect in the first several times of treatment, followed by a stable phase, which may last different time period for different individuals, as the new metabolic balance was being reconstructed by the acupuncture stimulation. Some TCM doctors consider that formula of the points as well as the reinforcing or reducing manipulations need to be regulated during this phase according to the syndrome differentiation, and then the third phase may come smoothly.

Mechanisms of acupuncture on weight loss

The mechanisms of acupuncture's therapeutic effects for simple obesity are not completely understood. TCM holds that obesity belongs to the mixture of root-deficiency (mainly qi deficiency) and symptoms excess (excess of phlegm-dampness). Acupuncture acts to strengthen the function of spleen, stomach, 1iver and kidney, supplement antipathogenic qi and remove pathogenic qi by stimulating points and regulating meridians. Syndrome differentiation is especially important for the treatment. According to the syndrome differentiation of TCM theory, combination of both chief and supplementory points canregulate qi and blood of meridians, correct yin-yang disorder of zang-fu organs and make lasting force with marked effect.

Although the exact mechanism by which acupuncture works is often unknown, the mechanism that helps to suppress appetite in patients who use acupuncture has been of interest to researchers. Ear has close relation with both meridians and zang-fu organs. There have been reports of reduced appetite or craving for food from subjects wearing auricular acupuncture devices (Dung,1986; Richards & Marley, 1998). The conclusion was usually obtained according to subjective reports rather than quantitative analysis of the food intake. Wang et al (2008), in their animal study, measured daily food intake of the rats, and found a reduction of food consumption in the electroacupuncture (EA) treatment group compared to control group subject to restraint only ($P<0.001$). This reduction was positively correlated with weight loss. It was noticeable that when rats were administered with 2 Hz EA every other day, a reduction of food consumption was observed only on the day of EA administration, suggesting that the effect of EA on appetite suppression lasted for <24 h. Considering the fact that most of the food consumption occurred in the dark phase, the researchers compared the effect of EA treatment delivered in the dark versus light. 2 Hz EA delivered just before the dark phase was more effective than that at the end of the dark phase, suggesting once again that the effect of EA in reducing food intake was immediate and short lasting.

Shiraishi et al. (1995) reported that auricular acupuncture applied to rats produced a reduction of the neural activity of lateral hypothalamus (LH, considered as the "feeding center") and an increase of the neural activity of ventral medial hypothalamus (VMH, considered as the "satiety center"). Because of this, it could help to control the sense of hunger. Furthermore, it has been determined that auricular acupuncture suppresses the appetite by stimulation of the auricular branch of nervous vagus, which has been shown to increase tone in the smooth muscle of the stomach, thus suppressing appetite (Richards& Marley, 1998).

Steyer and Ables（2009）reported that acupuncture affected the ventromedial nucleus of the hypothalamus. Rats that were stimulated with acupuncture needles demonstrated decreased levels of tyrosine and dopamine and increased levels of 5-hydroxytryptamine and 5-hydroxyindoline in this area of the brain. In a study by Wei and Liu in 2003, levels of tryptophan and 5-hydroxyindoleacetic acid were increased, and 5- hydroxytryptamine were decreased in the raphe nuclei of acupuncture treated rats. Thus, acupuncture appears to work on neurotransmitters within the brain to suppress appetite levels and thus help with weight loss.

Electroacupuncture stimulation of the somatic acupoints Zu-san-li and Nei-ting can also increase excitability of the satiety center in the ventral medial nucleus of the hypothalamus (Zhao et al., 2000). The arcuate nucleus of hypothalamus (ARH) is a crucial integrative center for modulation of food intake (Niswender & Schwartz, 2003; Cowley, et al. 2001). The ARH contains at least two populations of neurons that have opposite influence on food intake. One population expresses the anorexigenic peptide "alpha- melanocytestimulating hormone" (a-MSH). The other population expresses the orexigenic peptide "neuropeptide Y (NPY)". It was demonstrated that in obese rats withhyperphagia, the expression of a-MSH in ARH was significantly decreased (Lin et al, 2000; Tian et al., 2004). 2 Hz EA treatment produced an increase in the expression of mRNA encoding a-MSH as well as an increase of the peptide level of a-MSH (Tian et al., 2003). In the meantime, there was a downregulation of NPY expression (Tian et al., 2006). Thus, an increase of a-MSH expression and a reduction of NPY expression in the hypothalamic arcuate nuclei may constitute at least part of the mechanisms underlying the effect of 2 Hz EA for decrease of appetite and reduction of body weight.

Wenhe and Yucun (1981) observed that the level of serotonin (5-HT) in the central nervous system increased with acupuncture application. Serotonin has been implicated in the control of eating behavior and body weight. It is thought that an increase in the level of serotonin in the central nervous system with acupuncture application can provide weight loss, as it has a role in both reducing food intake and arranging the psychomotor balance.

It has been shown that uncoupling protein 3(UCP3) in the muscle accelerates the utilization of fatty acids as energy substrate and UCP3 mRNA expression is positively associated with energy expenditure (Costford et al., 2007). The expression of UCP3 could be up-regulated by the activation of 5'-AMP-activated protein kinase (AMPK) (Suwa et al., 2003), and a deletion of AMPKa resulted in a decrease of UCP3 expression in muscle [Jorgensen et al., 2005]. Several studies had implied that UCP3 may serve as a new target in reducing body weight by up-regulating energy expenditure (Schrauwen et al.,1999; Tiraby et al.,2007; Yoon et al.,2007). Wang et al. (2008) reported no significant change in the content of UCP3 protein in the muscle of obese rats was observed after 2 Hz EA treatment, and, in consistent with this, neither phosphorylated nor total protein level of AMPKa were changed. Therefore, the effects of 2 Hz EA in reducing body weight seem to be a result of decrease of food intake rather than an increase of energy expenditure through AMPK–UCP3 pathway.

In many studies it has been observed that electroacupuncture application caused an increase in the levels of beta endorphin both in serum and in the central nervous system (Jin et al., 1996; Takeshige et al., 1992, 1993; Fu, 2000; Petti et al., 1998). It also has been observed that low current frequency (2 Hz) electroacupuncture application increases the concentration of endomorphins, enkephalins, and beta endorphin but high current frequency (100 Hz) electroacupuncture application increased the concentration of dynorphin in the central nervous system (Han et al., 1999). Richter et al. (1983) investigated the lipolitic activity of beta endorphin in the isolated fat cells of rabbits in vivo. It was determined that as a result of the effect of beta endorphin on fat cells, the levels of free fatty acid and glycerol increased in the rabbit plasma. Vettor et al. (1993) studied the lipolitic activity of beta endorphin in isolated human fat tissue to observed that beta endorphin application caused the increase of glycerol secretion from isolated fat cells. According to the results obtained from these studies, it is thought that electroacupuncture, which increases the plasma beta endorphin levels, can contribute to the weight loss by increasing the lipolitic activity (Cabýoglu et al., 2006a).

As was reported before, diet-induced obese rats showed an increased level of plasma cholesterol and triglyceride. Sun (2005) applied ear- and body-acupuncture to obese subjects, and found a decrease in plasma levels of triglyceride, total cholesterol and LDL cholesterol as well as an increase in the HDL level. Since the manipulation of the needle according to traditional Chinese medicine is difficult to characterize, Wang et al. (2008) took advantage of using precisely identified frequency and intensity of the electrical stimulation applied on needles inserted into the acupoints. EA produced a reduction of plasma level of total cholesterol and triglyceride. In this respect, 100 Hz EA was more effective than 2 Hz EA. If it is verified that 2 Hz EA is more effective in body weight loss and 100 Hz EA more effective in decreasing plasma lipid content, it may be worthwhile to try the 2/100 Hz alternative mode of stimulation to cover both sides of the disorder.

Leptin is a peptide known to decrease the body weight and appetite. Kim et al. (2006) applied 100 Hz EA to ad libitum fed normal rats and revealed a significant increase of plasma leptin level. You et al. (2005) found that 100 Hz EA produced a significant decrease of plasma leptin in obese rats. These results suggest that the effect of EA in modulating plasma level of leptin depends on the energy balance state of the animal. On the other hand, the subject's sensitivity to leptin should be regarded as a more important factor in determining the occurrence of obesity than the plasma level of leptin (Kim et al., 2006; You et al. 2005). In other words, resistance to leptin is more important for the induction of obesity than the insufficient supply of leptin. Therefore, study should be proceeded to characterize whether the sensitivity of leptin can be improved by EA treatment.

In sum, the weight loss function of acupuncture might work through the following three means: 1) Regulating nerve system. It is believed that when needling certain acupoints, peripheral nerves were stimulated to regulate the autonomic nerve of the internal organs and make the intercoordination between sympathetic and parasympathetic nerve, which can inhibit gastric emptying and correct abnormal appetite on one side, and promote intestinal peristalsis and reduce the food absorption on the other side; It has been observed that acupuncture application causes changes in the concentrations of K^+, Na^+, and Ca^+ in the neurons (Deng, 1995), and the amount of neuropeptides like beta endorphin, leucine, encephalin, and neurotransmitters like aspartate in the central nervous system (Fu, 2000). Researchers strongly support the opinions that the effect of acupuncture is arranged by the brain (Futaesaku et al., 1995) and that EA application causes a great change in the action potential of nerve cells (Fu, 2000). 2) Regulating endocrine system. Actually endocrine disorder is both the cause and result of obesity. Acupuncture can restore normal endocrine by regulating the two systems of "hypothalamus-pituitary-adrenal cortex" and "sympathetic adrenal cortex" (Shi & Zhang, 2005). It has been determined that endomorphin-1, beta endorphin, encephalin, and serotonin levels increase in plasma and the central nervous system through acupuncture application. Encephalins as well as serotonin has an effect on feeling well, producing happiness, being pleased, producing a normal level of appetite, and achieving psychomotor balance. These effects play a role in the arrangement of psychological behaviors, including dietary behavior (Cabýoglu et al., 2006b). 3) Regulating lipid metabolism. It has been observed that the increases of endomorphin-1, beta endorphin, encephalin, serotonin, and dopamine cause lipolitic effects on metabolism. Needling certain points can reduce the content of lipid peroxide in the blood and accelerate the fat decomposition. In addition, acupuncture can regulate water and salt metabolism and thus correct the condition of water-salt retention (Sun et al., 1996).

CONTEMPLATION ON THE METHODOLOGY EMPLOYED IN RESEARCH INVESTIGATING THE EFFECTS OF ACUPUNCTURE ON WEIGHT LOSS

Most of the literature on acupuncture for the treatment of obesity is based on uncontrolled trials. Among the controlled trials with positive results, the interpretation of these results is limited by such methodological problems as short duration, inadequate placebo controls, and nonstandard treatment protocols. A recently conducted systematic review and meta- analysis of acupuncture for obesity by Lee et al. (2009), which included a total of 31 randomized controlled trials (RCTs) and 3013 individual cases,reported that: Compared to control of lifestyle, such as diet, exercise or *qigong*,acupuncture was associated with a significant reduction of average body weight (95% confidence interval, CI) of 1.72 kg (0.50–2.93 kg) and associated with an improvement in obesity (relative risk =2.57; 95% CI, 1.98– 3.34). Acupuncture significantly reduced a body weight of 1.56 kg (0.74–2.38 kg), on average, compared to placebo or sham treatments. Acupuncture also showed more improved outcomes for body weight (mean difference =1.90 kg; 1.66–2.13 kg), as well as for obesity (relative risk =1.13; 1.04–1.22), than conventional medication.

However, most studies have been of short duration, varying from 4 to 12 weeks. Noting that obesity is a chronic condition, it is likely to require longer periods of acupuncture treatment. Moreover, obesity may wax and wane with or without treatment, and thus a longer follow- up period with serial measurements of outcomes is suggested to determine the genuine effect of acupuncture as well as its long-term efficacy. Acupuncture may also be considered during the maintenance phase of weight loss programs to prevent relapse. In addition, attrition data at each phase of treatment would provide a more thorough evaluation of this alternative treatment (Lacey et al., 2003).

Concealing allocation of treatment vs control is uniquely challenging. When a nonacupuncture point ('sham') is used, it is important for the treatment to be blinded by all except the acupuncturist, since a needle is applied to the same depth and for the same duration as the treatment group but in a location that has no known effect. Although blinding of the therapist who applies acupuncture would be difficult, blinding of patients and other care providers, as well as outcome assessors should be attempted to minimize the performance and assessment bias of trials. Standardized controls should be used in all future clinical trials of acupuncture treatment in obesity. Nevertheless, there have been studies showing that up to 50% of individuals treated by 'sham acupuncture' processes show some physiological effect (Liang & Koya, 2010). The more recently developed placebo needle may be a more appropriate method to ensure validity in assessing the effectiveness of acupuncture. This placebo needle, with a nonpenetrating, blunt tip, held in place by a bandaid and plastic ring, was shown to be perceived by volunteers as similar to the true penetrating acupuncture needle (also held in place via bandaid and ring), and was significantly less effective (Liang & Koya, 2010). With or without appropriate blinding and placebo controls, expectations about the credibility (usefulness and efficacy) of unconventional methods such as acupuncture may influence outcome. Such moderating variables may obscure real differences between groups especially in small samples. One suggested mechanism to control for patients' expectations is the treatment credibility assessment, adapted from Borkovec and Nau (Lacey et al., 2003, as cited in Borkovec & Nau, 1982), which is a simple series of four questions designed to measure the individual's beliefin the efficacy of treatment. Ideally, the mean scale scores should be equivalent for both treatment and control groups to ensure that the groups are comparable.

Laser acupuncture is currently being used in double-blind studies (Liang & Koya, 2010). In this technique, a laser needle is rather fixed onto the skin than pricked into the skin, to deliver the laser power to the acupoints, and the precise power intensity can determined by using the intensity curve (Litscher & Schikora, 2002). In this way, the patient can hardly feel the stimulation and the operator may also be unaware of whether the laser needle system is active, and therefore true doubleblind studies in acupuncture research can be performed. In this regard, previous studies indicated that laser acupuncture applied to the placebo points did not produce marked cerebral changes compared with that applied to the acupoints (Litscher et al.,2004).

To assess accurately any potential benefits for treating obesity, the art of acupuncture must be effectively bridged with the science of evaluation. Standard algorithms need to be developed, based on principles used by practitioners, for example, criteria for selecting and changing point locations, and spacing of treatments (Lacey et al., 2003).

Several randomized controlled trials have suggested that acupuncture has a positive impact on short-term weight loss. These positive effects are typically not observed when acupuncture is used in the absence of dietary and/or behavioral interventions. Therefore, future studies should include a behavioral component across conditions in order to maximize success, provide an active treatment for the controls, and decrease attrition in the comparison groups (Lacey et al., 2003).

The 31 RCTs results in Lee's systematic review (Lee et al., 2009) do not make any consistent suggestions about which form of acupuncture may be the most effective for various types of obesity. So, in light of the evidence, it is a fair summary that acupuncture is an effective treatment for obesity, and further studies, especially rigorous long-term RCTs, are justified to overcome a number of challenges such as effective evaluation of acupuncture while meeting research standards required for evidence-based medicine, to provide conclusive evidence as to the efficacy of acupuncture for weight loss and which particular type of acupuncture should be offered in accordance with the syndrome differentiation of obesity.

Animal research studies are of great importance to identify the underlying mechanism of acupuncture in treatment of obesity. Originating in China centuries ago, acupoints were described in human body rather than in animals. Animal research of acupuncture was initiated in China in the early 1950s and various mammals such as monkey, horse, dog, mouse, rabbit and rat have been applied in acupuncture studies. Although there have been standardized acupoints in human body (WHO Regional Office for the Western Pacific,2008), no such acupoints have been defined in animals. Most animal studies applied acupoints corresponding anatomically to their original locations in humans. A recent report on transpositional acupoint location in mice and rats may be supplied as a reference (Yin et al., 2008).

Mapping the precise location of needles at specific acupoints, including insertion points, depth, direction and angle will definitely have impact on the effect of acupuncture. Further studies are needed to clearly map the site and depth of needle prick/insertion at acupoints (Liang & Koya, 2010).

Since the manipulation of the needle according to TCM is difficult to characterize, a wide spectrum of high-tech methods including Laser Doppler flowmetry and imaging, multidirectional transcranial ultrasonography, cerebral near infrared spectroscopy as well as functional magnetic imaging and a range of bioelectrical methods have been utilized for research in the field of modernization of acupuncture (Liang & Koya, 2010; Litscher et al.,2004; Litscher & Schikora, 2002). Today it is possible to perform transcontinental studies, for example, using teleacupuncture (Litscher, 2009). Slowly but surely the secrets of acupuncture will be demystified.

REFERENCES

1. Beifan, Z.; Cooperative Meta-analysis group of working group on obesity in China (2002). Predictive values of body mass index and waist circumference for risk factors of certain related diseases in Chinese adults: study on optimal cut-off points of body mass index and waist circumference in Chinese adults. *Asia Pacific Journal of Clinical Nutrition,* Vol.11, Suppl 8, (December 2002), pp. s685–s693, ISSN 0964- 7058
2. Cabýoglu,MT.; Ergene,N. & Tan,U. (2006a). The treatment of obesity by acupuncture. *International Journal of Neuroscience,* Vol.116, No.7, (February 2006), pp.165–175, ISSN 0020-7454
3. Cabýoglu,MT.; Ergene,N. & Tan,U. (2006b). The mechanism of acupuncture and clinical applications. *International Journal of Neuroscience,* Vol.116, No.2, (February 2006), pp.115–125, ISSN 0020-7454
4. Chen,HN.; Lin,JG.; Ying,LC.; Huang,CC. & Lin,CH. (2009). The therapeutic depth of abdominal acupuncture points approaches the safe depth in overweight and in older children. *The Journal of alternative and complementary medicine,* Vol.15, No.9, (October 2009), pp.1033-1037, ISSN 1075-5535
5. Cho,SJ.; Lee,JS.; Thabane,L. & Lee,J. (2009).Acupuncture for obesity: a systematic review and meta-analysis. *International Journal of Obesity,* Vol.33, No.2, (January 2009), pp.183–196, ISSN 0307-0565
6. Costford,SR.; Seifert,EL.; Bézaire,V.; Gerrits,M.; Bevilacqua,L.; Gowing,A. & Harper.ME. (2007). The energetic implications of uncoupling protein-3 in skeletal muscle. *Applied Physiology Nutrition and Metabolism,* Vol.32, No.5, (October 2007), pp.884- 894, ISSN 1715-5320
7. Cowley,MA.; Smart,JL.; Rubinstein,M.; Cerdán,MG.; Diano,S.; Horvath,TL.; Cone,RD. & Low,MJ. (2001). Leptin activates anorexigenic POMC neurons through a neural network in the arcuate nucleus. *Nature,* Vol.411, No.6836, (May 2001), pp.480–484, ISSN 0028-0836
8. Deng,QS. (1995). Ionic mechanism of acupuncture on improvement of learning and memory in age mammals. *American Journal of Chinese Medicine,* Vol.23, No.1, (January 1995), pp.1–9, ISSN 0192-415X
9. Dung,HC. (1986). Attempts to reduce body weight through auricular acupuncture. *American Journal of Acupuncture,* Vol.14, No.2, (May 1986),pp.117–122, ISSN 0091- 3960

10. Fu,H. (2000). What is the material base of acupuncture? The nerves! *Medical Hypotheses,* Vol.54, No.3, (March 2000), pp.358–359, ISSN 0306-9877

11. Futaesaku,Y.; Zhai,N.; Ono,M.; Watanabe,M.; Zhao,J.; Zhang,C.; Li,L. & Shi,X. (1995). Brain activity of a rat reflects apparently the stimulation of acupuncture. A radioautography using 2-deoxyglucose. *Cellular and Molecular Biology,* Vol.41, No.1, (January 1995), pp.161–170, ISSN 1165-158X

12. Han,Z.; Jiang,YH.; Wan,Y.; Wang,Y.; Chang,JK. & Han,JS. (1999). Endomorphin-1 mediates 2 Hz but not 100 Hz electroacupuncture analgesia in the rat. *Neuroscience Letters,* Vol.274, No.2, (October 1999), pp.75–78, ISSN 0304-3940

13. Integrated Chinese Medicine Holdings LTD.(Shen-Nong Info.a). How Does TCM View Obesity and Its Causes? Available from http://www.shen-nong.com/eng/lifestyles/tcmrole_obesityweight_ cause.html Integrated Chinese Medicine Holdings LTD.(Shen-Nong Info.b). Types of Obesity from a TCM Perspective. Available from http://www.shen-nong.com/eng/lifestyles/tcmrole_besityweight_ type.html Integrated Chinese Medicine Holdings LTD.(Shen-Nong Info.c). Body acupuncture therapy for Weight Loss. Available from http://www.shen-nong.com/eng/lifestyles/tcmrole_ obesityweight_ methods_ bodyacupuncture.html

14. Integrated Chinese Medicine Holdings LTD.(Shen-Nong Info.d). Techniques for Enhancing the Needling Stimulation. Available from http://www.shen-nong.com/eng/treatment/acupuncture_enhancing.html Integrated Chinese Medicine Holdings LTD.(Shen-Nong Info.e). Otopuncture & Aural Acu-points Stimulation for Weight Loss. Available from http://www.shen-nong.com/eng/lifestyles/ tcmrole_obesityweight_methods_ otopuncture.html

15. Jin,HO.; Zhou,L.; Lee,KY.; Chang,TM. & Chey,WY. (1996). Inhibition of acid secretion by electrical acupuncture is mediated via beta-endorphin and somatostatin.*American Journal of Physiology,* Vol.271, No.3(Pt 1), (September 1996), pp.G524–G530, ISSN 0002-9513

16. John,Z.; Nelson,M.; George,O.; Amy,S. & Derek,N. (2008). Effect of laser acupoint treatment on blood pressure and body weight—a pilot study. *Journal of Chiropractic Medicine,* Vol.7, No.4, (Decembe 2008), pp.134-139, ISSN 0744-9984

17. Jorgensen,SB.; Wojtaszewski,JF.; Viollet,B.; Andreelli,F.; Birk,JB.; Hellsten,Y.; Schjerling, P.; Vaulont,S.; Neufer,PD.; Richter,EA. & Pilegaard,H. (2005). Effects of alpha- AMPK knockout on exercise-induced gene activation in mouse skeletal muscle. *FASEB J,* Vol.19, No.9, (July 2005), pp.1146–1148, ISSN 8750-7587

18. Kanazawa,M.; Yoshiike, N.; Osaka, T.; Numba, Y.; Zimmet, P. & Inoue, S. (2002). Criteria and classification of obesity in Japan and Asia-Oceania. *Asia Pacific*

Journal of Clinical Nutrition, Vol.11, Suppl 8, (December 2002), pp. s732–s737, ISSN 0964- 7058

19. Kim,SK.; Lee,G.; Shin,M.; Han,JB.; Moon,HJ.; Park,JH.; Kim,KJ.; Ha,J.; Park,DS. & Min,BI. (2006). The association of serum leptin with the reduction of food intake and body weight during electroacupuncture in rats. *Pharmacology Biochemistry and Behavior*, Vol.83, No.1, (January 2006), pp.145–149, ISSN 0091-3057
20. Lacey,JM.; Tershakovec,AM. & Foster, GD.(2003). Acupuncture for the treatment of obesity: a review of the evidence. *International Journal of Obesity,* Vol.27, No.4, (April 2003), pp.419–427, ISSN 0307-0565
21. Liang,F. & Koya D. (2010). Acupuncture: is it effective for treatment of insulin resistance? *Diabetes, Obesity and Metabolism,* Vol.12, No.7, (July 2010), pp.555–569, ISSN 1462- 8902
22. Lin,S.; Storlien,LH. & Huang,XF. (2000). Leptin receptor, NPY, POMC mRNA expression in the diet-induced obese mouse brain. *Brain Research,* Vol.875, No.1-2, (September 2000), pp.89–95, ISSN 0006-8993
23. Litscher G, Schikora D. (2002). Near-infrared spectroscopy for objectifying cerebral effects of needle and laserneedle acupuncture. *Spectroscopy,* Vol.16, No.3-4, (July 2002), pp.335–342, ISSN 0887-6703
24. Litscher,G.; Rachbauer,D.; Ropele,S.; Wang,L.; Schikora,D.; Fazekas,F. & Ebner,F. (2004). Acupuncture using laser needles modulates brain function: first evidence from functional transcranial Doppler sonography and functional magnetic resonance imaging. *Lasers in Medical Science,* Vol.19, No.1, (August 2004), pp.6–11, ISSN 0268-8921
25. Litscher G. (2009). Modernization of traditional acupuncture using multimodal computer- based high-tech methods: Recent results of blue laser and teleacupuncture from the Medical University of Graz. *Journal of Acupuncture and Meridian Studies*, Vol.2, No.3, (September 2009), pp.202–209, ISSN 2005-2901
26. Liu,Z.; Sun,F. & Hu,K. (2004). Clinical study on treatment of simple obesity with acupuncture. *Journal of Acupuncture and Tuina Science*, Vol.2, No.2, (April 2004), pp. 10–13, ISSN 1672-3597
27. Mu,M. & Yuan, Y. (2008). Clinical study on simple obesity treated with abdomen acupuncture. *Journal of Acupuncture and Tuina Science*, Vol.6, No.3, (June 2008), pp. 165–168, ISSN 1672-3597
28. Niswender,KD. & Schwartz,MW. (2003). Insulin and leptin revisited: adiposity signals with overlapping physiological and intracellular signaling capabilities. *Frontiers in Neuroendocrinology*, Vol.24, No.1, (January 2003), pp.1–10, ISSN 0091-3022

29. Petti,F.; Bangrazi,A.; Liguori,A.; Reale,G. & Ippoliti,F. (1998). Effects of acupuncture on immune response related to opioid-like peptides. *Journal of Traditional Chinese Medicine,* Vol.18, No.1, (March 1998), pp.55–63, ISSN 0255-2922

30. Richards,D. & Marley,J. (1998). Stimulation of auricular acupuncture points in weight loss. *Australian Family Physician,* Vol.27, No.2, (July 1998),pp.S73–S77, ISSN 0300-8495 Richter,WO.; Kerscher,P. & Schwandt,P. (1983). Beta-endorphin stimulates *in vivo* lipolysis in the rabbit. *Life Sciences,* Vol.33, No.S1, (July 1983), pp.743–746, ISSN 0024-3205

31. Schrauwen,P.; Xia,J.; Bogardus,C.; Pratley,RE. & Ravussin,E. (1999). Skeletal muscle uncoupling protein 3 expression is a determinant of energy expenditure in Pima Indians. *Diabetes,*Vol.48, No.1, (January 1999), pp.146–149, ISSN 0012-1797

32. Shi,Y.; Zhao,C. & Zuo,XY. (2008). Clinical study on treatment of simple obesity due to spleen deficiency by acupuncture-moxibustion. *Journal of Acupuncture and Tuina Science,* Vol.6, No.6, (December 2008), pp.352–355, ISSN 1672-3597

33. Shi,Y. & Zhang,LS. (2005). Therapeutic Idea and Approaches to Obesity with Acupuncture. *Journal of Acpuncture and Tuina Science,* Vol.3, No.4, (August 2005), pp.54–57, ISSN 1672-3597

34. Shiraishi,T.; Onoe,M.; Kojima,T.; Sameshima,Y. & Kageyama,T. (1995). Effects of auricular stimulation on feeding-related hypothalamic neuronal activity in normal and obese rats. *Brain Research Bulletin,* Vol.36, No.2, (May 1995), pp.141–148, ISSN 0361-9230

35. Steyer,TE. & Ables,A. (2009). Complementary and alternative therapies for weight loss. *Primary Care: Clinics in Office Practice,* Vol.36, No.2, (June 2009), pp.395–406, ISSN 0095-4543

36. Sun,FM.; Liu,ZC. & Wang,YZ. (1996). The antiobesity effect of acupuncture and it's influence on water and salt metabolism. *Acupuncture Research,* Vol.21, No.2, (January 1996), pp.19–24, ISSN 1000-0607

37. Sun,H.(2008).Relationship between treatment course and therapeutic effect of acupuncture for female obesity of different types. *Journal of Traditional Chinese Medicine,* Vol.28, No.4, (November 2008), pp.258-261, ISSN 0255-2922

38. Sun,PH. (2005). Clinical observation on treatment of simple obesity with acupuncture. *Journal of Acupuncture and Tuina Science,* Vol.3, No.6, (December 2005), pp.26–28, ISSN 1672-3597

39. Suwa,M.; Nakano,H. & Kumagai,S. (2003). Effects of chronic AICAR treatment on fiber composition, enzyme activity, UCP3, and PGC-1 in rat muscles. *Journal of Applied Physiology,* Vol.95, No.3, (September 2003), pp.960–968, ISSN 8750-7587

40. Takeshige,C.; Nakamura,A.; Asamoto,S. & Arai,T. (1992). Positive feedback action of pituitary beta endorphin on acupuncture analgesia afferent pathway. *Brain Research Bulletin,* Vol.29, No.1, (July 1992), pp.37–44, ISSN 0361-9230

41. Takeshige,C.; Oka,K.; Mizuno,T.; Hisamitsu,T.; Luo,CP.; Kobori,M.; Mera,H. & Fang,TQ. (1993). The acupuncture point and its connecting central pathway for producing acupuncture analgesia. *Brain Research Bulletin,* Vol.30, No.1-2, (March 2003), pp.53–67, ISSN 0361-9230

42. Tian,DR.; Li,XD.;,Wang,F.; Niu,DB.; He,QH.; Li,YS.; Chang,JK.; Yang,J. & Han,JS. (2005). Up-regulation of the expression of cocaine and amphetamine-regulated transcript peptide by electroacupuncture in the arcuate nucleus of diet-induced obese rats. *Neuroscience Letters*, Vol.383, No.1-2, (July 2005), pp.17-21, ISSN 0304-3940

43. Tian,DR.; Li,XD.; Shi,YS.; Wan,Y.; Wang,XM.; Chang,JK.; Yang,J. & Han,JS. (2004). Changes of hypothalamic alpha-MSH and CART peptide expression in diet- induced obese rats. *Peptides*, Vol.25, No.12, (December 2004), pp.2147–2153, ISSN 0196-9781

44. Tian,DR.;Li,XD.;Niu,DB.;Shi,YS.;Chang,JK.&Han,JS.(2003).Electroacupuncture upregulated arcuate nucleus alpha-MSH expression in the rat of diet induced obesity.*Journal of PekingUniversity (Health Sciences)*, Vol.35, No.5, (October 2003), pp.458–461, ISSN 1671-167X

45. Tian,N.; Wang,F.; Tian,DR.; Zou,Y.; Wang,SW.; Guan,LL.; Shi,YS.; Chang,JK.; Yang,J. & Han,JS. (2006). Electroacupuncture suppresses expression of gastric ghrelin and hypothalamic NPY in chronic food restricted rats. *Peptides*, Vol.27, No.9, (September 2006), pp.2313–2320, ISSN 0196-9781

46. Tiraby,C.; Tavernier,G.; Capel,F. Mairal,A.; Crampes,F.; Rami,J.; Pujol,C.; Boutin,JA. & Langin,D. (2007). Resistance to highfat-diet-induced obesity and sexual dimorphism in the metabolic responses of transgenic mice with moderate uncoupling protein 3 overexpression in glycolytic skeletal muscles. *Diabetologia*, Vol.50, No.10, (October 2007), pp.2190–2199, ISSN 0012-186X

47. Vettor,R.; Pagano,C.; Fabris,R.; Lombardi,AM.; Macor,C. & Federspil,G. (1993). Lipolytic effect of beta-endorphin in human fat cells. *Life Sciences,* Vol.52, No.7, (July 1993), pp.657–661, ISSN 0024-3205

48. Wang,F.; Tian,DR. & Han,JS. (2008). Electroacupuncture in the treatment of obesity. *Neurochemical Research,* Vol.33, No.10, (August 2008), pp.2023–2027, ISSN 0364- 3190

49. Wei,B.(1992). Integrated Chinese and westem medieine criteria for diagnosis and therapeutic effects of simple obesity. *Chinese Journal of Integrated Traditional*

and Western Medicine, Vol.12, No.11, (November 1992), pp.690-691, ISSN 1003-5370

50. Wei,Q. & Liu,Z. (2003). Effects of acupuncture on monoamine neurotransmitters in raphe nuclei in obese rats. *Journal of Traditional Chinese Medicine*, Vol.23, No.2, (June 2003), pp.147–150, ISSN 0255-2922
51. Wenhe,Z. & Yucun,S. (1981). Change in levels of monoamine neurotransmitters and their main metabolites of rat brain after electric acupuncture treatment. *International Journal of Neuroscience,* Vol.15, No.3, (March 1981), pp.147–149, ISSN 0020-7454
52. WHO Regional Office for the Western Pacific. (2008). WHO standard acupuncture point locations in the western pacific region. 1 WPRO Nonserial Publication, ISBN-13 9789290613831 ISBN-10 9290613831
53. Xu,B.; Liu,ZC. & Zhang, ZC.(2004).Basic idea and approaches to treatment project design of obesity with acupuncture. *Chinese Acupuncture and Moxibustion,* Vol.24, No.2, (February 2004), pp.129–133, ISSN 0255-2930
54. Yin,CS.; Jeong,HS.; Park,HJ.; Baik,Y.; Yoon,MH.; Choi,CB. & Koh,HG.(2008). A proposed transpositional acupoint system in a mouse and rat model. *Research in veterinary science*, Vol.84, No.2, (April 2008), pp.159–165, ISSN 0034-5288
55. Yoon,Y.; Park,BL.; Cha,MH.; Kim,KS.; Cheong,HS.; Choi,YH. & Shin,HD. (2007). Effects of genetic polymorphisms of UCP2 and UCP3 on very low calorie diet-induced body fat reduction in Korean female subjects. *Biochemical and Biophysical Research Communications*, Vol.359, No.3, (August 2007), pp.451–456, ISSN 0006-291X
56. You,JS. & Hung,CC. (2005). Effect of electroacupuncture on plasma of leptin and insulin in diet-induced obese rats. *Journal of Chinese Medicine*, Vol.16, No.2-3, (September 2005), pp.101-109, ISSN 0143-8042
57. Zhang,X.(2008). A clinical survey of acupuncture slimming. *Journal of Traditional Chinese Medicine*, Vol.28, No.2, (May 2008), pp.139-147, ISSN 0255-2922
58. Zhao,M.; Liu,Z. & Su,J. (2000). The time–effect relationship of central action in acupuncture treatment for weight reduction. *Journal of Traditional Chinese Medicine*, Vol.20, No.1, (March 2000), pp.26–29, ISSN 0255-2922.

CHAPTER-13

SPERMATOGONIAL STEM CELLS AND THEIR ROLE IN ANIMAL TRANSGENESIS

INTRODUCTION: AN OVERVIEW

Spermatogonial stem cells (SSCs) play a crucial role in the field of animal transgenesis, which involves the introduction of foreign genetic material into the germ line of an animal to create genetically modified offspring. SSCs are unique in that they give rise to sperm throughout the entire life of a male animal, serving as a continuous source of genetic material. This feature makes SSCs an ideal target for genetic modification, as they can be manipulated to carry and transmit specific traits to the next generation. Researchers use techniques such as gene editing to introduce desired genes or modifications into SSCs, which can then be used to generate transgenic animals with the desired genetic traits. This approach has profound implications for biotechnology, agriculture, and biomedical research, as it enables the development of genetically modified animals for various purposes, including disease modeling, the production of valuable proteins, and the improvement of livestock and crop species.

Spermatogonial stem cells (SSCs) are unipotent adult stem cells responsible for the maintenance of the spermatogenesis throughout the entire life of the male. We could say that the mammalian spermatogenesis is a classic adult stem cell-dependent process, sustained by self renewal and differentiation of SSCs. They are the only germline stem cells in adults. These cells can be found in the seminiferous tubule, lying near to the basement membrane. The SSC may choose to self-renewal or generate a daughter cell committed to differentiation. Studying SSCs provides a model to better understand adult stem cell biology and decipher the mechanisms that control SSC functions. It was reported that these cells hold the ability to colonize the seminiferous tubules after transplantation, restoring spermatogenesis. Besides the biomedical potential to perform studies of infertility in many species, SSCs present a promising application in biotechnology in the production of transgenic animals. This alternative route for transgenesis is of interest because a single male will generate by regular mate a variety of transgenic progenies. The production of a transgenic gonad can overcome the obstacles faced with the sperm-mediated gene transfer (SMGT) due to the high specialization of sperm. The use of SSC for transgenesis relies on targeting a much more undifferentiated germ cell and the potential permanent modification of the germ line. In this manner, this chapter aims to review the following topics regarding SSCs: (1) Mammalian spermatogenesis and SSCs; (2) Characterization of SSCs; (3) Isolation and *in vitro* culture of SSCs; (4) Transplantation of SSCs and animal transgenesis.

SPERMATOGONIAL STEM CELLS AND SPERMATOGENESIS

Spermatogenesis is a highly organized and complex process that is responsible for sperm production in male individuals (Russell et al., 1990). Besides providing continuous source of spermatozoa, it is responsible for maintenance of its stem cell population by constant replication of SSCs. In mammals, millions of sperm cells are produced everyday from SSC (Meistrich & van Beek, 1993). In the testis, only SSCs hold the self-renewal ability, i.e. the ability to undergo a series of mitotic cycles without differentiating. In this manner, we can say that spermatogonial stem cells are at the foundation of spermatogenesis. They are the adult stem cell population of the testis, which is responsible for the maintenance of spermatogenesis throughout the entire life of the male. As observed in other tissue-specific stem cells, SSCs are rare, being only 0.03 percent present in an adult mouse testis (Tagelenbosch & de Rooij, 1993). They are present in the testis in such a small number due to the high density of differentiated germ cells, as differentiating spermatogonia, spermatocytes, spermatids and sperm, all originated from SSCs. We define SSCs as stem cells based on their ability to balance self-renewal and differentiation. The self-renewal sustains the stem cell pool at the testis because SSC undergo multiple mitosis producing new SSCs. These new SSCs hold the same self-renewal and differentiation potential as their precursors. Upon demand, SSCs start the cell division in order to produce a differentiated daughter cell. The balance of these two cell divisions maintains spermatogenesis, which can produce millions of sperm each day without causing depletion of cell source.

SSCs originate from primordial germ cells (PGC), which migrate from the embryonic ectoderm in the epiblast, through the allantois and hindgut until reaching genital ridges (Lawson & Pedersen, 1992; Clark & Eddy, 1975). Once PGCs colonize the genital ridge, they are enclosed by differentiating Sertoli cells, starting the formation of seminiferous cords, that will eventually give rise to seminiferous tubules (Byskov & Hø´yer, 1994). From this stage on, the germ cells are called gonocytes because they differ morphologic from PGCs (Clermont & Perey 1957; Huckins & Clermont 1968). In the late stage of gestation, the gonocytes undergo proliferation and become quiescent, i.e., arrested in the G0/G1 phase of the cell cycle (Clermont & Perrey, 1957). These cells remain in mitotic arrest until the peri- pubertal period, when they start proliferating again, this time to produce type A SSCs. Increasing levels of gonadotrophic hormones concentration triggers this massive proliferation of type A SSCs, marking the onset of spermatogenesis (Huckins & Clermont, 1968; Belveé et al., 1977).

It is known that in rhesus monkey (de Rooij et al., 2002) and human (Clermont, 1966), two subtypes of type A spermatogonia are morphologic distinguishable: Adark and Apale. The Adark spermatogonium act as a true SSC, forming the testis regenerative reserve while Apale has a progenitor role, constituting the functional reserve (Ehmcke et al., 2006). In rhesus monkey, these spermatogonia are followed by four generations of spermatogonia in different stages of differentiation (B1,B2, B3 and B4; de Rooij, 1986). In human, only one generation of type B spermatogonium can be observed before formation of spermatocytes (Clermont, 1966). In rodents, seven subtypes of type A spermatogonia have been reported: A simple (As), A pared (Apr), A aligned (Aal), A1, A2, A3 and A4 (Huckins, 1971a,b; Huckins & Oakberg, 1978). As spermatogonia are considered the SSCs. Although still undifferentiated, Apr and Aal produce expanded colonies of SSCs because they have already undergone mitosis. However, due to a formation of an intercellular bridge that connects their daughter cells, their division is considered incomplete (Zamboni & Merchant 1973). As a result, these cells no longer possess the self-renewal ability as subtype As spermatogonia. Subtypes A1-A4 spermatogonia also constitute expanded SSCs colonies, but differently from Aal and Apr, they are already synchronized with the cycle of the seminiferous epithelium. In this manner, it is possible to say that A1-A4 spermatogonia are already committed to differentiation into future spermatozoa. We refer as the cycle of seminiferous epithelium the synchronic evolution of germ cells from one stage of spermatogenesis to the next. In other words, the cycle of seminiferous epithelium is the completion of ordered events of cell association, divisions and stages in the seminiferous epithelium over time (Russell et al., 1990). In this cycle, the succession of spermatogonia, spermatocytes and spermatids from basementmembrane toward the lumen of seminiferous tubule is established in a stepwise manner during postnatal development.

In bovine, a livestock species, a similar classification was proposed in 1995 by Wrobel et al. According to this classification, there are basal stem cells, corresponding to type As and Apr spermatogonia in rodents, aggregated spermatogonial precursor cells, equivalent to Aal spermatogonia and finally committed spermatogonial precursor cells, equivalent to A1-A4 spermatogonia. It was suggested that type Apr spermatogonia also hold stem cell properties in bulls.

CHARACTERIZATION OF SPERMATOGONIAL STEM CELLS

SSC are the foundation of the productive spermatogenesis that results in the continuous production of spermatozoa in the postnatal life, but studies with SSCs are complicated because these cells are few in number and no unique identifying characteristics have been reported to date. Thus, little is known of their morphology, functional assay or biochemical characteristics and those evaluations become harder in postnatal tissue. Togelenbosch and de Rooij (1993) performed a quantitative study with spermatogonial cells in mouse testis comprising 1 in 3333 cells from adult mouse testis.

In the spermatogenic cycle with each division the number of cells theoretically double, but is important to remember that generally there is no divisions between Aal to A1 cells. Although morphological changes occur and A1 cells slightly resemble Aal cells. Only some spermatogonial cell types can be distinguished by morphologic characteristics and this may actually cause many disturbances in spermatogonial kinetics studies. In almost all species the type A have very similar morphologic characteristics when these cells are analyzed in whole seminiferous tubules. On the other hand, the type A (as a class), Intermediate and type B spermatogonia can be distinguished by minor morphological changes, using either light or electron microscopy (Russell et al., 1990).

The type A spermatogonia have two different surfaces: one flattened and another rounded. The first surface acquires this format because of its direct contact with basal lamina and the second surface is surrounded by Sertoli cells. In the nucleus is observed little presence of heterochromatin and the nucleolus is visible. The Intermediate spermatogonia typically show an ovoid nucleus, present more heterochromatin located close to the nuclear envelop compared to type A and have also a rounded and a flattened surface. Finally, type B spermatogonia has a rounded nucleus with a moderated quantity of heterochromatin allocated around nuclear edge. A smaller part of the cellular membrane is in contact with basal lamina then the most part of surface is rounded. Thus, main morphological aspects that are analyzed to distinguish the types of spermatogonial cells are, first, the amount of heterochromatin in the nucleus and its relation to nuclear membrane. The type A basically has no heterocromatin, Intermediate displays a moderate quantity and type B an abundant amount. The second important aspect is that spermatogonial cells are part of seminiferous epithelium and always have a flattened surface in contact with basal lamina, and rounded, in contact with Sertoli Cells (Russell et al., 1990). Approximately 300,000 cell (types As, Apr, Aal and Intermediated) were counted and characterized from mice seminiferous tubules (Togelenbosch & de

Rooij, 1993) being identified approximately 35,000 type As cells from each testis (Meistrich & van Beek, 1993). The morphological evaluation is an important toolfor permatogonial studies but it provides many disturbance in the analysis of data, mainly in cells that is analyzed outside of seminiferous tubules environment. Nevertheless, espermatogonial cells can be identified with functional assays or molecular techniques besides the morphological characterization.

In the functional assay, the presence of SSC, for example: from a new purification protocol, was checked by the transfer of progenitor germ cell to the testis of a recipient animal. Spermatogenesis of the recipient testis was previously depleted by the treatment with an alkylating agent, Busulfan or fractionated X-irradiation (local testicular doses of 1.5 and 12 Gy, 24 h apart; Aponte et al., 2005). After transplantation, SSC repopulate the recipient animal seminiferous tubules, that produces a spermatic cycle with donor progenitor cells from the same specie or not.

Whilst in the undifferentiated stage, SSCs express different proteins and genes. In both cases, they are not produced in greater differentiation (Type A1-A4, Intermediated, spermatocytes and spermatids; Caires et al., 2010). A key step in studying the biology of SSCs is to determine their gene expression profile. However, a scarce knowledge of molecular markers has been accumulated in recent years (Kokkinaki et al., 2010). Some research groups have demonstrated that glial cell-. derived neurotrophic factor (GDNF) is the most essential factor for SSC self-renewal and *in vitro* maintenance in rodents (Caires et al., 2010; Ryu et al., 2005; Meng et al., 2000; Kubota et al., 2004a,b; Kanatsu-Shinohara et al., 2005; Kanatsu-Shinohara et al., 2008; and Braydish-Stalle et al., 2005) and that GDNF receptor (GFRA1) is expressed by SSC/progenitor cells (Naughton et al., 2006; Hofmann et al., 2005 and He et al., 2007). The activation of GDNF pathway probably is related with other pathways that promote the proliferation and the self-renewal of SSCs (Caires et al., 2010; Jijiwa et al., 2004; Braydich-Stolle et al., 2007; Oatley et al., 2007 and Lee et al., 2006). Thereby, many molecular markers for SSC are associated with GDNF pathway.

SSC MARKERS AND DIFFERENCES AMONG SPECIES

The establishment of molecular signatures for SSCs are a complex and difficult process but some molecular markers have been defined for SSCs and undifferentiated spermatogonia (Caires et al., 2010). It is important to know that all these markers (Table 1) were established for different species using a pool containing undifferentiated germ cells.

One of the most important molecular marker for progenitor germ cells is the GRFA1 that is a co-receptor of RET for the GDNF (He et al., 2007). GDNF is related to neural development (He et al., 2007 and Garces et al., 2000) and renal morphogenesis (He et al., 2007 and Vega et al., 1996). In spermatogonial cells, this factor plays an important role in the regulation of proliferation and differentiation or undifferentiated (He et al., 2007, Naughton et al., 2006; Takakoro et al., 2002; Meng et al., 2000 and Hofmann et al., 2005). Others important markers are Nanog and Pou5f1 (Oct3/4). They are essential transcription factors for the maintenance of pluripotency (Goel et al., 2008). THY-1, a member of the Ig super family is highly expressed in rat stem cells and in SSCs from pre-pubertal bulls (Aponte et al., 2005, Ryu et al., 2004 and Reding et al., 2010), However, the role of THY-1 in the male fertility is still unknown (Aponte et al., 2005 and Barlow et al., 2002). PLZF (*Zfp145*) is a molecular marker for As, Apr and Aal spermatogonia and as GRFA1, it is related with self-renewal of SSC (Aponte et al., 2005; Buaas et al., 2004 and Costoya et al., 2004). NGN3 also is expressed in the same cell types as PLZF and acts in the differentiation of spermatogonia (Aponte et al., 2005 and Yoshida et al., 2004), but also is present in spermatocytes (Aponte et al., 2005 and Reverot et al., 2005).

Table 1. Spermatogonial cells molecular markers in different species

Molecular Marker	Author	Animal
Bcl6b	Oatley et al., 2006	mouse
	Izadyar et al., 2011	human
CD49f (alpha 6 integrin)	Maki et al., 2009	primate
	Alipoor et al., 2009	mouse
DBA	Izadyar et al., 2002	bovine
Etv5 (erm)	Oatley et al., 2007 and Schlesser et al., 2008	mouse
Gfra1 (Gfra1)	Naughton et al., 2006	mouse
GPR 125	Izadyar et al., 2011	human
	Seandel et al., 2007	mouse
Lhx1	Oatley et al., 2007	mouse
	Goel et al., 2008	swine
NANOG	Fujihara et al., 2011	bovine
	Sada et al., 2009	mouse
Neurog3 (Ngn3)	Yoshida et al., 2004	mouse
PGP 9.5	Goel et al., 2010	mouse
Pou5f1 (oct4)	Pesce et al., 1998	mouse
	Fujihara et al., 2011	bovine
Ret	Naughton et al., 2006	mouse
SSEA4	Izadyar et al., 2011	human
	Maki et al., 2009	primate

Molecular Marker	Author	Animal
THY1	Maki et al., 2009; Reding et al., 2010 and Herrid et al., 2007 Fujihara et al., 2011	primate bovine
UCHL1 (PGP9.5)	Fujihara et al., 2011	bovine
Utp14b	Boettger-Tong et al., 2000 and Shetty et al. 2006	mouse
VASA	Fujihara et al., 2011	bovine
Zbtb16 (Plzf)	Buaas et al., 2004	mouse
	Reding et al., 2010	bovine

ISOLATION AND *IN VITRO* CULTURE OF SPERMATOGONIAL STEM CELLS

Isolation techniques

As discussed earlier, SSCs are found close to the basement membrane and their presence in the adult testis is restricted to less than 0.1 percent of all germ cells (Togelenbosch & De Rooij, 1993). In this manner, the election of the most suitable technique to isolate them is very important to establish an *in vitro* culture of SSC. Nowadays, the two step enzymatic digestion is the most popular technique used to isolate SSCs. This technique is based on two incubations of testicular tissue fragments in the presence of enzymes to digest it. It was first proposed by Davis and Schuetz, (1975) in rats and Bellvé et al., (1977) in mice.

The enzymatic digestion have been adapted and applied to many other species since then. The isolation of SSCs is often followed by a purification or enrichment step, in order to increase the amount of SSCs in the cell culture. For that, many approaches have been reported, including the discontinuous Percoll density gradient (Van Pelt et al., 1996), differential plating (Izadyar et al., 2002), flow cytometry cell sorting and magnetic cell sorting using SSCs specific antibodies. We can highlight the Percoll gradient as the most popular enrichment protocol for SSCs. However, many modifications have been proposed to this technique regarding the Percoll density adopted and number of layers used to prepare the gradient. The differential plating consists on the overnight *in vitro* culture of freshly isolated SSCs followed by subculture of only non adherent cells. Germ cells tend to remain in suspension, while supporting cells and other testicular cells adhere to the culture dish. These two techniques are usually combined in order to enrich the population of SSCs.

Other important factor in SSCs isolation if the age of donor individuals at the moment of germ cell isolation. Kanatsu-Shinohara et al. (2004) isolated SSCs from newborn mice (0-2 days of age) because at this age, the most primitive types of spermatogonia are predominant in the testis. When Guan et al. (2006) isolated SSCs from mice with 4-6 weeks of age, they obtained a less pure population of SSCs. Similar results were obtained by Seandel at el. (2007) with 3-5 weeks mice. We believe the same happens in livestock species, as bovine. Izadyar et al. (2002) isolated 65-87% type A SSC population from 5-7 months calves.

In vitro culture of SSCs

In vitro culture of SSCs faces similar hurdles to those commonly observed in *in vitro* culture of adult stem cells. However, many advances have been achieved in this area, and we can find in the literature protocols with satisfactory outcomes.

It was first reported that SSCs can be *in vitro* cultured for months by Nagano et al. (1998). The same group later suggested the addition of GDNF is important to short-term *in vitro* maintenance of SSCs (Nagano et al., 2003). Aponte et al. (2006) demonstrated the importance of this growth factor in the *in vitro* culture of bovine SSCs. Kanatsu-Shinohara et al. (2003a,b) studied the dynamics of gonocytes throughout *in vitro* culture, assessing the cell number increasing. In this study, it was possible to observe a 10^{14}-fold increase in cell number. In 2005, Kanatsu-Shinohara et al. developed a serum-free culture condition, when germ stem cells were cultured *in vitro* over 6 months. Serum-free conditions to culture SSCs have been optimized in rodents (Kubota et al., 2004a,b and Ryu et al., 2005) in order to support long- term maintenance.

The co-culture of SSCs with monolayers of other cell types as mitotically inactivated murine embryonic fibroblasts (MEF) or STO feeder cells (Nagano et al., 1998) are still discussible, presenting different results so far. MEF is widely used in *in vitro* culture of embryonic stem cells (Evans & Kaufman, 1981) and has been applied as feeder cells to murine SSCs cultures (Kanatsu-Shinohara et al., 2004a). Oatley et al., (2002) cultured bovine SSCs over a monolayer of STO feeder cells and in 2004, the same group developed a lineage of bovine embryonic cells (BEF), which was shown to be effective in the *in vitro* maintenance of bovine SSCs. However, Lee et al. (2001) and Aponte et al. (2006) successfully cultured bovine SSCsunder feeder-free conditions. Izadyar et al. (2003a) adopted a lamimin based extracelular matrix to support bovine SSCs *in vitro*.

Cryopreservation of SSCs

There are two ways to preserve SSCs, long-term *in vitro* culture and cryopreservation. Culture and cryopreservation in combination could be used to immortalize a male's genetic line through the germ cells due to the spermatogonial stem cells ability to self-replicate (Oatley et al., 2004). Since it is still hard to maintain pure populations of SSCs for long periods under in vitro condition, the development of effective cryopreservation protocols have been considered of high interest. Izadyar et al., (2002) cryopreserved type A bovine SSCs using DMSO obtaining a survival rate of 50% after thawing cells. Kaul et al., (2010) also adopted DMSO as SSC cryoprotectant in caprine, also observing a 50% survival rate after cryopreservation. Due to the fact enrichment protocols for type A SSCs are still being improved, many groups adopted the cryopreservation of testicular tissue instead of isolated cells. It appears to provide good results in preserving germ cells (Sato et al., 2011; Kaul et al., 2010).

SSCS TRANSPLANTATION AND TRANSGENESIS

The development of male germ cells transplantation methods provided a powerful means to study the biology of SSCs and its role in spermatogenesis and opened a door to a new potential tool for transgenesis. In addition, the testis cells transplantation is considered the unique *in vivo* functional assay for SSCs. This technique was first used to verify the function of *in vitro* cultured SSCs in mice (Brinster & Zimmermann, 1994; Brinster & Avarbock, 1994). Testicular cells are isolated from a fertile donor and microinjected into the seminuferous tubule of an infertile recipient. It is expected to observe the resumption of spermatogenesis from a donor SSC-derived colonies in the recipient testes. These cell colonies rise from a single transplanted SSC, what allows the quantification of these clonal events (Brinster, 2002). Besides the possibility to study male infertility, the transplantation of SSCs also provides another way to conserve reproductive potential of genetically valuable individuals within or between species and, finally, can be used to produce transgenic animals after generation of transgenic sperm cells. This last application is especially important for species in which embryonic stem cell lines have not been established and other transgenic techniques present limited efficiency. Transgenic animals have huge applications from basic science such as the creation of animal models for human diseases, like Parkinson´s (Crabtree& Zhang, 2011) to production of recombinant pharmaceutic proteins in the animal's fluid: blood, milk (Houdebine, 2000a,b and Houdebine, 2002), egg white (Zhu et al., 2005; van de Lavoir et al., 2006 and Lillico et al., 2007) and seminal plasma (Dyck et al., 2003). Ever since the generation of the first transgenic animal, in 1980, through pronuclei microinjection in an embryo's pronuclei (Houdebine, 2009), this method has been used in other prolific species as rat, rabbit and pig (Houdebine, 2000a,b). However, along the years, many disadvantages of pronuclei microinjection were reported. One of the most important is the misplaced injection of DNA in the cell cytoplasm and not in its pronucleus. Additionally, the exogenous DNA interaction with the host cells genome is quite variable (Houdebine, 2009). Alternatively, other techniques were developed, such as: gene transfer with transposons, lentiviral vectors, sperm, pluripotent, stem and somatic cells. In 2002, Lavitrano et al. obtained a large number of transgenic pigs using sperm-mediated gene transfer (SMGT).

The authors reported that up to 80% of the animal had the exogenous gene integrated in the genome, thus SMGT was more efficient than other techniques previously described. SSCs of all mammalian species examined, including human, can replicate in mouse seminiferous tubules following transplantation, the growth factors required for SSCs self-renewal are probably conserved among mammalian species (Kubota et al., 2006).

Although most studies have been performed in rodents, germ cell transplantation has also been applied to non-rodent species as pigs, goats, cattle, monkeys and recently fish and chickens (Honaramooz et al., 2002a,b, 2003; Schlatt et al., 2002; Izadyar et al., 2003b; Takeuchi et al., 2003; Yoshizaki et al., 2005; Lee et al., 2006; Mikkola et al., 2006; Okutsu et al. 2006; Trefil et al. 2006), as reviewed by Dobrinski, (2008). The first hurdle found when germ cell transplantation was applied to livestock species was the differences in testicular anatomy and physiology. While in rodents it is possible to microinject germ cells directly into the seminiferous tubule via efferent ducts, the same is not feasible in larger animals. Thus, the alternative use of ultrasoung to guide the needle during the injection of cells was successfully reported by Kaul et al. (2010) in goats. In bovine, this ultrasound guided needle technique was successfully applied when an autologous transplantation (Izadyar et al., 2002) as well as a heterologous transplantation (Herrid et al., 2006) was performed. It has been demonstrated that germ cell transplantation can be performed more efficiently after suppression of spermatogenesis in recipient animals. The most popular chemical treatment consists of administering a DNA alkylating agent, Busulfan, that destroys proliferating cells. Busulfan is commonly used to suppress spermatogenesis in rodents. An alternative to Busulfan, irradiation of the testis (Creemers et al., 2002; Schlatt et al., 2002), being frequently adopted in studies with large animals as bulls (Izadyar et al., 2003). Despite all the described potential of SSCs to produce transgenic animals, until now, few groups have genetically modified these cells, being the efforts more restricted to laboratory species.

SUMMARY

As discussed in this chapter, SSCs, like every other adult stem cell in mammals, retains the ability of either self-renewal or differentiation. In this case, the differentiation process is known as spermatogenesis. Interest in spermatogonia has grown in recent years as a result of exciting developments in stem cell research in general and the development of new research tools allowing the isolation, culture and transplantation of these cells.

Because there are a low concentration of SSCs in mammal testis and isolation processes are difficult (Meachem et al., 2001), the assessment of biological activity and cell viability are essentials for the maintenance of the SSC (Potten & Loeffler,

1990; van der Kooy & Weiss, 200; Watt & Hogan, 2000). Brinster & Avarbock (1994) performed the first SSC transplantation and reported this technique to be a good functional assay. In this way, the progenitor germ cells would be in the correct environment having direct contact with somatic niches (Brinster, 2002). Because SSCs are capable of restoring spermatogenesis after transplantation into testes which spermatogenesis had been suppressed, their transplantation opened the door to many possibilities of usage. In this context, we can include the preservation and reestablishment of the reproductive potential of the animals. For example, and animal with desirable genetic traits which can no longer mate, cancontinue to spread its genetics through germ cell transplantation. In humans, cancer patients now have the opportunity to cryopreserve their SSCs during chemotherapy or radiation treatments. However, there is so much to study and to understand regarding the biology of SSCs before their transplantation in human becomes a routine procedure. When SSC culture becomes available for clinical use, efficient protocols for cryopreservation of these cells and testicular tissue will be of great value.

Finally, the most exciting potential usage of SSCs relies on their capability to transfer genetic modifications to the next generation in a fast manner. The potential use of SSCs in animal transgenesis has attracted the attention of many research groups all over the globe. As discussed, it is almost impossible to describe the value of a transgenic animal, since they can be used in basic science as animal models to human diseases. In addition, transgenic animals can serve as bioreactors, producing proteins of high interest in the human pharmaceutical industry.

In conclusion, SSCs has many future research perspectives, such as: infertility treatment, contraceptive strategy, *in vitro* spermatogenesis, the development of markers for identification of spermatogonial subtypes, innovative research using germ cell transplantation, preservation of fertility for cancer patients, generation of transgenic animals and preservation of valuable animals (Meachem et al., 2001).

REFERENCES

1. Alipoor, F. J., Ali, M., Gilani, S. (2009). Achieving high survival rate following cryopreservation after isolation of prepubertal mouse spermatogonial cells. J Assist Reprod Genet, v.26, p. 143–149.
2. Aponte, P. M., Maaikep, A. V. B., de Rooiji, D. G. & Van Pelt, M. M. (2005). Spermatogonial stem cells: characteristics and experimental possibilities. *APMIS*, v. 113, p. 727–42, 2005.
3. Aponte, P. M., Soda, T., van de Kant, H. J. & de Rooij, D. G. (2006). Basic features of bovine spermatogonial culture and effects of glial cell line-derived neurotrophic factor. *Theriogenology*, v. 65, n. 9, p. 1828-47, 2006.

4. Barlow, J. Z, Kelley, K. A., Bozdagi, O. & Huntley, G. W. (2002). Testing the role of the cell- surface molecule Thy- 1 in regeneration and plasticity of connectivity in the CNS. *Neuroscience,* v. 111, p. 837–52, 2002.

5. Bellvé, A. R., Cavicchia, J. C., Millette, C. F., O'Brien, D. A., Bhatnagar, Y. M. & Dym, M. (1977). Spermatogenic cells of the prepuberal mouse. Isolation and morphological characterization. *J Cell Biol*, v. 74, n. 1, p. 68-85, 1977.

6. Boettger-Tong, H. L., Johnston, D. S., Russell, L. D., Griswold, M. D. & Bishop, C. E. (2000). Juvenile spermatogonial depletion (jsd) mutant seminiferous tubules are capable of supporting transplanted spermatogenesis. *Biology of Reproduction*, v. 63, p. 1185– 1191, 2000.

7. Braydich-Stolle, L., Kostereva, N., Dym, M. & Hofmann, M. C. (2007). Role of Src family kinases and N-Myc in spermatogonial stem cell proliferation. *Developmental Biology*, v. 304, p. 34–45, 2007.

8. Brinster, R. L. & Avarbock, M. R. (1994). Germ-line transmission of donor haplotype following spermatogonial transplantation. *PNAS*, v. 91, p. 11303-11307, 1994.

9. Brinster, R. L. & Zimmerman, J. W. (1994). Spermatogenesis following spermatogonial male germ cell transplantation. *PNAS*, v. 91, p. 11298-11302, 1994.

10. Brinster, R. L. (2002). Germline Stem Cell Transplantation and Transgenesis. *Science*, v. 296, n. 5576, p. 217-2176, 2002.

11. Buaas, F. W., Kirsh, A. L., Sharma, M., McLean, D. J., Morris, J. L., Griswold, M. D., de Rooij, D. G. & Braun, R. E. (2004). Plzf is required in adult male germ cells for stem cell self-renewal. *Nature Genetics,* v. 36, p. 647–652, 2004.

12. Byskov, A. G. & Høyer, P. E. (1994). Embryology of mammalian gonads and ducts. In: *The Physio- logy of Reproduction*, E Knobil and JD Neill, pp. 487–540, 2nd edn, New York.

13. Caires, K., Broady, J., & Mclean, D. (2010). Maintaining the male germline: regulation of spermatogonial stem cells. *Journal of Endocrinology*, v. 205, p. 133-145, 2010.

14. Clark JM, Eddy EM. 1975. Fine structural observations on the origin and associations of primordial germ cells of the mouse. Dev Biol. 1975 Nov; v.47, n.1: p.136-55.1975.

15. Clermont, Y. & Perey, B. (1957). Quantitative study of the cell population of the seminiferous tubules in immature rats. *Am J Anat*, v. 100, p. 241–67, 1957.

16. Clermont, Y. (1966). Spermatogenesis in man. A study of the spermatogonial population. (1966). *Fertil Steril*, v. 17, n. 6, p. 705-21, 1966.

17. Costoya, J. A., Hobbs, R. M., Barna, M., Cattoretti, G., Manova, K. & Sukhwani, M., et al. (2004). Essential role of Plzf in maintenance of spermatogonial stem cells. *Nat Genet*, v. 36, p.653–9, 2004.

18. Crabtree, D.M. & Zhang, J. (2011). Genetically engineered mouse models of Parkinson's disease. *Brain Research Bulletin*, doi:10.1016/j.brainresbull.2011.07.01

19. Creemers, L. B., Meng, X., den Ouden, K., van Pelt, A. M. M., Izadyar, F., Santoro, M., de Rooij, D.G. & Sariola, H. (2002). Transplantation of germ cells from glial cell line- derived neurotrophic factor-overexpressing mice to host testes depleted of endogenous spermatogenesis by fractionated irradiation. *Biol. Reprod.*, v. 66, p. 1579–1584, 2002.

20. Davis J. C., Schuetz A. W. (1975). Separation of germinal cells from immature rat testes by sedimentation at unit gravity. Exp Cell Res, v. 1; n.91(1), p.79-86.

21. de Rooij, D. G., van de Kant, H. J., Dol, R., Wagemaker, G., van Buul, P. P., van Duijn- Goedhart, A., de Jong, F. H. & Broerse, J. J. (2003). Long-term effects of irradiation before adulthood on reproductive function in the male rhesus monkey. *Biol Reprod*, v. 66, n. 2, p. 486-94, 2002.

22. Dobrinski I. Germ cell transplantation and testis tissue xenografting in domestic animals.(2005). Animal reproduction science, v.89, n.1-4, p.137-45.

23. Dyck, M. K., Lacroix, D., Pothier, F. & Sirard, M. A. (2003). Making recombinant proteins in animals—different systems, different applications. *Trends Biotechnol*, v. 21, p. 394–9, 2003.

24. Ehmcke, J., Wistuba, J. & Schlatt, S. (2006). Spermatogonial stem cells: questions, models and perspectives. *Hum Reprod Update*, v. 12, n. 3, p. 275-82, 2006.

25. Evans, M. J. & Kaufman, M. H. (1981). Establishment in culture of pluripotencial cells from mouse embryos. *Nature*, v. 292, p. 154-156, 1981.

26. Fujihara, M., Kim, S. M., Minami, N., Yamada, M. & Imai, H. (2011). Characterization and in vitro culture of male germ cells from developing bovine testis. *The Journal of reproduction and development*, v. 57, n. 3, p. 355-64, 2011.

27. Garces, A., Haase, G., Airaksinen, M. S., Livet, J., Filippi, P. & deLapeyriere, O. (2000). GFRalpha 1 is required for development of distinct subpopulations of motoneuron. *J Neurosci*, v. 20, p. 4992–5000, 2000.

28. Goel, S., Fujihara, M., Minami, N., Yamada, M., Imai, H. (2008). Expression of NANOG, but not POU5F1, points to the stem cell potential of primitive germ cells in neonatal pig testis. Reproduction, v.135, n.6, p. 785-9.

29. Guan, K., Nayernia, K., Maier, L. S., Wagner, S., Dressel, R., Lee, J.H., Nolte, J., Wolf, F., Li, M., Engel, W. & Hasenfuss, G. (2006). Pluripotency of spermatogonial stem cells from adult mouse testis. *Nature*, v. 440, n. 7088, p. 1199-203, 2006.

30. He, Z., Jiang, J., Hofmann, M. C. & Dym, M. (2007). Gfra1 silencing in mouse spermatogonial stem cells results in their differentiation via the inactivation of RET tyrosine kinase. *Biology of Reproduction*,v. 77, p. 723- 733, 2007.

31. Herrid, M., Vignarajan, S., Davey, R., Dobrinski, I. & Hill, J. R. (2006). Successful transplantation of bovine testicular cells to heterologous recipients. *Reproduction*, v. 132, p. 617–624, 2006.

32. Herrid, M., Davey, R. J. & Hill, J. R. (2007). Characterization of germ cells from pre-pubertal bull calves in preparation for germ cell transplantation. *Cell Tissue Res*, v. 330, n. 2, p. 321-9, 2007.

33. Hofmann, M. C., Braydich-Stolle, L. & Dym, M. (2005). Isolation of male germ-line stem cells; influence of GDNF. *Dev Biol*, v. 279, p. 114–124, 2005.

34. Honaramooz, A., Megee, S. O. & Dobrinski, I. (2002a). Germ cell transplantation in pigs. *Biol Reprod*, v. 66, p. 21–28, 2002.

35. Honaramooz, A., Snedaker, A., Boiani, M., Scholer, H. R., Dobrin- ski, I. & Schlatt, S. (2002b). Sperm from neonatal mammalian testes grafted in mice. *Nature*, v. 418, p. 778–781, 2002.

36. Honaramooz, A., Behboodi, E., Blash, S., Megee, S. O. & Dobrinski, I. (2003). Germ cell transplantation in goats. *Mol Reprod Dev*, v. 64, p. 422–428, 2003.

37. Houdebine L. M. (2009). Production of pharmaceutical proteins by transgenic animals. Comp Immunol Microbiol Infect Dis, v.32, n.2, p.107-21.

38. Houdebine, L. M. (2000a). Production of pharmaceutical proteins by transgenic animals. Comparative Immunology. *Microbiologyand Infectious Diseases*, v. 32, p. 107–121, 2000.

39. Houdebine, L. M. (2000b). Transgenic animal bioreactors. *Transgenic Res*, v. 9, p. 305–12, 2000.

40. Houdebine, L. M. (2002). The methods to generate transgenic animals and to control transgene expression. *J Biotechnol*, v. 98, p. 145–60, 2002.

41. Huckins, C. (1971) The spermatogonial stem cell population in adult rats. II. A radioautographic analysis of their cell cycle properties. *Cell tissue kinet*, v. 4, p. 313- 334, 1971.

42. Huckins, C. (1971). The spermatogonial stem cell population in adult rats. I. Their morphology, proliferation and maturation. *Anat Rec*, v. 169, p. 533–558, 1971.

43. Huckins C., Clermont Y. 1968. Evolution of gonocytes in the rat testis during late embryonic and early post-natal life. Arch Anat Histol Embryol, v.51, n.1, p.341-54.1968.

44. Huckins, C. & Oakberg, E. F. (1978). Morphological and quantitative analysis of spermatogonia in mouse testes using whole mounted seminiferous tubules, I. The normal testes. *Anat Rec*, v. 192, n. 4, p. 519-28, 1978.

45. Izadyar, F., Spierenberg, G. T., Creemers, L. B., den Ouden, K. & de Rooij, D. G. (2002). Isolation and purification of type A spermatogonia from the bovine testis. *Reproduction*, v. 124, n. 1, p. 85-94, 2002.

46. Izadyar, F., Den Ouden, K., Creemers, L. B., Posthuma, G., Parvinen, M. & De Rooij, D. G. (2003a). Proliferation and differentiation of bovine type A spermatogonia during long-term culture. *Biol Reprod*, v. 68, p. 272–281, 2003.

47. Izadyar, F., Den Ouden, K., Stout, T. A., Stout, J., Co- ret, J. & Lankveld, D. P., et al.. (2003b). Autologous and homo- logous transplantation of bovine spermatogonial stem cells. *Reproduction*, v. 126, p. 765–74, 2003.

48. Izadyar, F., Wong, J., Maki, C., Pacchiarotti, J., Ramos, T., Howerton, K., Yuen, C., Greilach, S., Zhao, H. H., Chow, M., Chow, Y. C., Rao, J., Barritt, J., Bar-Chama, N. & Copperman, A. (2011). Identification and characterization of repopulating spermatogonial stem cells from the adult human testis. *Human Reprod*, v. 0, p. 1– 11, 2011.

49. Jijiwa, M., Fukuda, T., Kawai, K., Nakamura, A., Kurokawa, K., Murakumo, Y., Ichihara, M. & Takahashi, M. (2004). A targeting mutation of tyrosine 1062 in Ret causes a marked decrease of enteric neurons and renal hypoplasia. *Molecular and Cellular Biology*, v. 24, p. 8026–8036, 2004.

50. Kanatsu-Shinohara, M., Ogonuki, N., Inoue, K., Ogura, A., Toyokuni, S. & Shinohara, T. (2003a). Restoration of fertility in infertile mice by transplantation of cryopreserved male germline stem cells. *Hum Reprod*, v. 18, n. 12, p. 2660-7, 2003.

51. Kanatsu-Shinohara, M., Toyokuni & S., Shinohara, T. (2003b). CD9 is a surface marker on mouse and rat male germline stem cells. *Biol Reprod*, v. 70, n. 1, p. 70-5, 2003.

52. Kanatsu-Shinohara, M., Inoue, K., Lee, J., Yoshimoto, M., Ogonuki, N., Miki, H., Baba, S., Kato, T., Kazuki, Y., Toyokuni, S., Toyoshima, M., Niwa, O., Oshimura, M., Heike, T., Nakahata, T., Ishino, F., Ogura, A. & Shinohara, T. (2004). Generation of pluripotent stem cells from neonatal mouse testis. *Cell*, v. 119, n. 7, p. 1001-12, 2004.

53. Kanatsu-Shinohara, M., Miki, H., Inoue, K., Ogonuki, N., Toyokuni, S. & Ogura A, et al. (2005). Long termculture of mouse male germline stem cells under serum-or feeder- free conditions. *Biol Reprod*, v. 72, p. 985–91, 2005.

54. Kanatsu-Shinohara, M., Muneto, T., Lee, J., Takenaka, M., Chuma, S., Nakatsuji, N., Horiuchi, T. & Shinohara, T. (2008). Long-termculture of male germline stem cells from hamster testes. *Biology of Reproduction*, v. 78, p. 611–617, 2008.

55. Kokkinaki, M., Lee, T.-lap, He, Z., Jiang, J., Golestaneh, N., Chan, W.-yee, & Dym, M. (2010). *NIH Public* Access, v. 139, n. 6, p. 1011-1020, 2010.

56. Kubota, H., Avarbock, M. R. & Brinster, R. L. (2004a). Culture conditions and single growth factors affect fate determination of mouse spermatogonial stem cells. *Biol Reprod*, v. 71, p. 722–31, 2004.

57. Kubota, H., Brinster, R. L. (2006). Technology insight: In vitro culture of spermatogonial stem cells and their potential therapeutic uses. *Nat Clin Pract Endocrinol Metab*. v. 2, n. 2, p. 99-108, 2006.

58. Kaul, G., Kaur, J., Rafeeqi, T.A. Ultrasound Guided Transplantation of Enriched and Cryopreserved Spermatogonial Cell Suspension in Goats. Reprod Dom Anim doi: 10.1111/j.1439-0531.2009.01549.x ISSN 0936-6768.

59. Lavitrano, M., Bacci, M. L., Forni, M., Lazzereschi, D., Di Stefano, C., Fioretti, D., Giancotti, P., Marfé, G., Pucci, L., Renzi, L., Wang, H., Stoppacciaro, A., Stassi, G., Sargiacomo, M., Sinibaldi, P., Turchi, V., Giovannoni, R., Della Casa, G., Seren, E. & Rossi, G. (2002). Efficient production by sperm-mediated gene transfer of human decay accelerating factor (hDAF) transgenic pigs for xenotransplantation. *PNAS*, v. 99, n. 22, p, 14230-5, 2002.

60. Lawson K. A., Pedersen R. A. 1992. Clonal analysis of cell fate during gastrulation and early neurulation in the mouse.Ciba Found Symp .v.165, n. 3-21; discussion 21-6. 1992.

61. Lawson, K. A., Pedersen, R. A., Lee, J., Kanatsu-Shinohara, M., Inoue, K., Ogonuki, N., Miki, H., Toyokuni, S., Kimura, T., Nakano, T., Ogura, A. & Shinohara, T. (2007). Akt mediates self-renewal division of mouse spermatogonial stem cells. *Development*, v. 134, p. 1853–1859, 2007.

62. Lee, D. R., Kaproth, M. T. & Parks, J. E. (2001). In vitro production of haploid germ cells from fresh or frozen-thawed testicular cells of neonatal bulls. *Biol Reprod*, v.65, p. 873–878. 2001.

63. Lee, Y. M., Jung, J. G., Kim, J. N., Park, T. S., Kim, T. M., Shin, S. S., Kang, D. K., Lim, J. M. & Han, J. Y. (2006). A testis-mediated germline chimera production based on transfer of chicken testicular cells directly into heterologous testes. *Biol Reprod*, v. 75, p. 380– 386, 2006.

64. Lillico, S. G., Sherman, A., McGrew, M. J., Robertson, C. D., Smith, J. & Haslam, C., et al. (2007). Oviduct-specific expression of two therapeutic proteins in transgenic hens. *Proc Natl Acad Sci USA*, v. 104, p. 1771–6, 2007.

65. Maki, C. B., Pacchiarotti, J., Ramos, T., Pascual, M., Pham, J., Kinjo, J. & Izadyar, F. (2009). Phenotypic and molecular characterization of spermatogonial stem cells in adult primate testes. *Human reproduction*, v. 24, n. 6, p. 1480-91, 2009.

66. Meachem, S., M., von Schönfeldt, V., Schlatt, S. (2001). Spermatogonia: stem cells with a great perspectiveReproduction, v.121, p.825–834.

67. Meistrich, M. L. & van Beek, M. E. A. B. (1993). Spermatogonial stem cells, In: Cell and molecular biology of testis, Desjardins, C. and Ewing, L. L., pp. 266-295, Oxford University Press.

68. Meng, X., Lindahl, M., Hyvonen, M. E., Parvinen, M., de Rooij, D. G., Hess, M. W., Raatikainen-Ahokas, A., Sainio, K., Rauvala, H. & Lakso, M., et al. (2000). Regulation of cell fate decision of undifferentiated spermatogonia by GDNF. *Science*, v. 287, p. 1489–1493, 2000.

69. Mikkola M, Sironen A, Kopp C, Taponen J, Sukura A, Vilkki J, Katila T, Andersson M. (2006). Transplantation of normal boar testicular cells resulted in complete focal spermatogen- esis in a boar affected by the immotile short-tail sperm defect. Reprod Domest Anim, v.41, p.124–128

70. Nagano, M., Avarbock, M. R., Leonida, E. B., Brinster, C. J. & Brinster, R. L. (1998). Culture of mouse spermatogonial stem cells. *Tissue Cell*, v. 30, p. 389-397, 1998.

71. Nagano, M., Ryu, B. Y., Brinster, C. J., Avarbock, M. R. & Brinster, R. L. (2003). Maintenance of mouse male germ line stem cells in vitro. *Biol. Reprod.*, v. 6 p. 2207–14, 2003.

72. Naughton, C. K., Jain, S., Strickland, A. M., Gupta, A. & Milbrandt, J. (2006). Glial cell-line derived neurotrophic factor-mediated RET signaling regulates spermatogonial stem cell fate. *Biology of Reproduction*, v. 74, p. 314–321, 2006.

73. Oatley, J. M., de Avila, D. M., McLean, D. J., Griswold, M. D. & Reeves, J. J. (2002). Transplantation of bovine germinal cells into mouse testes. *J Anim Sci*, v. 80, n. 7, p. 1925-31, 2002.

74. Oatley, J. M., Avarbock, M. R, Telaranta, A. I., Fearon, D. T. & Brinster, R. L. (2006). Identifying genes important for spermatogonial stem cell self-renewal and survival. *Proceedings of the National Academy of Sciences of the United States of America*, v. 103, n. 25, 2006.

75. Oatley, J. M., Avarbock, M. R. & Brinster, R. L. (2007). Glial cell line-derived neurotrophic factor regulation of genes essential for self-renewal of mouse

spermatogonial stem cells is dependent on Src family kinase signaling. *Journal of Biological Chemistry*, v. 282, p. 25842–25851, 2007.

76. Okutsu, T., Suzuki, K., Takeuchi, Y., Takeuchi, T. & Yoshizaki, G. (2006). Testicular germ cells can colonize sexually undiffer- entiated embryonic gonad and produce functional eggs in fish. *Proc Natl Acad Sci U S A*, v. 103, p. 2725–2729, 2006.
77. Potten, C. S., Loeffler, M. (1990). Stem cells: attributes, cycles, spirals, pitfalls and uncertainties. Lessons for and from the crypt. Developmen, v.110, p.1001–1020.
78. Raverot, G., Weiss, J., Park, S. Y., Hurley, L. & Jameson, J. L. (2005). Sox3 expression in undifferentiated sperm- atogonia is required for the progression of sper- matogenesis. *Dev Biol*, v. 283, p. 215–25, 2005.
79. Reding, S. C., Stepnoski, A. L., Cloninger, E. W. & Oatley, J. M. (2010). THY1 is a conserved marker of undifferentiated spermatogonia in the pre-pubertal bull testis. *Reproduction*, v. 139, n. 5, p. 893-903, 2010.
80. Russell, L. D., Ettlin, R. A., Sinha Hikim, A. P., Clegg, E. D. (1990). Histopathological Evaluation of the Testis, In: *Histological and Histopathological Evaluation of the Testis*, pp. 1-40, Cache River Press, Clearwater, FL.
81. Ryu, B. Y., Orwig, K. E., Kubota, H., Avarbock, M. R. & Brinster, R. L. (2004). Phenotypic and functional charac- teristics of spermatogonial stem cells in rats. *Dev Biol*, v. 274, p.158–70, 2004.
82. Ryu, B. Y., Kubota, H., Avarbock, M. R. & Brinster, R. L. (2005). Conservation of spermatogonial stem cell self-renewal signaling between mouse and rat. *PNAS*, v. 102, p. 14302–14307, 2005.
83. Sada, A., Suzuki, A., Suzuki, H. & Saga, Y. (2009). The RNA-binding protein NANOS2 is required to maintain murine spermatogonial stem cells. *Science*, v. 325, p. 1394– 1398, 2009.
84. Schlatt, S., Foppiani, L., Rolf, C., Weinbauer, G. F. & Nieschlag, E. (2002). Germ cell transplantation into X-irradiated monkey testes. *Hum. Reprod.*, v. 17, p. 55–62, 2002.
85. Schlesser, H. N., Simon, L., Hofmann, M. C., Murphy, K. M., Murphy, T., Hess, R. A. & Cooke, P. S. (2008). Effects of ETV5 (ets variant gene 5) on testis and body growth, time course of spermatogonial stem cell loss, and fertility in mice. *Biology of Reproduction*, v. 78, p. 483–489, 2008.
86. Seandel, M., James, D., Shmelkov, S. V., Falciatori, I., Kim, J., Chavala, S., Scherr, D. S., Zhang, F., Torres, R., Gale, N. W., Yancopoulos, G. D., Murphy, A., Valenzuela, D. M., Hobbs, R. M., Pandolfi, P. P. & Rafii, S. (2007).

Generation of functional multipotent adult stem cells from GPR125+ germline progenitors. *Nature*, v. 20, n. 449(7160), p. 346-50, 2007.

87. Shetty, G., Weng, C. C., Porter, K. L., Zhang, Z., Pakarinen, P., Kumar, T. R. & Meistrich, M. L. (2006). Spermatogonial differentiation in juvenile spermatogonial depletion (jsd) mice with androgen receptor or follicle-stimulating hormone mutations. *Endocrinology*, v. 147, p. 3563–3570, 2006.

88. Sato, T., Katagiri, K., Gohbara, A., Inoue, K., Ogonuki, N., Ogura, A., et al. (2011). In vitro production of functional sperm in cultured neonatal mouse testes. *Nature*, v. 471, n. 7339, p.504-7.

89. Tadokoro, Y., Yomogida, K., Ohta, H., Tohda, A. & Nishimune, Y. (2002). Homeostatic regulation of germinal stem cell proliferation by the GDNF/FSH pathway. *Mech Dev*, v. 113, p. 29–39, 2002.

90. Takeuchi, Y., Yoshizaki, G. & Takeuchi, T. (2003). Generation of live fry from intraperitoneally transplanted primordial germ cells in rainbow trout. *Biol Reprod*, v. 69, p. 1142–1149, 2003.

91. Tegelenbosch, R. A. & de Rooij, D. G. (1993). A quantitative study of spermatogonial multiplication and stem cell renewal in the C3H/ 101 F1 hybrid mouse. *Mutat Res*, v. 290, p. 193–200, 1993.

92. Trefil, P., Micakova, A., Mucksova, J., Hejnar, J., Poplstein, M., Bakst, M. R., Kalina, J. & Brillard, J. P. (2006). Restoration of spermatogenesis and male fertility by transplantation of dispersed testicular cells in the chicken. *Biol Reprod*, v. 75, p. 575– 581, 2006.

93. van Pelt A. M., Morena A. R., van Dissel-Emiliani F. M., Boitani C., Gaemers I. C., de Rooij D. G., Stefanini M. (1996). Isolation of the synchronized A spermatogonia from adult vitamin A-deficient rat testes. Biol Reprod,v. 55, n.2, p.439-44.

94. van de Lavoir, M. C., Diamond, J. H., Leighton, P. A., Mather-Love, C., Heyer, B. S. & Bradshaw, R., et al. (2006). Germline transmission of genetically modified primordial germ cells. *Nature*, v. 441, p.766–9, 2006.

95. van der Kooy, D., Weiss, S. (2000). Why stem cells? Science, v.287, p.1439–1441.

96. Vega, Q. C., Worby, C. A., Lechner, M. S., Dixon, J. E. & Dressler, G. R. (1996). Glial cell line- derived neurotrophic factor activates the receptor tyrosine kinase RET and promotes kidney morphogenesis. *Proc Natl Acad Sci U S A*, v. 93, p. 10657–10661, 1996.

97. Watt, F. M., Hogan, B.L. (2000). Out of Eden: stem cells and their niches. Science, v.287, p.1427–1430.

98. Wrobel, K. H., Bickel, D., Kujat, R. & Schimmel, M. (1995). Configuration and distribution of bovine spermatogonia. *Cell Tissue Res*, v. 279, p. 277-289, 1995.

99. Yoshida, S., Takakura, A., Ohbo, K., Abe, K., Waka-bayashi, J. & Yamamoto, M., et al. (2004). Neurogenin3 de- lineates the earliest stages of spermatogenesis in the mouse testis. *Dev Biol*, v. 269, p. 447–58, 2004.

100. Yoshizaki, G., Tago, Y., Takeuchi, Y., Sawatari, E., Kobayashi, T. & Takeuchi, T. (2005). Green fluorescent protein labeling of primordial germ cells using a nontransgenic method and its application for germ cell transplantation in salmonidae. *Biol Reprod*, v. 73, p. 88–93, 2005.

101. Zamboni, L. & Merchant, H. (1973). The fine morphology of mouse primordial germ cells in extragonadal locations. *Am J Anat*, v. 137, p. 299–336, 1973.

102. Zhu, L., van de Lavoir, M. C., Albanese, J., Beenhouwer, D. O., Cardarelli, P. M. & Cuison, S., et al. (2005). Production of human monoclonal antibody in eggs of chimeric chickens. *Nat Biotechnol*, v. 23, p. 1159–69, 2005.

CHAPTER-14

UTILIZING GENE EXPRESSION MICROARRAYS FOR MICROGRAVITY RESEARCH WITH A FOCUS ON IDENTIFYING SIGNIFICANT FACTORS WITHIN THE UNITED STATES

INTRODUCTION: AN OVERVIEW

Gene Expression Microarrays play a vital role in advancing microgravity research, with a particular focus on identifying major contributors within the United States. These cutting-edge technologies allow scientists to comprehensively analyze and compare gene expression patterns in microgravity environments, providing valuable insights into how space conditions impact gene regulation. By scrutinizing the genetic responses of organisms under reduced gravity, researchers can uncover key factors influencing various biological processes. This research is not only crucial for understanding the fundamental effects of microgravity on living organisms but also holds significant implications for space exploration, healthcare, and biotechnology, offering the potential to improve our understanding of human health and adaptation to space travel.

Crew members of space flights commonly experience certain health condition changes such as immune system dysregulation, musculoskeletal changes (e.g., significant bone and muscle loss), and neurological alterations. Since the space exploration of the 1960s and 1970s, physiological changes in several organ systems due to weightlessness have been identified. Some of the adverse effects are a decline in cellular immune responses (Leach et al, 1990; Cogoli 1993; Pippia et al; 1996; Borchers et al, 2002;), cardiovascular deconditioning (Fritsch-Yelle, et al

1996), bone deterioration (Mack et al 1967; Vose et al 1974; Atkov 1992; Schneider et al, 1995; Collet et al, 1997) and muscular atrophy (Thomason and Booth 1990; Aubers et al, 2005; Trappe et al, 2009). Human exposure to microgravity has been demonstrated to be a major environmental factor during space flight (Cogoli et al 1993; Ullrich et al, 2008). Many of the adverse effects of microgravity have much in common with earthbound health problems related to low physical activity or less mechanical loading. For example, bone and muscle loss as well as immune system dysfunction are some of the main consequences common to both extended spaceflight and physical inactivity such as that associated with the aging population and people suffering from degenerative disorders. In a recent review article, mechanotransduction is attributed as the possible convergence point for all the "abnormalities" associated with aging and microgravity because human adaptation to microgravity has all the features of accelerated aging (reviewed by Vernikos & Schneider 2010). More recently, similarities between the clinical presentation (such as atrophy in muscle and bone, cardiovascular disturbances, and alterations in renal, immune and sensory motor systems) of individuals living with spinal cord injury (SCI) and those who experience prolonged gravity unloading (especially astronauts) are reviewed (Scott et al, 2011). It is evident that continued effort in microgravity research will deeper our understanding of space adaptation response and improve many of our health-related problems on earth. Thus, microgravity based research can further our understanding of human diseases such as SCI, diabetes, osteoporosis and premature aging that are related to physical inactivity. Most effective counter measures can then be formulated to ensure safe experience in microgravity and promote healthy beings especially at the senior level.

Despite over 50 years of manned space flight, there is still much to be learned about the consequences of living in space for extended periods of time. Microgravity exposure fromspaceflight has global effects on cells in virtually all organ systems in the body. Most cell types, ranging from bacteria to mammalian cells, are sensitive to the microgravity environment, suggesting that microgravity affects fundamental cellular activities. Studies at the cellular and molecular levels have been reported from both space flight and ground- based microgravity simulations. Ground-based gravity-simulation experiments at the cellular and molecular levels have gained much insight into the underlying molecular and cellular alterations induced by microgravity stress as well as the mechanisms of the microgravity effects (reviewed in Cogoli, 1993, 1996; Sonnenfeld & Shearer 2002; Sonnenfeld 2005). Environmental change as drastic as sudden gravity change is likely to alter the function and transcriptional activities of groups of genes. This is because any change in the physiological activity of a cell or an organism is most likely the result of changes in certain genes' expressions. Through the 1990s a number of gene expression studies were carried out to determine microgravity effect on organisms and these studies tended to focus on a few genes at a time. It was only with the advent of high-throughput genomic technology such as microarrays that large scale genome-wide studies have been performed. For most genes (especially structural genes), gene expression in response to an environmental change is mainly controlled at the level of transcription, which provides the base for the successful development of mRNA-based high-throughput assays such as DNA microarray technology. The integrated application of biotechnologies in microgravity research with high throughput microarrays for gene expression analysis and various ground-based simulated microgravity models makes it possible for well controlled experimental studies (Hammond& Hammond, 2001). In this chapter, I will first give a brief overview of the ground based simulated microgravity technology and microarray technology for space lifescience research. Then I will review the combined use of these biotechnologies in the study of microgravity effect on gene expression of mammalian cells with specific focus on the areas where most studies tend to focus on such as cells in the immune system, bone, and muscles. In addition, I will make an attempt toward the identification of major space genes by combing data from all the retrievable microarray-based microgravity studies for each of the specific areas as well as an overall combination of these areas with other less studied areas through bioinformatics analysis. Furthermore, I will discuss the initial list of candidate major space genes that are most frequently altered by microgravity environments through microarray based assays and cross-platform, cross-species bioinformatics analysis.

Two relevant biotechnologies in space bioscience research: Simulated microgravity and gene expression microarrays

As we depend on the unit gravity (1g) on earth for our daily lives, most of the cells in our body depend on the gravity for proper growth and function. When the cells are placed in reduced gravity environment, many of their functions are affected at various degrees. A clear understanding of microgravity effects on our genes of cells in our organ systems is essential for extraterrestrial health of space travelers. Because of the high-cost and low efficiency of space flown experiments, ground-based methods for simulating the microgravity environment have been developed. These simulated microgravity studies include head down bed rest for humans (LeBlanc et al, 2007), tail suspension for rodents (Sonnenfeld & Shearer 2002; Sonnenfeld 2005), cell and microorganism cultures with high aspect ratio wall vessel bioreactors (RWV) (Schwarz et al, 1992; Tsao et al 1992; Hammond & Hammond 2001; Nickerson et al, 2003) and random positioning machines (RPM) (Hoson et al 1993; Walther etal, 1998; Pardo et al, 2005), denervation (Nikawa et al, 2004), and diamagnetic levitation (Dai et al, 2009; Hammer et al, 2009). Bed rest with the head tilted down at ~ 6? has been found to induce physiological alterations similar to those experienced in the space environment. In a similar way, tail suspension for mice and rats presents physiological effects analogous to those observed in a microgravity environment. To date, most of the microarray based studies of microgravity effects on gene expressions have used the ground-based RWV and RPM bioreactors to simulate microgravity environment (Figure 1 & Table 1). The following section in the introduction gives a brief overview of the major kinds of simulated microgravity models.

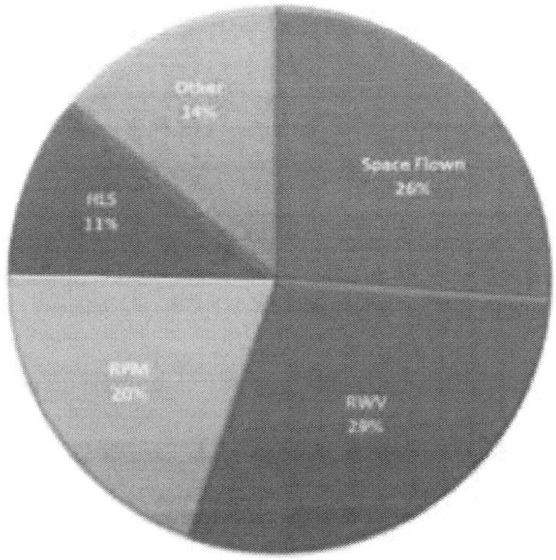

Fig. 1. The pie chart shows the percentage of each kind of microgravity used for the microarray studies discussed in this chapter.

Some ground-based simulated microgravity in bioscience research

Ground-based gravity-simulation experiments at the cellular and molecular levels have gained much insight into the underlying molecular and cellular alterations induced by microgravity stress as well as the mechanisms of the microgravity effects. In ground-based microgravity bioscience research, most simulation models simulate reduced gravity in the range of $10^{-4} \sim 10^{-6}$ g which is very small, close to "micro" (10^{-6}) g level (Klaus 2001). The most commonly used devices for simulating a microgravity environment are the RWV (Figure 2A) and the RPM (Figure 2B), which are also known as the 3D clinostat (van Loon 2007). The RWV and RPM bioreactors were developed to simulate microgravity by mimicking a functional weightless state. The RWV bioreactor rotates cells in a zero head space suspension culture that keeps the cells in a near free fall state (Figure 2A), which we have used for our studies (Clement et al 2007; 2008). The RPM is constructed of two independently rotating frames; one inside the other (Figure 2B). The frames are computer controlled and rotated at random rates (Hoson et al, 1997). This allows samples to continuously randomly position resulting in a vector-averaged simulation of near weightlessness (van Loon, 2007). Both of these systems have been designed to attempt to mimic the weightlessness experienced by objects in orbit around the Earth. Neither system eliminates gravity, but they do make a time-averaged g-vector close to zero (Klaus 2001). Both devices do not allow the cells to receive gravitational loads in any fixed direction.

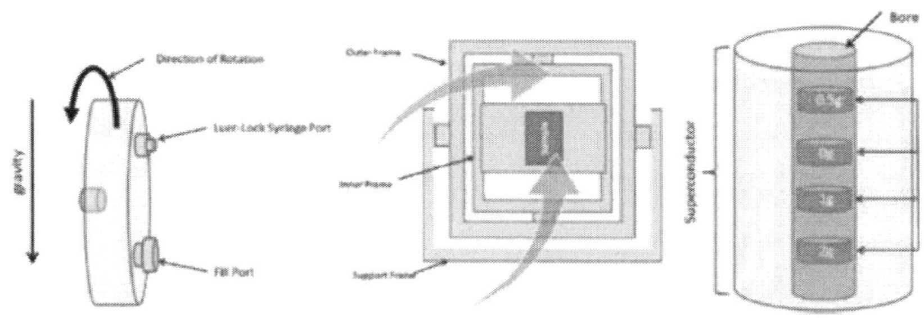

Fig. 2. A Simplified Schematic View of How the Three Ground-Based Simulated Microgravity Bioreactor Models Work. a). Rotating Wall Vessel bioreactor, rotating along horizontal axis. b) Random Positioning Machine: The blue arrow shows the direction of rotation of the outer frame and red arrows shows the rotation of the inner frame. c) Diamagnetic levitation model: A simple schematic of a variable magnetic force apparatus, a superconducting solenoid encased in a liquid nitrogen shielded liquid helium dewar with a room temperature bore passing through the center.

In addition to the RWV and RPM models, another ground-based simulated microgravity model that has been used less frequently is the diamagnetic levitation model (Figure 2C). Diamagnetic levitation is a method that uses magnetic force to create a near weightless state for ground based gravity studies. Such a variable magnetic force apparatus (VMF) is used to simulate a gravity environment from 0g to 2g (Valles & Guevorkian 2002; Coleman et al, 2007; Hammer et al, 2009). The device is essentially a superconducting solenoid encased in a liquid nitrogen shielded liquid helium dewar with a room temperature bore passing through the center (Figure 2C). Samples placed in the bore experience different gravitational force depending on the vertical position in the bore (Valles & Guevorkian 2002; Coleman et al, 2007).

GENOMIC TECHNOLOGY OF GENE EXPRESSION MICROARRAYS

Since the first microarray studies were published over 15 years ago, DNA microarrays have been used in many areas of biomedical sciences. Gene expression DNA microarrays have the potential to become key tools in space bioscience research because gene activity regulation is mostly controlled at the RNA level which is mainly determined by transcription initiation step. There are essentially four main manufacturing techniques for DNA microarrays: photolithography, contact printed, non-contact printed (inkjet), and bead arrays. Photolithography is a form of photochemical synthesis. A main advantage is the ability to put millions of features on one chip (Dalma-Weizhausz et al, 2006). Contact printing involves using robotically controlled print heads that spot or "print" the cDNA or oligonucleotides on a glass slide. Contact printed microarrays are what are commonly used for creating in-house microarrays. Non-contact printed works in a similar fashion to an inkjet printer. This technology does not have the capacity of arrays manufactured through photolithography, but it is improving every year. In 2006, around 180,000 features could be printed on one glass slide (Wolber et al, 2006). Today that has risen to close to 500,000 (Agilent website). With bead arrays the oligonucleotides are attached to 3µm silicon beadswhich are randomly deposited on a substrate such as a glass slide. The technology allows for hundreds of thousands of features to be attached to one slide (Fan et al, 2006). Both the photolithography and bead array are proprietary process of Affymetrix and Illumina, respectively. Adoption of these kinds of arrays, means adopting a complete system including the microarray scanner. In contrast, the contact and non-contact printed arrays are printed on glass microscope slides and can be used in a wide variety of microarray scanners and with a wide variety of software packages.

Although the high throughput gene expression microarray analysis is tremendously time efficient in that genes from the entire genome can be analyzed simultaneously in one experiment, there have been a number of concerns that have called into question the validity of microarray technology. With spotted (contact printed) microarrays there is evidence that cross-hybridization can happen (Handley et al, 2004), which reduces the specificity of the detection power. There have also been major concerns about the reproducibility of microarray data (Tan et al, 2003) as well as the bias and lack of interpretation that is present in many microarray studies (Richard, 2010). However, high fidelity tends to occur from experiments or studies performed using the same platform within a research laboratory. Tan et al reported a study that used the same samples and conditions, but three different microarray platforms. They found there was a high correlation (>0.9) between the data using the same platform, but when the data was compared between platforms it was as low as 0.47 and only reached 0.59 at best (Tan et al, 2003). Despite the justified concerns over the repeatability of results and cross-platform correlations, microarrays still hold tremendous potential for application to research into areas involving changes in environmental conditions such as microgravity research. In recent years, the MicroArray Quality Control (MAQC) consortium has shown that if standards are met and maintained, microarrays can yield a wealth of reliable data. They were able to show that if proper standards are set and followed inter-platform and intra-platform results are reproducible (MAQC Consortium, 2006). In addition to standardization of protocols, it is important to perform biochemical assays (such as Northern blotting, RT-PCR, etc.) to validate some of the microarray data obtained in individual research labs (Clement, 2010). Most microarrays that are currently being used are high-density, whole genome and multiplexed microarrays. Regardless of microarray platforms, the general procedure for a gene expression microarray assay is more or less the same towards evaluating mRNA abundance.

Once an experimental design has been decided upon, the procedure can be divided into bench work (wet lab) and desk work (computer analysis). A more detailed discussion of gene expression microarray and experimental design can be found from recent publications (Stekel et al 2003; Clement 2010).

THE INTEGRATION OF MICROARRAYS AND SIMULATED MICROGRAVITY BIOTECHNOLOGIES IN THE QUEST FOR GRAVITY-RESPONSIVE GENES

With the advent of high-throughput genomic technology such as gene expression microarrays, large scale genome-wide search for gravity sensitive genes have been carried out using mRNAs from a variety of organisms such as human, rat, mouse, xenopus, yeast,

- elegans, Drosophila, and several types of plants and microorganisms. The majority (about two thirds) of these studies have been on gene expression in cells grown in some form of simulated microgravity (Table 1).

To create the table of published microarray based microgravity studies, I started by doing two searches in PubMed using the search terms "microgravity and microarray" and "spaceflown and microarray". This yielded 47 and 36 citations respectively. I then analyzed the lists to remove redundancies, review articles, and articles that did not directly relate to analyzing the microgravity environment. This yielded a list of 49 published articles. After reviewing the literature, I was able to identify 7 more published microarray studies that did not appear in the PubMed searches. I added these to create the final list of 56 microarray based microgravity studies. Microarray data from these microgravity studies will be subjected to bioinformatics analysis towards the identification of major space genes.

The main aim of this chapter is therefore to review the current status of gene expression microarray technology in space bioscience research.

Table 1. Publications on Microarray Based Analysis of Microgravity Effects on Cells and Organisms

Organism/Cell or tissue type	Space/Microgravity	Duration	Microarray Platform	Vendor	Author	Year
Human						
Renal cortical cells	RWV/FTS	6d	cDNA	Incyte	Hammond et al	1999
Renal cortical cells	RWV/FTS	6d	cDNA	Incyte	Hammond et al	2000
HepG2	RWV	?	cDNA	Clontech	Khaoustov et al	2001
[illegible]	[illegible]	[illegible]	cDNA	[illegible]	[illegible] et al	2001
WI-38 fibroblast	[illegible]	[illegible]	cDNA	In house	[illegible]	2002
[illegible]	[illegible]	[illegible]	cDNA	In house	Vadrucci et al	2003
T-Cells	RPM	4h	oligonucleotide	Affymetrix	Boonyaratanakornkit et al	2005
[illegible]	RWV	24h	oligonucleotide	Affymetrix	Ward et al	2006
[illegible]	[illegible]	[illegible]	oligonucleotide	[illegible]	Clement et al	2006
EA.hy926 Endothelia	RPM	7d, 10d	multiplex array	Luminex	Grimm et al	2007
HepG2	RWV	[illegible]	oligonucleotide	Agilent	Clement et al	2007
hFOB Osteoblasts	RPM	[illegible]	cDNA	Clontech	Pardo et al	2007
[illegible]	RWV	[illegible]	oligonucleotide	Agilent	Clement et al	2008
[illegible]	[illegible]	[illegible]	oligonucleotide	In house	Chopra et al	2009
MG63 Osteoblasts	CL	24h	cDNA	Affymetrix	Qian et al	2010
[illegible]	[illegible]	[illegible]	oligonucleotide	[illegible]	[illegible] et al	2010
TK6 & WHC lymphoblastoid	ISS	8d	oligonucleotide	Agilent	Takahashi et al	2010
Mesenchymal Stem Cells	RWV	24h	oligonucleotide	Affymetrix	Sheyn et al	2010
Mouse						
[illegible]	RPM	[illegible]	oligonucleotide	Amersham	[illegible]	2005
[illegible]	RPM	[illegible]	oligonucleotide	Affymetrix	Patel et al	2007
[illegible] & gastrocnemius	HLS	24h	oligonucleotide	Agilent	[illegible] et al	2008
Brain	HLS	[illegible]	cDNA	In house	[illegible] et al	2008
Thymus [illegible]	STS/HLS	13d	oligonucleotide	Affymetrix	Allen et al	2009
[illegible]	RWV	[illegible]	oligonucleotide	Agilent	[illegible]	2009
[illegible]	[illegible]	[illegible]	oligonucleotide	Affymetrix	[illegible]	2009
Embryonic Stem Cells	RWV	72	cDNA	Roche	Blaber et al	2010
[illegible]	RWV	[illegible]	oligonucleotide	Agilent	[illegible] et al	2010
Thymus	STS	13d	oligonucleotide	Affymetrix	Lebsack	2010
Rat						
Soleus	HLS	[illegible]	cDNA	Clontech	[illegible] et al	2001
Soleus	HLS	[illegible]	oligonucleotide	Affymetrix	[illegible] et al	2003
tibialis anterior and gastrocnemius	STS	[illegible]	cDNA	Clontech	Taylor et al	2002
gastrocnemius	STS/HLS/DN	[illegible]	oligonucleotide	Affymetrix	Nikawa et al	2004
Soleus	HLS	[illegible]	cDNA	Clontech	[illegible]	2004
PC12	RWV	4h	cDNA	In house	[illegible] et al	2005
BMEC	RWV	[illegible]	oligonucleotide	CapitalBio	Dai et al	2007
Xenopus						
A6 Kidney	RPM	[illegible]	oligonucleotide	Agilent	Ikuzawa et al	2014
A6 Kidney	RPM	5d, 8d, 10d,13d	oligonucleotide	Agilent	Ikuzawa et al	2015
A6 Liver	RPM	[illegible]	oligonucleotide	Agilent	[illegible]	2017
A6 Liver	RPM	5d, 8d, 10d,13d	oligonucleotide	Agilent	Ikuzawa & Asashima	2008
Yeast						
S. cerevisiae	RWV	[illegible]	oligonucleotide	Affymetrix	[illegible] et al	2002
S. cerevisiae	RWV	[illegible] generations, [illegible] generations	oligonucleotide	Affymetrix	Sheehan et al	2007
Bacteria						
Salmonella enterica	RWV	[illegible]	cDNA	In house	Wilson et al	2002
Salmonella enterica	RWV	[illegible]	cDNA	In house	Wilson et al	2002
Salmonella enterica	RWV	[illegible]	cDNA	[illegible]	Chopra et al	2006
Salmonella typhimurium	STS	?	cDNA	In house	Wilson et al	2007
[illegible]	RWV	[illegible] generations	cDNA	[illegible]	[illegible] et al	2007
[illegible]	[illegible]	[illegible]	oligonucleotide	In house	[illegible] et al	2008
[illegible]	RWV/RTV	24	oligonucleotide	Affymetrix	Crabbé et al	2010
Plant						
Arabidopsis Thalia	RPM	?	cDNA	In house	Kittang et al	2004
Arabidopsis Thalia	RPM/SR	[illegible]	cDNA	In house	Martzivanou et al	2006
[illegible] (Fern)	STS	?	cDNA	In house	Salmi & Roux	2008
[illegible] (Barley)	[illegible]	?	oligonucleotide	[illegible]	[illegible] et al ([illegible])	2013
Drosophila/C.Elegans						
C.elegans	[illegible]		cDNA	Affymetrix	[illegible] et al	2006
C.elegans	[illegible]	[illegible]	oligonucleotide	Affymetrix	Selch et al	2008
Drosophila melanogaster	ISS/RTV	8.5d	oligonucleotide	Affymetrix	[illegible] et al	2010
Drosophila melanogaster	[illegible]	[illegible]	oligonucleotide	Affymetrix	[illegible] et al	2013

ISS = International Space Station
STS – Space Shuttle
RWV = Rotating Wall Vessel Bioreactor
RPM – Random Positioning Machine
HLS – Hind Limb Suspension
[illegible]

SR = Bed Rest
D – Denervation
B = Balloon
[illegible] – [illegible] Limb Suspension
SR – Sounding Rocket

UTILIZING MICROARRAYS TO EXAMINE IMMUNE SYSTEM CELLS EXPOSED TO MICROGRAVITY

The complex immune system evolved on earth has many windows of opportunity for a sudden switch to the space environment to dysregulate it. Hopefully, with dedicated research efforts of the space life sciences, especially with the application of advanced biotechnologies, a better understanding of immunology in the space environment will lead to effective countermeasures.

Study of the immune system is very important since it is known that astronauts have a much higher rate of infection during and after spaceflight (Sonnenfeld 1988; Sonnefeld 2005; Klaus and Howard 2006). It has been shown that inhibition of T cell activation in microgravity was a result of microgravity itself; factors other than microgravity can be excluded from the depressed activation of lymphocytes during spaceflight (Cogoli et al 1983; Ullrich et al, 2008). To understand the molecular mechanisms for the reduced activation of T cells during microgravity, many experiments have been documented from various laboratories using a variety of cell lines or animal models exposed to spaceflight and different simulated microgravity models (reviewed by Cogoli 1997; Sonnefeld 2005; Aponte et al, 2006). Although variable and contradictory results are common, overall evidence indicates that many functions such as signal transduction, cell-cell contact, cytoskeleton, and cell migration tend to be altered in microgravity. Experiments with human and mouse lymphocytes demonstrated a significant decrease in cell proliferation and in IL-2 and IL-Ra synthesis (Cogoli, 1997; Walther et al, 1998). An increase in urinary IL-6 excretion was observed on space flight and after landing (Stein and Schluter, 1994). IL-2 production decreased after space flight for three different T cell subsets ($CD3^+$, $CD4^+$ and $CD8^+$) and IFN-γ production decreased in the $CD4^+$ subset (Crucian et al, 2000). Spaceflight studies performed with rodents have found that microgravity caused rodents to be more susceptible to infection (Sonnenfeld et al, 1988), inhibited NK cell activity (Rykova et al, 1992), reduced capability for wound healing (Davidson et al., 1999), inhibited INF-γ production (Gould et al., 1987) and a reduction in lymphoid organ size (Congdon et al, 1996). A study using rhesus monkeys, microgravity inhibited IL-1 production and decreased response to colony stimulating factor (CSF) on bone marrow cells (Sonnenfeld et al, 1996). Similar to the in flight experiments, antiorthostatic suspension experiments on rodents have shown a increased levels of corticosterone, a reduced ability to clear bacteria from organs and a increased rate of mortality (Aviles et al, 2003). The important function of cytoskeleton in sensing of microgravity during spaceflight is reviewed (Hughes-Fulford, 2003). Over the past three decades, many more (thousands) publications have documented various results toward elucidation of microgravity mediated immune dysfunction. The most consistently observed effects of microgravity on the immune system have been a reduction in T cell and NK cell populations and functions, especially a reduction in cell-mediated immunity, altered cytokine production, as well as an increased susceptibility to infection under space flight conditions. Decreases in the reactivity of T cells, T cell cytotoxicity, and T cell helper activities have been documented for both spaceflight and ground based simulated microgravity studies.

Towards deciphering why T-cell activation is inhibited in microgravity, genome-wide microarray based analysis has become increasingly used within the last decade. Cell response to microgravity has been studied in relatively well-controlled clinostats and bioreactor cell cultures which are particularly convenient for time-course or multiplexed microarray analysis. The first report on cDNA microarray analysis of space flown T cells was documented in 2001(Lewis et al, 2001). In this study, human T cells (Jurkat, human acute leukemic T cell line, E6-1) were space-flown for 24h and 48h. cDNA microarray (GeneFilter™) analysis was performed to evaluate gene expression of 4,324 human genes at the 24h time point and 20,000 genes at the 48h time point. They identified differentially expressed genes that encode proteins for cytoskeletal organization, growth and metabolism, adhesion and signal transduction, transcription, apoptosis and tumor suppression (Lewis et al, 2001). The cDNA microarray (GeneFilter™) analysis of Jurkat cells flown on STS-95 in 1998 found that around 98% the genes examined had similar expression patterns when the space-flown cells were compared to ground control cells. They were the first to report that cytoskeletal genes were differentially regulated: calponin, dynamin, tropomodulin, keratin8, myosins, ankyrin, an actinlike protein, plectin, and C-NAP-1 were upregulated; gelsolin was downregulated. Their data indicated that the expression of genes functioning in interconnect cytoskeletal elements to each other and to cell membrane, regulate filament polymerization and microtubule organization centers were altered by spaceflight (Lewis et al, 2001).

Another earlier microarray based analysis of microgravity effects on T cell activation was reported by Meloni et al (2002). They intended to discriminate between effects of microgravity and cosmic radiations on the influence of microgravity on mitogenic activation of T cells and studied the effects of high cosmic radiations on the gene expression in human T cells boarded in a stratospheric balloon (22 hours flight). They used cDNA microarray hybridization technology for the gene expression analysis and found that activated cells react to the ionizing stress by activating genes involved in cell cycle check-point, oxidative stress response, heat shock protein production or by repressing genes involved in antigen recognition.

Aiming to examine the roles of early genes in initiating and maintaining T cell activity, Hughes-Fulford's lab performed a microarray analysis of simulated microgravity (4h RPM) effect on human peripheral lymphocytes and found that PKA was a key player in the loss of T cell activation in microgravity (Boonyaratanakornkit, et al. 2005). For this study they used Human Genome Focus Arrays (Affymetrix) and identified 91 down-regulated genes as a result of exposure to simulated microgravity. The expression of early genes regulatedprimarily by transcription factors NF-κB, CREB, ELK, AP-1, and STAT were impaired inmicrogravity, suggesting that microgravity either slows, impedes, or fully blocks keysignaling pathways in early T cell activation (Boonyaratanakornkit et al, 2005). They showed that IL-2 was among the down-regulated genes, which correlates well with previous non- array based reports (Cogoli et al, 1993; Pippia et al, 1996; Walther et al, 1998; Hughes- Fulford et al, 2005). Interestingly, IL-2 was also significantly decreased (and IL-10 expression was increased) in a recent real time PCR based analysis of gene expression in T-cells of mice after being flown in space aboard STS-118 for 13 days (Gridley et al, 2009).

Ward et al used Affymetrix Human U133A arrays to study activated human peripheral lymphocytes exposed to simulated microgravity in RWV (22 rpm) for 24h (Ward et al, 2006). From their triplicate experiments they identified 89 (10 up, 79 down) genes that were statistically significant ($P \leq 0.01$) and at least 1.5 fold up or down-regulated in an all of the arrays. A larger proportion of these affected genes are found to be players in fundamental cellular processes such as immune responses, signal transduction, DNA repair and apoptosis, and metabolic pathways (Ward et al, 2006).

A microarray based study of mRNA expression in murine thymus tissue extracted from C57BL/6NTac female mice that had been flown aboard the space shuttle Endeavour (STS118) for thirteen days was reported recently by Lebsack et al (2010). They used Affymetrix gene chips for this study and found 970 genes significantly differentially regulated (644 up and 326 down). Among the identified genes in stress response, RNA binding motif 3(RBM3) and cold inducible binding protein (CIRBP) were up regulated, while HSP90, HSP110, STIp1, FKBP4 were down regulated. More importantly, genes that regulate immune response were affected by space flight: CD44 and CXCL10 that promote T cell development were down regulated; whereas CTLA-4 (negative regulator of T cell activation) mRNA was upregulated. Overall, the genes identified in this study were involved in stress regulation, glucocorticoid receptor metabolism, and T cell signaling and activity (Lebsack et al, 2010).

The above five studies made good use of the biotechnologies in their respective microgravity studies. Although impressive findings are documented in each individual report, widely varied results in terms of the type of genes or trends in expressions are noted among these microgravity studies using microarrays in the immune system. For a systematic view of these studies, I attempted to compile the data into a tabulated form and was hampered by obstacles, mainly due to a lack of overlap in the identified gravity sensitive genes. The only "overlap" was two genes (STAT1 and XCL1) found to be differentially regulated in two studies: in one study they were both down-regulated (Boonyarantankornkit et al, 2005), while in the other they were both up-regulated (Ward et al, 2006). The large variation in the resulting gravity sensitive genes may in part be due to numerous variables and the overall complexity of the immune system itself (Gridley et al, 2003). In addition, variations of the types of microgravity, different cell lines or cell types, various microarray platforms, etc., could also contribute to the differences in the results. Furthermore, how the data was documented and reported can also contribute to the lack of consistency in resulting sensitive genes. Another contributing factor is the fact that these few studies span a large time frame (2001 ~ 2010) for the relatively young microarray biotechnology, the time period the technology itself undergoes development and is still in the process of standardization. With more standardization in the genome wide assays, meaningful compilation of data can be applied efficiently in a statistical analysis toward the identification of major space genes.

Nonetheless, it is still of interest to see which pathways or molecular functions the gravity sensitive genes from afore mentioned studies. To generate the KEGG Pathways for immune system related cells, I needed to prepare the data for analysis in DAVID (Database for Annotation, Visualization, and Integrated Discovery)(Huang et al, 2009a; Huang et al, 2009b). Out of the five microarray studies, four had gene tables, or lists of genes that were relatively easy to use. Three of these lists used human cells (Lewis et al, 2001; Boonyarantankornkit et al, 2005; Ward et al, 2006), one used mouse tissue (Lebsack et al, 2010). Since the majority of the studies were done with human cells, I chose those three gene lists to compile a master gene list by manually combining the genes from the three microarray studies and using the DAVID conversion tool to convert them all into the same format. This conversion is necessary because of inconsistency in the way gene tables are presented: some of the gene lists included gene bank accession numbers and gene symbols, some included gene symbols but no accession numbers, and others included accession numbers but no symbols, still others with no gene list at all. To limit confusion and for ease of references, both accession numbers and gene symbols should be included in gene lists whenever possible. The converted gene list resulted in 142 genes after redundant, unknown, or unable to define genes were eliminated. This combined gene list was uploaded to the DAVID Functional Annotation tool to identify the statistically significant KEGG Pathways (Table 2). Interestingly but not surprisingly, most of the functional pathways identified using the combined gravity sensitive genes are key pathways for innate and adaptive immunity (Table

2). This is indeed significant, despite the fact that such a large variation in the type and trends of gene expressions were found among these studies.

Table 2. Pathway Identification of Microarray Identified Gravity Sensitive Genes in Immune System. Legend: red means up regulated; green means downregulated; purple means opposite trends between studies; underlined means it appears in more than one study.

Term	Count	PValue	Genes
hsa04060:Cytokine-cytokine receptor interaction	14	4.27E-06	CSF2, CCL3, IL2RA, TNFSF14, CCL4, CXCL10, LIF, TNFRSF9, TNFSF10, CCL20, IFNG, XCL1, XCL2, LTA
hsa03050:Proteasome	7	1.14E-05	PSMB5, PSMB10, PSMC4, PSME2, PSMA3, IFNG, PSMB9
hsa04062:Chemokine signaling pathway	9	0.001014	CCL3, DOCK2, CCL20, NFKB1, XCL1, STAT1, XCL2, CCL4, CXCL10
hsa04612:Antigen processing and presentation	5	0.01353	PSME2, CREB1, HSPA1A, KIR2DL3, LTA
hsa04620:Toll-like receptor signaling pathway	5	0.02593	CCL3, NFKB1, STAT1, CCL4, CXCL10
hsa04640:Hematopoietic cell lineage	4	0.071181	CSF2, IL2RA, CD59, CD2
hsa00970:Aminoacyl-tRNA biosynthesis	3	0.076664	WARS, SARS, GARS
hsa04630:Jak-STAT signaling pathway	5	0.094726	LIF, CSF2, IL2RA, IFNG, STAT1

Does the cell use "many roads lead to Rome" approach to adapt to microgravity environment? It may be possible that major space pathways rather than specific major space genes are key determinants for adaptation to microgravity. More studies would be needed for statistically based cross-laboratory and cross-platform analysis of microarray based data, which will provide key insight into the molecular mechanism of microgravity mediated immune dysfunction. An altered immune response to microgravity is attributed to be a key factor for bone loss because the altered production and action of cytokines in the immune system could affect bone remolding (Zayzafoon et al, 2005).

MICROARRAY ASSESSMENT OF BONE CELLS EXPOSED TO MICROGRAVITY CONDITIONS

Physical inactivity or mechanical unloading to the skeletal system is an underlying cause for bone density loss in clinical disorders such as spinal cord injury, stroke, prolonged bed-rest, aging and osteoporosis as well as in spaceflight microgravity environments (LeBlanc et al, 2005; Beller et al, 2011). Therefore, continued effort in deciphering the mechanism and finding a remedy for bone density loss is beneficial to human health both in space and on earth. Spaceflight caused bone density reduction specifically involves weight-bearing bones (Mack et al, 1967; Vose et al, 1974; Schneider et al, 1995; Collet et al, 1997; Lang et al, 2006; Keyak et al., 2009). In the space environment, bone density in the lower extremities and spine of crew member is lost at a rate of 1% to 2% per month (LeBlanc et al, 2007; Amin, 2010). Microgravity disturbs the balance between bone formation and resorption in bone remolding process: it tends to increase bone resorption functions of osteoclast and decrease bone formation functions of osteoblast. A net loss of calcium in-flight of similar magnitude to that observed in earlier studies from Skylab as well as an increase in bone resorption markers (Smith et al, 2005). Studies of spaceflights showed decreased serum levels of bone formation markers such as alkaline phosphatase (ALP), osteocalcin (OCN), and the C-terminal peptide of pro-collagen I (Collet et al, 1997; Caillot-Augusseau et al, 2000). Increased resorption with little change in formation is the main finding in space flight caused uncoupled bone remodeling (reviewed by LeBlanc et al, 2005). The unbalanced bone resorption and formation resulted in increased Ca^{2+}secretion. Ground-based simulated microgravity studies showed decreased bone formation (Nakamura et al, 2003; Zayzafoon et al, 2004; Pardo et al, 2005; Patel et al, 2007; 2009). Although a gene-specific approach has identified several key genes involved in bone cell growth and development that are affected in microgravity, a comprehensive genome-wide search allows for the identification of more genes as well as possible pathways through which the bone loss occurs.

A number of microarray based analysis of microgravity effect on bone cells have been published relatively recently. Pardo et al showed that gene expression of 140 genes (88 down and 52 up) were significantly altered after exposure of 2T3 murine preosteoblast to 3 days simulated microgravity in RPM (Pardo et al, 2005). They used CodeLink Uniset Mouse 1 Bioarrays (Amersham Biosciences) and the median intensity of all the probes to normalize the intensity of the individual probes. In agreement with spaceflight data and simulated microgravity studies, their microarray data showed genes important for bone density such as alkaline phosphatase (ALP), runt-related transcription factor 2 (Runx2), osteomodulin, parathyroid hormone-related protein(PTHrP), parathyroid receptor 1 (PthR1), and platelet derived growth factor (PDGF) were significantly down-regulated. In contrast, cathepsin K (responsible for bone resorption in osteoclasts) is upregulated in the 2T3 cells in response to simulated microgravity. ALP is a known marker for bone formation, and Runx2 is involved in osteoblastic differentiation and skeletal morphogenesis. Both interact with secreted bone morphogenic proteins and with insulin like growth factor 1 (IGF1), which has been shown to regulate Runx2 in endothelial cells (Qiao et al 2004). In their data, IGF1 was shown to be slightly down-regulated although it did not pass the significance test (Pardo et al 2005). PthR1 promotes the release of Ca^{2+} and it has already been shown to be gravity sensitive in bone cells (Torday 2003). Overall, this study shed much light on the mechanism of microgravity mediated bone loss through the use of microarray analysis. Patel et al (2007), the same research group as theaforementioned study by Pardo et al, published a further study on 2T3 murine preosteoblast cells grown in RWV for 3 days. The cells were seeded on microcarrier beads and rotated on the RWV at 22 rpm. Microarray analysis of the simulated microgravity treated preosteoblasts showed that the microgravity downregulated 61 and upregulated45 genes by more than twofold compared to static 1 g controls. Comparison of the RWV/microarray data with the data from previous RPM/microarray analysis they found 14 mechanosensitive genes that were changed in the same direction. Once again ALP, runx2, PthR1, and PDGF were shown to be significantly down-regulated. Thus, the two different simulators of microgravity on the same cell line assayed by microarray kits from two different vendors produce similar results with regard to bone cell differentiation and osteoblast function. In a further non-microarray based study (Patel et al, 2009), they cultured 2T3 cells in SMG using RPM and PCR-based analysis. They again found that ALP, Runx2, PthR1 were down-regulated in simulated microgravity. They also found that low magnitude and high frequency (LMHF) mechanical loading (0.1 ~ 0.4 g at 30 Hz for 10 ~ 60 min/day) prevented a decrease in ALP, Runx2, PthR1, but static conditions had no effect (Patel et al, 2009).

Yamada et al studied osteoblasts (NOS-1 cells derived from a human osteosarcoma) that were exposed to 3 days of simulated microgravity in RPM. They used Atlas™ Human 3.8K microarray for the mRNA analysis. As with the previous studies, ALP activity was significantly reduced in the cells exposed to simulated microgravity. However, there is not a gene list for the microarray analysis. It is significant to find that the addition of chitosan (a natural polyaminosaccharide) significantly increased ALP activity in the cells exposed to simulated microgravity (Yamada et al 2007).

A recent study by Capulli et al (2009) used primary mouse calvarial osteoblasts grown in simulated microgravity (RWV) for 5 days and used Agilent microarrays for the analysis of simulated microgravity exposed bone cells. The cells were seeded to microcarrier beads and grown in RWV at 16 rpm to simulate microgravity. They found that 133 genes were differentially regulated, 45 genes were significantly up-regulated and 88 were down- regulated. The significantly differentially regulated genes were presented in tables of clusters and molecular function classifications. Among the differentially regulated genes are genes involved in osteoblast differentiation, function, and osteoblast–osteoclast cross-talk, genes of extracellular matrix, glycosaminoglycan/heparin-binding activity, and growth factor activity. The findings concerning FN1 are consistent with other studies in different cell lines (Dapp et al, 2004, Nikawa et al, 2004, Sheyn et al, 2010). The finding with CTGF is also consistent with other findings (Sheyn et al, 2010).

Qian et al (2009) examined gene expression profile changes of human osteoblast-like cell line MG-63 in response to 24 h simulated microgravity, highmagneto-gravitational environment (HMGE). They used 35 mm cell culture plates seeded with cells and placed them into the HMGE at special positions to achieve the gravity effects of 0, 1, and 2 g by the object stage. They used a self-made circulating water-bath as a control for temperature of 37±0.5°C. High-density human genome (HG) U133 Plus 2.0 Arrays (Affymetrix) were used for the gene expression analysis. Among the total of 54,613 gene probes examined with the microarray, they found 53 genes were statistically down-regulated and 55 genes were statistically up-regulated compared with the 0 g with the temperature control. They presented the identified genes in tables according to cellular functions. They specificallynoted that cytoskeleton-related genes such as WASF2 and WIPF1 genes were the most mechanosensitive.

Hammer et al used diamagnetic levitation to conduct experiments of how gravity affects MC3T3-E1 osteoblastic cell line. In the experiment, they exposed cells to 0g, 1g, and 2g gravity. The cells were exposed to the magnetic field for two days after which RNA was extracted and microarrays were run. They used Affymetrix Mouse genome arrays and the intensity of housekeeping genes for normalization, Robust Multichip Average (RMA) for normalization. The data was then scaled to a median of 100; genes below 50 were filtered out. They used a 3 fold cut-off to determine the gene list for further bioinformatics analysis. Based on this criterion they identified 2270 genes that were upregulated and 135 genes that were down-regulated when 1g samples were compared to 0g samples. The focus of the paper is more on testing the diamagnetic levitation biotechnology rather than on the genes that are differentially regulated (Hammer et al, 2009). .

Sambandam et al (2010) published a recent study of osteoclast grown in RWV SMG (16 rpm to simulate 0.0008g environment) for 24 hours. For the study they used Agilent whole genome arrays. They followed standard statistical procedures to identify their list of differentially regulated genes. They identified 3,404 differentially expressed genes. They have followed MIAME standards by depositing the microarray profile data in the GEO database. Some genes of interest that were up-regulated include CTSK, CTSL, and CTSB, as well as several MMPs, bone matrix degrading proteases. Their microarray data agrees with previous finding that stimulation of osteoclastogenesis in microgravity environment.

All seven studies also made good use of the biotechnologies in their respective studies. To further examine all the microarray-identified genes from bone cells exposed to microgravity, I performed pathway analysis using DAVID. As in the immune system section, the use of gene bank accession numbers and gene symbols are not standardized here also. In order to find out which pathways the altered genes are involved, I first used the DAVID conversion tool to convert them all into the same format. Out of the seven microarray studies involving bone cells, five studies have the gene lists published in the paper or supplemental tables. Since the majority (four out of five) were studies done in mice (Pardo et al, 2005; Patel et al, 2007; Capulli et al, 2009; Sambandam et al 2010) and one was in human (Qian et al, 2007), I chose to use those four gene lists to compile into one master list in the same format, and uploaded them to the DAVID Functional Annotation tool for pathway analysis. The list was run as species Mouse and Identified the statistically significant KEGG Pathways (data not shown).

Within the seven microarray analysis of microgravity exposed bone cells discussed above, there are certain consistency or overlap in the identified gravity sensitive genes among these different studies. Are there potential major space genes or bone-specific space genes among these studies? To this end, I performed further cross-laboratory, cross-species, cross- microgravity-platform, and cross-microarray-platform comparative analysis here. In order to identify the major gravity sensitive genes from these microarray/microgravity studies, I manually examined and picked out the genes that appeared to be significant in more than one of the five studies (two out of the seven studies do not have a gene list with the publications nor available in supplement). Essentially, any gene that was identified in two or more of the five studies were compiled into a table along with the data values (fold changes) from each specific source of origin for direct comparison (Table 3).

Table 3. Microgravity Sensitive Genes Identified by Microarray analysis of Bone Cells

	Mouse			Human	
Gene	Pardo et al, 2005	Patel et al, 2007	Capulli et al, 2009	Sambandan et al, 2010	Qian et al, 2009
Akp2	0.21	0.82			
Bmp4	1.03	0.40			
Col11a1	0.29	0.40			
Dcn	0.56	0.76			
Enpep	0.35	0.40			
Fn1			0.35		4.40
Gadd45g	0.49	0.40			
Has2	1.12	0.74			
Hspa1a		3.18	2.26		
Igf2	0.75		0.40		
Il1a	1.88	1.58			
Il6	0.96		4.44		
ITGB5				6.20	0.06
Ogn	0.38	0.23	0.21		
Omd	0.18	0.10	0.36		
POSTN			0.38		2.10
Pthr1	0.20	0.54			
Runx2	0.53	0.69			
Sod2	0.35		0.47		
Sfrp2	0.41	0.43	0.21		
Thbs1	1.42				3.00
Thbs2	0.72	0.39			
Wisp2		0.39	0.42		

It is more apparent now with this table that the data from the same research group tends to be more consistent (compare Pardo et al 2005 and Patel et al 2007).

The highest overlapped (in three of the five studies) genes were osteoglycin(OGN), osteomodulin (OMD), and secreted Frizzle-Related Protein 2 (SFRP2) and all three genes were down-regulated in all the microarray/microgravity studies that identified them (Pardo et al, 2005; Patel et al, 2007; Capulli et al, 2009). Osteoglycin(OGN) encodes a protein (a small proteoglycan) which induces bone formation in conjunction with transforming growth factor beta. Osteomodulin (OMD) is implicated in biomineralization processes. In addition to the reduction of expression of the two genes important in bone formation, the majority of the gravity sensitive genes shown in Table 3 are genes encode proteins in pathways involving ECM- receptor interaction and focal adhesion: Integrin β5 (ITGB5), thrombospondins (THBS1, THBS2), collagen (COL11A1), and fibronectin (FN1). In addition, four components in the TGF-β signaling pathway were also found in this table: bone morphogenic protein (BMP4), decorin (DCN), and the thrombospondins (THBS1, THBS2). Furthermore, the two cytokines, interleukin-1(IL-1) and interleukin-6(IL-6), that coordinate and regulate many cellular activities of the innate immunity were among the more frequently identified microgravity sensitive genes in bone (Table 3).

MICROARRAY ANALYSIS OF MICROGRAVITY EXPOSED MUSCLE CELLS

Muscle atrophy as a result of spaceflight has been a condition that was identified early on in the space program. As with bone cells and cells in the immune system, spaceflight also has a major effect on muscle cells. A study of an exercise program showed that crewmembers while aboard the International Space Station (ISS) for 6 months calf muscle volume and peak power decreased significantly, and there was a redistribution among the faster phenotypes, despite the fact that the crewmembers exercised regularly (Trappe et al, 2009). Microgravity effect on muscle has been one of the most studied areas in space life science research. Ground based analogs such as bed rest or unilateral leg suspension (ULLS) have been used in humans and HLS has been used in rats. Bed rest and ULLS have shown results that seem to be similar to spaceflight (Narici & deBoer, 2011). However, it has been cautioned that the HLS using rodent models may not be the best choice when extrapolating results to humans; the processes involved in muscle atrophy and disuse in rodent and humans are very different (Rennie et al, 2010). Currently, little is even known about mechanisms of muscle atrophy and disuse in general (Rennie et al, 2010). In studying microgravity effect on muscle cells, most commonly used model systems are rodents and space flight experiments have been performed extensively using both mice and rats flown in space. The most commonly used ground based simulated microgravity models are hind limb-suspension of rodents. These have been combined with microarray biotechnology for genome wide gene expression analysis of the microgravity effect. Here I will focus my discussion on the microarray based studies and apply DAVID bioinformatics tools to examine the genes that have been identified in these studies.

Taylor et al (2002) used Clontech Atlas DNA expression array to study the alteration of gene expression profiles in skeletal muscles (tibialis anterior and gastrocnemius) of male rats flown on the STS 90 Neurolab for 17 days. They found that 50 genes showed differential regulation: 38 genes were downregulated and 12 were upregulated. Genes related to cell proliferation and growth factor cascades (such as $p21^{cip1}$, Rb, cyclins G1/S, -E, -D3, MAP kinase 3, MDA3, and ras related protein RAB2) were down-regulated during spaceflight. The microarray data indicates that genes involved in regulation of muscle satellite cell replication are down regulated by microgravity (Taylor et al 2002). Thus, this experiment gives further insight into the mechanisms underlying muscle atrophy and diminished muscle repair capability associated with the space environment.

Another 2002 study (Wittwer et al 2002) using Clontech Atlas 1.2K Rat arrays identified 105 genes out of 1200 tested that were significantly differentially regulated after 35 days hind- limb suspension. Much of their microarray data agreed with similar studies reported previously. Data suggested a coordinated increase in the expression of genes (PFKM, ALDOA, and GAPDH) in the glycolic pathway involved in the cytoplasmic conversion of glucose to pyruvate in 35-day HS *m. solei*. In addition, the mRNA of an enzyme that controls glycolytic flux, adenylate kinase 1 (AK1), was also increased. mRNA of proteins (LDL receptor, SR-BI, FATP, and H-FABP) involved in the uptake and transport of fatty acids from the blood into the muscle fibers decreased. mRNAs of vesicle transport proteins (IRAP, M6P/IGFR2, and VAMP3) involving mitochondrial energy conversion were increased. These genes have been implicated in glucose uptake, mediated by the major insulin- mediated glucose transporter of skeletal muscle, glut-4. Data suggests the decreased fatty acid import proteins as a regulation towards reduced fatty acid uptake and transpot concomitant with adaptation toward greater glycogen utilization and a generally reduced energy demand in the atrophied muscle. The gene activities for intracellular protein degradation such as lysosomal protease mRNAs (cathepsin C, L, and D) and some enzymes involved in cytosolic protein degradation by proteasomes (i.e., TPPII, UBE2B, rP28 alpha, and carboxypeptidase D) were increased. Extracellular proteases (MMP-2, u-PA) as well as protease inhibitors (TIMP-2 and -3) were also increased at mRNA level. The findings gave much insight towards the understanding of skeletal muscle atrophy.

Stein et al (2002) studied soleus muscle isolated from 5 control and 5 hindlimb suspended rats (21 days) with the Affymetrix microarray system for assessing gene expression on fuel pathways within the muscle. Similar to the Wittwer et al report, they observed a consistent decrease in expression of genes involved in fatty acid oxidation and an increase in expression of genes in the glycolytic activity in the suspended group. Their microarray data further confirms that disuse atrophy is associated with a change in mRNA levels of enzymes involved in energy metabolism, a shift away from use of fat to use of glucose.

Dapp et al (2004) applied the mouse hindlimb suspension (HS) model for gene expressional alterations underlying loaddependent muscular adaptations. Gene expression was assessed from total RNA by a muscle-specific low-density cDNA microarray (custom designed Atlas cDNA arrays with 222 double spots on each nylon array). Immediate early genes such as FRA1, JUND, and JUN were induced as were IGFs, while myosin heavy chain was reduced in mouse soleus muscle after 7 days hind-limb suspension.

Nikawa et al (2004) published a study comparing gene expression patterns in gastrocnemius muscle (the largest muscle of the calf of the leg that acts by extending the foot and bending the knee) cells from rats flown in space, rats exposed to hind-limb suspension, and denervated rats. Using Affymetrix Rat Genome U34 GeneChips, they found that most gene expression changes were unique to spaceflight (Nikawa et al 2004). The DNA microarray data indicated that spaceflight specifically caused altered expression of some mitochondrial genes and cytosleletal genes (such as A-kinase anchoring protein and cytoplasmic dynein) as well as up-regulated ubiquitin ligase genes (MuRF-1, Cbl-b, and Siah-1A). Several oxidative stress-inducible genes were also upregulated and hightly expressed in the muscle of the spaceflown rats (Nikawa et al 2004). They proposed that mitochondrial dislocation during spaceflight may have caused atrophy in the form of insufficient energy supply and leakage of reactive oxygen species from the mitochondria. Although only a fraction of the genes are discussed in limited space of the original publication, comprehensive tables of microarray data are presented in the paper, which provides a very valuable source of reference information. The same research group published another research paper using human 20 day bed rest to study muscle disuse atrophy mechanism and used DNA microarray for their genome wide gene expression analysis (Ogawa et al, 2006). Their data suggested that Cbl-b or atrogin-1 mediated ubiquitination pathway could be important in unloading induced muscle atrophy in humans.

Mazzatti et al 2008 reported a study using gene arrays (Agilent Mouse Oligo Arrays) to determine the acute effects of short-term HLS on metabolic consequences of unloading. They used Agilent whole-genome arrays to examine mRNA expression in mouse soleus and gastronemius muscle cells after 24 hours hind-limb suspension and identified 600 genes with a FDR of 0.05 and at least 1.5 fold differential regulation in both cell types. Several proteins (PPARδ, UCP-3, AMPK, and CPT1/2) that have putative roles in the maintenanceof metabolic flexibility were upregulated. They also concluded that there was increased reliance on glucose as an energy source or a loss of metabolic flexibility. Muscle unloading appears to result in reduced fatty acid oxidation, decreased transcription of genes involved in lipid metabolism and increased expression of genes involved in glycogen synthesis (Mazzatti et al 2008). The findings that increased reliance on glucose fit with previous studies that have shown that atrophied muscle rely more on glucose for energy (Fitts et al, 2000; Henriksen and Tishler 1988; Stein et al, 2002; Martin et al, 1988).

A more recent study (Allen et al 2009) of murine skeletal muscles(gastrocnemius from mice) compared space flown mice (11-day, 19-h on STS-108 shuttle flight) with the ground-based unloading model of hindlimb suspension (one group of pure suspension and one of suspension followed by 3.5 h of reloading). They found that spaceflight causes differential regulation in genes involved in muscle growth and fiber type. Their microarray data showed that 272 mRNAs significantly differentially regulated by spaceflight, of which many genes were found to belong to pathways involved in muscle growth and adaptation. Spaceflight significantly altered the levels of mRNAs involved with the PI3- kinase/Akt/mTOR pathway: the PI3-kinase regulatory subunit polypeptide 1, pi3kr1/p85α, the forkhead box O1 (FoxO1), transcription factor, the muscle-specific ubiquitin ligase F-box only protein 32 (MAFbx/atrogin1), and the ubiquitin-conjugating enzyme E2 variant 2 mRNAs were increased; Insulin receptor substrate-1 (IRS-1) mRNA levels were decreased in space flown mouse gastrocnemius muscles. Genes in the TNF-α/NF-κB signaling pathway and the calcineurin/nuclear factor of activated T cells (NFAT) pathway were also affected: the TNF-α downstream target TNF-α-induced protein 2 mRNA and the NF-kB inhibitor nuclear factor B light chain gene enhancer in B cells inhibitor α(Nfkbia/IκBα) mRNA were significantly increased, whereas mRNA levels of the NFAT cytoplasmic, calcineurin- dependent 3 (Nfatc3) transcription factor were significantly decreased in SF gastrocnemius. In addition, mRNAs for three members of the CAAT/enhancer binding protein (C/EBP) family of transcription factors, C/EBP-α, C/EBP-β, and C/EBP-δ were also increased in SF gastrocnemius. They found that space flight increased myostatin (which limits muscle growth) mRNA and decrease the mRNA levels of myostatin inhibitor FSTL3. They also found that mRNA levels of the slow oxidative fiber-associated transcriptional coactivators, peroxisome proliferator activated receptor alpha (PPAR-α) and the PPAR-γ coactivator 1α (PGC1-α) decreased in space-flight. They concluded that spaceflight induced significant changes in mRNA expression of genes associated with muscle growth and fiber type toward a less oxidative phenotype (Allen et al 2009).

Chopard et al (2009) performed a genome-wide gene expression analysis of female skeletal muscles during 60 days of bed rest with and without exercise or dietary protein supplementation as countermeasures. They investigated the effects of long-term bed rest on the gene expression of soleus (SOL) and vastus lateralis (VL) muscles in healthy women using a customized microarray containing 6,681 muscles-relevant genes. They found clusters of genes involved in nucleic acid and protein metabolism were upregulated and that encoding components involved in energy metabolism were downregulated. Counter measures (exercise and nutrition) had some compensatory effects on gene expression profiles.

Reich et al (2010) examined the global gene expression patterns of the left vastus lateralis muscle in seven sedentary men following 48 h unloading via unilateral lower limb suspension and 24 h reloading. Microarray analysis of gene expression changes were usedfor the identification of enriched functions and canonical pathways. They found that the highest ranked canonical pathways were related to protein ubiquitination and oxidative stress response pathways. Gene functions related to mitochondrial metabolism were the most significantly downregulated. The increases in mRNA for ubiquitin proteasome pathway-related E3 ligase and stress response gene heme oxygenase-1 as well as extracellular matrix (ECM) component COL4A3 were confirmed by qRT-PCR. The unloading associated gene expression patterns were not reversed on reloading.

The ten studies discussed above made good use of microarray biotechnology in their respective studies. Out of the ten studies, we found seven that had gene lists of some form that could be further analyzed. Among the seven gene lists, a few genes were reported in more than one study. Fibronectin (FN1) was identified by several microarray studies in muscle. In space-flown rat muscles, FN1 was down-regulated (Nikawa et al, 2004) and it was down-regulated in HLS rats as well (Dapp et al, 2004). Interestingly, FN1 was up- regulated in the recovery time points in Dapp et al. FN1 was also up-regulated in women exposed to bed rest (Chopard et al, 2009). ACADVL was found to be down-regulated both in rat (Stein et al, 2002) and human muscles (Chopard et al, 2009). Another four genes were identified in the same directions between a study in mouse and a study in rat: MT1, MT2, and PIM3 were upregulated, whereas MARCKS was down regulated in both studies (Nikawa et al, 2004; Allen et al, 2009). In addition, CYR61 gene was identified to be upregulated by a study in mouse (Allen et al, 2009) and a study in rat muscles (Nikawa et al, 2004). I also did DAVID analysis for all the genes pooled from the ten studies and found that 32 pathways were involved (data not shown). This large variation could at least in part be due to the heterogeneity of the study conditions. The factors discussed in the immune and bone sections also apply.

MICROARRAY ANALYSIS OF MICROGRAVITY EXPOSED OTHER CELLS

Exposure of rat bone marrow mesenchymal stems cells (rBMSC) to HMGE simulated microgravity study found that BMSC proliferation and osteogenesis decreased and the cells were growth arrested in the G0/G1 phase of cell cycle (Dai et al 2007). Data from their microarray (CapitalBio Corporation) analysis confirmed that rBMSC proliferation and osteogenesis gene activities decreased under simulated microgravity. Insulin-like growth factor-I, epidermal growth factor, and basic fibroblastic growth factor that normally stimulated rBMSC proliferation had only a marginal effect in the simulated microgravity.

Sheyn et al (2010) using Affymetrix microarrays examined simulated microgravity (RWV) effect on human mesenchymal stem cells. The cells were seeded on microcarrier beads and placed in a RWV bioreactor, which was rotated at 16 rpm to simulate a microgravity environment. They identified 882 genes that were down regulated and 505 genes were up- regulated by 2 fold or above and with a P value of 0.05 or less. They identified a large number of gene clusters responded to microgravity. In agreement with previous studies, their microarray data showed a general trend of less osteogenesis and more adipogenesis when hMSCs cells were cultured in simulated microgravity. Most of the extracellular matrix related genes were downregulated. They identified many genes that were involved in the actin cytoskeleton such as COL1A1, COL1A2, FN1, SPARC, CTGF, and IGFBP3, were down-regulated. The mRNA of several growth and differention factors such as fibroblast growth factor, vascular endothelial growth factor, insulin-like growth factor-related proteins, andbone morphogenetic protein (BMP)6 were downregulated. Most osteogenic genes (such as BMP6, osteonectin, and collagen type I) were downregulated, whereas most adipogenic genes were upregulated.

A recent study using hematopoietic stem cells by Fridley et al (2010) was not a microgravity study per se. Instead it was a comparative study of stirred tank bioreactors and RWV bioreactors. They studied how these devices would work in the formation of embryoid bodies and the generation of hematopoietic progenitor cells. For the RWV study they ran the bioreactor samples at speeds ranging from 10 rpm to 40 rpm. They used Roche Nimblegene4 plex microarrays (72k) for the study. They did discuss some of the genes that were differentially regulated, but they did not publish a full gene list. One gene that was significantly down-regulated was E-Cadherin (Fridley et al, 2010).

Clement et al, using Agilent 22K Human 1A microarrays, examined skin cells (HEK001) for their response to simulated microgravity over the course of 3, 4, 9, and 10 days followed by recovery from SMG in normal cell culture conditions for 15, 50, and 60 days. A total of 162 genes ($P < 0.05$) were differentially regulated by at least 2 fold at some point during the time course. They also used Northern blotting to both qualitatively and quantitatively verify their data. The genes that were analyzed showed a statistical correlation with the microarray data ranging from 0.69 to 0.97. Interestingly, 80% of the differentially expressed genes recovered from 4 days SMG exposure after 15 days in normal cell culture conditions. Genes that were exposed to 9 days SMG required more than 50 days to recover to pre-exposure levels. Also of interest, HLA-G, a key gene in cellular immune response suppression, was found to be significantly up-regulated during the recovery phase (Clement et al 2008). This study indicates that longer term exposure to microgravity may have long term affects on the body even after returning to normal gravity. Further studies on microgravity exposure and subsequent recovery should be carried out.

The effect of microgravity on the nervous system and brain at the molecular level is still not well understood. Frigeri et al examined gene expression in mouse brain after 2 weeks hind- limb supesension (Frigeri et al 2008). Using 27K cDNA microarrays, they found 592 statistically significant genes with 1.5 fold or higher differential regulation. Hind-limb suspension also seems to affect the pathways involved in learning and memory as well as blood coagulation. They found that hind-limb suspension causes a more hyper-coagulative state and an increased risk of venous thrombosis (Frigeri et al 2008). In a recent study PC12 cells were exposed to 4 days SMG to analyze oxidation sensitive genes in microgravity (Kwon et al 2006). They found that 65 genes were up-regulated and 39 were down-regulated as a result of exposure to SMG. They found that genes involved in DNA repair and replication, cell proliferation, apoptosis, molecular transport, and oxidative phosphorylation were affected by exposure to SMG.

A 2002 study of human fibroblast grown onboard the space shuttle (STS-93) in 1999 identified changes in expression of 10 genes in the TNF or IL gene families. For this study, they chose 202 genes which were then spotted on to a nylon membrane. They did verify their results on some genes with semi-quantitative PCR with a correlation of 0.89. The genes identified are involved in the regulation of bone density or proinflammatory status (Semova et al 2002).

Some of the first microarray studies of space flown cells was done using renal cortical cells flown on STS-90 (Hammond et al, 1999). They applied a high throughput microarray systemto study gravity-induced gene-expression changes in human renal cells and found 1632 out of 10,000 genes changed as a result of exposure to microgravity (Hammond et al, 2000).

Kitamoto et al studied Xenopus A6 cells grown in RPM simulated microgravity environment and examined gene expression using customized Agilent 8K Xenopus laevis DNA microarrays (Kitamoto et al, 2004, Kitamoto et al, 2005). They found that 52 out of 8091 genes examined showed differential regulation. Their time course data showed interesting gene expression patterns of microgravity effect and cellular adaptation. They also found that SPARC was down-regulated Xenopus A8 cell lines (Ikuzawa and Asashima 2008). Their data combined with previous observations (Kitamoto et al, 2005) lead them to conclude that SPARC might play a key role in response to microgravity (Ikuzawa and Asashima 2008). Down-regulation of SPARC and SPARC related genes has also been observed in human mesenchymel stem cells grown in SMG (Sheyn et al, 2010).

Liver cells have also been studied in SMG. Both of the SMG studies on liver cells were carried out using HepG2 cells (Clement et al, 2007; Khaoustov 2001). Both studies showed that liver cells grown in a RWV caused altered gene expression patterns. The Clement study used Agilent 22K human 1A microarrays to study HepG2 cells. They identified 139 genes which were differentially regulated with a fold change of >1.5 and P<0.01. Khaoustov et al used 6K Human Array containing 6144 genes showed that 95 genes were differentially regulated. More systematic studies would be needed in further studying microgravity effects on liver cells.

TOWARDS THE IDENTIFICATION OF MAJOR SPACE GENES

A main focus of this chapter was to try to identify potential major space genes by reviewing gene expression profiles, compare and contrast the expression profiles of cells from different lineages of different organisms exposed to various microgravity conditions. My original hypothesis was that it could be possible that a common set of key microgravity sensitive genes in different cells would be preferentially altered in microgravity conditions. The identification of this common set of genes might lead to the identification of "major space genes" that together play a major check-and-balance role ultimately determining the outcome of a cell, or an organism such as an individual person, in the response to microgravity conditions. This is analogous to areas of cancer research resulting in the identification of growing numbers of major cancer genes (oncogenes and tumor suppressor genes) that are directly involved in the determination the fate of a cell---whether a cell becomes cancerous or not. I therefore believed that a systematic examination of molecular alterations of cells under microgravity conditions would enhance our knowledge of cellular responses to microgravity and the underlying mechanisms. Since cells of different organ or tissue systems may be affected by this environmental factor differently, studying and comparing the cellular response to such gravity changes using various cell lines of different lineage or tissue origin would generate a comprehensive database of cellular activity alteration associated with microgravity, and the identification of major space genes. This would provide scientific basis for therapeutic intervention aimed to prevent and correct abnormalities resulting from space flight.

Examining high throughput microarray data as well as other gene expression data from cells and organisms exposed to microgravity, there were a great many differentiallyregulated genes that showed little or no commonality across closely related studies. In most study areas such as immune, bone, and muscle response to microgravity, there was overall very little commonality or overlap among the reports at the individual gene expression level. However, when looked at gene pathways through DAVID analysis, relatively focused and very insightful pathways were identified for the microarray based analysis of microgravity effects on the immune and bone cells. Would it be that major space pathways rather than major space genes are playing key roles in microgravity response? Typically, there are many gene components in a pathway. It is possible that genes in a pathway can have various levels of activity in response to an environmental challenge and the level of activity of the pathway is optimized to be able to best adapt to environmental change by adjusting component genes activities. Whatever the case, it is too early to conclude from such an initial attempt of cross platform and cross species examinations.

Nonetheless, as part of the identification of space genes, I decided to make a further attempt to compare the gene lists from the microarray studies in mammals that were discussed in this chapter. I compared gene lists from these studies to determine what genes might be common among these studies. There were a large number of differences between the gene lists. This has been shown to be the case in other studies examining microarray experiments between labs and across platforms. These variations are not surprising considering the significant variations between experiments. First, the type of cells used for the studies were very different. These cells included rat muscle, mouse muscle, Xenopus renal, and human liver, renal, keratinocyte, fibroblasts, etc. The methods for studying microgravity were different. These included Random Positioning Machine (RPM), Rotating Wall Vessel (RWV) bioreactors, bed rest, hind limb suspension, as well as space flown. Finally, and perhaps more importantly, the microarray platforms had major variations. All of these variables make it particularly difficult to compare data. For the purpose of this discussion, we decided to include any gene on our final list of potential "space genes" if it appeared on two or more gene lists of the studies we examined. Using this loosely defined criterion, the initial potential candidates for "space genes" were subjected to bioinformatics analysis.

Using the DAVID Bioinformatics Resources, I processed these genes together to identify relevant KEGG pathways (Table 4). The KEGG Pathways and Gene Ontologies were included in the tables if they were considered to be statistically significant ($P \leq 0.05$).

The pathway table shows that Extracellular Matrix-Receptor Interaction and Focal Adhesion Pathways are two important pathways. Both have been identified as pathways that may be affected by Microgravity (Ingber 1998).

As part of the attempt to identify and create a list of potential space genes from the microarray based microgravity studies, I decided to analyze a variety of gene lists from these studies. First, I limited the scope of the bioinformatics analysis to mammalian cells. I then examined all of these studies to identify the ones that had useable gene lists. Out of a total of 36 mammalian studies, I identified 26 studies that could be used for this purpose. If a gene appeared as differentially regulated in at least four of these studies, it would be included in the candidate gene list. This yielded a list of 8 potential space genes, name Putative Space Genes (Table 5).

The initial candidate major space genes thus identified in Table 5 (for which I briefly discussed while reviewing each of the involved studies) will provide a clue for furtherscrutiny in future more systematic studies. I believe that systematic examination of molecular alterations of cells under microgravity conditions will enhance our knowledge of cellular responses to gravity changes and the underlying mechanisms.

KEGG Pathways	Genes	
ECM-receptor interaction	CD44, COL3A1, COL1A2, ITGB5, COL1A1, THBS1, THBS2, COL11A1, FN1	
Focal adhesion	ACTB, COL3A1, COL1A2, ITGB5, COL1A1, THBS1, THBS2, COL11A1, FN1	
Hypertrophic cardiomyopathy (HCM)	ACTB, ACTC1, IL6, TGFB3, ITGB5, TPM1	H,M,R,X
TGF-beta signaling pathway	BMP4, INHBA, TGFB3, DCN, THBS1, THBS2	H,M,R
Dilated cardiomyopathy	ACTB, ACTC1, TGFB3, ITGB5, TPM1	H,M
Prion diseases	IL6, HSPA1A, IL1A	H
Pathways in cancer	WNT5A, BMP4, IL6, TGFB3, CDH1, MMP1, FN1	
Glycolysis / Gluconeogenesis	HK2, PCK2, ENO1	
Bladder cancer	CDH1, THBS1, MMP1	
Pathogenic Escherichia coli infection	ACTB, CDH1, TUBB3	

Table 5. List of putative space genes.

Gene Symbol	Gene Description	Species*
4 studies		
CD44	antigen (homing funciton snd Indian blood group system)	H, M
MARCKS	Myristoylated alanine-rich protein kinase C substrate	R, M, X
5 Studies		
FN1	Fibronectin	H, R, M
TUBA1	Tubulin, Alpha 1	H, M, X
6 Studies		
CTGF	Connective tissue growth factor	H, M, X
CYR61	Cysteine-rich, angionic inducer, 61	H, R, M, X
MT2	Metallothionein 2	H, R, M
7 Studies		
MT1	Metallothionein 1	H, R, M, X

*H = human, M = Mouse, R = Rat, X = Xenopus

REFERENCES

1. Atkov, O. (1992). Some medical aspects of an 8 month's space flight. *Advances in Space Research.* Vol.12, No.1, pp. 343-345.
2. Aponte, V.M., Finch, D.S. & Klaus, D.M. (2006). Considerations for non-invasive in-flight monitoring of astronaut immune status with potential use of MEMS and NEMS devices. *Life Sciences.* Vol.79, pp. 1317-1333.
3. Amin, S. (2010) Mechanical factors and bone health: effects of weighlessness and neurological injury. *Current Rheumatology Reports.* Vol.12, No.3, pp. 170-176
4. Allen, D.L., Bandstra, E.R., Harrison, B.C., Thorng, S., Stodieck, L.S., Kostenuik, P.J., Morony, S., Lacey, D.L., Hammond, T.G., Leinwand, L.L., Argraves, W.S., Bateman, T.A. & Barth, J.L. (2009). Effects of spaceflight on murine skeletal muscle gene expression. *Journal of Applied Physiology.* Vol.106, pp. 582-595
5. Aubert, A.E., Beckers, F. & Verheyden, B. (2005). Cardiovascular function and basics of physiology in microgravity. *Acta Cardiologica.* Vol.60, No.2, pp. 129-151.
6. Aviles, H., Belay, T., Vance, M., Sun, B. & Sonnenfeld, G. (2004). Active hexose correlated compound enhances the immune function of mice in the hindlimb-unloading model of spaceflight conditions. *Journal of Applied Physiology.* Vol.97, pp. 1437-1444
7. Beller, G., Belavy, D.L., Sun, L. & Armbrecht, G. (2011). WISE-2005: Bed-rest induces changes in bone mineral density in women during 60 days simulated microgravity. *Bone.* doi:10.1016/j.bone.2011.06.021
8. Boonyaratanakornkit, J.B., Cogoli, A., Li, C.F., Schopper, T., Pippia, P., Galleri, G., Meloni, M.A. & Hughes-Fulford, M. (2005). Key gravity-sensitive signaling pathways drive T cell activation. *FASEB Journal.* Vol.19, pp. 2020-2022
9. Borchers, A.T., Keen, C.L. & Gershwin, M.E. (2002) Microgravity and immuneresponsiveness: implications for space travel. *Nutrition.* Vol.18, pp. 889-898.
10. Capulli, M., Rufo, A., Teti, A. & Rucci, N. (2009) Global transcriptome analysis in mouse calvarial osteoblasts sets of genes regulated by modeled microgravity and identifies a "mechanoresponsive osteoblast gene signature". *Journal of Cellular Biochemistry.* Vol.107, No.2, pp. 240-252
11. Carmeliet, G., Nys, G., Stockmans, I. & Bouillon, R. (1998). Gene expression related to the differentiation of osteoblastic cells is altered by microgravity. *Bone.* Vol.22, No. 5 Suppl., pp. 139S-143S
12. Clement, J.Q. (2010). Microarray profiling of genome-wide expression regulation in response to environmental exposures, In: *A Practical Guide to Bioinformatics Analysis,* G.P.C. Fung, pp. 22-40, Iconcept Press, ISBN 978-0-9807330-2-0, Brisbane

13. Clement, J.Q., Lacy, S.M. & Wilson, B.L. (2007). Genome-wide gene expression profiling of microgravity effect on human liver cells, *Journal of Gravitational Physiology.* Vol.14, pp.P121-122

14. Clement, J.Q., Lacy, S.M. & Wilson, B.L. (2008). Gene expression profiling of human epidermal keratinocytes in simulated microgravity and recovery cultures. *Genomics, Proteomics, and Bioinformatics.* Vol.6, No.1:8-28.

15. Coleman, C.G., Gonzalez-Villalobos, R.A., Allen, P.A., Johanson, K., Guevorkian, K., Valles, J.M. & Hammond, T.G. (2007). Diamagnetic levitation changes, growth, cell cycle, and gene expression of *saccharomyces cerevisiae. Biotechnology and Bioengineering.* Vol.98, No.4, pp. 854-863

16. Caillit-Augusseau, A., Vico, L., Heer, M., Voroviev, D., Sourberbielle, J., Zitterman, A., Alexandre, C. & Lafage-Proust, M. (2000). Space flight is associated with rapid decreases of undercarboxylated Osteocalcin and increases of markers of bone resorption without changes in their circadian variation: observations in two Cosmonauts. *Clinical Chemistry.* Vol.46, pp. 1136-1143

17. Collet, P., Uebelhart, D., Vico, L., Moro, L., Hartmann, D., Roth, M. & Alexandre, C. (1997). Effects of 1- and 6- month spaceflight on bone mass and biochemistry in two humans. *Bone.* Vol.20, No.6, pp. 547-551

18. Cogoli, A. (1993). Space flight and the immune system. *Vaccine.* Vol.11, No.5, pp. 496-503.

19. Collet, P. et al (1997). Effects of 1- and 6-month spaceflight on bone mass and biochemistry in two humans. *Bone.* Vol.20, pp. 547-551.

20. Congdon, C.C., Allebban, Z., Gibson, L.A., Kaplansky, A., Strickland, K.M., Jago, T.L., Johnson, D.L., Lange, R.D. & Ichiki, A.T. (1996). Lymphatic tissue changes in rats flown on Spacelab Life Sciences-2. *Journal of Applied Physiology.* Vol.81, No.1, pp. 172-177.

21. Crabbe, A., Pycke, B., Van Houdt, R., Monsieurs, P., Nickerson, C., Leys, N & Cornelis, P. (2010). Reponse of *Pseudomonas aeruginosa* PAO1 to low shear modeled microgravity involves AlgU regulation. *Environmental Microbiology.* Vol.12, No.6, pp. 1545-1564

22. Crucian, B.E., Cubbage, M.L. & Sams, C.F. (2000) *Journal of Interferon Cytokine Research.* Vol.20, No.6, pp. 547-556

23. Dai, Z.Q., Wang, R., Ling, S.K., Wan, Y.M. & Li, Y.H. (2007). Simulated microgravity inhibits the proliferation and osteogenesis of rat bone marrow mesenchymal stems cells. *Cell Proliferation.* Vol.40, pp. 671-684

24. Dalma-Weiszhausz, D.D., Warrington, J., Tanimoto, E.Y. & Miyada, C.G. (2006). The Affymetrix GeneChip® platform: an overview. *Methods in Enzymology.* Vol.410, pp. 3-28

25. Dapp, C., Schmutz, S., Hoppeler, H. & Fluck, M. (2004). Transcriptional reprogramming and ultrastructure during atrophy and recovery of mouse soleus muscle. *Physiologic Genomics.* Vol.20, pp. 97-107

26. Davidson, J.M., Aquino, A.M., Woodward, S.C. & Wilfinger, W.W. (1999). Sustained microgravity reduces wound healing and growth factor responses in the rat. *FASEB Journal.* Vol.13, No.2, pp. 325-329

27. Fan, J.B., Gunderson, K.L., Bibikova, M., Yeakley, J.M., Chen, J., Garcia, E.W., Lebruska, L.L., Laurent, M., Shen, R. & Barker, D. (2006) Illumina universal bead arrays. *Methods in Enzymology.* Vol.410, pp. 57-73

28. Fan, X., Lobenhofer, E.K., Chen, M., Shi, W., Huang, J., Luo, J., Zhang, J., Walker, S.J., Chu, T.M., Li, L., Wolfinger, R., Bao, W., Paules, R.S., Bushel, P.R., Li, J., Shi, T., Nikolskaya, T., Nikolsky, Y., Hong, H., Deng, Y., Cheng, Y., Fang, H., Shi, L. & Tong, W. (2010) Consistency of predictive signature genes and classifiers generated using different microarray platforms. *The Pharmacogenomics Journal.* Vol.10, pp. 247-257

29. Fitts, R.H., Desplanches, D., Romatowski, J.G. & Widrick, J.J. (2000). Physiology of a microgravity environment. Invited review: microgravity and skeletal muscle. . *Journal of Applied Physiology.* Vol.89, pp. 823-839

30. Fridley, K.M., Fernandez, I., Li, M.A., Kettlewell, R.B. & Roy, K. (2010). Unique differentiation profile of mouse embryonic stem cells in rotary and stirred tank bioreactors. *Tissue Engineering: Part A.* Vol.16, No.11, pp. 3285-3298

31. Frigeri, A., Iacobas, D.A., Iacobas, S., Nicchia, G.P., Desaphy, J.F., Camerino, D.C., Svelto, M. & Spray, D.C. (2008). Effect of microgravity on gene expression in mouse brain. *Experimental Brain Research.* Vol.191, pp. 259-300

32. Fritsch-Yelle, J.M., Charles, J.B., Jones, M.M. & Wood, M.L. (1996). Microgravity decreases heart rate and arterial pressure in humans. *Journal of Applied Physiology.* Vol. 80, No.3, pp. 910-914.

33. Gould, C.L., Lyte, M., Williams, J., Mandel, A.D. & Sonnenfeld, G. (1987) Inhibited interferon-gamma but normal interleukin-3 production from rats flown on the space shuttle. *Aviation, Space and Environmental Medicine.* Vol.58, No.10, pp. 983-986

34. Gridley, D.S., Nelson, G.A., Peters, L.L., Kostenuik, P.J., Bateman, T.A., Morony, S., Stodieck, L.S., Lacey, D.L., Simske, S.J. & Pecaut, M.J. (2003). Selected contribution: effects of spaceflight on immunity in the C57BL/6 mouse. II. Activation, cytokines, erythrocytes, and platelets. *Journal of Applied Physiology.* Vol.94, pp. 2095-2103.

35. Gridley, D.S., Slater, J.M., Luo-Owne, X., Rizvi, A., Chapes, S.K., Stodieck, L.S., Ferguson, V.L. & Pecaut, M.J. (2009). Spacefilght effects on T lymphocyte distribution, function and gene expression. *Journal of Applied Physiology.* Vol.106, pp. 194-202

36. Hammond, T.G. & Hammond, J.M. (2001). Optimized suspension culture: the rotating-wall vessel. *American Journal of Physiology- Renal Physiology* . Vol.281, pp. F12–F25

37. Hammond, T. G., Lewis, F.C., Goodwin, T.J., Linnehan, R.M., Wolf, D.A., Hire, K.P., Campbell, W.C., Benes, E., O'Reilly, K.C., Globus, R.K. & Kaysen, J.H. (1999). Gene expression in space. *Nature Medicine* Vol.5, No.4, pp. 359–359

38. Hammond, T.G., Benes, E., O'Reilly, K.C., Wolf, D.A., Linnehan, R.M., Taher, A., Kaysen, J.H., Allen, P.L. & Goodwin, T.J. 2000. Mechanical culture conditions effect gene expression: gravity-induced changes on the space shuttle. *Physiological Genomics* Vol.3, pp. 163-173

39. Handley, D., Serban, N., Peters, D., O'Doherty, R., Field, M., Wasserman, L., Spirtes, P., Scheines, R. & Glymour, C. (2004). Evidence of systematic expressed sequence tag IMAGE clone cross-hybridization on cDNA microarrays. *Genomics.* Vol.83, pp. 1169-1175

40. Henriksen, E.J. & Tishler, M.E. (1988) Glucose uptake in rat soleus: effect of acute unloading and subsequent reloading. *Journal of Applied Physiology.* Vol.64, pp. 1428-1432

41. Hoson,T.,Kamisaka,S.,Masuda,Y.,Yamashita,M.&Buchen,B.(1997).Evaluation of the three-dimensional clinostat as a simulator of weightlessness. *Planta.* Vol.203, pp. S187-S197

42. Hoson, T., Kamisaka, S., Miyamoto, K., Ueda, J., Yamashita, M. & Masuda, Y. (1993). *Microgravity Science and Technology.* Vol.6., No.4, pp. 278-281

43. Huang,D.W.,Sherman,B.T.&Lempicki,R.A.(2009a).Systematic and integrative analysis of large gene lists using DAVID Bioinformatics Resources. *Nature Protocols.* Vol.4, No.1, pp. 44-57.

44. Huang, D.W., Sherman, B.T. & Lempicki, R.A. (2009b). Bioinformatics enrichment tools: paths toward the comprehensive functional analysis of gene lists. *Nucleic Acids Research.* Vol.37, No.1, pp. 1-13

45. Hurley, M.M. & Lorenzo, J.A. (2004). Systemic and Local Regulators of Bone Remodeling, In: *Bone Formation,* Bonner, F. & Farach-Carson, M.C., pp. 44-70, Springer-Verlag, ISBN 1-85233-717-6, London

46. Hughes-Fulford, M. (2003). Function of the cytoskeleton in gravisensing during spaceflight. *Advances in Space Research.* Vol.32, No.8, pp. 1585-1593

47. Hughes-Fulford, M. (2002). Physiological effects of microgravity on osteoblast morphology and cell biology. *Advances in Space Biology and Medicine.* Vol.8, pp. 129-157

48. Ikuzawa, M. & Asashima, M. (2008). Global expression of simulated microgravity- responsive genes in Xenopus liver cells. *Zoological Science.* Vol.25, No. 8, pp. 828-837

49. Jagoe, R.T., Lecker, S.H., Gomes, M. & Goldberg, A.L. (2002). Patterns of gene expression in atrophying skeletal muscles: response to food deprivation. *FASEB.* Vol.16, pp. 1697- 1712

50. Keyak,J.H., Koyama, A.K., LeBlanc, A. & Lang, T.F. (2009). Reduction in proximal femoral strength due to long-duration spaceflight. *Bone.* Vol.44, pp. 449-453

51. Khaoustov, V.I., Risin, D., Pellis, N.R. & Yoffe, B. (2001). Microarray analysis of genes differentially expressed in HepG2 cells cultured in simulated microgravity. *In Vitro Cellular & Developmental Biology – Animal.* Vol.37, No.2, pp. 84-88Hammer, B.E., Kidder, L.S., Williams, P.C. & Xu. W.W. (2009). *Microgravity Science and Technology.* Vol.21, No.4, pp. 311-318.

52. Kitamoto, J., Fukui, A. & Asashima, M. (2004). Global and temporal regulation of gene expression in Xenopus kidney cells in response to presumed microgravity generated by 3D clinostats. *Biological Sciences in Space.* Vol.18, pp. 152-153

53. Kitamoto, J., Fukui, A. & Asashima, M. (2005). Temporal regulation of global gene expression and cellular morphology in Xenopus kidney cells in response to clinorotation. *Advances in Space Research.* Vol.35, pp. 1654-1661.

54. Klaus D.M. (2001). Clinstats and bioreactors. *Gravitational Space Biology Bulletin.* Vol.14, No2., pp. 55-64

55. Klaus, D.M. & Howard, H.N. (2006). Antibiotic efficacy and microbial virulence during space flight. *Trends in Biotechnology.* Vol.24, pp. 131-136

56. Kwon, O., Sartor, M., Tomlinson, C.R., Millard, R.W., Olah, M.E., Sankovic, J.M. & Banerjee, R.K. (2006). Effect of simulated microgravity on oxidation-sensitive gene expression in PC12 cells. *Advances in Space Research.* Vol.36, No.6, pp. 1168-1176

57. Lang, T.F., LeBlanc, A.D., Evans, H.J. & Lu, Y. (2006). Adaptation of the proximal femur to skeletal reloading long-duration spaceflight. *Journal of Bone and Mineral Research.* Vol.21, No.8, pp. 1224-1230.

58. Leach, C.S. (1990). Medical considerations for extending human presence in space. *Acta* Astronautica. Vol.21, pp. 659-666.

59. LeBlanc, A.D., Spector, E.R., Evans, H.J. & Sibonga, J.D. (2007). Skeletal responses to space flight and the bed rest analog: a review. *Journal of Musculoskeletal and Neuronal Interactions.* Vol.7, No.1, pp. 33-47

60. Lebsack, T.W., Fa, V., Woods, C.C., Gruener, R.. Manziello, A.M., Pecaut, M.J., Gridley, D.S., Stodieck, L.S., Ferguson, V.L. & DeLuca, D. (2010). Microarray analysis of spaceflown murine thymus tissue reveals changes in gene expression regulating stress glucocorticoid receptors. *Journal of Cellular Biochemistry.* Vol.110, pp. 372-381

61. Lewis, M.L., Cubano, L.A., Zhao, B., Dinh, H.K., Pabalan, J.G., Piepmeier, E.H. & Bowman, P.D. (2001). cDNA microarray reveals altered cytoskeletal gene expression in space- flown leukemic T lymphocytes (Jurkat). *FASEB Journal.* Vol.15, No.10, pp. 1783- 1785

62. Mack, P.B., LeChange, P.A., Vose, G.P. & Vogt, F.B. (1967). Bone demineralization of foot and hand of gemini-titan IV, V and VII astronauts during orbital flight. *American Journal of Roentgenology, Radium Therapy and Nuclear Medicine.* Vol.100, No.3, pp. 503-11

63. MAQC Consortium (2006). The MicroArray quality control (MAQC) project shows inter- and intraplatform reproducibility of gene expression measurements. *Nature Biotechnology.* Vol.24, No.9, pp. 1151-1161

64. Marcu, O., Lera, M.P., Sanchez, M.E., Levic, E., Higgins, L.A., Shmygelska, A., Fahlen, T.F., Nichol, H. & Bhattacharya, S. (2011). Innate immune responses of *Drosophila Melanogaster* are altered by Spaceflight. *PLoS One*, Vol.6, No.1, e15316

65. Martin, I.P., et al 1988. Influences of spaceflight on rat skeletal muscle. . *Journal of Applied Physiology.* Vol.65, pp. 2318-2325

66. Mazzatti, D.J., Smith, M.A., Oita, R.C., Lim, F.L., White, A.J. & Reid, M.B. (2008). Muscle unloading-induced metabolic remodeling is associated with acute alterations in PPAR□ and UCP-3 expression. *Physiological Genomics.* Vol.34, pp. 149-161

67. Meloni, M.A., Galleri, G., Carta, S., Negri, R., Costanzo, G., de Sanctis, V. Cogoli, A. & Pippia, P. (2002). Preliminary study of gene expression levels in human T-cells exposed to cosmic radiations. *Journal of Gravitational Physiology.* Vol.9, No.1, pp. P291-292.

68. Nakamura, H., Kumei, Y., Morita, S. Shimokawa, H., Ohya, K. & Shinomiya, K. (2003) Suppression of osteoblastic phenotypes and modulation of pro- and anti-apoptotic features in normal human osteoblastic cells under a vector-averaged gravity condition. *Journal of Medical and Dental Sciences.* Vol.50, No.2, pp. 167-176

69. Narici, M.V. & de Boer, M.D. (2011). Disuse of the musculo-skeletal system in space and on earth. *European Journal of Applied Physiology.* DOI 10.1007/s00421-010-1556-x McPhee, J (2006). Life sciences research standardization. *Journal of Gravitational Physiology.* Vol.13, pp. 59-72.

70. Nath, R., Kuman, D., Li, T. & Singal, P.K. (2000). Metallothioneins, oxidative stress and the cardiovascular sytem. *Toxicology.* Vol.155, pp. 17-26

71. Nickerson, C.A., Ott, C.M., Wilson, J.W., Ramamurthy, R., LeBlanc, C.L., Honer zu Bentrup, K., Hammond, T. & Pierson, D.L. (2003). Low-shear modeled microgravity: a global environmental regulatory signal affecting bacterial

gene expression, physiology, and pathogenesis. *Journal of Microbiological Methods.* Vol.54, No.1, pp. 1-11.

72. Nikawa, T., Ishidoh, K., Hirasaka, K., Ishihara, I., Ikemoto, M., Kano, M., Kominami, E., Nonaka, I., Ogawa, T., Adams, G.R., Baldwin, K.M., Yasui, N., Kishi, K. & Takeda, S. (2004). Skeletal muscle gene expression in space-flown rats. *FASEB Journal.* Vol.18, pp. 522-524

73. Pardo, S.J., Patel, M.J., Skyes, M.C., Platt, M.O., Boyd, N.L., Sorescu, G.P., Xu, M., van Loon, J.J.W.A., Wang, M.D. & Jo, H. (2005). Simulated microgravity using the random positioning machine inhibits differentiation and alters gene expression profiles of 2T3 preosteoblasts. *American Journal of Cell Physiology.* Vol.288, pp. C1211-C1221

74. Patel, M.J., Liu, W., Sykes, M.C., Ward, N.E., Risin, S.A., Risin, D. & Jo, H. (2007). Identification of mechanosensitive genes in osteoblasts by comparative microarray studies using the rotating wall vessel and the random positioning machine. *Journal of Cellular Biochemistry.* Vol.101, pp. 587-599

75. Patel, M.J., Chang, K.H., Skyes, M.C., Talish, R., Rubin, C. & Jo, H. (2009). Low magnitude and high frequency mechanical loading prevents decreased bone formation responses of 2T3 preosteoblast. *Journal of Cellular Biochemistry.* Vol.106, pp. 306-316

76. Pippia, P., Sciola, L., Cogoli-Greuter, M., Meloni, M.A., Spano, A. & Cogoli, A. (1996) Activation signals of T lymphocytes in microgravity. *Journal of Biotechnology.* Vol.47, pp. 215-222

77. Qian, A., Di, S., Gao, X., Zhang, W., Tian, Z., Li, J., Hu, L., Yang, P., Yin, D. & Shang, P. (2009). cDNA microarray reveals the alterations of cytoskeleton-related genes in osteoblasts under high mangneto-gravitational environment. *Acta Biochimica et Biophysica Sinica.* Vol.41, No.7, pp. 561-577

78. Qiao, M., Shapiro, P., Kumar, R. & Passaniti, A. (2004). Insulin-like growth factor-1 regulates endogenous RUNX2 activity in endothelial cells through phosphatidylinositol 3- kinase/ERK-dependent and Akt-independent signaling pathway. *Journal of Biological Chemistry.* Vol.279., No. 41, pp. 42708-42718

79. Reich, K.A., Chen, Y.W., Thompson, P.D., Hoffman, E.P. & Clarkson, P.M. (2010). Forty- eight hours of unloading and 24h of reloading lead to changes in global gene expression patterns related to ubiquitination and oxidative stress in humans. *Journal of Applied Physiology.* Vol.109, No.5, pp. 1404-1415

80. Rennie, M.J., Selby, A., Atherton, P., Smith, K., Kumar, V., Glover, E.L. & Philips, S.M. (2010). Facts, noise and wishful thinking: muscle protein turnover in aging and human disuse atrophy. *Scandinavian Journal of Medicine and Science in Sports.* Vol.20, pp. 5-9

81. Rykova, M.P., Sonnenfeld, G., Lesnyak, A.T., Taylor, G.R., Meshkov, D.C., Mandel, A.D., Medvedev, A.E., Berry, W.D., Fuchs, B.B. & Konstantinova, I.V. (1992). Effect of spaceflight on natural killer cell activity. *Journal of Applied Physiology*. Vol.73, pp. 196S-200S

82. Sambadan, Y., Blanchard, J.J., Daughtridge, G., Kolb, R.J., Shanmugarajan, S., Pandruvada, S.N.M., Bateman, T.A. & Reddy, S.V. (2010). Microarray profile of gene expression during osteoclast differentiation in modeled microgravity. *Journal of Cellular Biochemistry*. Vol.111, pp. 1179-1187Schneider, V. (1995). Bone and body mass changes during space flight. *Acta Astronaut.* Vol.36, pp. 463-6.

83. Schwarz, R.P., Goodwin, T.J. & Wolf, D.A. (1992). Cell culture for three dimensional modeling in rotating-wall vessels: an application in microgravity. *Journal of Tissue Culture Methods.* Vol.14, No.2, pp. 51-58

84. Scott, J.M., Warburton, D.E., Williams, D., Whelan, S. & Krassioukov, A. (2011). *Spinal Cord.* Vol.49, No.1, pp. 4-16

85. Semova, A., Semova, N., Lacelle, C., Marcotte, R., Petroulaski, E., Proestou, G. & Wang, E. (2002). Alterations in TNF- and IL-related gene expression in space-flown WI38 human fibroblasts. *FASEB Journal.* Vol.16, No.8, pp. 899-901

86. Sheehan, K.B., McInnerney, K., Purevdorj-Gage, B., Altenburg, S.D. & Hyman, L.E. (2007). Yeast genomic expression patterns in response to low-shear modeled microgravity. *BMC Genomics.* 8:3

87. Sheyn, D., Pelled, G., Netanely, D., Domany, E., Gazit, D. (2010) The effect of simulated microgravity on human mesenchymal stem cells cultured in an osteogenic differentiation system: a bioinformatics study. *Tissue Engineering: Part A.* Vol.16, No.11, pp. 3403-3412

88. Shi, L. (2006) Executive Summary: the MicroArray Quality Control (MAQC) Project, *Food and Drug Administration,* accessed July 8, 2011, available from: <http://www.fda.gov/downloads/ScienceResearch/BioinformaticsTools/MicroarrayQualityControlProject/UCM132150.pdf >

89. Smith, S.M., Wastney, M.E., O'Brien, K.O., Morukov, B.V., Larina, I.M., Davis-Street, J.E., Oganov, V. & Shackelford, L.C. (2005) *Journal of Bone and Mineral Research.* Vol.20, No.2, pp. 208-218

90. Sonnenfeld, G. (1999). Space flight, microgravity,stress, and immune responses. *Advances in Space Research.* Vol.23, pp. 1945-1953

91. Sonnenfeld, G. (2005). The immune system in space, including Earth-based benefits of space- based research. *Current Pharmaceutical Biotechnology.* Vol.6, pp. 343-349

92. Sonnenfeld, G., Gould, C.L., Williams, J. & Mandel, A.D. (1988). Inhibited interferon production after space flight. *Acta Microbiologica Hungarica.* Vol.35, No.4, pp. 411-416.

93. Sonnenfeld, G. & Shearer, W.T. (2002). Immune function during space flight. *Nutrition.* Vol.18, pp. 899-903.

94. Sonnenfeld, G. 2005. Use of animal models for space flight physiology studies, with special focus on the immune system. *Gravitational Space Biology Bulletin.* Vol.18, pp. 31-35

95. Stein, T.P. & Schluter, M.D. (1994) Excretion of IL-6 by astronauts during spaceflight. *American Journal of Physiology.* Vol.266, No.3, pp. E448-E452

96. Stein, T.P., Schluter, M.D., Galante, A.T., Soteropoulus, P., Tolias, P.P., Grindeland, R.E., Moran, M.M., Wang, T.J., Polansky, M. & Wade, C.E. (2002). Energy metabolism pathways in rat muscle under conditions of simulated microgravity. *Journal of Nutritional Biochemistry.* Vol.13, pp. 471-478

97. Stekel, D. (2003) *Microarray Bioinformatics.* Cambridge University Press, ISBN 0-521-52587-X, Cambridge, UK

98. Tan, P.K., Downey, T.J., Spitznagel, E.L., Xu, P., Fu, D., Dimitrov, D.S., Lempicki, R.A., Raaka, B.M. & Cam, M.C. (2003). Evaluation of gene expression measurements from commercial microarray platforms. *Nucleic Acids Research.* Vol.31, No.19, pp. 5676-5684

99. Taylor, W.E., Bhasin, S., Lalani, R., Datta, A. & Gonzalez-Cadavid, N.F. (2002). Alteration of gene expression profiles in skeletal muscle of rats exposed to microgravity during spaceflight. *Journal of Gravitational Physiology.* Vol.9, No.2, pp. 61-70

100. Tsao, Y.D., Goodwin, T.J., Wolf, D.A. & Spaulding, G.F. (1992). Responses of gravity level variations on the NASA/JSC bioreactor system. *Physiologist.* Vol.35, No.1 Suppl., pp. S49-S50.

101. Thomason, D.B. & Booth, F.W. (1990). Atrophy of the soleus muscle by hindlimb unweighting. *Journal of Applied Physiology.* Vol. 68, No.1, pp. 1-12

102. Trappe, S., Costill, D., Gallagher, P. Creer, A., Peters, J.R., Evans, H., Riley, D.A. & Fitts, R.H. (2009). Exercise In Space: Human Skeletal Muscle After 6 Months Aboard The International Space Station. *Journal of Applied Physiology.* Vol39, No.4, pp. 463-471

103. Torday, J.S. (2003). Parathyroid hormone related protein is a gravisensor in lung and bone cell biology. *Advances in Space Research.* Vol32, pp. 1569-1576

104. Ullrich, O., Huber, K. & Lang, K. (2008). Signal transduction in cells of the immune system in microgravity. *Cell Communication and Signaling.* 6:9

105. Valles, J.M. & Guevorkain, K. (2002). Low gravity on earth by magnetic levitation of biological material. *Journal of Gravitational Physiology.* Vol.9, No.1, pp. p11-p14

106. van Loon, J.J.W.A. 2007. Some history and use of the random positioning machine, RPM, in gravity related research. *Advances in Space Research.* Vol.39, pp. 1161-1165

107. Vernikos, J. & Schneider, V.S. (2010). Space, gravity, and the physiology of aging: parallel or convergent disciplines? A mini-review. *Gerontology.* Vol.56, No.2, pp. 157-166.

108. Vose, G.P. (1974). Review of roentgenographic bone demineralization studies of the Gemini space flights. *American Journal of Roentgenology, Radium Therapy and Nuclear Medicine.* Vol121, No.1 , pp. 1-4

109. Walther, I., Pippia, P., Meloni, M.A., Turrini, F., Mannu, F. & Cogoli, A. (1998). Simulated microgravity inhibits the genetic expression of interleukin-2 and its receptor in mitogen-activated T lymphocytes. *FEBS Letters.* Vol.436, No.1, pp. 115-118.

110. Ward, N.E., Pellis, N.R., Risin, S.A. & Risin, D. (2006). Gene Expression Alterations in Activated Human T-Cells Induced by Modeled Microgravity. *Journal of Cellular Biochemistry.* Vol.99, pp. 1187-1202.

111. Wilson, J.W., Ott, C.M., Ramamurthy, R., Porwollik, S., McClelland, M., Pierson, D.L. & Nickerson, C.A. (2002a). Low-shear modeled microgravity alters the Salmonella enterica serovar typhimurium stress response in a RpoS-independent manner. *Applied Environmental Microbiology.* Vol.68, No.11, pp. 5408-16

112. Wilson, J.W., Ramamurthy, R., Porwollik, S., McCelland, M., Hammond, T., Allen, P., Ott, M.C., Pierson, D.L. & Nickerson, C.A. (2002b). Microarray analysis identifies Salmonella genes belonging to the low-shear modeled microgravity regulon. *Proceedings of the National Academy of Science USA.* Vol.99, No.21, pp. 13807-13812

113. Wittwer, P., Fluck, M., Hoppeler, H., Muller, S., Desplanches, D. & Billeter, R. (2002). Prolonged unloading of rat soleus muscle causes distinct adaptations of the gene profile. *FASEB Journal.* Vol.16, pp. 884-886

114. Wolber, P.K., Collins, P.J., Lucas, A.B., De Witte, A. & Shannon, K.W. (2006). The Agilent *in situ*-synthesized microarray platform. *Methods in Enzymology.* Vol.410, pp. 28-57

115. Yamada, S., Ganno, T., Ohara, N. & Hayashi, Y. (2007). Chitosan monomer accelerates alkaline phosphatase activity on human osteoblastic cells under hypofunctional conditions. *Journal of Biomedical Materials Research Part A.* Vol.83, pp. 290-295

116. Yu, Z.B., Zhang, L.F. & Jin, J.P. (2001). A proteolytic NH2-terminal truncation of cardiac troponin I that is up-regulated in simulated microgravity. *Journal of Biological Chemistry.* Vol.276, No.19, pp. 15753-15760

117. Zhang, P, Hamamura, K. & Yokota, H. (2008). A brief review of bone adaptation to unloading. *Genomics, Proteomics, and Bioinformatics.* Vol6, No.1, pp. 4-7

118. Zayzafoon, M., Gathings, W.E. & McDonald, J.M. (2004). Modeled microgravity inhibits osteogenic differentiation of human mesenchymal stem cells and increases adipogenesis. *Endocrinology.* Vol.145, No.5, pp. 2421-2432

CHAPTER-15

SECURING HUMAN HEALTH THROUGH BIOTECHNOLOGY PATENTS

INTRODUCTION

Biotechnology patents serve as a critical safeguard for human health by incentivizing innovation in the field of biotechnology. These patents encourage researchers and companies to invest in the development of novel medical treatments, diagnostics, and therapies. By protecting their intellectual property, inventors can secure the resources necessary to bring groundbreaking discoveries to the market. As a result, the availability of new biotechnological advancements can lead to improved healthcare, the development of more effective pharmaceuticals, and the early diagnosis of diseases. This not only enhances our overall well-being but also helps address some of the most pressing global health challenges. The safeguarding of human health through biotechnology patents underscores the importance of fostering innovation and ensuring access to cutting-edge medical solutions for people around the world.

Health related problems affect every human being in an interconnected way, between generations and between societies, through spatial and temporal transitions. This timeless and all-pervasive aspect of health makes biotechnology unique among all technologies. While the science that drives biotechnology has far to go before it reaches a comparable level of maturity of eighteenth century physics, nevertheless biology is now deeply rooted in science; it has taken huge strides from its humble beginnings as a classification science to cell biology, to molecular biology, and now modestly to quantum biology. Erwin Schrödinger, a pioneer of quantum mechanics, was among the first scientists to suggest, in his book *What is Life*? (Schrödinger, 1944), that quantum mechanics can provide deep insights into life's mechanisms. However, our current understanding of how such quantum phenomena as superposition, entanglement, collapse of the wavefunction, etc. affect the chemistry of life is nascent (Ball, 2011).

The links in the supply chain that support biotechnology products and services include knowledge creation in R&D laboratories, product creation in biotechnology and pharmaceutical industries, and the ultimate receiver of therapeutic remedies—a human patient, *inter alia*, communally bound by morality and ethics. The fact that this supply chain, in principle, must cater to every human being on our planet, demands that it be protected with utmost care. Perhaps the most vulnerable link in this chain is the patentable knowledge created through privately funded R&D, which, unless diligently protected, is easy prey to infringement and theft. The maintenance of this chain is astronomically expensive and complex as it must balance some extreme needs: huge funding for exploration-intensive, curiosity-driven, 'blue-sky' R&D; highly risky upfront investments by industry before going to market; enormous funds to protect its intellectual property, if necessary, through litigation; and the need to provide safe and affordable remedies to very large populations of indigent people in the world to keep them healthy.

This chapter is written for biotechnologists who wish to get an understanding of the role patents could play in protecting and advancing their research output in the service of mankind when commercial applications of that research is the optimal means of doing so. Here we discuss general principles of biotechnology patents and related issues, rather than country specific ones.

THE PROMISE OF BIOTECHNOLOGY

Biotechnology took roots more than 10,000 years ago. The 'old or traditional' methods of biotechnology were mainly fermentation (through the unwitting use of microbes) to produce such products as beer, wine, and cheese, and cross-breeding (through the unwitting use of genetic material) to modify plants and animals through progressive selection for desired traits. These methods were developed empirically, patiently, and over countless years (Darwin, 1872; Smith 2009). It was only during 1857 and 1876 that the fermentative ability of microorganisms was demonstrated by Louis Pasteur (Smith, 2009). The discovery by Alexander Fleming of the antibiotic penicillin in 1929 and its large-scale production in the 1940s created major advances in fermentation technology. Since then the technology has advanced rapidly not just in the production of antibiotics but in many other biochemical products including organic acids, polysaccharides, enzymes, vaccines, and hormones. Modern breeding methods now selectively move genes within the same species or between species.

The 'new or modern' biotechnology that emerged in the 1970s is (Lilly, 1997) "the application of scientific and engineering principles to the processing of materials by biological agents to provide goods and services." Its methods use microbial, animal or plant cells or enzymes for the purposes of breaking down, synthesizing, or transforming materials. The scientific foundation of biotechnology was laid in a remarkable paper by James D. Watson and Francis H. C. Crick in *Nature* (Watson & Crick, 1953), which elucidated the double-helix structure of cellular DNA[1] (deoxyribonucleic acid). It gave birth to molecular biology and paved the way for developing recombinant DNA and cell fusion techniques along with scientific versions of 'old' biotechnological processes of modern biotechnology. Advances such as the transformation of *Escherichia coli*[2]; cutting and joining DNA strands (recombinant DNA technology) (Cohen, et al, 1973)[3]; the rapid cloning of DNA strands (PCR technique) (Mullis, et al, 1986); the ability to make monoclonal antibodies (hybridoma technique) (Köhler & Milstein, 1975)[4], etc. have made possible the creation of genetically engineered life forms capable of manufacturing new and improved drugs, such as, human insulin, interferons, vaccines, and treatments for a host of human afflictions such as septic shock, anemia, diabetes, AIDS, cancer, hepatitis, and heart attack. Genetically engineered transgenic animals such as the Harvard mouse[5] and the SCID mouse[6] play an immensely important role in cancer and immunology research, respectively. Many other transgenic life forms such as bacteria, cows, pigs, goats, etc. play a crucial role inthe development of therapies and the manufacture of pharmaceuticals. Thus one of the main objectives of biotechnology is to find means of scaling-up biological processes.

The mapping of the human genome independently by the Human Genome Project and by Celera Genomics in 2001 was a remarkable breakthrough in data collection to aid studies of the human body.[7] A crucial step in providing personalized medical care was thus taken. The breakthrough creation of a bacterial cell controlled by a chemically synthesized genome was reported by Craig Ventor's team in May 2010 in *Science* (Gibson, et al, 2010). The team reported synthesizing the genome of the bacterium *Mycoplasma mycoides*, comprising some 1.1 million base pairs as a proof of principle that cells can be produced based upon genome sequences designed in the computer. There is, of course, much to be learnt before one can construct and transplant whole computer-designed genomes of higher life forms. Important limiting factors are insufficient scientific knowledge of gene structure and function, and of microRNAs.

The potential curative abilities of stem cells come from their remarkable ability to renew themselves through cell division, sometimes even after long periods of inactivity, and to develop into many different cell types in the body. When a stem cell divides, each new cell, under appropriate circumstances, may either remain a stem cell or become a specialized cell such as a muscle cell, a red blood cell, or a brain cell. This unique regenerative ability of stem cells offers new opportunities for treating diseases such as diabetes and heart disease. However, much remains to be understood about them before reliable cell-based therapies can be designed to treat diseases.

The strength of modern biotechnology comes from its interdisciplinary nature and the interactions it orchestrates between various parts of biology and engineering. It draws insights and knowledge from a wide range of fields: biochemistry, microbiology, molecular biology, cell biology, immunology, protein engineering, enzymology, breeding techniques, chemical engineering, mechanical engineering, computational methods, mathematical simulation, bioinformatics, etc. The products and processes it spawns are the results of intense R&D, astronomical funding, and the unique entrepreneurial spirit of the biotechnology community in converting R&D results into therapies and cures, diagnostic tools and tests for disease detection, etc.

At present, biotechnology produces a range of embryonic enabling technologies for which some applications are known and many more expected. It is sustained by an enormous faith that suites of these enabling technologies, when further refined and augmented, will eventually find vast new applications of tremendous value to society. So the main benefits essentially lie in the future. The expected benefits include novel personalized pharmaceutical drugs and therapies for many diseases based on individual genomic information, genetically engineered healthier food with longer shelf-life, and new energy- efficient techniques for protecting the environment. The biotechnology industry'sdependence on multi-nation patent protection for survival in the marketplace is therefore not at all surprising. In fact, it is imperative that every biotechnology researcher understands the circumstances when acquiring and protecting his research results by patents is crucial. Breakthrough R&D results by themselves are not enough; to serve society they must lead to commercially viable products and processes or find philanthropic hosts or find federal support. Of particular research interest are genes and their corresponding proteins as they are believed to represent the future of diagnostic and therapeutic medicine.

FUNDAMENTAL PRINCIPLES OF PATENT LAW

A patent is a limited period monopoly intellectual property right granted to an inventor of any country by a Government of any other or same country for an invention that fulfills prescribed statutory requirements of the granting country. Patent monopoly differs from market monopoly; a patent is a right to exclude, a right to prevent trespassing. In this sense it is no different from, say, the right to keep our house or car or any other personal possession free from trespassers. A patent grants inventors the right to exclude others from making, using, selling or offering to sell, and importing the claimed invention in the country of grant; it does not confer any right to practice the invention. This is because in practicing the invention, one may well need complementary patents held by others unwilling to cooperate or there may be other laws, rules or regulations that prevent its practice.

Patents are issued only to the first inventor (or group of joint inventors) of an invention who files a legally valid patent application; all others are barred, even if they independently created the invention. Consequently, those other inventors must get a license from the first inventor if they wish to practice the invention. Patents granted by a country, like its laws, have no extraterritorial effect; hence patents are unenforceable, if infringed, in another country where the invention in question is not patented. If multi-country patent protection is required, the invention must be patented in each desired country. There is no such thing as a "world-wide patent".

The modern concept of patents dates back to the year 1421, when the Italian city-state Florence granted the first recorded patent to Fillippo Brunelleschi, for the design and use of a ship, the Badalone (seagoing monster; it was used to carry marble along the Arno river), for three years.[8] The Venetian Senate passed the first patent law on March 14, 1474, granting limited duration monopoly for original devices. That same Venice in 1594 granted Galileo[9] a "privilege" (what we know as a patent) on a machine which he had invented[10] "for raising water and irrigating land with small expense and great convenience," on the condition that it had never before been thought of or made by others. In his petition for the privilege he said, "it not being fit that this invention, which is my own, discovered by me with great labor and expense, be made the common property of everyone," adding further, that if he were granted the privilege, "I shall the more attentively apply myself to new inventions for universal benefit." Clearly, even the great scientist Galileo was not willing to divulge his invention for free exploitation by others without just compensation for his efforts. The Venetian Council granted Galileo a "privilege" for 21 years.

An invention is the creation of a new technical idea *and* of the physical means to accomplish or embody it. An idea *per se* is not an invention; a useful and successful implementation of an idea is. Four types of inventions are eligible for patents: process, machine, manufacture, or composition of matter, collectively known as statutory subject matter. They are subject to certain limitations that vary from country to country. However, there is universal agreement among nations that abstract ideas (e.g., mathematical formulas), laws of nature, natural phenomena, and products of nature are ineligible, but their application to a known structure or process may be eligible. What to exclude from patent monopoly is a national prerogative, largely derived from government policy decisions that accrue from the prevailing socio- economic conditions it must manage, and international treaty obligations.

Inventions that qualify as statutory subject matter must then face additional stringent statutory tests of substantial and credible utility (industrial application) in the eyes of an expert (such as a patent examiner in a patent office) in the field of the invention, novelty with respect to prior art (state-of-the-art) as it exists on the date of filing the patent application (in the United States there is some relaxation available), and nonobviousness, i.e., the invention has an inventive step that is unlikely to have been made by a person having ordinary skill in the art (PHOSITA), if required, assuming he would make the effort to study relevant prior art.

There is a *quid pro quo* attached to the grant of patents. To get a patent the inventor must put the invention in the public's possession. He must therefore fully describe his invention (written description requirement) in the patent application *before* the invention is formally examined by a patent examiner (examination typically takes two or more years). This description must be so clear and detailed as to enable a person skilled in the technologies related to the invention in question (an expert) to independently reproduce the invention (enablement requirement) without undue extra-solution activity, such as further research, data gathering, etc. on his part. In fact, this description should leave no doubt that the patent applicant was in possession of the claimed invention at the time of filing his application. Patents may be granted on improvements over existing inventions.

Patent laws of a country do not over-ride its other laws that might regulate the invention's use. For example, a new pharmaceutical cannot be marketed without the approval of appropriate authorities. Patent laws of a country may take into account moral, cultural, ethical, social, environmental, or scientific concerns of society.

In most countries with a patent regime, a pending patent application is placed in the public's possession 18 months after the first "priority" filing date[11] of the application or the date of patent grant, whichever is earlier, by means of publication in print and world-wide- web enabled media. This gives an opportunity to others to improve upon the invention and possibly patent improvements (or focus on something else), without unduly stifling innovation. Limited period monopoly (usually 20 years from the priority filing date) is meant to prevent undue concentration of economic power, yet allow inventors (a rare breed) an opportunity to recover costs and earn profit from their long and expensive effort, not otherwise possible if others could reverse engineer the invention (often a far less expensive process) and duplicate it without penalty. The goal has been to get as many useful inventions into the public domain and in free use as soon as possible and thus enrich society as a whole without being unfair to the inventor. Patent protection is therefore a bargain struck by society on the premise that, in its absence there would be insufficient invention and innovation. Patent and other laws do not forbid an altruistic inventor (unless bound by a legal contract, say, to his employer) from freely placing his patentable inventions in the public domain without patent protection.

Patents are granted to inventors. The rights attached to a patent may be exercised by the patentee, his or her heirs or assigns during the term of the patent. A patent may be assigned (e.g., to one's employer) or licensed, with or without conditions attached, to one or more legal entities. A patent license to a licensee is an agreement that the patent owner will not enforce certain or all rights of exclusion against the licensee. Anyone else infringing the patent can be sued in a court of law by the patent's owner.

Limited period patent monopoly may provide an enormous first mover advantage to an entrepreneur, especially if it involves new technology. Alexander Graham Bell's two telephone patents—"Improvement in Telegraphy" (U.S. 174,465), granted on March 7, 1876, provided a monopoly on the basic principle of telephony, and "Improvement in Electric Telegraphy" (U.S. 186,787) granted on January 30, 1877, provided a monopoly on the telephone hardware—are outstanding examples. By the time the patents expired, American Bell (later to become AT&T) had acquired a "natural monopoly" in the telephone business.

An alternative to patent protection is to keep the invention a trade secret, which lasts as long as the secret is kept. This works if the invention's independent discovery is so unlikely that it can be monopolized indefinitely. Otherwise, independent discoverers of the secret can practice their invention with impunity, and worse, one of them may patent it and deny all others the use of the invention if not licensed from him. If one is keen to commercialize the invention, patent protection is much safer than trade secret, especially during negotiations with investors when detailed exposure of the invention may be necessary. At times, keeping marginal improvements of a patented invention as trade secrets may be preferred, especially if constrained by patenting costs.

Submitting and Advancing a Patent Application

To get a patent one must file a patent application at an appropriate office designated for the purpose, usually the patent office of the selected country. Each country has its own rules and regulations for filing and these must be strictly followed. To claim priority over others for an invention, it is necessary to be the "first to file" the application in accordance with the country's statutory requirements, which may include statutory grace periods. To claimpriority, one may file a provisional application which, at the minimum, fulfills the written description and enablement requirements, but it must be followed, within a year, by a proper application for the same invention as described in the provisional application, else the priority date is lost. Patent offices act only on proper applications. Provisional applications remain dormant during their life.

Prosecution is the process by which a (proper) patent application is defended before a patent office. The process often lasts several years. The application includes a complete description of the invention, a list of claims on statutory subject matter sought to be protected, and the requisite filled-in patent office forms, along with a filing fee. As prosecution proceeds, there may be other fees to be paid at various stages. For filing and prosecution details, visit the website of the desired patent office.

Patent claims

The legal core of a patent application is the list of claims. Each claim in this list covers and secures a process, a machine, a manufacture, a composition of matter, or a design, but never the function or result of either, nor the scientific explanation of their operation. The claims define the scope of a patent grant and function to forbid not only exact copies of an invention but also products that go to the heart of the invention but avoid the literal language of the claim by making a non-critical change. (See Section 4 below.) Whether a claim is allowed by a patent office is judged on the basis of novelty, nonobviousness, and utility (industrial applicability) of the invention being considered. Of course, claims are interpreted in light of the description of the invention provided in the patent application and information elicited during prosecution from the inventor, prior art and other sources. Almost all litigation related to patent infringement centers on the validity and scope of the claims.

In biotechnology, claims may be product claims, process claims, or product-by-process claims. Product claims may include such things as novel protein products, known but purified protein products, DNA sequence of a gene that encodes a particular protein, etc. Process claims may include preparation or use of recombinant DNA, the use of bacteria or cultured cells transformed with vectors containing DNA encoding a desired protein product, methods of use for proteins, methods for production or use of monoclonal antibodies, etc. Product-by-process claims deal with products that are too complex to be described conventionally (e.g., with reference to its composition, structure or some other testable parameter) and hence are described by the process with which it is made. By such claims it is not possible to use a new process to claim an old product. The focus here is the patentability of the product itself, not on the process used to describe it since the reference to a process serves only the purpose of defining the product.

There have been attempts by biotechnology inventors to get "reach-through claims" granted. Such claims seek to protect things which may not have been identified by the applicant in his patent application but which *may be* identified in the future by others by carrying out the invented process. This is different from the product-by-process claims as the products claimed in reach-through claims are speculative and hence do not fulfill the statutory requirements of disclosure and enablement. The purported justification for such broad extra-legal claims is that a pioneering invention paves the way for subsequent inventions and hence its inventor is "entitled" to capture some of the follow-on value based on the relative contribution of his pioneering invention. (Christie & Lim, 2005; IPO, 2009)

INFRINGEMENT

Protecting an active patent when infringed can be a nightmare. It is time consuming, and hugely expensive (usually measured in millions of U.S. dollars) if it involves litigation. Alleged infringers, if challenged, are quite likely to counter-challenge by questioning the validity of the disputed patent. It is therefore imperative, especially in biotechnology where patents underpin business, that patent applications are prepared by experienced patent attorneys and that inventors work closely with them to minimize litigation possibilities. Considerations that go into the preparation of a fortified patent application include the doctrine of equivalent, prosecution history estoppel, reverse doctrine of equivalents, prior art or state-of-the-art, and the anticipated profile of the imaginary PHOSITA.

Doctrine of equivalents

Literal infringement of a valid active patent where the alleged infringer exactly or nearly exactly copies an invention without a licence from the patent owner is understandably rare. Generally, one tries to work around a patented invention by introducing differences and variations to avoid infringement. The question then is whether the modified product or process is remote enough from the patent that it will not infringe. Inadvertent infringement may arise if a product or process is invented in ignorance of an active patent whose existence is discovered only later, say, after a business commitment has been made to produce the product or use the process.

Such situations are partially dealt with by the judicially created doctrine of equivalents. This is a rule of claim interpretation under which a product or process, although not a literal infringement, is an infringement if it performs substantially the same function in substantially the same way to obtain the same result as a patented product or process. This doctrine, which has universal appeal, expands patent protection beyond the literal language of the claim. To determine what counts as an equivalent one must find a balance between two opposing public policies:

(1) the importance of providing public notice as to what infringes by requiring clear and distinct claims, and (2) the need to prevent an infringer from avoiding liability by merely playing semantic games or by making only minor changes in the accused product or process to avoid the literal language of the claims (Belvis, 2003). In litigation, courts may seek expert opinion as to scientific or engineering facts and the decision may well depend on the most believable expert. Note that things that are equivalent for one purpose may not be so for other purposes. The Supreme Court of the United States sums it succinctly in *Graver Tank*[12]:

What constitutes equivalency must be determined against the context of the patent, the prior art, and the particular circumstances of the case. Equivalence, in the patent law, is not the prisoner of a formula and is not an absolute to be considered in a vacuum. It does not require complete identity for every purpose and in every respect. In determining equivalents, things equal to the same thing may not be equal to each other and, by the same token, things for most purposes different may sometimes be equivalents. Consideration must be given to the purpose for which an ingredient is used in a patent, the qualities it has when combined with the other ingredients, and thefunction which it is intended to perform. An important factor is whether persons reasonably skilled in the art would have known of the interchangeability of an ingredient not contained in the patent with one that was.

In *Warner-Jenkinson*[13] the same Court then clarified and restricted the application of the doctrine of equivalents, holding that: Each element contained in a patent claim is deemed material to defining the scope of the patented invention, and *thus the doctrine of equivalents must be applied to individual elements of the claim, not to the invention as a whole.* It is important to ensure that the application of the doctrine, even as to an individual element, is not allowed such broad play as to effectively eliminate that element in its entirety. [Emphasis added]

This restriction on the doctrine of equivalents serves to eliminate one of the great mischiefs that could be played in patent law. Absent this rule, one could attempt to use the doctrine of equivalents to subvert patent claims. Rather than focusing on specific claim language and elements of the claim, the case could be tried based on how the accused device was equivalent to that claim as a whole. The Court further held that the equivalence determination was to be made at the time of the alleged infringement and not at the time the patent issued. It is likely that the less certain and more complex the courts perceive a scientific field underlying a technology to be (as is the case with biotechnology), the less scope will be given to patents under the doctrine of equivalents. If the patent is a pioneer in a whole new field, it will generally receive a broader range of equivalents than one for a narrow improvement to existing technology (Blenko, 1990). There are a few other restrictions that circumscribe the doctrine of equivalents: prosecution history estoppel, the reverse doctrine of equivalents, and prior art.

Prosecution history estoppel

During prosecution, quite likely, one or more claims will be rejected or require amendment to become narrower and detailed, in view of prior art. If a claim is allowed after being narrowed to avoid prior art, the patentee is barred from asserting the narrowed claim in its earlier broader sense under the doctrine of equivalents. This means that broad claims that have to be amended during prosecution can be difficult to enforce, if infringed. In short, rejected or narrowed claims cannot be expanded to their earlier scope under the doctrine of equivalents. In fact, such claims practically forego any benefit that could have accrued under the doctrine of equivalents in infringement cases.

Reverse doctrine of equivalents

A further restriction on the doctrine of equivalents is the *reverse doctrine of equivalents*. As noted by the Supreme Court of the United States in the *Graver Tank* case:

> The wholesome realism of this doctrine [of equivalence] is not always applied in favor of a patentee but is sometimes used against him. Thus, where a device is so far changed in principle from a patented article that it performs the same or similar function in a substantially different way, but nevertheless falls within the literal words of the claim, the doctrine of equivalents may be used to restrict the claim and defeat the patentee's action for infringement. [Citations omitted]

Thus, where an invention relies on the fundamental concept embodied in a patent but is more sophisticated than the patented device due to "a significant advance," the accused device does not infringe by virtue of the reverse doctrine of equivalents. Once a patentee establishes literal infringement, the burden is on the alleged infringer to establish non- infringement under the reverse doctrine of equivalents. This is an untested area of patent law but may become important in biotechnology with respect to certain pioneering technologies, such as, synthetic cell technology.

Prior art

Prior art or state-of-the-art is all information, available in any form, in the public domain. It does not include secret information, such as trade secrets. The existing reservoir of ideas and their expression form the foundation on which new intellectual property is built. Normally, prior art does not include unpublished work or mere conversations (although in the European Patent Convention, oral disclosures do form prior art[14]). There is a continuing effort by various countries to document their respective traditional knowledge, such as medicinal properties of plants, and make that knowledge available as searchable prior art. The doctrine of equivalents excludes whatever is already prior art.

The PHOSITA in biotechnology

In examining a patent application, the patent examiner faces an immediate problem. How to define the relevant PHOSITA? In patent law the PHOSITA is a legal fictional character or a team of characters analogous to the "reasonable person" in the common law of torts. The PHOSITA is a statistical concept in the sense that there is a very high probability that no one from the community of ordinarily skilled persons in the relevant technical field(s) will be able to come up with the invention in question or its close equivalent or a superior one if the community was required to do so. So the PHOSITA, by definition, is neither a genius nor a layperson, but one possessing normal skills and knowledge in the required technical field. In this sense he serves as a litmus test for deciding if an invention is nonobvious or involves an inventive step. If the PHOSITA is deemed capable of coming up with the invention by applying his mind, knowledge, skill, and common sense, that particular invention is deemed unpatentable. In short, a "person of ordinary skill is also a person of ordinary creativity, not an automaton."[15]

Unfortunately, "ordinary skill" must be determined on a case-by-case basis, depending on the sophistication and technological features of the invention. Clearly, the ordinary skills ofa nuclear physicist are different from those of a chef or a cobbler or a molecular biologist. Factors used in determining ordinary skill include the time frame of the invention; education level of the inventor, education level of active workers in the field of the invention, and the type of problems generally encountered in the field; prior art solutions relevant to the invention; rapidity with which innovations are made in the field; sophistication of the technology; etc. Further, with time, the profile of a PHOSITA, in advancing technologies, will only improve due to the infusion of new knowledge and skills. Therefore, in rapidly advancing fields, such as biotechnology, determining the profile of a PHOSITA requires great skill and frequent revision.

The PHOSITA's role is crucial in several places—in the enablement requirement, the nonobviousness requirement, the utility requirement (the invention must operate as described if he is to be enabled), and the written description requirement, as compliance with these requirements is measured from his perspective. Therefore, claims must be written so that a PHOSITA would understand the bounds of the patent, including the territory covered by the doctrine of equivalents. A fundamental test for the doctrine of equivalents is whether a PHOSITA would reasonably interchange the elements in a claim at issue in an infringement case. What is not clear is that as higher education spreads and the PHOSITAs learn to solve problems at conceptual levels, how that will affect the doctrine of equivalents.

The enablement and non-obviousness questions arise before the issuance of a patent while the question of interchangeability arises at the time of infringement.

Note that while biotechnology patent examiners are experts in biotechnology, infringement and validity cases are decided by judges who are not. So there is often a misalignment of the PHOSITA's profile separately conjured by the examiner and the judge in any given biotechnology patent case. In fact, it is rather difficult for courts to insert, in their decisions, the role of "common sense" a PHOSITA might routinely employ in his day-to-day work.

In infringement cases, the cut-off date chosen to ascertain prior art and the PHOSITA's profile can become a critical factor even when the dates differ by only a few weeks. Scientific breakthroughs and pioneer inventions suddenly appearing on the scene around the cut-off date can complicate matters tremendously. In the fast changing world of biotechnology, what is nonobvious today may well be obvious next year or next week!

REQUIREMENTS FOR PATENT ELIGIBILITY IN THE BIOTECHNOLOGY FIELD

Large scale patenting of living matter is recent. Indeed, prior to 1980, few patents had been granted on non-living biological matter and biologically pure cultures of micro-organisms as they did not exist in nature in their pure form; they could only be produced in carefully controlled laboratory environments. Patent laws around the world till then had assumed that higher life forms were not patentable as they were deemed products of nature. An abrupt change in legal thinking occurred when the Supreme Court of the United States in its June 16, 1980 decision in *Diamond* v. *Chakrabarty* held that "a live, human-made micro-organism is patentable subject matter" under the U.S. Patent Act of 1952. Recall that recombinant DNA technology was already well known in 1980. The Court reasoned that Chakrabarty's microorganism was a "nonnaturally occurring manufacture or composition of matter—a product of human ingenuity" worthy of liberal encouragement under thepatent system. It declared that "the relevant distinction was not between living and inanimate things, but between products of nature, whether living or not, and human-made inventions." The floodgates for biotechnology patents were thus opened in the United States,[16] and eventually, using similar reasoning, patents on living matter were allowed in other countries. In 1988, the United States Patent and Trademark Office (USPTO) issued the first transgenic animal patent on the now famous Harvard mouse[17], a mouse genetically engineered to be particularly susceptible to tumor growth. Patents have since been issued on many other genetically engineered plants and animals.

The nature of biotechnology and its close working association with bioinformatics and molecular biology has added a new and complex dimension in patenting. The DNA is both a material molecule as well as a literal embodiment of coded information (the book of life). The courts are still trying to understand this deep two-facedness of DNA. For example, can artificially created DNA or gene sequences be copyrighted?

Finally, when filing a biotechnology patent application, particular attention should be paid to meeting the legal requirements of (1) statutory subject matter, (2) utility, (3) novelty, (4) nonobviousness, and (5) specification (description, enablement, and claims). The vast majority of litigation cases revolve around these statutory requirements.

Statutory subject matter

Biotechnology deals with bio-matter itself (including products of biotechnology living or non-living) and processes of making bio-matter. Examples of non-living bio-matter are amino acids, peptides, proteins, fats and fatty acids, and nucleic acids. They are all chemical compounds and are usually better known in the form of antibodies, hormones, enzymes, antibiotics, steroids, cholesterol, DNA molecules, etc.

The primary entity in living bio-matter is the cell, the smallest reproducible unit of life. A wide range of biotechnology product inventions, e.g., proteins, antibodies, intracellular components of plant and animal cells (DNA fragments, DNA constructs, DNA promoters, plasmids, vectors, RNAs, ribosomes, chloroplasts, mitochondria, Golgi bodies, etc.) and living matter *per se*, such as cell lines, fused cells, plant seeds, tissue cultures, microorganisms, plants and nonhuman animals, are patentable subject matter.

Biotechnology process inventions include processes for sequencing DNA, RNA or proteins; processes for genetically manipulating cells, plants or animals; processes for recovering proteins produced by cell lines or animals; processes for detecting and characterizing mutagenic agents; processes for culturing tissue or cells; processes for diagnosing or detecting biological states; fermentation, chemical and diagnostics methods; methods of treating human or animal bodies; methods of controlling pests; etc. Biotechnology processes also provide the potential for creating genetically altered bio-matter itself. In the early to mid-1980s researchers were already creating genetically altered transgenic mice, hamsters, rats, hogs, poultry, cattle, sheep, and fish. In 2010, the first synthetic cell capable ofcontinuous self-replication was created, a cell that was completely controlled by a computer designed synthetic chromosome.

The utility requirement (genetic materials)

A patent examiner will accept a utility asserted by an applicant unless there is evidence or sound scientific reasoning against it. Clinical trials of a new pharmaceutical are not required to establish its utility[18]. Transgenic animals are generally created with a specific use in mind, so their utility is usually obvious. For gene sequences a nontrivial utility of the protein it will produce must be shown. Citing generic useful functions such as a marker, probe, or primer for various genetic researchers may likely be considered trivial given the present state-of-the-art in gene research.

New processes related to genetic materials must show utility for the product of a process as well as the process itself, otherwise one could end in patenting a process which yielded an unpatentable product. "Until the process claim has been reduced to production of a product shown to be useful, the metes and bounds of that monopoly are not capable of precise delineation. It may engross a vast, unknown, and perhaps unknowable area."[19] Less this constraint, the patentable field would be too broad. Therefore, to assert utility, say, for a process for making a protein, one must establish that the protein itself has substantial and specific utility in currently available form.

General utility is disallowed because it would embrace a broad class of an invention. For example, regarding ESTs (expressed sequence tags), "a claim to a polynucleotide whose use is disclosed simply as a 'gene probe' or 'chromosome marker' would not be considered to be *specific* in the absence of a disclosure of a specific DNA target."[20] An EST does not explain the purpose and use of the gene. Therefore an EST patent "would amount to a hunting license"[21] for performing research that may lead nowhere. Likewise, cDNA fragments used as probes for finding full-length genes lacks specific utility because, "[a]ny partial nucleic acid prepared from any cDNA may be used as a probe in the preparation and or identification of a full-length cDNA." [22] Biotechnology patents must present a higher degree of utility than for most other types of patents, say, in mechanical or electrical engineering.

The novelty requirement

Non-naturally occurring life forms such as transgenic animals that are "man-made" or "man-altered" for the first time satisfy the novelty requirement. Gene sequences, either artificially created or purified and altered from their natural state, say, by deleting the introns and retaining the protein coding exons, may fulfill the novelty requirement.

However, the current state-of-the-art in human gene sequencing and its rate of advance are such that in future litigation courts may well set a more stringent criterion for novelty. This is perhaps inevitable given that many of the products of interest to the biotechnology industry are synthetic versions of substances that already exist in nature, and creating those synthetic versions is within the capabilities of a PHOSITA. Under these circumstances, can a synthetic version be called "new"? While methods of use of "new" sequences may be novel, claiming those sequences as new compositions may not be easily allowed.[23]

The nonobviousness requirement

The nonobviousness requirement is an important policy lever by which governments can efficiently control the transfer of intellectual wealth to promote industrial products and processes. This is particularly important when dealing with innovative medical products and diagnostic tools being produced and mass marketed by multinational pharmaceutical and biotechnology firms.

The core of a researcher's activity is hypotheses testing. This is what many biotechnology PHOSITA do routinely. Scientific inventions in biotechnology rarely come about *de novo*. Thus how much of the experimental research or testing conducted in the lead-up to an invention is attributable to a PHOSITA is central to the obviousness test. Routine, ordinary, logical, or workshop activity is not deserving of patent monopoly. "The results of ordinary innovations are not the subject of exclusive rights under patent law;" otherwise "patents might stifle rather than promote the progress of useful arts."[24] Thus routine testing in the lead-up period to invention in anticipation of reasonable expectation of success should be expected of an ordinarily creative PHOSITA.

Setting obviousness standards for gene patents is difficult because scientists use similar techniques to isolate different gene sequences, even though the gene may be new. A related question immediately arises. If homology-based utility satisfies the requirement of utility, would the invention be considered obvious? The USPTO's view[25], obviously in the context of U.S. patent law, is that nonobviousness and utility requirements are separate. This is because even though a claim to a nucleic acid is supported by a homology-based utility over a set of nucleic acids, that utility is not *prima facie* obvious. Homology-based deductions may provide a reason or motivation to make the claimed composition, but it would still be necessary to establish a fact-intensive comparison of the claim with the prior art rather than the mechanical application of one or another *per se* rule. In short, "obvious-to-try" and obviousness is not always the same thing (rules of thumb are not rules of law). The mere fact that something is "obvious to try" in view of known prior art does not automatically imply that the invention resulting therefrom is obvious. This is especially true where the prior art does not contain any suggestion or teaching that might suggest how the invention might be accomplished or any basis for reasonable expectation that beneficial results will accrue by proceeding along the lines taken by an inventor.

To assess nonobviousness requires profiling a PHOSITA, who in biotechnology usually holds a PhD and is therefore an expert in common perception. Moreover, this PHOSITA is more likely to be a team of experts rather than a "mythical individual". So how does one reasonably determine this mythical PHOSITA at a point in time in an area of technology which is advancing so rapidly that the profile would need to be updated, at times, on a weekly basis? How are questions related to the doctrine of equivalents to be handled, especially if the invention is a synthetic version of a 'product of nature'? After all, Nature too is experimenting constantly with its own creations, including the destruction and creation of new species in the predator-prey game of "survival of the fittest" or "natural selection". These are extremely difficult questions to deal with in litigation.

The specification requirement

A specification is targeted at an expert in the field of the invention. Therefore, it is unnecessary for an applicant to spell out every detail but only enough to convince an expert that the inventor possessed the invention as of the filing date, and to enable a PHOSITA to make and use the invention without undue experimentation[26]. For example, in a gene related patent, a written description doesn't need a recitation or incorporation by reference of genes and sequences that are well documented in the prior art. An adequate description of the invention therefore depends on the nature and scope of the invention, not the description's length. Furthermore, an actual reduction to practice is *not* required. An invention can be "complete" even without an actual reduction. The Court of Appeals for the Federal Circuit in the United States, in *Falkner v. Inglis*, 448 F.3d 1357, 1366, 79 USPQ2d 1001, 1007 (Fed. Cir. 2006)[27] has succinctly stated: "(1) examples are not necessary to support the adequacy of a written description; (2) the written description standard may be met ... even where actual reduction to practice of an invention is absent; and (3) there is no per se rule that an adequate written description of an invention that involves a biological macro- molecule must contain a recitation of known structure."

Our current knowledge depicts living matter as incredibly complex (almost like a black-box) and therefore not describable either completely or accurately as required under the written description requirement. Inventions involving biological materials such as cell lines, cloning vectors, hybridomas, plasmids, microorganisms, etc., are sometimes impossible to describe adequately in words and reproducing them is not always a completely repeatable process. This problem is addressed by depositing appropriate biological materials with a recognized repository which provides permanence and availability to other researchers on demand (Berns, et al, 1996). This removes any uncertainty regarding the precise characterization of the material, such as a microorganism or cell line claimed in the invention, while ensuring that others will be able to practice the invention completely. In cases where the deposit requirement applies to higher-life forms, such as transgenic animals, the requirements may be satisfactorily fulfilled if a deposit of the lowest common denominator of a higher life form, e.g., the sperm, egg, fertilized egg, embryo, etc. is deposited. As on June 03, 2011, 75 Contracting Parties had signed the Budapest Treaty on the International Recognition of the Deposit of Micro-organisms for the Purposes of Patent Procedure[28] (done at Budapest on April 28, 1977, and amended on September 26, 1980). The Treaty specifically requires the following:

Contracting States which allow or require the deposit of microorganisms for the purposes of patent procedure shall recognize, for such purposes, the deposit of a microorganism with any international depositary authority. Such recognition shall include the recognition of the fact and date of the deposit as indicated by the international depositary authority as well as the recognition of the fact that what is furnished as a sample is a sample of the deposited microorganism. (Article 3(1)(a))

As far as matters regulated in this Treaty and the Regulations are concerned, no Contracting State may require compliance with requirements different from or additional to those which are provided in this Treaty and the Regulations. (Article 3(2))

The satisfaction of the specification requirement is largely a procedural matter and depends on the skill of the patent counsel and the inventor's cooperation in preparing the patent application. On the other hand, fulfilling the requirements of utility, novelty, and non- obviousness depend more on the substantive merits of the invention itself.

TREATIES AND AGREEMENTS PERTAINING TO PATENTS

Grant of patents and their enforcement, if infringed, rests with national governments. Differing national economic and geopolitical needs have resulted in wide differences among national patent systems that, at times, have led to odious disharmonies in patent enforcement and flow of trade and commerce. There has been a long-felt need for harmonized patent laws, especially by those who need their inventions protected concurrently in major world markets. Since the 1880s, limited harmonization among groups of nations, mainly related to procedural matters, has been achieved through various international treaties. The important ones are: (1) Paris Convention for the Protection of Industrial Property (1883),[29] (2) Patent Cooperation Treaty (1970),[30] (3) Agreement on Trade- Related Aspects of Intellectual Property Rights (TRIPS) (The Agreement is Annex 1C of the Marrakesh Agreement Establishing the World Trade Organization, signed in Marrakesh, Morocco on 15 April 1994),[31] and (4) The Trilateral Cooperation (1983)[32] agreement among the patent offices of the United States, Europe, and Japan.

The Paris Convention

The Paris Convention (1883), now administered by the World Intellectual Property Organization (WIPO)[33], has shaped the patent laws of various countries, especially those of its member States called Contracting Parties. It was the first important international treaty designed to help people of one country obtain protection in another for their intellectual creations in the form of inventions (patents), trademarks, and industrial designs. As on July 15, 2011, there were 173 member States[34]. The Convention does not allow Contracting Parties to discriminate between their own nationals and nationals of other Contracting Parties as regards the protection of industrial property. *Inter alia*, the Convention lays down the common basic structure for patent protection to which the Contracting Parties are bound. This basic structure does not unduly trespass on the sovereign rights of Contracting Parties or compromise their national interests. In fact, Article 19, of the Convention states that Contracting Parties "reserve the right to make separately between themselves special agreements for the protection of industrial property, in so far as these agreements do not contravene the provisions of this Convention." Because of this, patent laws of respective member States share a substantial common core.

The Paris Convention forms the foundation for two other important treaties related to patents—the Patent Cooperation Treaty (1970) and the Agreement on Trade-Related Aspects of Intellectual Property Rights (TRIPS) (1995).

Patent Cooperation Treaty (PCT)

The *Patent Cooperation Treaty* (PCT), administered by WIPO, became operational on June 1, 1978. As of September 23, 2011, the PCT had 144 signatories[35]. While the Paris Convention provides a means of access into different countries' patent systems, a patent application once filed, must be prosecuted through each national patent system. Under the PCT, patent applicants from Contracting States enjoy a relatively simple way of commencing patent applications in a number of countries simultaneously. This provision has since been encoded in the patent laws of most Contracting States.

The PCT provides a centralized mechanism for filing patent applications, prior art search and preliminary examination of the patent application; it does *not* grant patents. PCT Contracting States are bound by Chapter II of the PCT relating to the international preliminary examination of patent applications. An applicant can designate specific countries or regional conventions for grant of patent by filing an international patent application in the appropriate receiving office. After an international search report and a non-binding preliminary opinion on patentability has been provided, the applicant must still apply separately and individually to each jurisdiction where patent protection is required. While the search and the preliminary opinion might reduce subsequent search- related workload of national patent offices examining the patent application, the main and substantial workload still belongs to the national patent office. Every biotechnologist should become familiar with the process of filing a patent application under the PCT.

TRIPS

Of all the treaties in force, TRIPS is the most ambitious. Ratification of TRIPS is a prerequisite for a country to become a member of the World Trade Organization (WTO)[36]. As on October 4, 2011, WTO had 153 members. TRIPS entered into force on January 1, 1996, and covers various forms of intellectual property rights, including patents. It introduced intellectual property law into the international trading system for the first time. It was negotiated at the end of the Uruguay Round of the General Agreement on Tariffs and Trade (GATT) in 1994. TRIPS nudged signatory countries towards a level of uniformity, which most are still struggling to cope with even though it was sweetened with some concessions for developing and underdeveloped countries. For example, art. 1.1 leaves member states "free to determine the appropriate method of implementing the provisions of this Agreement within their own legal system and practice" and a November 2005 decision of the Council for TRIPS allowed least-developed country members to postpone implementation of many TRIPS obligations until 2013.[37] The difficulties faced by developing countries are not just due to their inferior stages of technological advancement but also due to social, administrative, infrastructural, and other costs incurred in implementing TRIPS. This is particularly visible in the case of pharmaceutical products.[38]

Under the TRIPS Agreement, member countries are required to make patents available "for any invention, whether products or processes, in all fields of technology" without discrimination, subject to certain legal requirements being met. These requirements are that they must fulfill the member country's legislated criteria for novelty, inventiveness, and industrial applicability. Further, once patent rights are granted, the owner of the patent should be able to enjoy those rights in the member country without discrimination as to the place of invention and whether products are imported or locally produced. The above is subject to three exceptions: (1) the invention should not be contrary to *ordre public* or morality; (2) inventions related to diagnostic, therapeutic and surgical methods for the treatment of humans or animals may be excluded from being patented; and (3) inventions related to plants and animals other than micro-organisms and essentially biological processes for the production of plants or animals other than non-biological and microbiological processes may be excluded. However, when such exclusions are made in the patent system for plant varieties, an effective *sui generis* system of protection must be provided. Subsequent to the Doha Round which took several years to negotiate, TRIPS permits countries to issue compulsory licenses to meet the health needs of nations unable to produce locally needed medicines.[39] This, however, means little to countries which lack the ability to manufacture pharmaceuticals locally.

Several TRIPS articles remain open to wide interpretation to allow each member country freedom to tailor its patent system according to its domestic needs, present state of development, and growth potential. For example, while TRIPS lists an "inventive step" as one of the requirements for patentable subject matter (art. 27(1)), it does not define the term. Likewise it defines the scope of a patent in terms of the nature of the rights conferred (art. 28), but does not set out the breadth of the technological terrain a patent must cover. This allows member states to supply their own definitions of "inventive step" and determine the scope of patent protection.

The Trilateral Cooperation

The Trilateral Cooperation was set up in 1983 between the USPTO, the European Patent Office and the Japan Patent Office (collectively known as the Trilateral Offices) to overcome certain problems arising due to a dramatic rise in the number of patent filings in the early 1980s. These Offices process the greater part of all patent applications filed worldwide including PCT applications. Under the Cooperation, the Offices focus on addressing global patent workload challenges, e.g., decreasing pendency and examination backlogs, improving patent quality, and leveraging IT solutions to accelerate processing of patent applications. Through work sharing arrangements the Offices leverage work done earlier by another Office to improve their own search and examination practices. One of their goals is to eventually develop a paperless administration of the patent procedure, the exchange of documents, and electronic filing of applications.

Hurdles in the path of harmonization

- Attempts to harmonize different national patent systems face major hurdles: the standards to be followed for utility, novelty, and nonobviousness; defining circumstances when research exemption and compulsory licensing are appropriate; setting objective standards for analyzing infringement and award of relief; etc. These issues are dealt with in widely differing ways by different countries. Any debate on global harmonization must also consider alternative mechanisms for encouraging technological innovation, not just the patent system, and account for the fact that different countries are at different stages of transition—from the industrial age to the information age. A recent paper (Reichman and Dreyfuss, 2007) succinctly notes:

- [T]he worldwide intellectual property system has entered a brave new scientific epoch, in which experts have only tentative, divergent ideas about how best to treat a daunting array of emerging new technologies. The existing system has become increasingly dysfunctional because it operates with a set of rudimentary working hypothesis that have not kept pace with technical change.
- Any attempt to push harmonization beyond TRIPS would require great care. At the least, individual nations must be clear about the patent system that would best serve their interests in the new knowledge economy. The daunting nature of the task becomes evident from the experience of the United States, which after six years of feet-dragging and several aborted attempts at reforming its Patent Act, finally enacted the Leahy-Smith America Invents Act, 2011[40] on September 16, 2011. It is seen as "a jobs creation bill." The Act, most importantly, changes the earlier "first-to-invent" system to a "first-to-file" system to make it compatible with the rest of the world, raises patentability standards, makes injunctions and damages harder to obtain, provides new options for challenging bad patents, provides enhanced funding of PTO operations, provides for expedited examination of patent applications (for a fee), etc. Full implementation of the new law will take several months. A lack of political will to sink differences to bring about change was evident throughout.

THE INFLUENCE OF BIOTECHNOLOGY PATENTS ON SOCIETY

Since the 1970s the spotlight has shifted from technological advances of the industrial revolution (driven by Newton's laws of motion and Maxwell's laws of electromagnetism) to advances in biotechnology (driven by molecular biology and biochemistry). Towards the end of the twentieth century, conventional wisdom asserted that while that century belonged to physics and chemistry, which led to huge industrialization and consequent megacities, the twenty-first century will belong to biology and associated technologies. Its impact on society is expected to be phenomenal, affecting every inhabitant on our planet in an intimate way. The nature of the impact will crucially depend on how society accepts or rejects new technical innovations in biotechnology (Smith, 2009). Even though, since the 1980s, biotechnology has been recognized and welcomed as a highly promising strategic technology by many industrialized nations (Bera, 2009c), it has not resulted in automatic acceptance by society. The rate at which R&D results can be assimilated will depend less on scientific or technical considerations but more on such factors as availability of venture capital, the ability to acquire and protect patents, marketing skills, the efficacy and cost- effectiveness of new technologies, and possibly of far greater importance, public perception and acceptance. Perhaps no other industry is as heavily dependent on patents as the biotechnology industry and on public trust.

The emergence of the modern biotechnology industry in the 1970s as an intermediate sector between academic research institutions and Big Pharma was novel. Academic researchers played an important role in the founding of many biotechnology companies; some participated in both worlds, some turned into entrepreneurs. University-industry collaboration became a critical factor in commercial success as did the foresight of some venture capitalists who were willing and able to support inexperienced companies entering a market with a seven- to ten-year product development cycle. Indeed, without patent rights in areas such as isolation and purification of proteins, DNA sequences, monoclonal antibodies, transgenic organisms and gene expression systems, etc., many biotechnology companies could not exist (Bera, 2009b; Williams, 2005).

New technologies spark patenting debates

Historically, the birth of each new technology tends to spark a patenting debate. In the early 20th century the debate was whether agricultural inventions could be patented on the grounds that agriculture was not an industry. In the 1970s, it was argued that pharmaceutical patents were unethical. In the 1980s, the biotechnology industry faced hostility over "patenting of life". Even now, bioethical, social, and legal questions related to biotechnology patents are far from being over, as are issues related to intellectual property, scientific integrity, and conflicts of interest in research. For a satisfactory resolution, the debate must involve experts from diverse fields: science, engineering, theology, and philosophy. Finally, since the 1990s, the computer revolution and the Internet have produced many controversies related to software and business method patents (Poynder, 2000). An apparently persuasive argument against biotechnology patents is the field's rapid pace of development. For example, a few decades ago, finding a gene may have taken ten years, but now one can be found within seconds using a computer search and gene maps (Demaine & Fellmeth, 2002). If invention is inevitable, does it merit reward of exclusivity? If yes, for how long? A long period may spur innovation but also limit the spread of new and useful products and processes and make them more expensive. On the other hand, rapid discovery of genes does not imply rapid availability of useful and safe applications based on those genes. Those applications come from creative geniuses. Should they not be rewarded with patents? For inventions, such as vaccines for public health, the balancing act is never easy, given the enormous R&D costs of developing them and the crucial role of Nobel class researchers who make them possible. As Todd Dickinson (former director of the USPTO) once remarked, "there are so many chemicals in the human body that, if we ruled them all off limits to patenting, we would rule out an extraordinary number of valuable and important inventions. ... Without the funding and incentives that are provided by the patent system, research into the basis of genetic diseases and the development of tools for the diagnosis and treatment of such diseases would be significantly curtailed." (Dickinson, 2000)

Another objection is that the "current model rewards particular kinds of creative effort, namely those which result in commercial gain. It is therefore likely to hinder innovation of products that have limited market value, but which have huge social benefit."[41] This flawed argument overlooks the obvious fact that intellectual property laws *were* meant to encourage commercial gain. There are other laws which encourage social benefit and one does not criticize those laws for hindering innovation that lead to commercial gains. The correct approach, if providing social benefit which have limited market value is the objective, is to elect governments that will act more enthusiastically in providing social benefits (of course, the government will have to increase taxes to do so), encourage the general population to contribute to philanthropic activity by donating time and service to community activities, including creating intellectual property. There are no laws against such philanthropic activities, but there is a huge lack of enthusiasm on the part of the general population to help itself. That same population works more energetically when it gets a share of "commercial gains" in terms of employment opportunities and wages. Commenting on the allegation that the global intellectual property regime denies poor people access to drugs, Alasdair Poore said, "Without an effective patent system, who would have made the necessary investment to discover and manufacture those drugs? It's politics and economics that block access to drugs for the world's poor, not the IP system." (Prowse, 2009)

On closer inspection, one finds that the broadest debates concern ethical and societal aspects of patenting genetic materials, the perceived rights of indigenous communities that have shaped their environment and its organisms and thus the genetic resource embodied therein, and the manner in which the bioindustry prospects (or allegedly pirates) biological resources of poor countries and commercializes the products it derives through patents (Koopman, 2003). At another level, while patenting of biotechnology inventions is being criticized, it is really the science behind it that the opponents seem to be against.

Knowledge is commercial power

An important, although not the only, measure of a technology's success is its embodiment in products and processes that generate a profitable commercial market through public acceptance. Public acceptance is a factor *only* if there is a supplier willing to assume business risks and enter the market. One might then assume that if a suite of patentable products and processes, paid for out of the public purse, were owned by the government and made available free or at a nominal cost for commercial exploitation, business risks would be lowered. That such is not the case was the genesis of the Bayh-Dole Act[42] of 1980 in the United States. What the government found was that discoveries made in the universities with federal funds were grossly underutilized[43] because government policy required that it take title to all such inventions and license them non-exclusively. The vast majority of university discoveries, as expected, were early stage discoveries that required substantial additional investment to turn them into a marketable product. It was estimated in 2002 that a "dollar's worth of academic invention or discovery require[d] upwards of $10,000 of private capital to bring to market."[44] (The government funds the 'inspiration' while the private sector funds the 'perspiration'!) "New drug development costs have risen from $0.8 billion (1997) to an expected $1.9 billion (2013)."[45] Without the protection of an exclusive license, companies were reluctant to invest huge sums when the resulting products could easily be appropriated by competitors.[46]

The Bayh-Dole Act was a bold, against-the-grain, initiative meant to rejuvenate the U.S. economy. Under Bayh-Dole, the government relinquished its intellectual property rights on the outputs of federally funded research and permitted universities and small businesses to acquire title to inventions created with federal funds. It also allowed exclusive licensing of patents thus acquired, to industry since, without it, companies were wary of investing in the further development of university developed technologies. In addition, descriptions of inventions were given legal protection from public dissemination and from requests under the *Freedom of Information Act*[47] for a reasonable period to enable patent applications to be filed. In return, the government retained a royalty-free, non-exclusive license to practice the patented inventions coming out of federally funded research throughout the world (including use by government contractors) and held 'march-in rights', which allowed the government to take back the title if the patent owner failed to commercialize the invention. However, the exercise of march-in rights was made substantially difficult and appealable in courts. The march-in rights were basically introduced to prevent companies from licensing university patents with the sole intention of blocking rival companies from doing so. Therights were meant to ensure fair competition and to meet the needs of U.S. citizens; they were not meant for the government to set prices, as some have tried to claim.[48]

So, post-Bayh-Dole, we now witness the hitherto unimagined situation where innovations, already paid for by the public, can be brought to the market only if those innovations are privatized and resold to the public via patents acquired by commercial entities. Otherwise certain markets will likely vanish on their own because the risks are too high! The biotechnology industry, through university-industry collaboration, has shown that knowledge is a phenomenal commercial power. The emulation of the Act by other countries, although common now, will not necessarily have the same impact that has been visible in the U.S. because of dissimilar national circumstances, or the absence of world-class research universities.

Patent laws never anticipated that together "blue sky" research *and* living matter would play such a fundamental role in late 20th century commerce with such speed, and economists and policy makers never imagined the deep post-1980 university-industry (or generally, public-private) collaborations that rapidly gained momentum in the biotechnology sector following three signal events in 1980 in the United States: the Bayh- Dole Act, grant of the Cohen-Boyer patent, and the Chakrabarty court decision (Bera, 2009d). These rapid changes have bewildered lawmakers, patent offices, the judiciary, and relevant enforcement agencies. To older generations it is sacrilege that "even the pure quest for knowledge is subverted by the need for profit."[49] To the new generation it is the welcome emergence of a new and refreshing paradigm where pure knowledge is rapidly converted to applications to serve consumers through conventional market mechanisms of demand and supply. The future may see the emergence of other market mechanisms, whose advent no economist will likely anticipate, because they will occur to accommodate needs triggered by innovation alongside need inspired innovations.

The writing on the wall is clear; the times are changing, and so must the way we teach, create, use, and protect knowledge and the innovations they spawn. Initially, important chunks of that knowledge in biotechnology coming from universities will be tacit and scarce, hence university-industry collaboration will be crucial for technology transfer and commercial success. The relative youth of the biotechnology industry and its dependence on scientific breakthroughs means that star scientists—their accessibility, location, motivation to collaborate at the bench-science level with scientists in industry in converting basic scientific knowledge into commercially viable products and processes—will be crucial in determining the pace of diffusion of tacit scientific knowledge (Zucker & Darby, 1996). To remain relevant in an economically global world, the social role of universities and government research laboratories must change as must our understanding of morality, ethics, and citizenship.

It is mainly due to the university-industry collaboration example set by the United States that universities elsewhere are now expected to transform themselves into engines of economic growth, rather than remain as not-for-profit ivory towers. This is an enormous social transformation, and an enormous opportunity for universities to help the world settle down in the new era of a knowledge-intensive global economy. In this world, university-industry collaborative research is a natural means of providing continuing education to knowledge workers throughout their professionally productive life. It is also a natural means of mutual technology transfers between academic researchers and applied industry researchers, especially of tacit knowledge. Donald Kennedy of Stanford University was spot on when he said, "Technology transfer is the movement of ideas in people."[50] This movement in biotechnology must frequently happen under the protective cover of patents. One of the outstanding examples of technology transfer between university and industry is the licensing policy adopted by Stanford University with respect to the Cohen-Boyer patents[51] (Bera, 2009a). No doubt, other technology transfer models to fulfill emerging needs will evolve as the biotechnology sector matures.

Modern science-based industries (SBIs) critically depend on monetary funds, star scientists (human capital), and protected intellectual property (intellectual capital). Where and when star scientists publish also has a determining effect on the commercial adoption of new technologies. Geographically, SBIs tend to nucleate in close proximity of universities hosting a star group of scientists active in the relevant science, as it greatly improves mutual accessibility of both people and research facilities. This has generally been the case for biotechnology, especially in the United States (Zucker & Darby, 1996). Governments must bear this in mind when framing policies for economic and industrial growth and providing infrastructure. Once nucleation is complete and substantial diffusion of tacit knowledge of the stars has taken place, further expansion of the industry can spread to far-off places, especially of manufacturing units and support R&D groups.

Gene patents (unresolved issues)

A recent gene patent case, *Association for Molecular Pathology* v. *the USPTO and Myriad Genetics*, No. 10-1406 (Fed. Cir. 2011)[52], decided by the United States Court of Appeals for the Federal Circuit, has gained extraordinary attention. In this case, the validity of a series of patents that claim isolated DNA compositions and methods for testing the presence of genetic mutations that are correlated with an increased risk of certain breast and ovarian cancers (the BRCA1 and BRCA2 genes) was challenged. Myriad owns or is the exclusive licensee of these patents. Opponents of the patents argued that the patent claims encompass patent-ineligible subject matter, e.g., products of nature. At issue were such fundamental questions as to whether isolated DNA should be eligible for a patent, and whether the patenting of genes promotes or stifles innovation and development of new diagnostics and therapies. The Court stated expressly that an isolated partial DNA fragment, not just cDNA has a "markedly different structure to native DNA" and so reaffirmed that isolated gene sequences are patentable. Patentability of DNA sequences as diagnostics remains uncertain. Whether the Supreme Court of the United States will entertain an appeal in the Myriad case is not yet known.

Quite independent of how this case eventually ends, patent law will need to revisit the grant of gene patents. We now know that the one-gene-one-protein assumption of yore is no longer true. The proteome is larger than the genome—there are more proteins than genes due to alternative splicing of genes and post-translation modifications like glycosylation and phosphorylation. The cause of most disorders and diseases is a combination of genetic and environmental factors and this raises important questions about the adequacy, scope and purpose of patent law in view of rapidly advancing knowledge in biotechnology and related fields. It is amply possible that under present laws, a single gene or a short DNA sequence, if patented, could result in a near monopoly on diagnostic tests and treatments for widespread and serious ailments, such as, diabetes, cancer, multiple sclerosis, and Alzheimer's disease. Can such patents be considered as serving human society if the patent owner cannot pursue all known downstream opportunities and blocks others from pursuing those or new ones? If such patents are inevitable, what steps should be taken to ensure that they do not obstruct others ready to pursue opportunities not pursued by the patent owner. Even otherwise, courts and administrative agencies continue to struggle with issues raised by gene patents and their predecessors—chemical patents—as to when and how patents should be granted on biochemicals in their natural and modified states under existing patent laws.

The fact that creation of transgenic humans is, in principle, possible, inevitably raises questions of human dignity, and moral and ethical issues. There appears to be a general consensus that transgenic humans are not patentable. Yet, no unambiguous definition of a transgenic human exists. Given that the genomic DNA differences between human and chimpanzee is only about 1.2%, the possibility of creating a patentable transgenic talking chimpanzee that can communicate with humans is not a fantasy. Such a chimpanzee might actually be able to speak and be capable of making connections between human words, objects, and even emotions. For the first time we may then be able to establish verbal communication with another species and derive remarkable insights about the animal kingdom. Should this possibility be denied to the human race because the transgenic chimpanzee is also a transgenic human? (Bera, 2009d)

The owner of a gene patent does not "own" any organism containing that gene. Thus, a person whose body contained a patented gene would not infringe the patent. However, if a gene, patented or not, is inserted into a living organism that organism may become patentable and then commercially exploited. Ownership and commercial exploitation of plants and animals, such as buying and selling them, is widely accepted in our society, but not of humans in *today's* world. Finally, animal breeding is not new and has been practiced since virtually the beginning of agriculture. Human breeding through marriage customs is also not new. Clearly, when one discusses moral issues related to patents, it is not the invention that is morally repugnant but its use in certain unintended ways. No one in the patent system—inventors, patent examiners, judges, or even legislatures (representing the people) can anticipate all uses of a particular patent that may eventually turn out to be, on a statistical balance, detrimental or beneficial to society. Any premeditated restrictions on the grant of patents must carefully consider the possibility that such restrictions may undermine the patent law's primary objective of promoting technical innovation.

CONCERNS OVER BIOPIRACY

A paradigm shift inevitably entails shifts of power. The rise of biotechnology has sharpened the divide between the science-based industrial nations, and the genetically endowed but less-developed nations whose genetic resources are prospected by the former. The conflicting issues being debated include the proprietary character of natural genetic material and the nature of commercial exploitation of the value added to those materials through R&D. Natural genetic resources abound in developing countries with a tropical climate, e.g., Brazil, Peru, Costa Rica, and India, in the form of gene pools, organisms, and ecosystems. Biotechnologists are obviously interested in these resources and related 'traditional knowledge' held by indigenous communities as inputs to research while the biotechnology industry is interested in prospecting those resources for potential commercial exploitation.

The methodology and approach of traditional knowledge is holistic and applied according to notions of biocentricism (a political or ethical stance which asserts the value of non- human life in nature), co-evolution and equality; it does not rely on empirical verification, rather it seeks connections between the physiological characteristics of organisms (visible phenotypic properties) with their spiritual ones. In contrast, modern biotechnology concentrates on biochemical genotypic properties. Nevertheless, traditional knowledge can be a valuable starting point for biotechnologists by indicating to them specific organisms and their known medical usage (Koopman, 2003). The enormous gap in terms of effectiveness and use between products and processes derived from traditional knowledge and from modern biotechnology must be filled by very expensive R&D, which is clearly outside the capabilities of indigenous communities. These are uncontested facts.

The biopiracy debate then essentially revolves around ethical and societal values as viewed from two widely different cultures over the patenting of genetic material whose natural inputs were prospected in and transported from indigenous communities on the basis of their traditional knowledge, with next to nothing in return in terms of acknowledgement or affordable products and processes or sharing of R&D knowledge. In short, the indigenous communities see this as blatant biopiracy. This is a clash between two cultures—of shared community rights against privately held individual rights. Indigenous communities seldom recognize the concepts of individual ownership, exclusion and competition that underlie the Western concepts of property law regimes.

Such irreconcilable differences have found palatable compromises in the form of the Convention on Biological Diversity[53] and TRIPS, where each culture makes concessions to the other. Countries providing access to genetic resources or traditional knowledge are permitted, and some have implemented, *sui generis* systems where they provide access on certain conditions, such as, getting prior informed consent of a national office dealing with such matters, benefit sharing arrangements (e.g., sharing of proceeds derived from commercial exploitation, training in R&D, transfer of technology under 'fair and most favorable terms' consistent with the 'adequate and effective protection of intellectual property rights'), and treating certain violations of the statutes as criminal offenses, in exchange for biological samples and traditional knowledge. In a subtle way, these *sui generis* systems are enforcing "reach through claims" on others for advantages nature has endowed indigenous communities with and the traditional knowledge they have developed long ago. The debate never mentions the tremendous unpatentable scientific knowledge of Newton, Maxwell, Einstein, and others which has been freely bestowed on the world without seeking rents, and which has allowed such dreams as putting a man on the moon possible.

SUMMARY

Our new understanding of living matter is leading us to uncharted territories. Recombinant DNA technology, transgenic animals, synthetic cells are just the tip of the iceberg. Creation of computer designed, engineered life is no longer science fiction but a potential reality. Will surreptitiously created transgenic humans one day enslave natural humans and rule the world. Can such an event be stalled? What will a world dominated by transgenic humans be like? Will it be more humane than ours? Will they rule more by the head and less by the heart or the other way round? We have no way of knowing.

While patent laws forbid patenting of abstract ideas, thought processes, laws of nature, natural phenomena, and products of nature so as not to stifle advancement of knowledge, the laws do recognize, as they did for Galileo, that certain down-to-earth inventions, conjured through human ingenuity by applying these forbidden things, go beyond philosophical musings and have potential commercial value, because of their utility to humans. In such cases, a *quid pro quo* system that provides limited period monopoly with commercial advantage in the form of patents, in exchange for a full public disclosure of the invention not later than the date of patent grant, encourages further creation of new inventions or improvements over old ones. Patent laws were meant to encourage commercial gain in an equitable manner. There are other laws and practices that encourage material and spiritual contributions to society through raised taxes, philanthropy, free social service, open-source, etc., which are not motivated by commercial gain. Patent laws do not interfere with these other laws and practices. Outside of contractual obligations to, say, his employer, a biotechnologist can choose to patent or not patent. Patent laws do not operate in isolation. Their purpose is to benefit society on the whole so that the positives outweigh the negatives in a statistical sense. There is no denying that countries with a thriving patent system, with all its faults, have advanced technologically more rapidly than all other countries and provided a better quality of life to their citizens. It is hoped that this chapter will help the reader decide when patenting is appropriate in light of other alternatives.

REFERENCES

1. Ball, P., The dawn of quantum biology, *Nature*, Vol. 474, 16 June 2011, pp. 272-274.
2. Belvis, G. (2003). Chapter 2 Overview of the Doctrine of Equivalents and § 112, 6 Equivalents, In: *Intellectual Property Law Update*, G. Belvis, pp. 35-66, Aspen Publishers, Retrieved from http://www.brinkshofer.com/files/102.pdf
3. Bera, R.K. (2009a). The Story of the Cohen-Boyer Patents. *Current Science*, Vol. 96, No. 6, (25 March 2009), pp. 760-763.
4. Bera, R.K. (2009b). The Changing role of Universities and Research Institutions in a Global Economy: Lessons Drawn from the U.S. Biotechnology Sector. *Current Science*, Vol. 96, No. 6, (25 March 2009), pp. 774-778.
5. Bera, R.K. (2009c). Intellectual Property Fuels a Global Sense of Competitiveness. *Current Science*, Vol. 96, No. 7, (10 April 2009), pp. 898-903.
6. Bera, R.K. (2009d). Post-1980 World of Biotechnology Patents in the U.S. *Current Science*, Vol. 96, No. 10, (25 May 2009), pp. 1343-1348.
7. Berns, K.I., Bond, E.C., & Manning, F.J. (Eds.). (1996). *Resource Sharing in Biomedical Research*, National Academy Press, Washington.
8. Blenko, Jr., W.J. (1990). The Doctrine of Equivalents in Patent Infringement, *JOM*, Vol. 42, No. 5, (May 1990), p. 59.
9. Cohen, S. N., Chang, A. C. Y., Boyer, H. W., & Helling, R. B. (1973). Construction of Biologically Functional Bacterial Plasmids In Vitro, *Proc. Nat'l Acad. Sci.*, Vol. 70, No. 11, (November 1973), pp. 3240-3244.

10. Christie, A. F. & Lim, A. (2005). Reach-through Patent Claims in Biotechnology: An Analysis of the Examination Practices of the United States, European and Japanese Patent Offices. *Intellectual Property Quarterly*, Vol. 3, (2005), pp. 236-266.

11. Darwin, C. (1872). *The Origin of Species*, 6th edition, Modern Library, ISBN 0451625587.

12. Demaine, L.J., & Fellmeth, A.X. (2002). Reinventing the double helix: a novel and nonobvious reconceptualization of the biotechnology patent, *Stanford Law Review*, Vol. 55, No. 2 (November 2002), pp. 303-462.

13. Dickinson, T. (2000). Gene Patents and Other Genomic Inventions: Hearing before the Subcommittee on Courts & Intellectual Property of the Committee on the Judiciary, House of Representative, *106th Congress*, (July 13, 2000), Retrieved from http://commdocs.house.gov/committees/judiciary/hju66043.000/hju66043_0f.htm

14. Gibson, D.G., *et al.* (2010). Creation of a Bacterial Cell Controlled by a Chemically Synthesized Genome, *Science*, (20 May 2010), pp. 1-12.

15. IPO (2009). Examination Guidelines for Patent Applications relating to Biotechnological Inventions in the Intellectual Property Office, *IPO*, U.K., (April 2009).

16. Koopman, J. (2003). Biotechnology, Patent Law and Piracy: Mirroring the Interests in Resources of Life and Culture, *Electronic Journal of Comparative Law*, Vol. 7.5, (December 2003), pp. 1-19, Available from http://www.ejcl.org/ejcl/75/art75-7.html

17. Köhler, G, & Milstein C. (1975). Continuous cultures of fused cells secreting antibody of predefined specificity, *Nature*, Vol. 256, Issue 5517, (07 August 1975), pp. 495-7.

18. Lilly, M.D. (1997). The development of biochemical engineering science in Europe, *Journal of Biotechnology*, Vol. 59, Issues 1-2, (17 December 1997), pp. 11-18.

19. Mullis, K.B. et al. (1986). Specific enzymatic amplification of DNA in vitro: the polymerase chain reaction, *Cold Spring Harbor Symp. Quant. Biol.*, Vol. 51, pp. 263-73.

20. Poynder, R. (2000). Internet sparks patenting controversy, *IP Matters*, (30 April 2001), Available from http://www.richardpoynder.co.uk/internet_sparks.htm

21. Prowse, P. (2009). Patent profession welcomes Manchester Manifesto on science but slams 'misleading' comments on IP, (November 27, 2009), Retrieved from http:// www.cipa.org.uk/pages/press/article?D5C2CBED-894B-488B-ACD2- 07B01E204A06

22. Reichman, J.H. & Dreyfuss, R.C. (2007). Harmonization without Consensus: Critical Reflections on Drafting a Substantive Patent Law Treaty, *Duke Law Journal*, Vol. 57, No. 1, pp. 85-130.

23. Schrödinger, E. (1944). What is Life? Cambridge University Press, Cambridge, Retrieved from http://whatislife.stanford.edu/LoCo_files/What-is-Life.pdf

24. Smith, J.E. (2009). *Biotechnology*, (5th edition), Cambridge University Press, Cambridge. Watson, J.D., and Crick, F.H.C. (1953). Molecular structure of nucleic acids, *Nature*, Vol. 171, No. 4356, (April 25, 1953), pp. 737-738.

25. Williams, A. (2005). The New Innovation: Rethinking Intellectual Property for the ONE Big Idea, New Paradigm Learning Corporation, (June 2005).

26. Zucker, L.G., & Darby, M.R. (1996) Star scientists and institutional transformation: Patterns of invention and innovation in the formation of the biotechnology industry, *Proc. Natl. Acad. Sci. USA*, Vol. 93, November 1996, pp. 12709-12716.

PART-IV
GENERAL BIOTECHNOLOGY

CHAPTER-16

BIOTECHNOLOGY VIRTUAL LABS: ENABLING ON-DEMAND LABORATORY ACCESS FOR CLASSROOM EDUCATION

INTRODUCTION: AN OVERVIEW

Biotechnology virtual labs have emerged as powerful tools for enhancing the educational experience by providing students with the opportunity to engage in laboratory experiments anytime and anywhere. These virtual environments simulate real-life laboratory settings, enabling students to conduct experiments, analyze results, and gain practical insights into biotechnological processes. This accessibility is particularly beneficial for remote and online learning, offering a solution to the limitations posed by physical laboratory constraints. Biotechnology virtual labs not only democratize access to quality education but also foster a deeper understanding of biotechnological concepts, preparing the next generation of scientists and innovators for the dynamic field of biotechnology.

Biotechnology is becoming more popular and well identified as a mainline industry. Students have shown greater interest in learning the techniques. As a discipline, biotechnology has led to new advancements in many areas. Criminal investigation has changed dramatically thanks to DNA fingerprinting. Significant advances in forensic medicine, anthropology and wildlife management have been noticed in the last few years. Biotechnology has brought out hundreds of medical diagnostic tests that keep the blood safe from infectious diseases such as HIV and also aid detection of other conditions early enough to be successfully treated. Medical kits for diabetes, blood cholesterol and home pregnancy tests are also biotechnology diagnostic products. Industrial biotech applications have led to cleaner processes that produce less waste and use less energy and water in such industrial sectors as chemicals, pulp and paper, textiles, food, energy, and metals and minerals. Laundry detergents produced in many countries contain biotechnology-based enzymes making them nature friendly and safer. Agricultural biotechnology benefits farmers, consumers and the environment by increasing yields and farm income, decreasing pesticide applications and improving soil and water quality, and providing healthful foods for consumers. Biotechnology has created more than 200 new therapies and vaccines, including products to treat cancer, diabetes, HIV/ AIDS and autoimmune disorders.

This rise in application has led to an increased rise in the number of students undertaking University-level biotechnology courses. However, biotechnology education requires an eclectic approach of combining various sub-disciplines. Biology courses and chemistry courses in biotechnology have diversified the approach of the topic. Most common courses that biotechnology degree programs focus at the University level in India consist of cell biology, molecular biology, microbiology, immunology, ecology, statistics and biophysics.

A brief description of the courses will be sketched so a better picture can be understood on the university-level curriculum at most places in India and abroad. Cell biology is a course that focuses on theoretical fundamentals behind the structure, function, and biosynthesis of cellular membranes and organelles; cell growth and oncogenictransformation; transport, receptors, and cell signaling; the cytoskeleton, the extracellular matrix, and cell movements; chromatin structure and RNA synthesis. Molecular biology course covers a detailed analysis of the biochemical mechanisms that control the maintenance, expression, and evolution of prokaryotic and eukaryotic genomes. The topics also include gene regulation, DNA replication, genetic recombination, and mRNA translation. In particular, the logic of experimental design and data analysis is emphasized. Microbiology course introduces students to the principles of infectious agents. Fundamental techniques in microbiological researches, such as sterilization, isolation, morphological observation, and cultivation are usually covered. Immunology courses focus on the mechanisms which govern the immune response. This will usually include the cells, organs and molecules that mediate the innate and adaptive aspects of the immune system as they apply to infection, tumor recognition, autoimmune diseases, immunodeficiency, cancer and hypersensitivity. Population ecology courses introduce students to major concepts in population ecology including topics such as mathematical models of population growth, population viability analysis, habitat fragmentation and meta-populations, dispersal, population harvesting, predation and population cycles, competition, and estimation of population parameters in the field. Biochemistry course explores the roles of essential biological molecules focusing on protein chemistry, while covering lipids and carbohydrates. It provides a systematic and methodical application of general and organic chemistry principles. Students examine the structure of proteins, their function, their binding to other molecules and the methodologies for the purification and characterization of proteins. Enzymes and their kinetics and mechanisms are covered in detail. Metabolic pathways are examined from thermodynamic and regulatory perspectives. A typical course in biochemistry provides the linkage between the inanimate world of chemistry and the living world of biology. Biophysics is a course that usually links to the study of underlying physical phenomenon in biology and their function. Biophysics course usually cover techniques, methods and applications besides molecular structure and function. Topics in biophysics covered will include an introduction to cell and molecular biology, biorheology, Brownian motion, molecular interactions in macromolecules, protein and nucleic acid structure, physics of biopolymers, chemical kinetics, mechanical and adhesive properties of biomolecules, molecular manipulation techniques, cell membrane structure, membrane channels and pumps, molecular motors, neuronal biophysics and related biophysical mechanisms. A very significant yet seemingly unrelated course is biostatistics. A single introductory course in biostatistics involves an emphasis on principles of statistical reasoning, underlying assumptions, and careful interpretation of results. Topics covered include descriptive statistics, graphical displays of data, introduction to probability, expectations and variance of random variables, confidence intervals and tests for means, differences of means, proportions, differences of proportions, chi-square tests for categorical variables, regression and multiple regressions, an introduction to analysis of variance.

Using software technologies for education has become a new trend. Computer-based technologies developed by academic institutions as well as industries worldwide are revolutionizing the educational system. A new field involving the use of virtual reality techniques is becoming the training environment. Through virtual labs, a new interdisciplinary field of science brings together biologists and physicists to tackle this grand challenge through quantitative experiments and models. Using several pro-learning evendistance education courses have started using virtual laboratories to enable students to access equipment since they are independent from opening hours and the work schedule of the staff. In many engineering courses within India, simulation is the most effective tool in training students in the use of sophisticated as well as complicated instruments that are routinely employed in modern biological and chemical laboratories. For the life sciences, this also circumvents the use of expensive and hazardous biological and chemical agents which toxic to the experimentalists as well as to the environment. Above all, the virtual lab technology is cheap as well as cost effective.

Education in many universities and research institutes include their own virtual laboratories on the web, which are accessible to people around the world. Although some laboratory practice requires getting one's hands 'dirty', it has already been established that the Virtual Lab enables the students to understand the underlying principles and the theory behind laboratory experiments. E-learning plays and will play an important role in diverse regions such as India where the traditional lab facilities at Universities are not very well localized to suit requirements of all sub-regions. With multi-campus scenarios as in some Universities such as ours, offering cross-disciplinary courses needs to exploit the use of extensive e- learning facilities (Bijlani et al., 2008).

Biotechnology lab courses richly rely upon new up-to-date content and various techniques that require a new synergy of knowledge and experimental implementation. Hence a new kind of experimental science that can be brought as a virtual simulation based laboratory is necessary. The developments of the virtual labs include mathematical techniques in biology to study, to hypothesize and to demonstrate complex biological functions. However virtual labs in heavy engineering topics such as analyzing nanomaterials with high-power microscopes and lab courses in biotechnology or biology will also have to exploit multiple techniques besides simulators alone as many scenarios cannot be reproduced mathematically while retaining the "real" lab-like feel.

In this chapter, we focus on the development and use on the virtual biotechnology laboratory courses through a combination of techniques to try completing the learning experience as that of a regular University laboratory.

WHY VIRTUAL LABS?

There are many main reasons to focus on creating virtual labs for University education (Auer et al., 2003). Among the primary reasons include the cost and lack of sufficient skill- set for facing the current growth in biotechnology sector. The setup cost of laboratories puts a large overhead on the educators. The Universities also need to setup laboratories to educate sufficient target group with the details of common biotechnological techniques and protocols (O'donoghue et al., 2001).

Another new motivation is the need to introduce and focus well-explored potential virtual lab areas which use computational methods, mathematical modeling and biophysics, computational biology and computational neuroscience. Computational biology and biophysics are upcoming areas and most techniques derive basis from real laboratory experiments. Another intention of using virtual labs via a computational approach is to train young scientists in the field of the mathematical thinking for life sciences and related environments. Main goals of cross-disciplinary sciences include the need to ensure thatthe students will be able to integrate different exhaustive models into a larger framework,i.e. in the perspective of comprehensive biological systems such as cells and biological networks. Such a role will also give an overview of the modeling approaches that are most appropriate to describe life-science processes. For the everyday biologist, the major use of virtual labs will also be in the learning perspective of advanced but common-to-use simulation tools.

Virtual labs and use of virtual tools should lead to an increase the awareness of a crucial need for standard model descriptions. Most simulators and common-use tools require various formats and schema and with the explosion of data, the use of virtual labs across the country or across multiple countries is also intended to unite educators to work towards common model descriptions and standardization of their data.

For the biotechnology sector, a highly favoring motivation for the shift to the virtual lab paradigm is the explosion of data-rich information sets, due to the genomics revolution, which are difficult to understand without the use of analytical tools. Also, recent development of mathematical tools such as chaos theory to help understand complex, nonlinear mechanisms in biology seems to push the need for information-rich virtual labs in simulation domain.

To aid further, an increase in computing power which enables calculations and simulations to be performed that were not previously possible, have set a new trend in the concept and use of computing. Simulations in the past that needed more intensive computers now can plainly be run through long battery-life laptops (Aycock et al., 2008), given that in many cases laptops today even host servers.

A slightly different reason that also pushes the concept of virtual labs for undergraduate and master level education at the Universities also seems to be an increasing interest in in silico experimentation due to ethical considerations, risk, unreliability and other complications involved in human and animal research.

Given all above reasons and motivation, virtual labs are today's experimental approach towards a newer trend in future education. However the virtual lab environments are still under severe testing and newer models seems to switch to more intelligent and adaptive platforms that can yield efficient knowledge dissipation. One such common model is the adaptive learning system (ALS) currently employed by many e-learning applications strewed on the internet.

OTHER VIRTUAL LABS AND ONLINE COURSES IN BIOSCIENCES

Very little work has been actually done in the biology sector. There are online "dissections" of frog tutorials by Mable Kinzie developed in 1994 and an improved version of the same was hosted in 2002 (http://curry.edschool.Virginia.EDU/go/frog/menu.html). Quick "movies": http://www.bio.unc.edu/faculty/goldstein/lab/movies.html Virtual "experiments": Biology Labs On-Line (BLOL) is a collaboration of the California State University system Center for Distributed

Learning and Addison Wesley Longman, with partial funding provided by the National Science Foundation (http://biologylab.awlonline.com). A project titled "BIOTECH Project" developed by University of Arizona, with aim of supporting Arizona teachers to conduct molecular genetics (DNA science) experiments with their students and assists teachersin developing new activities for their classroom (http://biotech.bio5.org/home). "Protein Lab" by A.J. Booth, is a computer simulation of protein purification. These labs are extremely helpful for beginners in the art of protein purification. It gives them a chance to get beyond the details of individual techniques and get a sense of the overall process of a protein purification strategy. (http://www.booth1.demon.co.uk/archive).To enhance education, there is a great need for individualized courseware to provide educational content that fits to the learner's learning style and knowledge base. University of Utah's genetic science learning center has its very animated genetics labs at http://learn.genetics.utah.edu/gslc. The labs were developed with the mission in making science easy for everyone to understand. Similar projects at Howard Hughes http://www.hhmi.org/biointeractive/vlabs and at Pearson's http://www.phschool.com/science/biology_place/labbench have been useful as virtual education websites.

Online biotechnology courses are available through several leading universities around the world, including the Massachusetts Institute of Technology (MIT), Osaka University and the Open University. OpenCourseWare (OCW) from TUFTS and MIT offer courses on the Web that containing all or some of the materials from the university's original on- campus classrooms. Many biotechnology courses on OCW make use of several different learning materials available online or by download, lab notes, assignments, lectures on scientific communications and study materials. Online biotechnology courses are known to be very helpful for students to study/prepare for the positions as lab technicians, research assistants and quality assurance analysts in such fields as agriculture, pharmaceuticals and manufacturing.

AMRITA VL

Amrita University's Virtual and Accessible Laboratories Universalizing Education (VALUE) initiative was initially targeted towards making biotechnology, physics and chemistry courses virtually accessible for undergraduate and postgraduate education. The project led to the development of 14 labs in biotechnology and 13 labs in physical and chemical sciences. The schema of virtual labs was based on one of our studies.

An average survey of the VL framework software was performed and the tests were shown (see Table 1 in Diwakar et al., 2011). The developed virtual labs are available for public use (See http://amrita.edu/virtuallabs). Any user may login with an open-id or Google's gmail account and access the authentication-compulsory regions such as the remote-panel, simulator and animations. The website uses the name and email address that provider gives only to set up an user account.

METHODS: ANIMATION, SIMULATIONS, AND REMOTE-INITIATED EXPERIMENTS

The key learning component in many biological laboratories is the complexity of the procedure and details of the step-by-step protocol carried out in the laboratory. Although some of these biological processes can be replaced by mathematical equations modeling the system, most of the "feel" is in performing the detailed procedure which is not derived from sets of equations. Graphical animations deliver a high degree of the reality to the virtual labs through their seeming closeness to the appearance and feel of the lab. Graphical animations also cut out the complexity of the modeling process by increasingthe "feel" of experiment. Like the proverb goes, "a picture is worth a thousand words", animations reveal better information that cannot be easily conveyed via text alone or static illustration.

In our biotechnology virtual labs, the animation type of experiments include the use of 2D flash based animations for illustrating detailed procedures such as wet lab protocols and heavy engineering techniques that are out of scope for simulation due to various reasons like complicated equations, numerical issues in simulation, lack of modeling data etc. Besides animation, another common technique in our virtual labs included engineering- based approaches such as remote-triggered experiments or remote-controlled experiments.

The very common and research-inspiring approach is the use of mathematical simulators to model biological and biotechnological processes or sub-processes. Although mathematics has long been intertwined with the biological sciences, an explosive synergy between biology and mathematics seems poised to enrich and extend both fields greatly in the coming decades. Among the various scenarios to study biology and disseminate information effectively and efficiently, includes the use of e-learning as a medium to offer courses.

Applying mathematics to biotechnology for virtual lab creation has recently turned into an explosion of interest in the field. The NASA virtual laboratory or the HHMI virtual labs at Howard Hughes Medical Institute or the Utah genetics virtual laboratory are some examples.

For our labs, a combination of user-interactive animation, mathematical simulations, remote-trigger of actual equipment and the use of augmented perception haptic devices are used to deploy effectively the real laboratory feel of a biotech lab online.

MODELS IN BIOLOGY – AS VIRTUAL LABS

Design of simulation labs requires basic mathematical models. Some models that were used to develop the virtual labs are listed below.

Neurophysiology and neuronal biophysics

In order to understand neuronal biophysics and simulations on voltage clamp and current clamp in detail, we modeled a section of excitable neuronal membrane using the Hodgkin- Huxley equations (Hodgkin and Huxley, 1952) that can be accessed a graphical web-based simulator. Various experiments using this simulator deal with the several parameters of Hodgkin-Huxley equations and will model resting and action potentials, voltage and current clamp, pharmacological effects of drugs that block specific channels etc. This lab complements some of the exercises in the Virtual Neurophysiology lab.

Population ecology

As part of population ecology virtual labs, we developed a set of mathematical ecology models to understand the basic dynamics and behavior of population in various aspects. Some models include:

- Exponential growth with continuous and discrete rate of growth. If a population has a constant birth rate through time and is never limited by food or disease, it has what is known as exponential growth. With exponential growth the birth rate alone controls how fast (or slow) the population grows. The objectives include the study the growth pattern of a population if there are no factors to limit its growth, to understand the various parameters of a population such as per capita rate of increase(r), per capita rate of birth (b) and per capita rate of death (d) and to understand how these parameters affect the rate of growth of a population. A case study on tiger population will indicate the applicability of exponential models as classroom tools.
- Leslie matrix is a discrete, age-structured model of population growth that is very popular in population ecology. It (also called the Leslie Model) is one of the best known ways to describe the growth of populations (and their projected age distribution), in which a population is closed to migration and where only one sex, usually the female, is considered. This is also used to model the changes in a population of organisms over a period of time. Leslie matrix is generally applied to populations with annual breeding cycle.
- Study of meta-populations using Levin's model shows a simple model to understand population changes. Meta population is a population

in which individuals are spatially distributed in a habitat to two or subpopulations. Populations of butterflies and coral-reef fishes are good examples of metapopulation. A virtual lab using Levin's model explains how to understand the basic concepts and dynamics of metapopulation and population stability with the help of mathematical models. In addition it is a study on how variations affect the population dynamics and how the initial number of patches occupied in a system affects the local extinction after a few years.

- Lotka-Volterra Predator Prey interactions (Wangersky, 1978) and logistic growth functions.

Biochemistry, cell biology, microbiology, immunology and molecular biology

Simple linear equations were used to understand molecular mass flow in AGE, PAGE exercises. No differential equations were used in biology oriented virtual labs where the focus was on the look and feel. In many cases, animation played a major role in these areas rather than mathematical simulations. As in the case of realistically animating experiments there are a lot of advantages; although it cannot be considered as a complete replacement of real labs due to its limitations. One solution was to provide the necessary details of the instruments we were using for the lab. Per say, if we use cooling centrifuges for an experiment in the virtual lab, one may not fully show all details corresponding to the operating methods of the centrifuge. But in the case of a real laboratory the student gets an opportunity to have a hands-on experience on the equipment while doing the experiment. Also, many of the experiments require instrumentation facilities. Also instruments from different companies have slight differences in design and operating mechanisms, which may not be shown in the virtual labs. Thus even though virtual lab meets the major target, it shadows the minor details of the experiments. Not all parameters such as changes in temperature during an experiment especially (where small changes do not matter) may not be included in the virtual lab for the sake of simplicity. In a real lab, curious students canperform these kinds of interesting experiments but to do the same in virtualized experiments is difficult.

MAJOR CHALLENGES

Setting and developing AMRITA virtual labs (see Fig. 2) as a complete learning experience has not been an easy task. Amongst the major challenges we faced included usage/design scalability, deliverability efficiency, network connectivity issues, security and speed of adaptability to incorporate and update changes into existing experiments.

Owing to the scientific domain, biotechnology lends the following challenges to establishing virtual labs:

- The development of analytical solutions in the arena is limited as biological processes are typically non-linear and are coupled systems of differential equations in various forms.
- The mathematics behind models is hidden by their complexity and appears refined through simulation platforms.
- Most simulation platforms need direct hands-on experience between teachers and students.
- The number of students that can be catered at any given time is restricted.
- Besides, such courses also need simultaneous theoretical explanations which may need classroom-like scenarios with video presentations, white-board and other tools. We could overcome the issue here using a collaborative suit, AVIEW (Bijlani et al., 2008).
- There are not many courses in India developed for this scenario.

In order to address some of these issues and to overcome restrictions, we deployed virtual lab experiments as web-client based animations or simulators besides remote triggered experiments. The virtual lab was based on a website that was designed for favorable use within intranets and internets. However, efficiency depended on the internet bandwidth and connectivity. Our target was any campus with a download link of 256kbps should suffice. To retain this compatibility the animations had to be size-delimited. To overcome the problem, longer experiments had to be sliced to smaller portions, each loading in sequence. This was possible as we maintained the virtual lab experiments as flash animations (Adobe, USA). Having labs in flash environments allowed the scalability and access although flash based action script programming needed additional programmers and training.

Other e-learning issues such as student-teacher collaboration via chat, video interfacing etc. were overcome via AVIEW-like environment (Bijlani et al, 2008). The intention of the virtual labs was also to extend the facility to develop an applied computational laboratory.

METHODOLOGY

Amongst others, the focus of having and designing virtual labs was also based on John Keller's ARCS model of motivation. Design of courses, simulations and models for computational approaches in biology will be the highlight. A lot of attention was on courses whose content will be applicable to the existing P.G. programs.

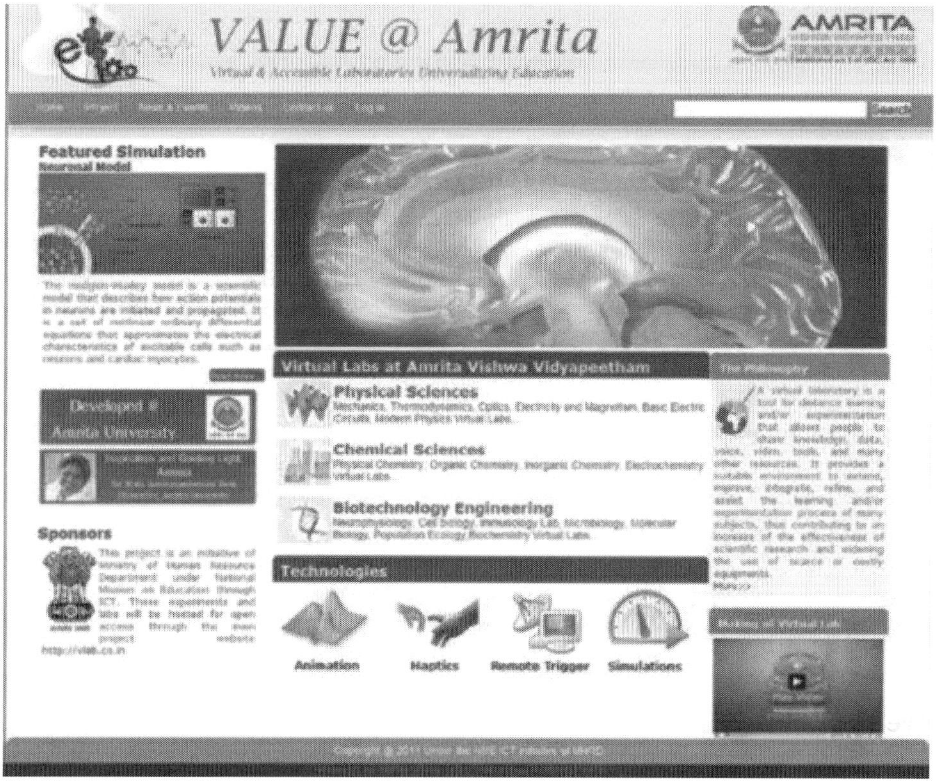

Fig. 1. Sakshat Amrita virtual labs. Accessible at http://amrita.edu/virtuallabs

For all biotech virtual labs, we had set the following lab-level objectives as general guidelines.

- Virtual labs should be adaptive. An adaptive e-learning system is a system in which modifies its behavior (the learning process) in response to the changes in the learners input data and information gathered from various teaching process. It should be able to incorporate data and user changes as and when possible.
- Introduce and focus virtual lab areas in core computational and protocol-based biotechnological sciences.
- To train young scientists in the field of the mathematical thinking for life sciences and related environments.
- To ensure that they will be able to integrate different exhaustive models into a larger framework, in the perspective of a comprehensive biological systems such as cells and biological networks.
- To give an overview of the modeling approaches most appropriate to describe life- science processes.

- To give a practical introduction to advanced but common-use simulation tools.
- To increase the awareness of a crucial need for standard model descriptions.

The implementation of animation and simulation based virtual labs was mainly done in Action Script 3 in Adobe flash in order to bring better definition to 2-D graphics. Action script allowed flash swf files as output thereby allowing both a better look-and-feel and an enhanced interactivity with the software. The physics simulator tools worked reasonably well. We did not use java as a programming medium in our learning tool to make sure we have complete cross-OS, cross-browser compatibility, to reduce initial loading time and also to consider support for the commercial operating systems such as Microsoft's Windows platform that support flash better than Java plug-in.

We used a new VLCOP platform (Nedungadi et al., 2011) in its full functionality for the virtual labs. The minor intention was to deploy preliminary platform with a learning environment and later render the environment adaptive and intelligent as per the user-audience. The main reason to precursor with such a test was cost-efficiency. Cost-efficiency of e-learning programs has been increasingly important because some institutions have failed due to the lack of well- thought out financial plans (Wentling et al, 2002; Morgan, 2000).

Virtual Labs use self-assessment based on questionnaire to evaluate user's experience. Although not implemented, an advanced form of the lab is being planned to include teacher's assessment, peer-assessment and collaborative assessment. Teacher assessment will actually have a "real" instructor on the deployment site to evaluate the lab user/student. Peer-assessment will include any student or teacher to assess another. Collaborative assessment will include both the instructor and the student to perform assessment on the completion of an experiment.

For our installation and deployment, we focused to reduce internet downtime. A 2004 study indicated that overall downtime costs companies an average of 3.6% of annual revenue (internet sources, see www.sentinelbussiness.it) indicating leading causes for downtime being software failure and human error. Through our studies, we managed to reduce unnecessary events and maintain downtime to less than 27 minutes for 6 months (not as in Amrita Learning software, see Table I in Diwakar et al., 2011). However, this could be because of our lack of full incorporation of the complex adaptive learning system as it was done for the schools where it was tested. However a test on real-time upgrade to such a model based on our previous experiences (data not shown) with Amrita learning (Nedungadi and Raman, 2010) indicated that overall loss of virtual lab in terms of downtime will be significantly less.

FEEDBACK AND ASSESSMENT

Feedback is usually not used as an evaluator but an assessment tool for student quality. With that in mind, the virtual lab evealuation criterion was focussed on measuring and estimating the student's involvement in the particular experiment of a particular lab. A way to increment the quantity and timing of feedback is to provide enough detail. Through animation, we have also increased evaluatory criterion and details in the virtual environments. It was noted that in more than 95 experiments performed by more than 30 people within a particular time-window there were more than 91% of appreciation (further statistics pending, data not shown) when two experiments, one with detail oriented interactive animation and other without interaction were delivered to assess the involvement of the students in terms of their self-assessment.

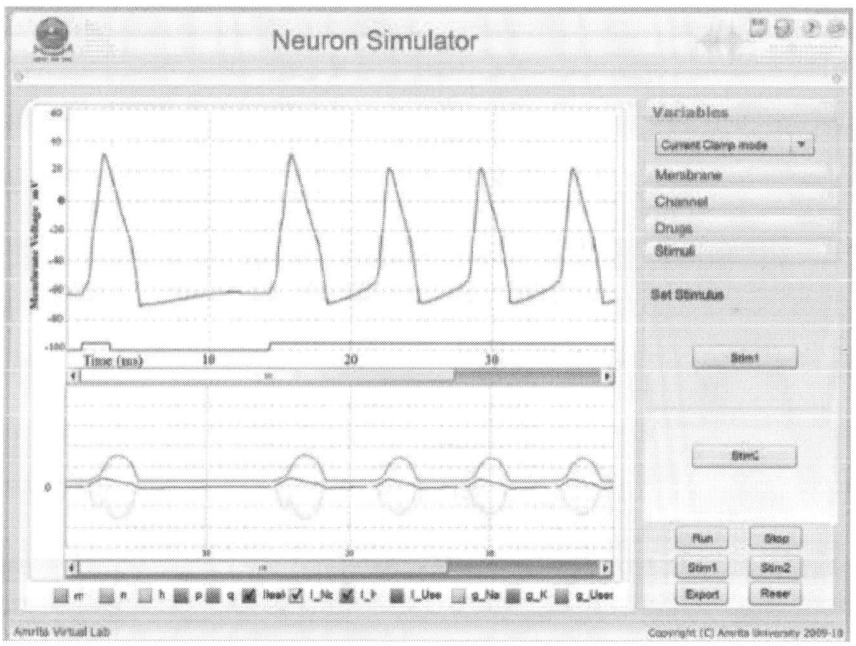

Fig. 2. Neuron simulator. The Neuron simulator lab uses Hodgkin-Huxley equations to study and analyze the action potential properties. The simulator allows some pharmacological studies and complements the neurophysiology virtual lab.

CASE STUDY: VIRTUAL NEUROPHYSIOLOGY LABORATORY

Our preliminary studies in the biotech sector were on neurophysiology techniques. The virtual neurophysiology laboratory provides an opportunity for students to substitute classroom physiology course with detailed techniques and protocols of a real laboratory. Besides the material like chemicals, physiology demands extensive knowledge and experience from the instructor. For example, rat brain slicing protocol which is the first experiment (in the virtual lab) takes approximately 6-10 hours to complete training and about 2-3 weeks to train one student in a real laboratory.

With the focus on time (Rohrig et al., 1999) and learning know-how, we adapted the usual lab experimental protocols as user-interactive animations of the neurophysiology lab experience. The work involved both animators and programmers. For some experiments such as brain slice preparation, animations were sufficient whereas for some others such as Hodgkin- Huxley neuronal model (Hodgkin et al., 1952, see Fig. 3.) for demonstrating behavior of single neurons, we used Java based simulator. The same simulator was embedded into other experiments such as voltage clamp protocol and current clamp protocol to allow the student to see the corresponding behavior as seen in real neurons (Koch, 1999).

A new set of experiments developed included the use of electronic resistance-capacitance (RC) circuits that could be remotely triggered as mimicking the electrical dynamics of a passive neuronal membrane. Passive neuronal membranes are modeled as RC-circuits with high resistance and low capacitance (for more details see Koch, 1999). In the simulation lab that was developed to complement the exercises of the VL, a model detailed study was added. Some of the main objectives and experiments using a neuron simulator included:

- Modeling action potential
- Modeling resting potential
- Modeling sodium ion channel and its effect on neural signaling
- Modeling delayed rectifier potassium channels
- Modeling passive membrane properties
- Current clamp protocol
- Voltage clamp protocol
- Understanding pharmacological implications of ionic currents
- Capacitive transients using Voltage Clamp
- Effect of temperature on neuronal dynamics
- Plotting F-vs-I curve
- Plotting V-vs-I curve

Also as part of the labs, we follow a particular formatting for each experiment within the lab. The goal was to allow the student to study the theory, the approach and do a self-test before actually going into the simulator or the virtual experiment. Covering some explanations and incorporating the same theory into the actual "lab" part of the experiment has been one of the primary goals. Each experiment in the labs (especially in Biotechnology) opens by default with the textual theory, which can also be randomly accessed by clicking on the icon "theory".

All the control and experimental parameters are explained in the "manual". The instructor and the student are informed on how various parameters change in the experiment in the very context of the virtual experimental lab procedure. For those experiments that have both an animation learning component and simulator component, each of the user controls and the variable parameters are explained. Also included in the manual is a help that actually explains the usage of radio controls and icons covered by the experiment. The intention was to evaluate the basic info that once the student completes the familiarization process by going through the theory and manual sections, he/she can take a "self-evaluatory" quiz module that chooses to test the student on some questions based on the theory background of the experiment.

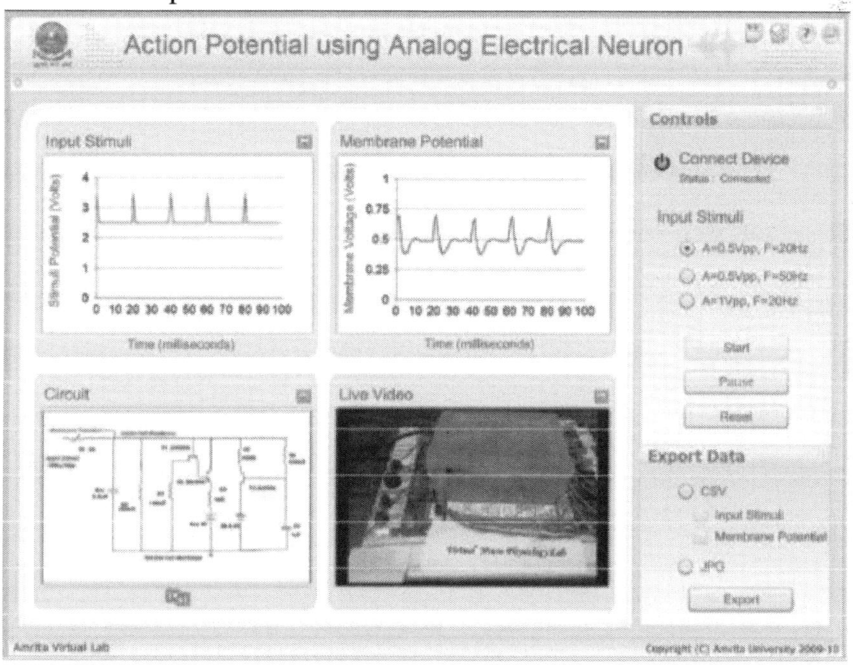

Fig. 3. Remote-triggered Experiment. Remote panel is also made with re-configurable panels and control options. This experiment emulates action potential generation using analog neurons.

The "simulator" tab actually leads to the experiment workbench. "Protocol for brain slicing" that is actually a detailed lab process that would take 6-10 weeks for post-master's student to learn and about 3-10 hours per procedure. That experiment we have virtualized by means of an interactive action script based animation. The second neurophysiology experiment concerns the modeling of a neuronal cell. In this case we have used a Flash based learning component along with a HH-simulator of a biophysical neuron.

The "assignment" icon is the lab experiment question with which intention the student performs the experiment. An instructor version of the assignment will include a modelsolved question or key tips in case of a protocol-like experiment. Additional reading material and reference information and other details will be found in our "misc info" icon.

Among the various methodologies the lab covers simulation-based, animation-based and remote-triggered experiments. The simulator was that of a bio-realistic model cell and was combined with an interactive animation-based learning-tool made using Flash. Maldarelli et al. (2009) report the advantage of virtual lab demonstration as an effective lab tool. The remote-triggered experiments were based on real electronic circuits that mimic the phenomenon observed in neuronal cells. The basic behavior of Resistance- capacitance circuits that can be modified remotely by a user to study and imitate real neuronal circuits as he/she does in a neuronal biophysics laboratory on a patch of a neuronal membrane.

We also tested the virtual lab via a questionnaire-based feedback for overall quality. Among the major questions, several virtualizations related questions were presented in the questionnaire. The general developer/designer related questions included in the lab were to rate the experiment that was most recently completed, extent of control on the interface, closeness to lab environment and feel, measurement and analysis of data, user-manual quality, adequacy of bibliography and references, results interpretation, whether any clear information was gained by using the virtual lab, any problems faced, how helpful the lab was and overall motivation.

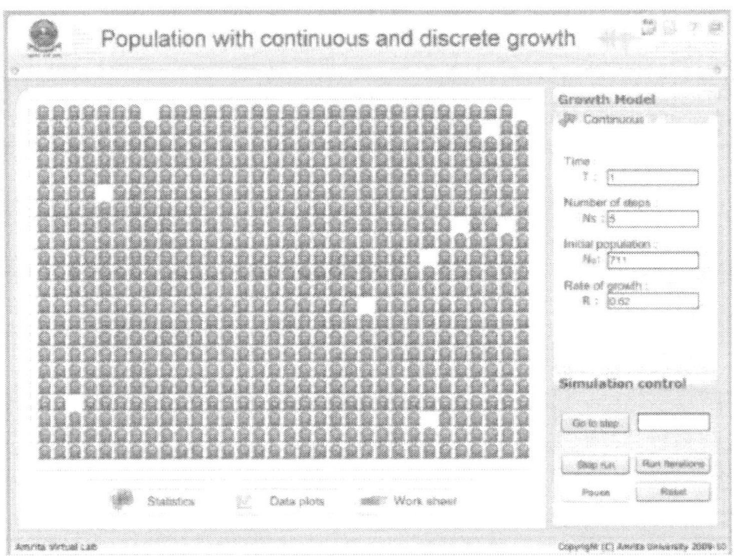

Fig. 4. Tiger population study. Using exponential growth patterns to predict tiger population in India.

TAKING PROJECT TIGER TO THE CLASSROOM: A VIRTUAL LAB CASE STUDY

Using a virtual lab was not our only objective. We wanted to test a real scenario and see if the virtual lab could be used a as research tool. Tiger population study uses virtual labs to take India's Project tiger to the classroom. Half of the tiger population in the world is in India. Due to reduction in their population in large numbers, from 1969 onwards the 'tiger' was declared as an endangered species (by CITES). Educating about tiger populations is vital. Typically courses in population ecology deal about population variations. In this section, we suggest on the applicative use of population ecology simulators as classroom models to complete the learning experience for a population ecology laboratory course. This section also reports the analysis, interpretation and some preliminary predictions in variations of tiger population in India.

Here we used the exponential growth model experiment in population ecology lab 1 (at http://amrita.edu/virtuallabs). First step is to select an experiment followed by selecting amathematical model which is described in the experiment from the set of experiments (see Fig. 4). Some existing data tested indicated the validation of the technique. (Fig 5B).

Data collection

Statistical data for this study was collected from Project Tiger which includes the tiger population from 1972 – 2002 of various tiger reserves (see Table 1). And the second data set was the crime reported for the numbers of tiger that have been killed in past few years were from WPSI's Wildlife Crime Database (14. WPSI's Tiger Poaching Statistics, http://www.wpsi-india.org/statistics/index.php).

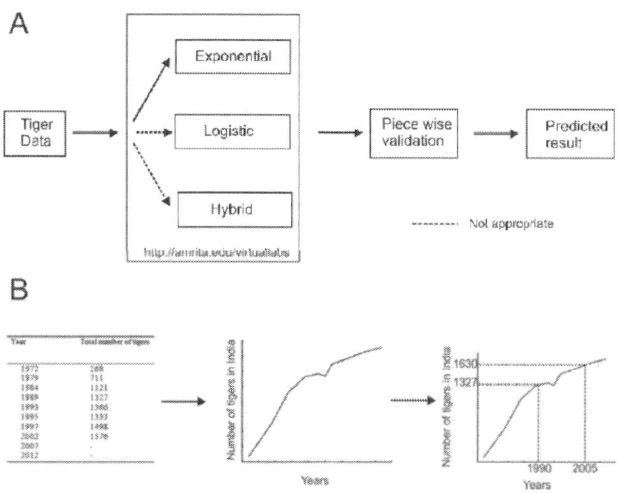

Fig. 5. Virtual Lab model for tiger population study. Note that exponential model was chosen based on decline in populations. Predictions with other models such as Lotka- Volterra and Logistic growth were inappropriate or had errors.

Growth rate has been calculated by using the formula, Growth rate $g(t) = \frac{(t+1)-t}{N_{(t+1)}}$

$N_{(t1)}$ is the total number of individuals at t+1, t′ represent the time in years.

On-screen methods

We have used an adaptive growth rate for different periods as shown in Table 1. Simulator's viewable window contain three main tabs, 1) Statistics button will show the growth of population while the simulation is running 2) Data plots button, will Population size Vs Time, 3) Worksheet button is an implementation of the model in excel.

Assumptions with tiger populations and growth model

Population ecology models include several assumptions, in order to realistically apply the model on data. Growth of prey population is exponential in absence of predators;

- Tiger population grows/declines exponentially within a short duration (10 years)
- The rate of change of tiger population is proportional to its size.
- During the process, the environment does not change in favour of one species and the genetic adaptation is sufficiently slow.

Although the assumptions make it difficult to actually call the simulation 'realistic', the validations showed a realistic trend and hence for this model, a simple exponential growth simulator was used.

Table 1. Shows the statistical data for tiger population from 1972 – 2002 and extended (prediction column) the curve to 2012 using continuous growth model simulator (data collected from Tiger project India at http://projecttiger.nic.in/populationinstate.asp).

Year	Total number of tigers	Total number of tigers predicted by the model	Growth rate
1972	268	-	-
1979	711	-	0.6230
1984	1121	-	0.3657
1989	1327	-	0.1552
1993	1366	-	0.0285
1995	1333	-	-0.0247
1997	1498	1521	0.1236
2002	1576	1586	0.0409
2007	-	1664	0.0468
2012	-	1718	0.0314

Calculated average growth rate was = 0.1545

RESULTS

Taking real data to class rooms have been very difficult with population ecology due to its high unpredictability and model-related unreliability. However, using a simple growth rate simulator and using patterns from a short period, the model shows promise.

The simulation showed predictability in the growth of tiger population in India for the years 2003- 2012 by extending the behavioral pattern of tiger population in India for last 30 years (see Fig.6). Predicted tiger population for the year 2007 was 1664. According to National

Tiger Conservation Authority (on 2008), the total tiger population reaches 1,411 (i.e. ranging between 1,165 and 1,657). The difference in number of tigers from predicted to this statistical report may be because of some environmental factors, number of tigers that have been killed in past few years and the census by National Tiger Conservation Authority was only partially included West Bengal.

SUMMARY

The predicted data (Table.1, third column) for tiger population in India showed standard deviation of 10% from real data. With some assumptions, it was possible to use simple models like exponential growth models for studying tiger populations. For a very short duration (such as in the data shown in Table 1), basic growth show a slowly saturating exponential and hence data matched the predictions (see Fig. 6). Online population ecology experiments developed on the basis of mathematical equations could help students to get a deeper understanding on model dynamics by exploring the parameter space provided by the model. Also it is always feasible for the user to supply the real data as input and observe the corresponding dynamics. The possibility to study such experiments has value. Biotechnology studies often include data collection and such models allow building simple hypothesis based on the dynamics. This new e-learning environment engaged and motivated the students to practice and explore the parametric space provided for the population ecology experiments.

Newer studies for analyzing fish populations and deer populations are being developed as part of the ongoing process. Such data will be made available as a virtual lab for continued use and study. We also noticed that the undergraduate and postgraduate students show an increased attention to details when we trained them on virtual labs instead of plainly explaining the theory. There was a 23% (metric not shown) improvement in interest to critically analyze population models among students who were introduced to population ecology studies directly virtual labs.

COST OF VIRTUAL LABS

In order to estimate the true financial cost of our virtual lab project, we had to include both project development and delivery and maintenance costs. As indicated by Kruse (see http://www.e-learningguru.com/articles/art5_2.htm), design of courseware needed more initial costs than instructor-led learning but delivery and maintenance is affordably cheaper. We estimate, based on Amrita learning software experience that there will be negligible costs for maintain web-based experiments. The main post-deployment costs included administration and maintenance. The administration and maintenance estimates included tracking of user-behavior, technical support, content updates and technology updates. Student material development, instructor costs and subject expert costs were included in the development expenses.

SOME EVALUATORY SETBACKS AND ASSOCIATED FEEDBACK

What we know from the Virtual Lab studies performed is that user-involvement in assessment is vital for improving the knowledge-experience for the user. Self-assessment hints preliminary results but are not comprehensive. Users tend to show implicit behavior patterns indicating favor of the tool rather than the experiment for their choice of vote. Interactional voting behavior is also dependent on age and other characteristic learner attitudes. In our studies, younger students mostly at the undergraduate level evaluated the tool using mid-range scores compared to the varying yet favorably high votes of the Master's and Graduate students in the feedback assessment. Although this may need further testing, we believe that scores from the higher age-experience level indicated statistically relevant reliability much more than undergraduates (data not shown due to pending experiments).

Overall, 27 Master's level students who helped in intensive evaluation of the Virtual lab platform as part of their regular class-room course, appeared predominantly positive about the value of virtual labs in e-learning, but anxieties were also expressed about the potential for e-labs to replace face-to-face teaching and labs in the economically challenged regions of the country apart an indication either of the value on the personal and face-to-face tutoring through an expressed preference for it. Students who were positive about their experiences of virtual labs indicated that they had received appropriate introductions and felt supported by staff, indicating the importance of sound inductions into the use of institutional systems and technologies.

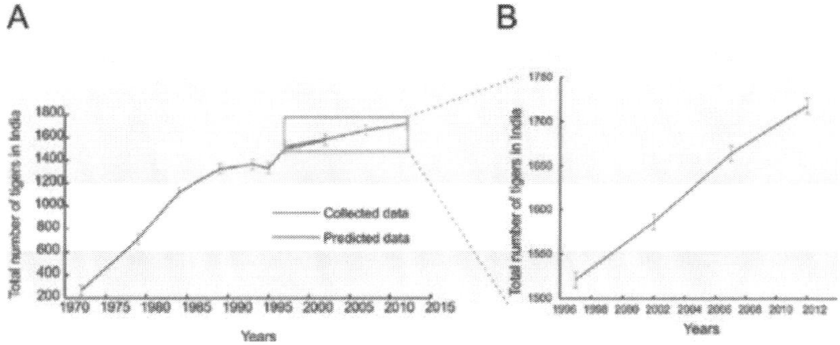

Fig. 6. Growth of tigers with predictions. A. The plot shows the nature / pattern of statistical data for tiger population in India from 1972 – 2002 (blue line) and an extended prediction (red line) of the curve to 2012 using continuous growth model simulator. The model assumes a 10% standard deviation shown by the error bar. B. The plot shows the enlarged curve of predicted piece-wise continuous growth of tiger populations in India

AT A GLANCE

Education using VL has been the new venture to better education and provide extensible laboratory experience to University students. The virtual lab protocols for neurophysiology and related sciences have been a successful complement to the usual theoretical education that happens at our school of biotechnology at the level of masters and undergraduate education. Although the elements can be improved, our approach to virtualization has answered many key results in establishing the virtual lab features such as teacher- independent/teacher-friendly approach to e-learning.

We tried to avoid the most usual failures in e-learning labs (Romiszowski, 2004) by focusing to avoid the common failures. Our design issues were based on a successfully tested e-learning software environment (Nedungadi and Raman, 2010) and included a clear identification and analysis of the real problem associated with University laboratory courses. Each virtual lab included overall strategic design decision such as structure of the courses, technologies employed and mode of experiment. Each experiment and the lab was considered with instructional design and elements that were evaluated so to motivate the learner experience. Such elements included the choice of graphical front-ends and authoring tools. We had also previously estimated issues related to dissemination for rapid, efficient and cost-effective usability taking into consideration both pedagogical and infrastructural complicacy.

Large scale tests will be needed to analyse and provide the assessment. These tests will also require both learners and educators (lab faculty) to use the software platform. Some tests in biotechnology are already underway via the VALUE initiative (Diwakar et al. 2011).

Several users raised the issue of how to support learners using VL. In real-world labs, learners work in the same place at the same time so there is teacher or peer support available. This kind of support is not immediately available to remote learners.

From our experience, the most vital requirement for each virtual lab is that of technical coordinators and subject matter experts whose inputs improve the lab's knowledge bank and usability. The Virtual lab project is already online for public preview via http://amrita.edu/virtuallabs or the National mission site http://vlab.co.in

REFERENCES

1. Auer M., Pester A., Ursutiu, D., Samoila C., Distributed virtual and remote labs in engineering, IEEE International Conference on Industrial Technology, Vol. 2, 2003, pp. 1208-1213.
2. Aycock J., Crawford H., deGraaf R., Innovation and technology in computer science education (Proceedings of the 13th Annual Conference on Innovation and Technology in Computer Science Education, Madrid(Spain), 2008, pp. 142-147.
3. Bijlani K., Manoj P., Rangan V., VIEW: A Framework for Interactive eLearning in a Virtual World, Proceedings of the Workshop on E-Learning for Business Needs 2008/BIS, Innsbruck(Austria),2008, pp. 177-187.
4. Hodgkin A.L., Huxley A.F., A quantitative description of membrane current and its application to conduction and excitation in nerve. Bull Math Biol, 1990, Vol 52, pp. 25-71.
5. O'donoghue J., Singh G., Dorward L., Virtual education in universities: A technological imperative, British Journal of Educational Psychology, 32(5), 2001, pp.511-523.
6. Rohrig C., Jochheim A., The Virtual Lab for controlling real experiments via Internet, Proceedings of the 1999 IEEE International Symposium on Computer Aided Control System Design, 1999, pp. 279 – 284.
7. Nedungadi P and Raman R. Effectiveness of Adaptive Learning with Interactive Animations and Simulations. Proceddings of the 2nd International Conference on Computer Engineering and Applications (ICCEA 2010), Bali, Indonesia, March 2010.
8. Wentling T and Park J. Cost Analysis of E-learning: A Case Study of A University Program, Proceedings of the AHRD, University of Illinois at Urbana-Champaign, p.1-11, 2002.

9. Morgan BM. Is distance learning worth it? Helping to determine the costs of online courses. Eric number 446611, 2000.

10. Maldarelli GA, Hartmann EM, Cummings PJ, Horner RD, Obom KM, Shingles R and Pearlman RS. Journal of microbiology & biology education, pp. 51-57, May 2009.

11. Romiszowski A. How's the E-learning Baby? Factors Leading to Success or Failure of an Educational Technology Innovation Educational Technology, 44(1), Jan-Feb 5-27, 2004.

12. Koch, C. Biophysics of Computation: Information Processing in Single Neurons,Stryker, M. (ed.) Oxford University Press, 1999.

13. Wenger E. Artificial Intelligence and Tutoring Systems: Computational and Cognitive approaches to the communication of knowledge. Morgan Kaufman Ed., 1987.

14. Diwakar S, Achuthan K, Nedungadi P and Nair B. Enhanced Facilitation of Biotechnology Education in Developing Nations via Virtual Labs: Analysis, Implementation and Case-studies, International Journal of Computer Theory and Engineering vol. 3, no. 1, pp. 1-8, 2011.

15. Nedungadi P, Raman R, Achuthan K, Diwakar S. Collaborative & Accessibility Platform for Distributed Virtual Labs, Proceedings of 2011 IAJC-ASEE International Conference, University of Hartford, NY, USA, April 29-30, 2011.

16. Wangersky P. Lotka-Volterra poulation models, Ann. Rev. Ecol Systems, 1978, 9:189-218.

CHAPTER-17

BIOTECHNOLOGY ATTITUDES: THE INTERSECTION OF GENDER, KNOWLEDGE, AND SCIENTIFIC EXPERTISE IN SHAPING PERCEPTIONS THROUGH TECHNOLOGY AWARENESS

INTRODUCTION: AN INTRODUCTION

This study investigates the relationship between gender, knowledge, scientific expertise, and public attitudes toward biotechnology. It emphasizes the importance of understanding technological salience and how knowledge is harnessed to form and influence attitudes. By delving into these factors, the research seeks to uncover the complex interplay between gender-based perspectives, individual knowledge levels, and the impact of scientific expertise on how people perceive and interact with biotechnological advancements. The findings of this study can shed light on the societal dynamics that affect the acceptance and apprehension of biotechnology, contributing to a more comprehensive understanding of public perceptions in this rapidly evolving field.

Since the advent of recombinant DNA techniques in the early 1970s, biotechnologies have received much attention both for their potential to help and to harm individuals and society. Through the development of gene therapies, stem cell technologies, reproductive technologies and genetically modified crops, biotechnologies have promised to help the sick, the barren, and the poor live more fulfilling lives. At the same time governments, the general public, and scientists themselves have recognized the risks involved with biotechnologies. Since the very beginning of modern biotechnological techniques, scientists have warned that there is "serious concern that some of these artificial recombinant DNA molecules could prove biologically hazardous" (Berg et al., 1974: 303). Risks include the unknown consequences of consuming organisms that have been genetically manipulated, and the release of novel genetic material into wild populations (Torgersen et al., 2002). But aside from the technical risks involved with biotechnologies, concerns have also arisen over the ethical implications of manipulating basic life processes. Critics have rebuked scientists for "taking a technological and reductionist perception of life itself" (Torgersen et al., 2002: 39), and the seemingly imminent development of human cloning and "designer babies" (McGee, 2000; Hughes, 2004) were added to the already controversial issues of embryonic stem cell research and genetically modified organisms. The potential to simultaneously do great help, and great harm, has made biotechnology one of the most important science policy issues of our time.

Often the views of scientists, industry, government, and the public have been at odds. Because of their enormous commercial potential, industry has largely downplayed the risks and emphasized the benefits of biotechnologies (see, for example, Rampton's and Stauber's (1998) exposé of Monsanto's attempts to suppress evidence that one of its products, recombinant bovine growth hormone, causes cancer). National governments, recognizing the importance of biotechnology as a key technology in the post-industrial marketplace, have simultaneously sought to protect industry from an overzealous public, yet appear responsive to public concerns. The general public has been especially critical ofbiotechnologies, though the level of skepticism varies by region and the particular issue at stake (Allum et al., 2002).

In the case of biotechnology, public concern has turned out to have significant influence. In the arenas of politics and economy, public concerns about biotechnology have translated into real transformations. Bauer and Gaskell observe that,In contemporary times, public opinion is not merely a perspective "after the fact"; it is a crucial constraint, in the dual sense of the limitations and opportunities for governments and industries to exploit the new technology. Whereas the biotechnology industry assumed that regulatory processes were the sole hurdle prior to commercialisation, it is now apparent that a second hurdle, national and international public opinion, must be taken into account (2002: 1).

Both by voting with their dollars, and by making it difficult for biotech firms to bring products to market (Weber et al., 2009), "public opinion" has affected the development and distribution of biotechnologies (for a more extended discussion of the efficacy of public opinion see Page and Shapiro, 1992). Furthermore, governmental bodies at the national level, as well as the European Union, have heard concerns and adjusted policy accordingly (Torgersen et al., 2002). Clearly, if the future of biotechnology is to be grasped, public opinion, and the processes by which it is formed, must be understood.

One issue that is frequently referred to when considering the public's attitudes is knowledge of biotechnology. Despite initial hesitance, "an increasing number of scientists concluded that the risks had been exaggerated" (Torgersen et al., 2002: 34). Yet public attitudes have become more negative, and more ambivalent, over time (Shanahan et al., 2001). Despite some studies that suggest a link between genetically modified organisms and health problems (e.g., a recent study that found a connection between the consumption of genetically modified corn and organ failure in rats; Vendômois et al., 2009), scientists have largely dismissed public outcry over biotechnologies as reactionary and a consequence of a public that remains uninformed (see, for example, McHughen, 2007). This very basic hypothesis – that people will become more accepting of biotechnology once they better understand it – has received some support in studies of public opinion (Allum et al., 2008; Bak, 2001; Wright & Nerlich, 2006). However, not all people use knowledge in the same way. The *contexts* in which knowledge is called forth have been shown to play a crucial role in how it is used to formulate opinions toward science and technology. Studies that test for interaction effects show that the effect of knowledge on attitudes varies by region (Allum et al., 2002), levels of political knowledge (Sturgis & Allum, 2004), and by gender (von Roten, 2004; Simon, 2010, 2011).

Gender differences in the use of knowledge to formulate attitudes are particularly intriguing because of the persistent gender gap in support for science and technology. Whether asked about environmental issues (Stern et al., 1993; Hayes, 2001; McCright, 2010), nuclear power (Freudenburg & Davidson, 2007; Davidson & Freudenburg, 1996; Krannich & Albrecht, 1995), biotechnologies (Qin & Brown, 2007; Bryant & Pini, 2006; Simon 2010, 2011) including reproductive technologies (Napolitano & Ogunseitan, 1999), or science in general (Trankina, 1993; Barke et al., 1997; von Roten, 2004; Mallow et al., 2010; Breakwell & Robertson 2001; Hayes & Tariq 2000), women express more skepticism toward science and technology than men do. Gender is an important determinant of people's attitudes toward science andtechnology, and the fact that there is some evidence that indicates that men and women use knowledge differently to form attitudes suggests this is of critical importance for understanding public attitudes toward science and technology.

In a previous study of gender differences in attitudes toward biotechnology (Simon, 2010), I found that the more knowledge that men had on the subject, the less inclined they were to be pessimistic about its effects on society. The effect I found for men clearly supports the knowledge deficit hypothesis: the more men know, the more likely they are to support biotechnology. However, the effect of knowledge on attitudes proved to be the *opposite* for women: women became more *critical* of biotechnology with more knowledge. The results of that study suggest scientific knowledge persuades men to be supportive of biotechnology, but that same knowledge causes women to be critical of it.

What is causing men and women to use knowledge in such radically different ways? One hypothesis is that science and technology, and biotechnologies in particular, have radically different implications for men and women. Because of gender roles associated with childbearing and childrearing, women are more likely to be directly affected by some of the negative consequences of science and technology. As Dorothy Nelkin suggests regarding attitudes toward nuclear power, women are its "most active and outspoken critics" because of "the special effects radiation has on the health of women and on future generations" (1981: 15). This perspective is echoed by ecofeminists such as Mallory (2006) who argue for a parallel between the exploitation of nature characteristic of science and technology and the exploitation of women, and Bryant and Pini who theorize "a relationship between chemicals and women's reproductive bodies" (2006: 268). Biotechnologies may be especially relevant to women because the implications for bodily contact and harm are an integral element of the controversies surrounding biotechnology (DuPuis, 2000). With respect to gender differences in attitudes toward biotechnology, Napolotano and Ogunseitan remark, "many of the applications [of biotechnologies] towards human health issues will likely affect fetuses, mothers, and young children more than adult males and non-childbearing female members of society" (1999: 202).

The implication of these arguments is that biotechnology is more *salient* for women than it is for men. I submit that the gender difference in *technological salience* produces the circumstances that facilitate a unique set of evaluative criteria for women that are not utilized by those with more distance from biotechnology. The hypothesis is that, for women, biotechnology is *personal* in a way that it is not for men, and the more women understand about biotechnology the better they are able to grasp this.

The technological salience explanation of gender differences in attitudes toward science and technology (developed with my colleague Katherine Johnson; Johnson and Simon, n.d.) suggests certain hypotheses. The theory states that the criteria that people will use to form attitudes toward science and technology will be determined by how salient the particular issue is to their own lived experiences. The hypotheses I intend to explore have to do with how men and women use knowledge differently with respect to their views on the public's role in science policy. It is intuitive that when people think the public should decide science policy, their own knowledge of a topic should be an important determinant of their attitude toward it. However, when people think experts should decide science policy, it is unclear how knowledge might relate to attitudes toward biotechnology. If biotechnology is really a personal matter for women, a matter relevant to their bodies, values, and roles in a way it isnot for men, then knowledge of it should be absolutely crucial for determining their attitudes under nearly all circumstances. Even when women are willing to abdicate responsibility to scientific experts or government to regulate other forms of science and technology, because biotechnology has an especial salience for women, they should be unlikely to disregard their own knowledge when evaluating it. On the other hand, I predict that men's use of knowledge to generate attitudes will be conditional upon their willingness to let scientific experts and government regulate science and technology. Because biotechnology is not especially salient for men compared to women, they should categorize it alongside other issues in science and technology that are not especially salient for them. When men abdicate responsibility for science policy to experts, their own knowledge of biotechnology should be less important in determining their attitudes compared to men who think science policy should be decided by the public.

In technical terms, I am hypothesizing a three-way interaction of gender, knowledge, and willingness to leave science policy to experts in determining attitudes toward biotechnology. For women, the effect of their knowledge on their attitudes should be the same no matter what their opinions on the role of experts are (Hypothesis 1), because even if they are generally willing to let experts decide science policy, the especial salience of biotechnology for women will make their understanding of it relevant under any circumstances. But because biotechnology is not especially salient for men, when the responsibility for determining a positive or negative attitude is externalized to experts, the importance of one's own knowledge about the topic should be less important because responsibility is transferred to an external source (Hypothesis 2). Because of the particular salience of biotechnology for women, the effect of their knowledge on their attitudes should be unconditional. Because biotechnology is less salient for men, the effect of their knowledge on their attitudes should be more likely to be affected by other circumstances; in particular, it should become less relevant when they believe the public should stay out of scientific affairs. Stated formally:

H1: Controlling for other relevant variables, the effect of knowledge on attitudes toward biotechnology will be *unconditional* upon the opinion that experts should decide science policy issues for females (i.e., the knowledge-experts interaction will *not* significantly predict attitudes for females).

H2: Controlling for other relevant variables, the effect of knowledge on attitudes toward biotechnology will be weaker when males hold the opinion that experts should decide science policy issues, and stronger when males hold the opinion that the public should decide science policy issues. (i.e., the knowledge-experts interaction *will* significantly predict attitudes for males, and it will be negative).

DATA AND METHODS

Sample

The Eurobarometer 63.1 is used to test these hypotheses because it features items on attitudes toward biotechnology, the role of experts in scientific decision-making, and a twelve-question quiz designed to measure scientific knowledge. The Eurobarometer 63.1 is a representative sample of thirty-two European nations and includes household respondents aged 15 and over. These nations include 25 European Union member countries: Austria, Belgium, Republic of Cyprus, Czech Republic, Denmark, Estonia, Finland, France, Germany, Greece, Hungary, Ireland, Italy, Latvia, Lithuania, Luxembourg, Malta, Netherlands, Poland, Portugal, Slovakia, Slovenia, Spain, Sweden, and the United Kingdom, plus the EU candidate countries of Bulgaria, Croatia, Romania, Turkey, as well as three European Free Trade Association countries: Iceland, Norway, and Switzerland (Papacostas, 2006).

Data for the Eurobarometer 63.1 were collected between the 3rd of January and the 15th of February 2005 by TNS Opinion and Social on behalf of the European Commission. It utilized a probabilistic, stratified sample design in which each nation was stratified first by region, then municipality. In each country, a number of sampling points were drawn with probability proportional to population size and to population density. Finally, households were chosen within each municipality. The survey was administered at the homes of the respondents in face-to-face interviews (Papacostas, 2006).

Dependent variable

The purpose of this study is to investigate gender differences in attitudes toward biotechnology. Hence, the dependent variable is derived from a question asking respondents whether they think biotechnology "will have a positive, negative, or no effect on our way of life over the next twenty years." This variable was recoded so that all respondents indicating that they think biotechnology will have a negative effect "on our way of life" were coded as 1 and all other respondents were coded as 0. The dependent variable is therefore a measure of *pessimism* about biotechnology.

Independent variables

Independent variables of interest are gender, scientific knowledge, and a variable derived from the question, "Which of the two following views is closest to your own? Decisions about science and technology should be based primarily on the advice of experts about risks and benefits involved, or on the general public's views of risks and benefits?" All respondents who indicated that decisions should be left to experts were coded as 1 and all respondents who indicated that the general public should make decisions were coded as 0.

To create the gender variable used in the analysis, all males were coded 1, and all females were coded 0.

The knowledge variable is a scale constructed from a twelve question factual test on scientific knowledge administered with the Eurobarometer. This test asked questions about science in general, not about biotechnology specifically. The items were recoded so that correct responses were equal to 1, and incorrect responses were equal to 0. Each respondent's score on the knowledge variable is equal to the proportion of correct answers they made on the knowledge test (e.g., someone who earned eight correct answers would be coded as .67). The knowledge scale obtained an alpha = .72.

Control variables included variables capturing political attitudes, age, education, and religiosity. The political attitudes variable was derived from responses to the question: "In political matters people talk of 'the left' and 'the right'. How would you place your views on this scale?" Respondents were asked to rate their political attitudes on a scale of 1 to 10, withlower values indicating that they are more "left", and higher values indicating that they are more to the "right". Respondents' age was coded in years. The Eurobarometer's education item does not directly measure respondents' educational attainment; instead respondents are asked to give the age at which they left school. This variable was recoded into dummies that roughly correspond to stages in the educational career: respondents who left school before high school age ("Edu. 0-14"), those of high school age ("Edu. 15-18"), those of college age but who have not reached the typical age of college graduation ("Edu. 19-21"), and those who left school after the age of 22 ("Edu. 22+"), assumedly college graduates. Another dummy was created for respondents who had not yet left school ("still studying"). This education measure is not ideal, but previous work (Simon, 2010, 2011) with the Eurobarometer's education measure has showed that it possesses criterion validity when predicting attitudes toward biotechnology. Respondents' religiosity was measured with an ordinal item asking respondents, "Apart from weddings or funerals, about how often do you attend religious services?", with responses ranging from "never" to "more than once a week". Higher values indicate more frequent attendance of religious services, and lower values indicate less frequent attendance.

Analytic sample

The total number of respondents included in the Eurobarometer 63.1 is 31,390. Missing values were addressed using pairwise deletion. After deleting missing values the sample was reduced to 26,621. Because of the complex sampling design, the data were weighted with a weight designed to accommodate analyses that make use of all the nations in the study. Weighting the data further reduced the analytic sample to 24,630. Splitting the sample by gender resulted in an $n = 12{,}379$ for males, and an $n = 12{,}251$ for females.

Analytic approach

Hypothesis 1 predicts that the experts-knowledge interaction will not be significant in the female model. Hypothesis 2 predicts that the experts-knowledge interaction will be significant and negative in the male model. To test the hypotheses, the sample was split by gender, and separate logistic regression models predicting attitudes toward biotechnology (with "it will have a negative effect on our way of life" coded as "1") were run for males and females, including a test of the "experts-knowledge" interaction in both models.

RESULTS

Table 1 shows means (or proportions) and standard deviations for all variables by gender. A few items are of note. First, while slightly less than 20 percent of males thought biotechnology will have a negative effect "on our way of life over the next twenty years", slightly more than 22 percent of females had the same opinion. While the gender difference in attitudes is slight, it is consistent with previous research that has shown females to be more skeptical of biotechnology than males. Also consistent with previous research, females possess less scientific knowledge than males. (averaging .59 on the knowledge test vs. .66, respectively, a difference of nearly one entire correct answer). Approximately three-quarters of both males and females believe that experts, not the general public, should decide science policy issues.

Table 1. Mean or proportion and standard deviation by gender.

	Males		Females	
	Mean or Proportion	Std. D.	Mean or Proportion	Std. D.
Pessimism	.20	.40	.22	.41
Knowledge	.66	.22	.59	.23
Experts Decide	.74	.44	.73	.45
Politics	5.30	2.23	5.22	2.08
Age	43.33	17.82	45.60	18.64
Edu 0-14	.18	.39	.23	.42
Edu 15-18	.38	.49	.38	.49
Edu 19-21	.15	.36	.16	.36
Edu 22+	.17	.38	.14	.35
Edu Still Studying	.12	.32	.10	.29
Religiosity	3.53	2.33	4.01	2.36

Males: $n = 12{,}379$
Females: $n = 12{,}251$

Table 2 features results of a series of logistic regressions predicting attitude toward biotechnology for females only. Odds ratios are shown. Model 1 features only the knowledge scale as an independent variable; it is positive and significant. Without controlling for any other variables, each additional correct answer on the knowledge test increases the likelihood of being pessimistic about biotechnology by about 33 percent. Model 2 adds the experts variable as a predictor; it is negative and significant. Controlling for level of scientific knowledge, feeling that experts and not the public should be responsible for science policy decreases the odds of being pessimistic toward biotechnology by about 31 percent. Model 3 retains the knowledge and experts variables as predictors, but adds the knowledge-experts interaction term. Recall that Hypothesis 1 predicted that, for females, the effect of knowledge on attitudes toward biotechnology would be about the same regardless of the respondent's opinion about the role experts should play in science policy issues. Model 3 reveals that the effect of knowledge on attitude is not affected by opinion about the role of experts in science policy. Model 4 shows that this is true even when controlling for political attitude, age, religiosity, and education, confirming Hypothesis 1.

Table 3 shows logistic regression results for the male sample only. Model 1 uses only the knowledge scale to predict pessimism toward biotechnology. Without accounting for any other variables, each additional correct answer on the knowledge test results a reduction in the odds of being pessimistic about biotechnology of about 20 percent for males. However, adding the experts variable reduces the effect of knowledge to insignificance. Like their female counterparts, when males leave policy decisions up to experts, they are much less likely to oppose biotechnology. Model 3 adds the interaction term. The effect is strong, negative, and highly significant. The correlation between knowledge and attitudeis much weaker for males who think science policy should be decided by experts compared to males who think it should be decided by the public. Model 4 includes the control variables. The experts-knowledge interaction effect on attitude is slightly reduced when controlling for political ideology, age, religiosity, and education, though it is still strong, negative, and significant, confirming Hypothesis 2. This is a much different scenario than for females, for which the experts-knowledge interaction did not predict biotechnology attitude.

To further illustrate how opinion on the role of experts in science policy tempers the effect of knowledge on biotechnology attitude for males but not for females, each male and female sample was further split by whether the respondent thought science policy decisions should be made by experts or by the public. Table 4 shows results of logistic regressions predicting biotechnology attitude by gender and science policy opinion. Model 1 displays the coefficients only for females who believe that science policy decisions should be left to experts. With all of the control variables in the equation, each additional correct answer on the knowledge test results in a 53 percent increase in the odds of being pessimistic about biotechnology. Model 2 shows results only for females who think the public should decide science policy issues. For these women, greater knowledge also leads greater odds of rejecting biotechnology. Model 3 features the same regression model, but this time only for males who put their faith in experts. Controlling for political views, age, religiosity, and education, the effect of knowledge on attitude is strong and negative: with each additional correct answer on the knowledge test, the odds of being pessimistic about biotechnology are reduced by about 47 percent.

Table 2. Logistic regressions predicting whether or not biotechnology will make things worse, female sample only.

	Model 1	Model 2	Model 3	Model 4
Knowledge Experts	1.325**	1.539*** .686***	1.354 .609**	1.421* .698
KnowXExperts Politics Age			1.208	1.029 .964** 1.003
Religiosity				1.017
Edu 0-14				.956
Edu 15-18				1.134
Edu 19-21 Edu Still Studying				1.068 .753*
(Edu 22+ ref cat)				

*sig. at p.<.05
**sig. at p.<.01
***sig. at p.<.001

However, the effect of knowledge on attitude is drastically different for males who think science policy should be the domain of the public. For these males, the effect of knowledge is strong and *positive*, nearly doubling the odds of being pessimistic about biotechnology with each additional correct answer on the knowledge test. For females, more knowledge leads to a greater likelihood of rejecting biotechnology no matter who they think should be making science policy decisions, and in both cases the effect of knowledge on attitude is about the same. For males, the effect of knowledge on attitude cannot be predicted unless the respondent's opinion on science policy decisions is known.

Table 3. Logistic regressions predicting whether or not biotechnology will make things worse, male sample only

	Model 1	Model 2	Model 3	Model 4
Knowledge	.800*	.818	2.087**	1.619*
Experts		.717***	1.782**	1.453
KnowXExperts			.257***	.353***
Politics				.951***
Age				.997
Religiosity				.993
Edu 0-14				1.388**
Edu 15-18				1.179*
Edu 19-21				1.323**
Edu Still Studying				1.254*
(Edu 22+ ref cat)				

*sig. at p.<.05
**sig. at p.<.01
***sig. at p.<.001

DISCUSSION

This paper made two predictions about how males and females use knowledge differently to generate attitudes toward biotechnology based on the idea of *technological salience.* It was predicted that because biotechnologies are more salient to females compared to males, the effect of their knowledge on their attitudes should hold no matter what their opinions on the role of experts are, because even if they are generally willing to let experts decide science policy, the especial salience of biotechnology for women will make their understanding of it relevant under a more robust set of circumstances.

But because biotechnology is not especially salient for men, when the responsibility for determining a positive or negative attitude is externalized to experts, the importance of one's own knowledge about the topic should be less important because responsibility is transferred to an external source. Because of the particular salience of biotechnology for women, the effect of their knowledge on their attitudes should be unconditional. Because biotechnology is less salient for men, the effect of their knowledge on their attitudes should be more likely to be affected by other circumstances; in particular, it should become less relevant when they believe the public should stay out of scientific affairs. Hypothesis 1 predicted that the experts-knowledge interaction would not significantly predict biotechnology attitude for females; that is, the effect of knowledge on attitude should be the same regardless of their opinion about who should make science policy decisions. Table 2 showed that, indeed, the experts-knowledge interaction was not significant for females. Hypothesis 2 predicted that the effect of knowledge on attitudes toward biotechnology should be weaker when males hold the opinion that experts should decide science policy issues, and stronger when males hold the opinion that the public should decide science policy issues; that is, the knowledge-experts interaction should be *negative* and significant for males. The evidence in Table 3 supports this hypothesis. When males do not trust the general public to make good policy decisions, their own knowledge is less important in determining their attitude toward biotechnology. Females, however, seem equally insistent upon using the knowledge they have under any circumstances.

Females Males

Table 4. Logistic regressions predicting whether or not biotechnology will make things worse, by gender and opinion on expertise.

	Experts	Public		Experts	Public
Knowledge	1.563*	1.171*		.526***	1.925*
Politics	.967*	.961		.952***	.950*
Age	1.000	1.012***		.998	.996
Religiosity	1.027	.988		.981	1.021
Edu 0-14	1.091	.702		1.290*	1.590*
Edu 15-18	1.155	1.071		1.178	1.166
Edu 19-21	1.105	1.000		1.343**	1.257
Edu Still Studying	.606***	1.314		1.477**	.784
(Edu 22+ ref cat)					

*sig. at p.<.05
**sig. at p.<.01
***sig. at p.<.001

Table 4 lends further support to the theory of technological salience, and complicates the way we understand how men's and women's knowledge differently affects their attitudes toward biotechnology. Consistent with previous research on how males and females use knowledge to generate attitudes (Simon, 2010), this study found that women become more likely to reject biotechnology with greater knowledge of it, and this relationship holds regardless of their opinion about who should make decisions about science in society. It is intuitive that when people think science should be controlled by the public, their own knowledge of a topic should be an important determinant of their attitude toward it. However, for women, even when they believe that science is best run by the experts, their knowledge is just as important for determining their attitude toward biotechnology, and in both cases they become more skeptical of it the more they understand. Table 4 also shows the dramatic difference in the way males who trust experts use knowledge compared to males who believe science policy decisions should be a public affair. Though Hypothesis 2 – that knowledge would be less important in determining attitude for males who abdicate to experts compared to males who think the public should be responsible for science policy decisions – was supported in by the results displayed in Table 3, Table 4 revealed that knowledge still significantly predicted attitude for males who defer to experts. While the effect of knowledge on attitude is stronger for males who think science policy should be decided by the public, Table 4 shows another telling gender difference. Males who trust experts become *less* likely to reject biotechnology with increasing levels of knowledge; males who insist that science should be controlled by the public become *more* likely to reject biotechnology with increasing levels of knowledge.

I propose that the explanation that best accommodates these findings is the differential salience of biotechnologies for women compared to men. Previous research has suggested that because of gender roles associated with childbearing and childrearing, women are more likely to be directly affected by some of the negative consequences of biotechnologies (Napolitano & Ogunseitan 1999; Bryant & Pini 2006). The more women understand biotechnology, the more they see the personal implications, and the more skeptical they become. When a technology is particularly salient to one's lived experiences, it is likely to be viewed from a much more personal perspective, one in which opinions about the role of experts in policy decisions become of secondary importance compared to how the technology is likely to affect one's own life.

Because biotechnologies are not as salient to males, they use knowledge to generate their attitudes in a different way. This study has suggested that when males award experts responsibility for making important decisions about science in society, the way they use knowledge largely conforms to the deficit model (Allum et al., 2008; Bak, 2001; Wright & Nerlich, 2006): the more they know, the more supportive they become. This seems to typify the kind of male frequently described in the literature on gender and science, one that has much invested in hegemony of scientific institutions, and the one described so eloquently by Hayes:

Because men have historically commanded the technoscientific components of society, they have not only acquired the necessary scientific and technological knowledge to dominate nature but have also been socialised to unecological attitudes toward the environment. Denied access to the technoscientific realm, not only havewomen been traditionally prevented from acquiring this knowledge but they have also been socialised to the more ecologically benign roles of mother and nurturer as reflected in their reproductive and greater child-rearing responsibilities within society (2001: 658).

However, not all males fit this stereotype. Although only about 27 percent of males in this study thought that science policy should be decided by the public, those who did became more critical of biotechnology with increasing knowledge of it, an effect of knowledge typical of females.

This study has several implications for the future of biotechnology and the study of its relations with society. If public opinion has been as instrumental in the development of biotechnologies as Bauer and Glaskel (2002) insist that it has been, then it seems that gendered experiences of biotechnologies have been instrumental in equal measure. In Bauer and Glaskel's edited volume on public perceptions of biotechnology – the most comprehensive and influential single text on this subject to date – gender is hardly mentioned at all, and when it is it is utilized as a control variable, not a focus of sustained analysis. If the future of biotechnology depends on its public reception, then it is also dependent upon gender roles and experiences, and how they relate to biotechnologies.

What do these results tell us about the relationship between understanding biotechnology and its acceptance? As Priest remarks, "It remains tempting for the scientific community and those who speak in public on its behalf to assume that dissent generally represents ignorance, and that it therefore can be reduced or eliminated by education" (2001: 98). But charges by scientific experts that resistance to biotechnologies is a consequence of an ill- informed public are only half true. It seems that, rather than providing the general public with accurate information, advocates of biotechnologies would have an interest in keeping women ignorant. Proponents of the "public understanding of science" have largely missed the point when it comes to women because they tend to see science and technology in terms of an objective cost-benefit analysis (e.g., Wertz et al., 1986) rather than as objects of lived experience. Embryonic stem cell techniques may sound innocuous to men, who may view the experimental material simply as a collection of cells. But when you have had a fetus in your own body, or even simply have the potential to carry a child, then embryonic stem cell techniques become more about powerful social institutions doing things to your body, and less about the greater good for humanity.

While this study is suggestive, it is also limited in some respects. The argument rests on claims about women's knowledge of biotechnology, though the knowledge variable used in the analysis measures general scientific knowledge, not knowledge about biotechnology specifically. It is assumed that general scientific knowledge can be taken as a faithful proxy for knowledge of biotechnology. There are surely arguments that could be summoned to challenge this assumption. However, it should be noted that the gender-knowledge interaction found here is similar to that found in a previous study (Simon, 2010) that made use of a knowledge test focused exclusively on genetics and biotechnologies. Another limitation is that the salience of biotechnology to respondents was not directly tested, but merely inferred from previous research on gender and technology and the gender differences in attitude formation observed from analyses of the Eurobarometer 63.1. I have elsewhere begun to perform more direct tests of technological salience with Katherine Johnson (Johnson and Simon, n.d.), but more research needs to be done to rule out alternative hypotheses. Perhaps there is some other gender division that is causing men and women to use knowledge in such disparate ways. The technological salience hypothesis is consistent with the findings presented here, but more direct tests need to be performed before it can be accepted with confidence.

SUMMARY

This chapter has sought to utilize the technological salience hypothesis to better understand gender differences in attitudes toward biotechnology. The analyses presented here support the idea that the salience of technologies to one's lived experiences engender unique criteria with which to evaluate those technologies. Based on previous research that has suggested that biotechnologies are especially salient for women compared to men, it was predicted that women would be less likely to discount their own knowledge in forming attitudes than men. This prediction was affirmed. Further analyses of the Eurobarometer 63.1 indicated that for men who concede science policy to experts, higher levels of scientific knowledge lead to a lesser likelihood of being pessimistic of biotechnology, while for men who insist that science policy should be the public's responsibility, higher levels of scientific knowledge lead to a greater likelihood of being pessimistic about biotechnology. While the analyses support the technological salience hypothesis, further research should use more direct measures of salience, rather than assuming gender automatically determines salience, as I have done.

REFERENCES

1. Allum, N., Boy, D., & Bauer, M.W. (2002). European regions and the knowledge deficit model. In: *Biotechnology: The Making of a Global Controversy*, edited by Bauer, M.W., & Gaskell, G., pp. 224-243. ISBN: 052177439. Cambridge, UK: Cambridge University Press.

2. Allum, N., Sturgis, P., Tabourazi, D., & Brunton-Smith I. (2008). Science knowledge and attitudes across cultures: A meta-analysis. *Public Understanding of Science*, Vol. 17, No. 1, pp. 35-54. ISSN: 0963-6625.
3. Bak, H. (2001). Education and public attitudes toward science: Implications for the "deficit model" of education and support for science and technology. *Social Science Quarterly*, Vol. 82, No. 4, pp. 779-795. ISSN: 1540-6237.
4. Barke, R.P., Jenkins-Smith, H., & Slovic, P. (1997). "Risk perceptions of men and women scientists." *Social Science Quarterly*, Vol. 78, No. 1, pp. 167-176. ISSN: 1540-6237.
5. Bauer, M.W., & Gaskell, G. (2002). "Researching the public sphere of biotechnology." In *Biotechnology: The Making of a Global Controversy*, edited by Bauer, M.W., & Gaskell, G., pp. 1-17. ISBN: 052177439. Cambridge, UK: Cambridge University Press.
6. Berg, P., Baltimore, D., Boyer, H.W., Cohen, S.N., Davis, R.W., Hogness, D.S., Nathans, D., Roblin, R., Watson, J.D., Weissman, S., & Zinder, N.D. (1974). "Potential biohazards of recombinant DNA molecules." *Science*, Vol. 185, No. 4148, p. 303. ISSN: 0036-8075.
7. Breakwell, G.M., & Robertson, T. (2001). The gender gap in science attitudes, parental and peer influences: Changes between 1987-88 and 1997-98. *Public Understanding of Science*, Vol. 10, pp. 71-82. ISSN: 0963-6625.
8. Bryant, L., & Pini, B. (2006). "Towards an understanding of gender and capital in constituting biotechnologies in agriculture." *Sociologia Ruralis*, Vol. 46, No. 4, pp. 261-279. ISSN: 1467-9523.
9. Davidson, D.J., & Freudenburg, W.R. (1996). Gender and environmental risk concerns: A review and analysis of available research. *Environment and Behavior*, Vol. 28, pp. 302-339. ISSN: 0013-9165.
10. DuPius, E.M. (2000). Not in my body: RBGH and the rise of organic milk. *Agriculture and Human Values*, Vol. 17, No. 3, pp. 285-295. ISSN: 0889-048X.
11. Freudenburg, W.R., & Davidson, D.J. (2007). Nuclear families and nuclear risks: The effects of gender, geography, and progeny on attitudes toward a nuclear waste facility. *Rural Sociology*, Vol. 72, No. 2, pp. 215-243. ISSN:1549-0831.
12. Hayes, B.C. (2001). Gender, scientific knowledge, and attitudes toward the environment: A cross-national analysis. *Political Research Quarterly*, Vol. 54, No. 3, pp. 657-671. ISSN: 1065-9129.
13. Hayes, B.C., & Tariq, V.N. (2000). Gender differences in scientific knowledge and attitudes toward science: A comparative study of four Anglo-American nations. *Public Understanding of Science*, Vol. 9, pp. 433-447.ISSN: 0963-6625.

14. Hughes, J. (2004). *Citizen Cyborg: Why Democratic Societies Must Respond to the Redesigned Human of the Future.* ISBN: 9780813341989. Cambridge, MA: Westview Press.

15. Johnson, K.M., & Simon, R.M. (n.d.). What affects faith in medical science? Contextualizing women's attitudes toward science and technology. Unpublished manuscript.

16. Krannich, R.S., & Albrecht, S.L. (1995). Opportunity/threat responses to nuclear waste disposal facilities. *Rural Sociology*, Vol. 60, No. 3, pp. 435-453. ISSN: 1549-0831.

17. Mallory, C. (2006). Ecofeminism and forest defense in Cascadia: Gender, theory, and radical activism. *Capitalism, Nature, Socialism*, Vol. 17, No. 1, pp. 32-49. ISSN: 1045-5752.

18. Mallow, J., Kastrup, H., Bryant, F.B., Hislop, N., Shefner, R., and Udo, M. (2010). Science anxiety, science attitudes, and gender: Interviews from a binational study." *Journal of Science Education and Technology*, Vol. 19, pp. 356-369. ISSN: 1059-0145

19. McCright, A.M. (2010). The effects of gender on climate change knowledge and concern in the American public. *Population and Environment*, Vol. 32, pp. 66-87. ISSN: 0199- 0039.

20. McGee, G. (2000). *The Perfect Baby: Parenthood in the New World of Cloning and Genetics.* ISBN: 9780847697595. Oxford, UK: Rowman & Littlefield.

21. McHughen, A. (2007). Public perceptions of biotechnology. *Biotechnology Journal*, Vol. 2, pp. 1105-1111. ISSN: 1860-7314

22. Napolitano, C.L., & Ogunseitan, O.A. (1999). Gender differences in the perception of genetic engineering applied to human reproduction. *Social Indicators Research*, Vol. 46, No. 2, pp. 191-204. ISSN: 0303-8300.

23. Nelkin, D. (1981). Nuclear power as a feminist issue. *Environment*, Vol. 23, No. 1, pp. 14- 20. ISSN: 0013-9157.

24. Page, B.I., & Shapiro, R.Y. (1992). *The Rational Public: Fifty Years of Trends in Americans' Policy Preferences.* ISBN: 9780226644783. Chicago, IL: University of Chicago Press.

25. Papacostas, A. (2006). *Eurobarometer 63.1: Science and Technology, Social Values, and Services of General Interest, January-February 2005.* Conducted by TNS Opinion & Social, Brussels, Belgium.

26. Priest, S.H. (2001). Misplaced faith: Communication variables as predictors of encouragement for biotechnology development. *Science Communication*, Vol. 23, No. 2, pp. 97-110. ISBN: 1075-5470.

27. Qin, W., & Brown, J.L. (2007). Public reactions to information about genetically engineered foods: Effects of information formats and male/female differences." *Public Understanding of Science*, Vol. 16, pp. 471-488. ISSN: 0963-6625.

28. Rampton, S., & Stauber, J. (1998). Monsanto and Fox: Partners in censorship. *PR Watch*, Vol. 5, No. 2. Retrieved July 8, 2011. http://www.prwatch.org/?q=prwissues/1998Q2/foxbgh.html.

29. Shanahan, D.S., & Lee, E. (2001). Trends: Attitudes about agricultural biotechnology and genetically modified organisms. *Public Opinion Quarterly*, Vol. 65, No. 2, pp. 267- 281. ISSN: 0033-362X.

30. Simon, R.M. (2011). Gendered contexts: Masculinity, knowledge, and attitudes toward biotechnology." *Public Understanding of Science*, Vol. 20, No. 3, pp. 334-346. ISSN: 0963-6625.

31. Simon, R.M. (2010). Gender differences in knowledge and attitude towards biotechnology. *Public Understanding of Science*, Vol. 19, No. 6, pp. 642-653. ISSN: 0963-6625.

32. Stern, P.C., Dietz, T., & Kalof, L. (1993). Value orientations, gender, and environmental concern. *Environment and Behavior*, Vol. 25, No. 3, pp. 322-348. ISSN: 0013-9165.

33. Sturgis, P., & Allum, N. (2004). Science in society: Re-evaluating the deficit model of public attitudes. *Public Understanding of Science*, Vol. 13, No. 1, pp. 55-74. ISSN: 0963-6625.

34. Torgersen, H, Hampel, J., von Bergmann-Winberg, M., Bridgman, E., Durant, J., Einsiedel, E., Fæstad, B., Gaskell, G., Grabner, P., Hieber, P., Felsøe, E., Lassen, J., Marouda- Chatjoulis, A., Nielsen, T.H., Rusanen, T., Sakellaris, G., Seifert, F., Smink, C., Twardowski, T., & Kamara, M.W. (2002). Promise, problems, and proxies: Twenty-five years of debate and regulation in Europe." In *Biotechnology: The Making of a Global Controversy*, edited by Bauer, M.W., & Gaskell, G., pp. 21-94. ISBN: 052177439. Cambridge, UK: Cambridge University Press.

35. Trankina, M.L. (1993). Gender differences in attitudes toward science. *Psychological Reports*, Vol. 73, pp. 123-130. ISSN:0033-2941.

36. Vendômois, J.S.D., Roullier, F., Cellier, D., & Séralini, G. (2009). A comparison of the effects of three GM corn varieties on mammalian health. *International Journal of Biological Sciences*, Vol. 5, No. 7, pp. 706-726. ISSN: 1449-2288.

37. von Roten, F.C. (2004). Gender differences in attitudes toward science in Switzerland. *Public Understanding of Science*, Vol. 13, No. 2, pp. 191–199. ISSN: 0963-6625.

38. Weber, K., Rao, H., & Thomas, L.G. (2009). From streets to suites: How the anti-biotech movement affected German pharmaceutical firms. *American Sociological Review*, Vol. 74, No. 1, pp. 106-127. ISSN: 0003-1224.

39. Wertz, D. C., Sorenson, J.R., & Heeren, T.C. (1986). Clients' interpretation of risks provided in genetic counseling. *American Journal of Human Genetics*, Vol. 39, No. 2, pp. 253- 264. ISSN: 0002-9297.

40. Wright, N., & Nerlich, B. (2006). Use of the deficit model in a shared culture of argumentation: The case of foot and mouth science. *Public Understanding of Science*, Vol. 15, No. 3, pp. 331-342. ISSN: 0963-6625.

CHAPTER-18

UTILIZING STRUCTURAL BIO-INFORMATICS IN THE FIELD OF PROTEIN ENGINEERING

INTRODUCTION: AN OVERVIEW

The application of structural bioinformatics in the domain of protein engineering represents a pivotal advancement in the development of novel proteins with tailored functions. By leveraging computational techniques and insights from the three-dimensional structures of proteins, scientists can design and engineer proteins for a wide array of applications, ranging from pharmaceuticals and biotechnology to industrial and environmental processes. Structural bioinformatics enables researchers to understand the intricate relationships between a protein's structure and its function, which is crucial for precise engineering. This technology not only accelerates the design process but also contributes to the creation of proteins with enhanced capabilities, such as increased specificity or stability. Overall, the integration of structural bioinformatics into protein engineering promises to revolutionize the way we approach protein design and holds immense potential for addressing diverse scientific and practical challenges.

Proteins are amongst the most abundant and functionally diverse macromolecules present in all living cells, and are linear heteropolymers synthesized using the same set of 20 amino acids found in all organisms from the archebacteria to more complex forms of life. By virtue of their vast diversity of amino acid sequences, a single cell may contain a huge variety of proteins, ranging from small peptides to large protein complexes with molecular weights in the range of 10^3-10^6 Daltons. Consequently, proteins show impressive diversity in biological function which includes enzyme catalysis, signal transmission via hormones, antibodies, transport, muscle contraction, antibiotics, toxic venom components and many others. Among these proteins, it may be argued that enzymes present the greatest variety and specialization, since virtually all the chemical reactions in the cell are catalyzed by enzymes (Aehle, 2004).

Due to their biotechnological and biomedical potential, enzymes and antimicrobial peptides have been the focus of extensive research efforts. This interest has been driven by a growing market for enzymes for various applications such as food supplements for humans and animals, enzymes to replace organochlorine compounds in some industrial processes, production of biofuels, vaccines, drugs, etc. (Aehle, 2004). Furthermore, the applications of enzymes in health, food, energy, materials and environment is likely to increase.

The elucidation of the three-dimensional structure of a protein is a fundamental step to understand its biological function. The 3D structure can be obtained by experimental techniques such as X-ray crystallography, nuclear magnetic resonance spectroscopy (NMR) and more recently by cryo-electron microscopy. As of mid-May 2011 there were more than 73,000 protein structures deposited in the Protein Data Bank (PDB - www.rcsb.org/pdb), of which approximately 90% have been determined by X-ray diffraction, 9% by NMR and 1% by other techniques.

Despite the availability of established experimental methodologies for the determination and analysis of three-dimensional structures of proteins to elucidate their structure-activity relationships, there are inherent limitations which can restrict their application. X-ray diffraction determines the spatial electron density distribution in the biomolecule and thereby provides the Cartesian coordinates of all atoms in the protein, except hydrogens. NMR spectroscopy is limited to low molecular weight proteins, generally smaller than 30kDa, although with the introduction of TROSY-based experiments (transverse relaxation- optimized spectroscopy) the molecular weight limitation has been extended to approximately 800kD (Fernández & Wider, 2003). Cryo-electron microscopy is a low- resolution technique that is a powerful tool for the determination of the structure of large proteins or protein complexes, such as virus structures (Mancini *et. al.*, 2000).

Conformational changes and dynamics are key processes to understand biological mechanisms and functions. Many biological processes involve functionally important changes in the protein 3D structure, however techniques for structure determination do not describe (or describe poorly) protein dynamics. Examples of the importance of protein dynamics in structural biology include the conformational changes associated with protein folding, catalytic functions of enzymes and signal transduction. In this context, bioinformatics techniques, such as molecular dynamics (MD) simulation, together with homology and statistical analysis of protein structures have become useful tools in structural biology not only for understanding protein function (van Gunsteren et. al., 2008), but also for protein engineering (Vieira & Degrève, 2009). Growing interest in bioinformatics techniques and advances in experimental methodologies have led to many successes in engineering enzyme function (Aehle, 2004).

The purpose of this chapter is to introduce the use of MD simulations in a biological context to target proteins of biotechnological interest, such as xylanases, antibody/antigen interactions and antimicrobial peptides. Using relevant models we discuss the use of MD simulations to study properties such as enzyme thermostability, protein conformational changes and functional movements, intra and intermolecular interactions through hydrogen bonding analysis, protein-protein interaction and modelling of peptide- membrane interactions. The main goal is to demonstrate how the three-dimensional structure, function, interaction potential and dynamics are highly correlated and together comprise a valuable data set to explore the broad research field of structure/function relationships of proteins.

MOLECULAR DYNAMICS SIMULATIONS: AN INTRODUCTION

Molecular dynamics (MD) simulation is one of the main computational methodologies which can provide detailed microscopic information at an atomic and molecular level for a variety of systems. The behaviour of matter can be understood by the structure and dynamics of its constituent atoms or atomic groups, and is considered as an *N*-body problem. Due to the lack of general analytical algorithms, a solution to the classical *N*-body problem is exclusively a numerical task. Therefore, the study of matter at this level requires the computational resources to allow the investigation of the movements of individual particles or molecules, and the MD approach aims to reproduce the properties of a real system using a microscopic model system. The continual increase in computational power enables the investigation of increasingly large model systems, and consequently to approach broader and more complex biological questions. MD simulation is not limited to biological and biochemical systems, and MD has a broad range of applications in material science, physics, chemistry and engineering (Leach, 2001).

MD simulations are applied to systems that are in states in which quantum effects are disregarded, which imposes a limit on the types of problem that can be addressed. For example, chemical reactions and transitions between different electronic states present a quantum nature due to their dimensional scale and their high energies, and cannot be treated by MD. MD simulation employs the classical equations to describe the motion of atoms and molecules as a function of the time under determined conditions (pH, temperature and pressure) taking into account an intermolecular interaction potential. Hybrid methodologies (Quantum Mechanics/ Molecular Mechanics, QM/MM) are capable of treating quantum phenomena in MD simulations to describe, for example, chemical reactions using Valence Bond MD.

Other well established hybrid methods are Carr-Parrinelo and *ab initio* MD (Leach, 2001), however these are limited by the intensive computing power required.

The use of MD simulation allows the study of the temporal evolution of molecular motions, generating a series of complete atomic coordenates (ie. Cartesian *x,y,z* coordinates for each atom) that permit the prediction of microscopic and macroscopic properties of the system from Statistical Mechanics analysis (McQuarrie, 1976). Starting from a pre-defined atomic arrangement that defines the initial configuration of the system, the individual atoms move under the influence of their intermolecular potentials. Given the atomic positions and velocities at a given time (t), the resulting forces can be calculated for each particle in the system permitting calculation of the atomic positions and velocities at a later time ($t+dt$). This procedure generates molecular trajectories for the whole system over the total simulation time.

The choice of the number of atoms in a system depends on the properties that are to be studied with the available computational capacity, and must be representative of the macroscopic real system. The current computational power with present methods allows simulations of systems constituted by up to 10^8 atoms. Recently, a 1-milliseconds all-atom MD simulation of a folded protein has been reported (Shaw et. al., 2010). The equilibrium properties of the system are computed as time averages over given time intervals that must be somewhat longer than the corresponding event observed at the atomic scale. Temperature, radial pair distribution functions, potential and kinetic energies are examples of fast convergence properties, while surface tension, diffusion coefficient and pressure are properties of slow convergence. An important characteristic of MD simulation is that information on the system is gained on a time scale that is often not easily obtained by experimental approaches (Rapaport, 2005).

Molecular Dynamics and Statistical Mechanics

Statistical mechanics deals with ensemble averages, and is well suited to analysis of the data produced by MD methods. During MD simulations, the mechanical, thermal and chemical thermodynamic variables must be fixed and simulations are usually conducted under constant NPT, NVT, NVE or μNT conditions. (where N is the number of atoms, P the pressure, T the temperature, V the volume, E the internal energy and μ the chemical potential). An ensemble is a large set of replicas generated under the defined external conditions (McQuarrie, 1976), and in the NVT (or canonical) ensemble, N, V and T are fixed external parameters and therefore the instantaneous temperature (obtained from the kinetic energy) must be controlled to a fixed value by an thermostat algorithm, which multiplies the instantaneous atomic velocities by a coefficient defined by the difference between instantaneous and desired temperatures. Several algorithms can be used to control the temperature including the Berendsen, Nosé-Hoover and V-rescale thermostats (Allen & Tildesley, 1992; Rapaport 2005). *NVT* and *NPT* ensembles are widely used in thermal treatments such as annealing in order to promote thermal denaturation of proteins in unphysical conditions (Rocco et. al.; 2008). *NPT* or isobaric-isothermal ensembles present better characteristics than the canonical ensemble, since the lack of the thermodynamic information about the equilibrium conditions are circumvented by the adjustment of the pressure to the imposed external conditions. This allows the volume to fluctuate by scaling the dimensions of the system by a coefficient defined by the instantaneous and desired pressures.

The equilibrium average of a given property *A* is expressed by:

$$\langle A \rangle_{ensemble} = \frac{\iint A(\vec{p}^N, \vec{r}^N) e^{-\beta E(\vec{p}^N, \vec{r}^N)} d\vec{p}^N d\vec{r}^N}{Q} \quad (1)$$

where p^N and r^N are the momenta and positions of the *N* particles at time t, $E(p^N, r^N)$ is the sum of the potential, $_{\rightarrow N}\ U(r)$, and kinetic energy, $_{\rightarrow N}\ K(p)$, and ? = $1/kT$. Th denominator term, *Q*, is the partition function. For a system of *N* identical particles the partition function for the canonical ensemble is:

$$Q_{NVT} = \frac{1}{N!}\frac{1}{h^{3N}} \iint e^{-\beta E(\vec{p}^N, \vec{r}^N)} d\vec{p}^N d\vec{r}^N \quad (2)$$

where *h* is the Planck's constant. The partition function is central to the statistical mechanics analysis since it contains all the information to determine the macroscopic thermodynamics properties, which are crucial for correlating the microscopic and macroscopic levels (McQuarrie, 1976). Generally, the partition function cannot be computed since it depends on the interatomic forces and positions.

The most fundamental axiom of statistical mechanics, the ergodic hypothesis, states that the ensemble average equals the time average.

$$\langle A \rangle_{ensemble} = \langle A \rangle_{time} \quad (3)$$

This theory is based on the fact that if one allows the system to evolve indefinitely in time it will pass through all possible energy states, therefore the MD simulation should generate enough configurations in time to ensure that the equality is satisfied. The average in eq. (1) is an average over all the possible positions of the particles. During a MD simulation, the averages are calculated over the configurations obtained as a function of time so that in accord with the ergodicity hypothesis:

$$\langle A \rangle_{time} = \lim_{\tau \to \infty} \frac{1}{\tau} \int_{t=0}^{\tau} A[\vec{p}^N(t), \vec{r}^N(t)]dt \approx \frac{1}{M} \sum_{m=1}^{M} A_m(\vec{p}^N, \vec{r}^N) \quad (4)$$

where M is the number of $A_m\ (p^N, r^N)$ values produced by the MD simulation.

A variety of thermodynamic properties can be calculated from the configurations generated by the MD simulations. Comparison with experimental results is an important way to check the accuracy of the simulation and the validity of the model and method. Simulation techniques can also be used for prediction of thermodynamic properties for which there is no experimental data. For example, the internal energy can be calculated as follows:

$$\langle E \rangle = \frac{1}{M} \sum_{i=1}^{M} E_i \quad (5)$$

and the heat capacity at constant volume, C_V, as given by the internal energy fluctuation is obtained from:

$$C_V = \frac{\langle E^2 \rangle - \langle E \rangle^2}{kT^2} \quad (6)$$

where $\langle E^2 \rangle$ and $\langle E \rangle^2$ can be calculated at the end of simulation.

Heat capacity is an important quantity to investigate phase transitions, because it shows a characteristic dependence upon temperature. A first-order phase transition shows an infinite heat capacity at the transition temperature and a discontinuity for a second-order phase transition. The heat capacity can also be calculated from a series of simulations at different temperatures followed by the differentiation of the energy with respect to the temperature.

Molecular Interactions and Equations of Motion

The main influence of a molecular simulation is the potential energy, which determines how the particles of the system interact. The simplest model of a particle is a sphere, and the simplest model of interaction is that between pairs of atoms. Such simple models are capable of reproducing two main features of real systems, *(i)* the resistance to compression (short range) and *(ii)* the capability of keeping atoms together in a condensed phase since atoms are attracted to each other over longer ranges (Rapaport, 2005). MD techniques use analytical expressions for the energy potential functions that are obtained from experimental data or quantum calculations. These analytical expressions are generally effective pair interaction potentials that allow the molecular systems to exhibit the correct characteristics of real systems (Maitland et. al., 1981).

The most widely used of these interaction potentials between monoatomic molecules was initially proposed to reproduce structural properties of liquid argon (Maitland & Smith, 1971), written in the form of the Lennard-Jones (LJ) potential:

$$U(r_{ij}) = 4\varepsilon\left[\left(\frac{\sigma}{r_{ij}}\right)^{12} - \left(\frac{\sigma}{r_{ij}}\right)^{6}\right] \tag{7}$$

where r_{ij} is the distance between atom i and j, $-\varepsilon$ is the depth of the potential well and σ the radius of the particles. An intermolecular potential energy must reproduce the attractive (long-range) and repulsive (short-range) behaviours, and in the case of LJ potential the repulsive and attractive behaviours are represented by the first and second terms respectively. More complete expressions describing the total intra and intermolecular potential energies for complex molecules will be described in section 2.5.

The intermolecular interactions are generally cut off at the separation distance r_C , i.e.

$$U(r_{ij}) = 0 \qquad r_{ij} \geq r_c \tag{8}$$

The force between the atoms i and j is given by:

$$\vec{F}_{ij} = -\frac{\partial U(r_{ij})}{\partial r_{ij}} \tag{9}$$

Special care must be taken with the treatment of the long range interactions, since the distance cut-off may introduce a discontinuity in the potential energy and the force near the given value. This artefact creates problems due to the requirement for energy conservation in MD simulations, and efficient mathematical approaches have been suggested to circumvent this problem, such as the use of switching functions (Allen & Tildesley, 1992).

The coordinates of the atoms of the system are obtained by solving the equation for Newton's second law;

$$m_i \frac{d^2\vec{r}_i}{dt^2} = m_i \vec{a}_i = \vec{F}_i = \frac{1}{2}\sum_{\substack{j=1 \\ j\neq i}}^{N} \vec{F}_{ij} \tag{10}$$

where F_i is the force acting on the atom i of mass m_i and acceleration a_i. Each pair of atoms need only be identified once because $\vec{F}_{ij} = -\vec{F}_{ji}$ (Newton's third law).

The differential equation (10) is solved by integration using standard numerical techniques, such as finite difference or predictor-corrector. Finite difference methods are widely employed in MD simulation to integrate the equation of motion, for example the Verlet and leap-frog algorithms (Allen & Tildesley, 1992).

The Verlet algorithm is based on positions $r(t)$ and accelerations $a(t)$. The positions at times $t \pm dt$ are obtained by Taylor expansion of $r(t)$:

$$\vec{r}_i(t+dt) = \vec{r}_i(t) + \vec{v}_i(t)dt + \vec{a}_i(t)\frac{dt^2}{2!} + ... \tag{11}$$

$$\vec{r}_i(t-dt) = \vec{r}_i(t) - \vec{v}_i(t)dt + \vec{a}_i(t)\frac{dt^2}{2!} - .... \tag{12}$$

where $v_i(t)$ is the velocity of the atom i, dt is the time step and, $r_i(t+dt)$ and $r_i(t-dt)$ are the later and earlier positions relative to $r_i(t)$. By addition of equations (11) and (12) and taking into account equation (10), we obtain the equation for advancing the positions:

$$\vec{r}_i(t+dt) = 2\vec{r}_i(t) - \vec{r}_i(t-dt)dt + \frac{dt^2}{m_i}\vec{F}_i(t) \tag{13}$$

Although the velocities are not necessary to compute the trajectories, they are useful to estimate the kinetic energy and hence the total energy of the sytem. The atomic velocities can be calculated from:

$$\vec{v}_i(t) = \frac{\vec{r}_i(t+dt) - \vec{r}_i(t-dt)}{2dt} \tag{14}$$

Other integration algorithms are available (Allen & Tildesley, 1992). The numerical integration of the motion equations encompasses numerical approximations so that the time step, dt, must be chosen to be at least one order of magnitude smaller than the high- frequency motions and generally of the order of a few femtoseconds. The high-frequency motions, such as bond stretching involving hydrogen atoms that can impair the use of $dt \approx 1fs$, are avoided by maintaining constant bond lengths, since generally these bonds are not interesting and have a negligible effect on the overall stability of the system.

Initial Configuration of the System

The MD simulations must be capable to sample the regions of the phase space. A MD simulation must be insensitive to the initial state, therefore any appropriate initial state is permissible. A simple method is to build the system from a random distribution of particles or from a regular lattice obeying the numerical density (*N/V*) of the real system. The initial velocities can be also assigned in different ways, either randomly or by fixing the value based on temperature (Maxwell-Boltzmann distribution). Velocities are adjusted to ensure that the center of mass of the system is at rest. In MD simulation of biomolecules, the initial structure must be known and the solvent molecules and ions are added to fulfil the system and neutralize it. Initial protein structures can be found in the PDB data bank or using specific methods of bioinformatics that predict and validate three-dimensional structures.

The set-up of the simulation box is the important first step in the definition of the system. The limits of the simulation box should not produce interface effects in homogeneous systems, and for this reason periodic boundary conditions (PBC) are applied to extend the system to the thermodynamic limit. PBC consists of building replicas of the around of the main box in order to obtain a system that mimics a macroscopic thermodynamic system. All the particles in the replica move identically so that the motion of the atoms is not limited by the walls of the box. During the course of a simulation, when a molecule moves in the central box, all its images move in the same manner (Allen & Tildesley, 1992). In other words, a system which is bounded but free of physical walls can be simulated by the application of PBC.

Five shapes of cells produce periodic images: the cube, the parallelepiped, the truncated octahedron, the rhombic dodecahedron and the hexagonal prism. The choice of the simulation box should be made based on the geometry of the system of interest, for example a globular protein is suitable to be simulated in a cubic or dodecahedron periodic cell, but inadequate for a parallelepiped.

Controlling the Simulations

Many changes occur in the structure of the system during a simulation process. As the initial structure undergoes modifications as a function of time, the MD data allow examination of the changes at each step and consequently, the evaluation of the equilibrium conditions. The equilibrium condition, *i.e.* small fluctuations near mean values of a given property, is fundamental to control the MD simulations. Stabilization conditions can be identified by accompanying determined properties that are calculated during the simulation. Many properties can be used to control the MD simulations such as the energy profiles, the stability of the temperature, etc. After establishing that the system is stabilized, the calculation phase can be initiated in which the properties of interest can be computed.

In a system constituted by a protein in water, the stability is established by a delicate energetic balance involving intramolecular and intermolecular interactions and kinetic energy. Despite the importance of the intramolecular interactions, the stability of biomolecules in general is strongly influenced by protein-water interactions. The solvent molecules make significant contributions to the conformational changes of the protein, and at equilibrium the protein/solvent interactions are completely relaxed. An important tool to detect equilibrium conformations in MD simulation of proteins is the Root Mean Square Deviation (RMSD), which is defined by;

$$RMSD(t) = \sqrt{\frac{\sum_{i=1}^{n} |\vec{R}_i - \vec{r}_i(t)|^2}{n}} \tag{15}$$

w here n is the number of atoms (frequently only the Cα atoms) used in the calculation. $|\vec{R}_i - r_i(t)|$is the distance between the atom i at time t and the same atom in the reference structure. The reference structure can be the experimentally determined initial structure or any other structure, for example the last structure generated by the simulation. The RMSD calculation is made after the superposition of the entire structures or may be applied to more limited groups of atoms or residues. The RMSD provides information about the structural stability as a function of time and the changes of the structures during the simulation process. Figure 1 shows two different cases of the RMSD of a protein which is not stabilized even after 100.0 ns as compared with the RMSD of a protein that reaches equilibrium in only 5.0 ns. These results illustrate the important point that the equilibration time is inherent to each protein, to its environment and to the quality of the initial three- dimensional structure.

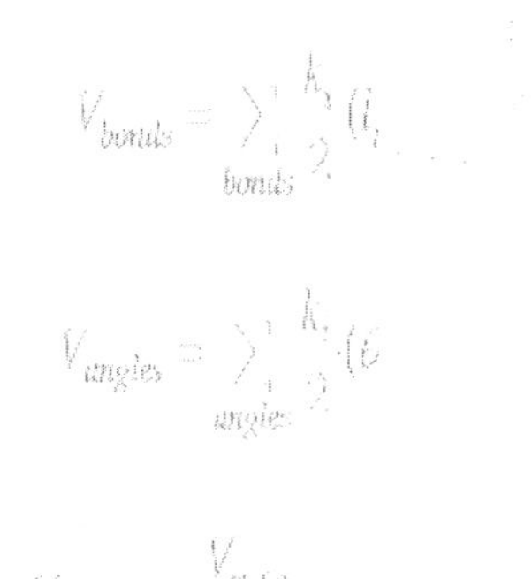

Fig. 1. Root mean square deviations (RMSD) for two different MD simulations.

The black line corresponds to a protein not stabilized in 100.0 ns and the gray line to the protein that shows a fast equilibration.

Force Fields and Molecular Simulations Software Packages

The Force Field in MD simulations is the name given to the set of parameters that define the potential interaction energies. These parameters are adjusted to fully describe the molecular behaviour and its interactions. Intra- and intermolecular potential interactions can be expressed in terms of relatively reduced number of types of components. The interaction between bonded atoms can be subdivided into bond stretching, angle bending and dihedral- angle bending (torsion) components, while the interactions between non-bonded atoms are by electrostatic and van der Waals interactions. The full expression can be described by:

$$V_{bonds} = \sum_{bonds} \frac{k_i}{2}(l_i - l_{i,0})^2 \quad (16)$$

$$V_{angles} = \sum_{angles} \frac{k_i}{2}(\theta_i - \theta_{i,0})^2 \quad (17)$$

$$V_{torsions} = \frac{V_n}{2}[1 + \cos(n\omega - \gamma)] \quad (18)$$

$$V_{non-bonded} = \sum_{i=1}^{N} \sum_{j=i+1}^{N} \left[\left(\frac{A_{12}}{r^{12}_{ij}} - \frac{B_6}{r^{6}_{ij}} \right) + \frac{q_i q_j}{4\pi\varepsilon_0 r_{ij}} \right] \quad (19)$$

$$V(\vec{r}^N) = V_{bonds} + V_{angles} + V_{torsions} + V_{non-bonded} \quad (20)$$

$V(r^N)$ denotes the total potential energy as a function of the positions *r* of *N* particles or atoms. Bonds and angles are modelled by harmonic potentials (eq. 16 and 17) that give the change in energy as x_i ($x \equiv l,?$) deviates from the reference value, (l_i ,0 ,?$_i$,0) . The third term, eq. 18, is the torsional contribution that evaluates how the energy changes in the vibrations of structures of four atoms (planar, like aromatic rings or angular such as in the sp^3 hybridization) and finally, the fourth term is the interaction between non-bonded atoms, including van der Waals and electrostatic contributions, as calculated between the non- bonded pairs of atoms in the same or different molecules. The terms A_{12} and B_6 in eq. 19 are the Lennard-Jones parameters which consider the size of the molecules and the distance between atoms at zero energy potential.

Electrostatic interactions are important long-range forces that influence protein stability and present difficulties in their treatment in MD simulations, and are integral parts of MD simulation software packages. Some methods can be cited as the most important for protein systems, such as the Reaction Field method and the Ewald Sum. Excellent books are available for consultation on this issue (Allen & Tildesley, 1992; Leach, 2001). Several Force Fields are available and all of them are, or at least should be equivalent, independently of the methodology used to compute their parameters.

An increasing number of force fields and MD software packages are available, many of which are free and can be obtained from the web sites of the research group responsible for their development. MD software packages have been employed in research in theoretical chemistry and biology, protein engineering and industry (pharmaceutical and materials). Some of the most important and referenced MD packages are listed below, and all are well documented and many offer parallelization options.

GROMACS (Groningem Machine for Chemical Simulation): Currently available in version 4.5.3 (2011) at the website www.gromacs.org. It refers to both the MD software package and the GROMACS Force Field that is updated like other force fields. It is parameterized to several solvent, proteins, nucleic acids and carbohydrates. AMBER (Assisted Model Building with Energy Refinement): refers to both the Force Field and the MD program specially developed for biomolecules, and can be obtained at symbolic price. CHARMM (Chemistry at Harvard Macromolecular Mechanics): is also a Force Field and MD program which although is not fully available for academic purposes, can be obtained from Accelrys. Inc through a commercial license. NAMD (Not Another Molecular Dynamics): has as its main feature the high degree of scalability in multi-processed computational systems, up to around one hundred processors with no loss performance. Q Program: is a MD software package designed for free energy calculations in biological systems. It is fully available for academic purposes from the Åqvist group website (Åqvist, 2007). Other MD simulation packages include TINKER, LAMMPS, AMMP, BOSS, CERIUS2, CPMD, INSIGHT2, ORAC and OPLS.

APPLICATIONS IN BIOTECHNOLOGY

Xylanases

Xylanases (EC 3.2.1.8) are enzymes produced by a wide variety of bacteria and fungi that hydrolyze the 1,4-β-D-xylosidic linkages of xylans, a mixed family of plant cell wall polysaccharides that are abundant in nature. The enzymatic hydrolysis of xylan has attracted considerable biotechnological interest with applications in the food engineering, bio-ethanol fuel production and cellulose pulp industries (Beg et al., 2001). Ongoing efforts have been directed toward both the identification of new xylanases and the improvement of the catalytic properties of existing enzymes with the goal of enhancing their compatibility for industrial applications.

Based on aminoacid sequence similarity and three-dimensional structural homology, xylanases are classified as family GH10 or GH11 hydrolases. GH11 xylanases present compact globular three-dimensional structures with a conserved scaffold comprised of a single α-helix and two extended pleated β-sheets forming a jellyroll fold. The major surface feature is a long cleft that spans the entire molecule and which contains the active site. The overall shape of GH11 xylanases resembles a right-hand, and as shown in Figure 2 the various structural features have accordingly been denominated as fingers, palm, thumb, cord and helix regions.

The globular structure and relatively small size (~200 amino acids) of the GH11 xylanases are good models for molecular simulation which can be used to investigate important physicochemical and biochemical properties, such as thermostability.

Thermostable proteins are of biotechnological interest because they can be employed in industrial processes which use high temperatures, such as cellulose pulp bleaching. Several strategies have been attempted with the aim of enhancing thermostability of proteins, including the introduction of disulfide bridges (Wakarchuk et al., 1994), increasing the hydrophobic contacts by addition of aromatic interactions (Georis et al., 2000), optimization ofthe electrostatic surface potential (Torrez et al., 2003), and optimization of intramolecular and protein/ solvent hydrogen bonding (Viera & Degrève, 2009). Due to the important enthalpic contribution of hydrogen bonding in biological systems, hydrogen bond optimization is a particularly interesting strategy to improve protein thermostability. Formally, hydrogen bonds are rigorously evaluated by quantum mechanics methodologies using the Morokuma Analysis that deconstructs the energy of hydrogen bonds into five components: electrostatic, polarisation, exchange repulsion, charge transfer and mixing (Morokuma, 1977). In contrast, molecular simulation obtains a reasonable estimative of hydrogen bond occurrence and stabilization by evaluating geometric criteria. For example, intramolecular hydrogen bonds are identified by interactions with a donor-receptor distance shorter than 0.24 nm, the (A-H···O) angle larger than 110º (A≡ N or O) and the fractional occurrence larger than 20% of the total simulation time. The intermolecular hydrogen bonds can be detected by two criteria; *(i)* based on the radial distribution functions, *g(r)*, and *(ii)* distribution of the interaction energies between the protein atoms and the solvent, F(E), (Degrève et al., 2004).

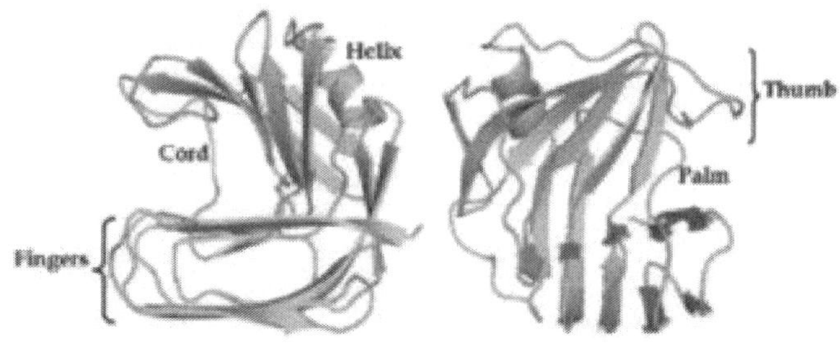

Fig. 2. Two different views of three-dimensional structure of the GH11 xylanase from *Bacillus subtilis*, PDB code 1XXN. The regions based in the right-hand analogy are indicated.

The dynamics and energetic properties of MD simulations are well suited to monitor interatomic distances and angles over time, and can characterize all atom pairs involved in hydrogen bonds, hydrophobic clusters or salt bridges. Figure 3 show the hydrogen bonding network in a three-dimensional structure of the *Bacillus subtilis* xylanase from a MD simulation identified by the geometric criteria describe above.

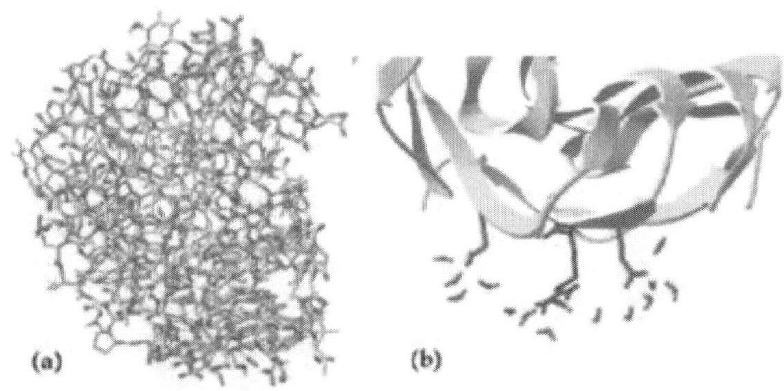

Fig. 3. (a) A snapshot at 10.0ns of GH11 xylanase from the MD simulation in canonical ensemble at physiological pH. The intramolecular hydrogen bonding network is showed as the green dotted lines. (b) The intermolecular hydrogen bonding interactions of the first solvation shell of two Glu residues (red) and one Arg residue (blue).

Intermolecular hydrogen bonds may be identified by considering the first solvation shell of a given residue. As shown in Figure 3(b), the $g(r)$ $g_{AW}(r)$ function must display a peak in an appropriate region, approximately 2Å, defined by the nature of the protein atom. Furthermore, the $F(E)$ should display a peak, or a shoulder, in the attractive energy region (negative energies) as observed in the example shown in Figure 4.

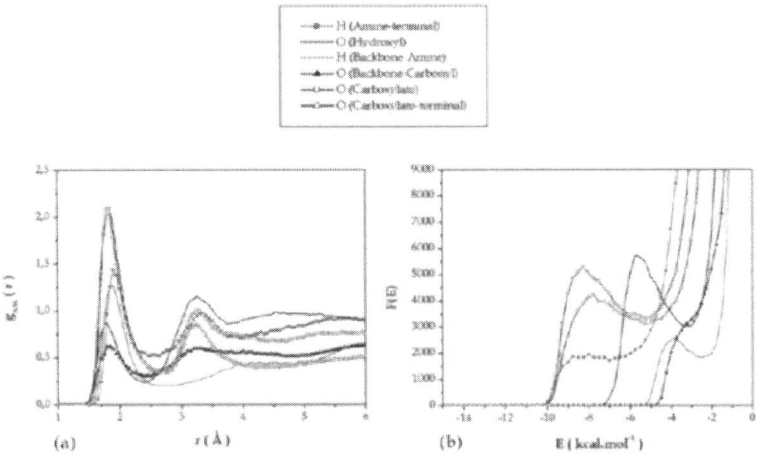

Fig. 4. (a) Radial distribution functions $gAW(r)$, where A is the protein atoms and W the water molecule. (b) Distribution of the interaction energies between atoms of protein and the water molecules. Protein atoms are defined in the legend.

As shown in Figs 4(a) and 4(b), each atom or a group of atoms has a defined influence on the protein energetics. These interactions are unique to each protein, and provide an individual map of the protein enabling us to make predictions with respect to the thermostabilization mechanism adopted by proteins. Recently, Vieira & Degrève have published an MD simulation study of a thermophilic-mesophilic pair of GH11 xylanases, in which a difference of 20°C in thermostability could be interpreted in terms of different patterns of intra and intermolecular hydrogen bonds exhibited by these proteins (Vieira & Degrève, 2009).

MD studies by Vieira & Degrève (2009) estimate the total interaction potential energy for the thermophilic xylanase to be around $-600 kcal.mol^{-1}$ more stable than its mesophilic counterpart, and approximately 20% of this difference is due to intermolecular hydrogen bonds energies. Although the backbone of the thermophilic xylanase is more rigid than the mesophilic enzyme (as determined by the intramolecular hydrogen bonds and salt bridges), the increased thermostability seems to be a consequence of the greater degree of solvation. Although both xylanases are highly solvated, the intermolecular hydrogen bonding network in the thermophilic enzyme is energetically more attractive than those at mesophilic counterpart. In other words, the highly solvated surface characterized by optimization of strong protein-water intermolecular hydrogen bonds is clearly a thermostabilization factor. The main residues which contribute to the thermostability of thermophilic xylanase were located in the finger region and have been targeted for site-directed mutagenesis. This insight as to a major contribution to protein thermostability would not be available from the analysis of a structure determined by X-ray crystallography, in which water molecules are poorly resolved. Therefore, MD simulations can complement the experimental studies in the development and of design process of new proteins.

Antibody engineering

Antibodies or immunoglobulins (Ig) are globular proteins that play an important role in protecting the host organism against infectious diseases. The primary function of an antibody is to bind molecules that are foreign to the host (antigens), and understanding the interactions between an antibody and its antigen is important for the design and development of new or improved antibodies. The host typically produces several types of antibody that have similar structures, consisting of four polypeptides (two heavy chains and two light chains) that are linked by disulphide bonds to form a molecular complex. Five heavy chain isotypes are known, giving rise to five different Ig classes (IgG1-4, IgA1-2, IgD, IgM, IgE), each with a distinct function in humans. Each class of heavy chain can combine with one of two light chain isotypes (kappa and lambda). As a result of these molecular combinations, antibodies differ in size, electric charge, amino acid composition and carbohydrate content (Roitt, 1997).

The most common class of IgG antibodies present a basic structure of two light chains (each of which has two domains) and two heavy chains (each having four domains). The two N-terminal domains of the heavy chain are linked by disulphide bonds to the two domains of the light chains to form a "Y"-shaped structure. The N-terminal domain of both the heavy and the light chain show pronounced amino acid sequence (and are appropriately called the variable domains, VL and VH), while the other domains are called constant domains (CL, CH1, CH2 and CH3) (Alberts et al, 1989). The variable domains have three regions of hypervariability in the amino acid sequence and are termed Complementarity Determining Regions (CDRs). These regions are responsible for the recognition and binding of the antibodies to a specific antigen, and determine the affinity and specificity to the antibody-antigen interaction. The "Y"-shaped antibody structure can be divided into three fragments: Fv (variable fragment), Fab (antigen binding fragment) and Fc (crystallization fragment). An engineered antibody may be obtained from separate segments of Fv heavy and light chains joined by a flexible peptide linker to form a single-chain Fv (scFv).

Molecular modeling by homology uses experimentally determined protein structures to predict the three-dimensional conformation of a protein with a similar amino acid sequence whose structure is unknown. Molecular modeling can be used with reasonable success if the amino acid sequences of the known and the unknown proteins share at least 40% identity, and within these limits more than

90% of the atoms in the main chain can be modeled with an accuracy of about 0.1 nm (Sanchez & Šali, 1997). The program *Modeller* (Šali & Blundell, 1993) can be used for homology modeling in conjunction with the programs *PROCHECK* (Laskowski et al., 1993) and *VERIFY_3D* (Luthy et al., 1992) that can be used to assess the quality of the generated structures. These tools can be used to model both the variable regions of antibodies and antigen.

Starting from the three-dimensional structures of antibody and antigen, the next step is to identify the epitopes on the antigen molecule that are recognized by the antibody. Docking techniques can be used to explore the regions of best fit between antibody and antigen, and the more information about the epitopes better the final solution generated by the docking program. For identification and optimization of regions of the antibody that enter in contact with the antigen, the use of available structural, biological and biochemical information is essential to obtain the most reliable results. Furthermore, the docking approach should not simply focus on solving a problem of rigid body fitting of geometric shapes between the two components, but should recognize that interatomic forces in the context of complex dynamics can give rise to structural changes that can maximize the contacts at the interface of the antibody-antigen complex.

Starting from predictions of antibody-antigen interfaces derived from docking studies, MD simulation can be used for the evaluation of the structures of the proposed antibody-antigen complexes. MD is a powerful tool that can be used to study the structural movements of antibody and antigen and to monitor interatomic interactions in a physiological aqueous environment. With this strategy, a new scFv (Ab) for a particular antigen (Ag) has been developed at the Brazilian company, Cientistas Associados (Cientistas Associados, 2008). It has been demonstrated that the binding between Ag and Ab is dynamic, with structural variations observed in both molecules, which can be improved by changing the contact specificity through the mutation of key residues at the Ag|Ab interface.

Crotoxin is a phospholipase A2 neurotoxin, and is a major component of the venom of the rattlesnake *Crotalus durissus terrificus*, and the lethal effects are due to blockage of neuromuscular transmission. Crotoxin is comprised of two non-identical subunits (CA and CB), which separately present present low toxicity. The CA subunit (acidic protein) and CB (basic phospholipase) spontaneously associate in a 1:1 complex, giving rise to the particularly lethal neurotoxin (Faure et al., 1991). Studies reported that three scFv could bind crotoxin (Cardoso et al., 2000). These scFvs recognized distinct epitopes close to CB, however only one scFv was able to effectively neutralize the activity induced by crotoxin.

In order to study this antibody|antigen complex, the structures of crotoxin and scFv were subjected to homology modelling. The template for the crotoxin was the phospholipase A2 from *Crotalus atrox*, which has 50% identity when compared to CB. In the case of a template for the scFv, a search was made of the "framework" (BLAST) and the structure of the CDRs with highest resolution and structural identity was selected. The Ag|Ab complex was obtained with the program HEX (Ritchie, 2003), and since the binding site of the antibody (CDR) is known through epitope mapping (Choumet et al. 2003), the crotoxin was in the correct orientation in relation to the CDRs.

Four MD simulations were performed with the GROMACS software package using a time step of 2 fs at 298 K in the NVT ensemble. A cut-off radius of 1.3 nm was applied. The simulated systems were electrically neutralized by the addition of counter ions and the simulation boxes were designed to avoid overlap of the cut-off radius through the sides of the boxes. The systems include a free Ab in solution (System 1), a free Ag in solution (System 2), Ag|Ab complex (System 3) and mutate Ab in Ag|Ab complex (System 4). In System 3, the antigen and antibody are initially separated by 0.7 nm. The initial structure of the Ab in the System 4 was mutated in four positions as compared to the Ab in System 3 to enhance the Ag|Ab interaction.

The Ag|Ab complex was formed by superposition of the final simulated antibody and antigen structures by docking. In the System 3, formation of the complex was monitored through the intermolecular interaction potential (IIP) between the Ag and Ab calculated during the simulation (Figure 5).

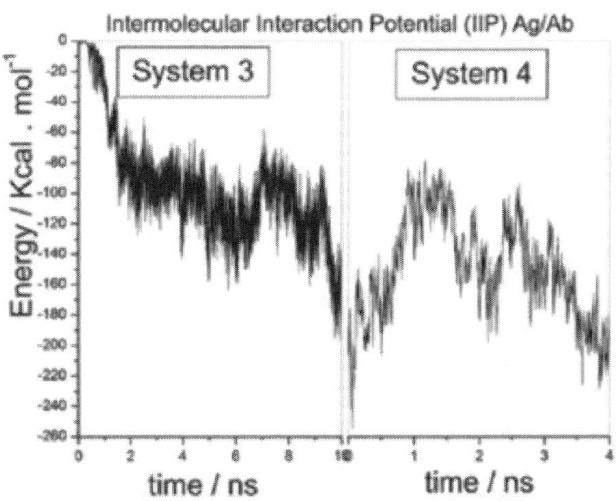

Fig. 5. Intermolecular Interaction Potential between the antigen and the antibody. A) System 3 and B) System 4 (see text for details).

The IIP shows that when Ag and Ab are separated by 0.7 nm (Figure 6a, time = 0 ns), the potential is almost zero and there is no effective interaction. Time around 1.0 ns and IIP = - 60 kcal. mol^{-1} marks the start of the Ag|Ab complex formation (Figure 6b), and from that moment the attractive interaction at the Ag|Ab interface increases, and the potential decreases to approximately -170 kcal.mol^{-1} in the equilibrium structure at the end of the simulation (Figure 6c). The Ag|Ab complexes obtained from MD simulation are presented in Figure 7. It is evident that the Ag|Ab interface can be extended since the Tyr and Thr residues are distant from the antigen.

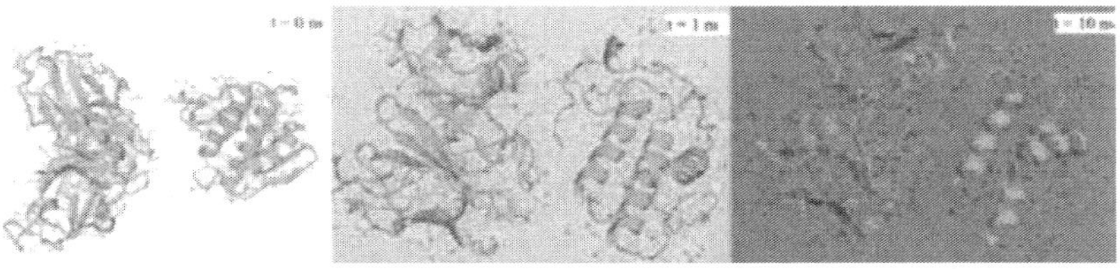

Fig. 6. Configurations of System 3 complex formation between the antibody (left) and antigen (right), at times 0, 1 and 10 ns.

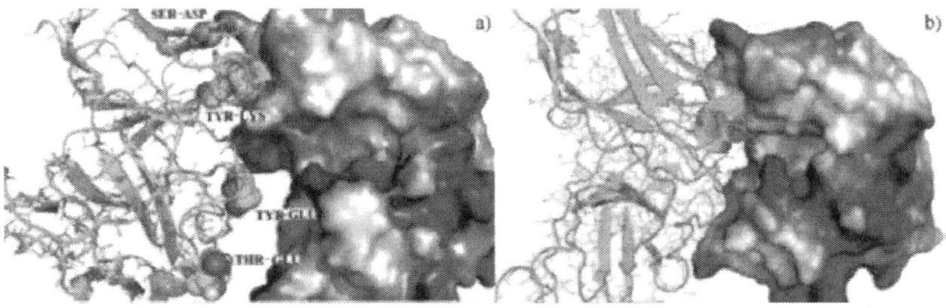

Fig. 7. Final configurations of Systems 3 and 4 from MD simulations, showing the Ag|Ab complex (left and right molecule respectively). Potential isocontours shown at +1 (kT/e) (in blue) and -1 (kT/e) (in red) for permittivities equal to 2 (solute) and to 78.5 (solvent). a) System 3 and b) System 4.

To study the contribution of electrostatic and van der Waals interactions of individual residues to the stability of the interface, four amino acids in the antibody were mutated with the aim of enhancing the Ag|Ab interaction. The substitutions were chosen to optimize the interface by choosing amino acids with opposite charges to those observed in the potential isocontours of the antigen. Figure 8 shows the residues mutated; Thr/Glu, Tyr/Lys, Tyr/Asp and Glu/Ser (System 4). Except the Tyr/Glu mutation, all other residues show a strong residue complementarity, and the polar Ser and Thr were replaced by negatively charged residues.

The most significant results of reduction of the IIP were observed for Thr/Glu and Ser/Asp mutations, which resulted in the addition of two charged residues. IIP for Thr/Glu and Ser/Asp mutations presented a reduction of approximately -10 and -20 kcal.mol^{-1}, respectively. Panels C and G, in Figure 8, show that there was a local change in the potential isocontours of the antigen in System 4. The final structures collected for the System 4 show that the complex is more compact, with a more complementary interface in relation to System 3 (Figures 7a and 7b).

The phenolic side chain of the Tyr residues have both aromatic and polar characteristics, which result in different interaction potential. The electrostatic potential is predominant in the hydroxyl group interaction, while in the aromatic ring the intermolecular interaction is predominantly van der Waals. In the case of Tyr/Glu mutation, the IIP increased as a result of repulsive interaction on the site. In System 3, the IIP of two Tyr residues are attractive, approximately -20 kcal. mol^{-1}, a sufficient energy for Tyr hydroxyl group to form hydrogen bonds. In this context, the mutations and Tyr to Lys and Tyr to Glu proved to be inefficient, since the IIP for the Lys is similar to that of Glu and Tyr shows positive IIP as a result of a local repulsion.

We conclude that uncharged residues at the Ag|Ab interface, such as Tyr residues (System 3), may result in more attractive interactions in relation to charged residues when in a favorable conformation with a good fit, showing that van der Waals forces play an important role in interface stabilization. In addition, charged residues on the antibody interact with antigen residues over longer distances, strengthening the binding between antigen and antibody. Therefore, the formation of Ag|Ab complex is mediated by residues whose interactions are dominated by van der Waals interactions and which are important for specificity (antigen recognition), while the charged residues are important for the enhancing affinity (binding). A MD strategy can therefore be used in antibody engineering for the evaluation of the proposed changes both locally to evaluate each residue for specificity, and globally to evaluate the effect of residue changes on affinity.

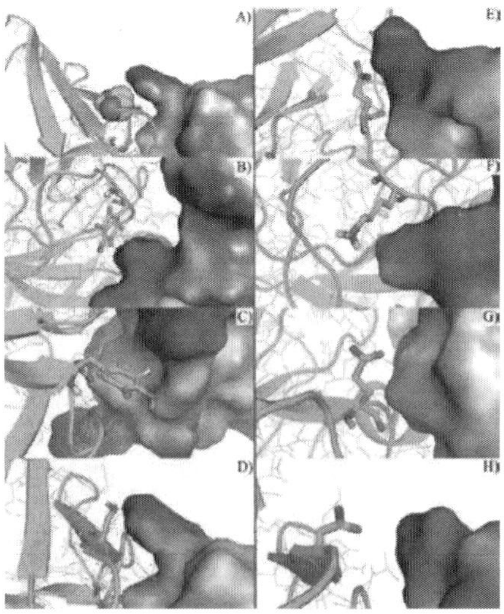

Fig. 8. Final configurations from MD simulations of Systems 3 and 4, showing the local environments of each residue of the antibody (left) compared to the antigen (right), represented as potential isocontours. Panels A, B, C, and D show the antibody residue in System 3 and panels E, F, G and H show the mutated residues in System 4.

Antimicrobial peptides

Antimicrobial peptides (AMP) play an important role in the innate immune defense system in all organisms and display activity against a wide variety of microorganisms. AMP are generally comprised of fewer than 50 amino acid residues, and are characterized both by an overall positive charge due the presence of multiple lysine and arginine residues and by a large proportion (at least 50 %) of hydrophobic residues. The Antimicrobial Peptide Database (APD) (Wang & Wang, 2004) contains more than 1750 AMPs, despite this experimental effort, knowledge with respect to the structures of these peptides remains limited since 3D structures are available for only 229 (13% of the total) of these AMPs. Therefore, it is evident that the investigation of AMPs structure and dynamics will yield knowledge that is important for the understanding of the mode of action of these biomolecules.

During the last decade, the accumulation of a large body of experimental data has demonstraed that AMPs act by predominantly affecting the integrity of cell membranes through their interaction with phospholipids. The cytoplasmic membranes of multicellular organisms and bacteria have distinct lipid compositions, and the specidicity of AMPs activity is therefore determined not only by the physico-chemical properties of the peptide but also by the composition of the target cell membranes. An understanding of the interactions between AMP and cell membrane at a molecular level is therefore of importance for the development of AMPs as therapeutic agents, which ideally would have a potent antimicrobial activity with low toxicity against host cells. The cytoplasmic membrane of the Gram negative bacteria *Escherichia coli* contains 70 to 80% neutral lipids with phosphatidylethanolamine (PE) head groups, 20 to 25% with phosphatidylglycerol (PG) head groups, which imparts a negative charge to the membrane, and other lipids that are present in smaller quantities (Dowhan, 1997). In eukaryotic cells, the extracellular leaflet of the cytoplasmic membrane is composed predominantly of lipid with phosphatidylcholine (PC) head groups (Mateo et al., 2006). It is crucial that these differences in lipid composition between eukaryotic and bacterial cell membranes be taken into account for investigations with respect to the toxic and antimicrobial activities of AMPs.

AMPs with altered amino acid sequences have been synthesized with the aim of decreasing toxicity and increasing cell selectivity, for example, the indolicidin (Selsted et al., 1992). Indolicidin (IND) is a peptide containing 13 amino acid residues (ILPWKWPWWPWRR-NH_2) and contains a large portion of tryptophan residues (39%). The IND is amidated at its C-terminus, displays activity against a broad range of microorganisms however is toxic to lymphocytes and erythrocytes. Two mutants of IND, a Pro/Ala mutant (CP10A ILAWKWAWWAWRR-NH_2) and the mutant CP11 (ILKKWPWWPWRRK-NH_2) that contains two extra positive charges, have been the target of several functional studies (Zhang et al., 2001; Halevy et al., 2003). This set of peptides is of interest as a model system since the toxic activity of the CP10A mutant is higher against human erythrocytes when compared to the native IND, while the CP11 has the lowest hemolytic activity among the three peptides (Halevy et al., 2003). Conversely, the antimicrobial activity is greatest for the CP11 and lowest for the CP10A (CP11>IND>CP10A) (Zhang et al., 2001). The functional properties of these mutants is therefore influenced by the physico-chemical characteristics of the peptide, and provides an excellent model system to study the peptide|membrane interaction using MD simulations.

MD studies were carried out with the IND and the two mutants CP10A and CP11 were used as model peptides to study the interaction of the peptides with two membrane models: one containing dipalmitoylphosphatidylcholine (DPPC) with 64 lipids in each layer, and the other containing a mixed bilayer with 64 lipids in each layer formed by 75 % of dipalmitoylphosphatidylethanolamine (DPPE) and 25 % of dipalmitoylphosphatidylglycerol (DPPG). Both bilayer systems were solvated and electrical neutrality of the mixed bilayer was maintained by the addition of one sodium ion for each DPPG molecule. Six systems were studied by introducing one of the peptides (IND, CP10A or CP11) into the bulk solution within a distance of about 4 nm from the center of the mass (CM) of one of the bilayers (DPPC or DPPE/DPPG). The simulations were initiated by energy minimization using the steepest descent algorithm, in order to eliminate bad contacts and undesirable forces. For each system, an initial simulation was performed for 0.5 ns applying restrictions to the peptides and bilayer atomic coordinates to equilibrate the systems using a $dt = 0.5\ fs$. The restrictions were removed and the systems were initially simulated by 30 fast heating and cooling cycles of simulated annealing (Fuzo et al., 2008) with $dt = 2\ fs$. In the next stage, the systems were submitted to 20 ns NpT simulations with dt of $2\ fs$, with 50 ns total simulation time for each system. Full details of the simulation parameters are given in Fuzo & Degrève (2009, 2011).

At the start of the simulations, the peptides were positioned in the solvent phase in order to allow a free interaction with the bilayer. During the simulation the peptides diffused through the solvent toward the membrane, before associating with the bilayers as observed in Figure 9. Differences were observed when the peptides were inserted into the two types of bilayers, and the mean position of the peptide CMs in relation to the bilayer CMs, given by a value d as measured along the axis perpendicular to bilayer plane, and which is lower when the peptide is inserted into the bilayer.

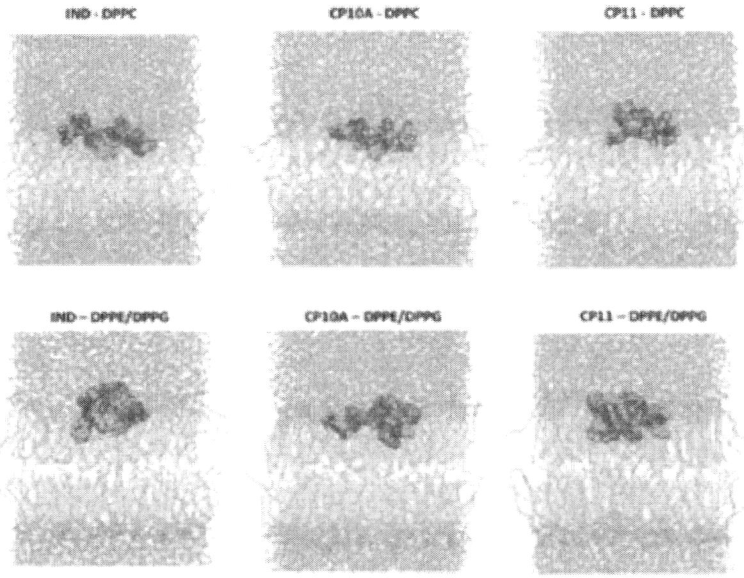

Fig. 9. Final configurations of systems containing peptides inserted into DPPC (top) and DPPE/DPPG (bottom) bilayers.

Table 1 shows that in the case of DPPC bilayers, all *d* values were less than the average position of the phosphorus atoms of DPPC (1.8 nm) in the order CP11>IND>CP10A, showing that the CP10A peptide inserted almost to the centre of the bilayer, the CP11 peptide remained close to the interface of the bilayer with the aqueous phase. The behavior of the IND peptide was intermediate between the two extremes. In the DPPE/DPPG systems the peptides also inserted into the bilayer giving *d* values that were lower than the phosphorous atoms, however it was observed that the order of insertion of the peptides was altered (ie. CP11>CP10A>IND) as compared to the DPPC bilayer. In all cases, the peptide that inserts deeper into a given bilayer is that which has the highest activity against a cell membrane of the same composition, thus in the systems studied with the DPPC bilayer the peptides that show a deeper insertion are those that have the highest hemolytic activity (Halevy et al., 2003), whereas for the DPPE/DPPG systems those peptides with higher antimicrobial activity were more deeply inserted (Zhang et al., 2001). This correlation between membrane activity and peptide insertion may have an important impact for the design of new AMPs, since more effective mutants can be designed rationallybased on the information obtained from MD simulation.

Table 1. Distances between the CM of peptides and bilayers along the axis perpendicular to bilayer plane (*d*), and the number of hydrogen bonds between the peptides and bilayers (*NHB*).

Peptide	d (nm)		N_{HB}	
	DPPC	DPPE/DPPG	DPPC	DPPE/DPPG
IND	1.43 ± 0.06	2.02 ± 0.05	24.4 ± 1.6	(13.8 ± 1.8)/(4.1 ± 0.8)
CP10A	1.24 ± 0.08	1.84 ± 0.08	20.8 ± 1.6	(8.1 ± 1.5)/(10.0 ± 1.6)
CP11	1.61 ± 0.07	1.70 ± 0.06	25.7 ± 2.1	(12.5 ± 1.5)/(4.0 ± 1.0)

Two important differences were obseved between the interactions of the peptides and the two different bilayers. The first was a greater number of hydrogen bonds (NHB) between the peptides and oxygen atoms of the DPPC bilayers (Table 1), which was due to a stretched conformation of the phosphotidylcholine groups of the DPPC molecule (Figure 10A**)** in which the distance between the CM of the phosphate and choline groups was 0.47 ± 0.06 nm, which compares with the phosphotidylethanolamine groups in the **DPPE/DPPG bilayer** where the distances between the phosphate and amine groups was 0.37 ± 0.01 nm. The proximity between the amine and phosphate groups in the DPPE molecules makes causes greater difficulty for the positively charged groups in peptides to approach the lipid headgroup phosphate group due to electrostatic repulsion with the positively charged amino group, thereby hindering the formation of HBs between the peptides and oxygen atoms. The second difference was the presence of cation-π interactions between choline groups of the DPPC and the side-chains of some Trp residues of the peptides. The cation-π interactions were evaluated by determining the radial distribution functions between the CM of the tryptophan side chains and nitrogen atoms of the choline groups, where the presence of the first peak (Dmax) at distances smaller than 0.5 nm and a number of choline groups greater than one (NChol) is indicative of cation-π interactions. These cation-π interactions were found with 1 to 2 choline groups for some tryptophan residues for all three peptides (shown in Table 2). Examples of the proximity of the choline headgroups for each peptide are shown in Figures 10C, 10D, and 10E for IND, CP10A, and CP11, respectivelly. The cation-π interactions observed between the choline groups and the tryptophan residues do not occur in lipid bilayers containing PE due the geometry of this group, in which the amine and phosphate groups are closer, thereby impeding the approximation of the tryptophan side chains to the amine groups to form cation-π interactions.

The simulations of the three peptides have shown that cation-π interactions between the choline headgroup and the tryptophan make an important contribution to the recognition of eukaryotic membranes by these peptides, thereby indicating that this type of interaction must be considered when designing new AMPs.

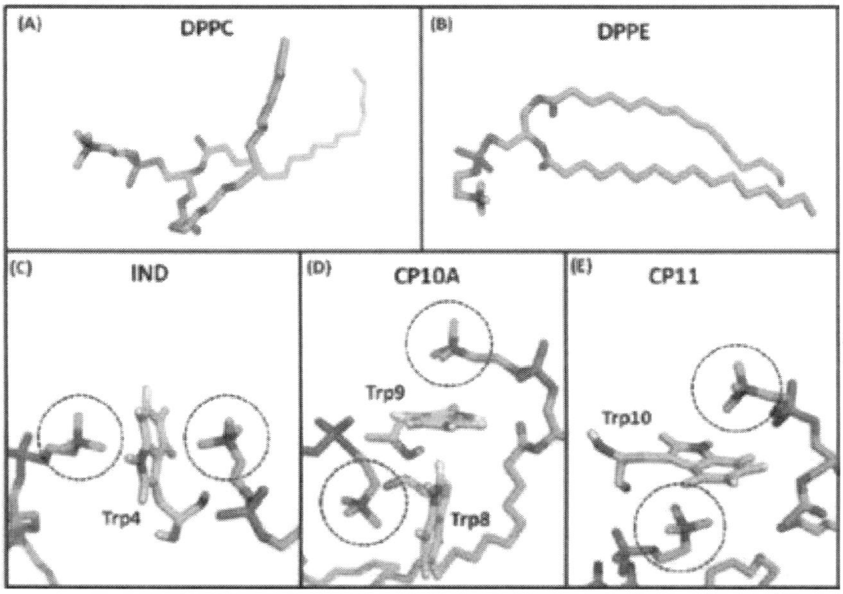

Fig. 10. Typical geometries of DPPC (A) and DPPE (B) molecules obtained from simulations. In (C), (D), and (E) are shown some examples of choline group neighbors of Trp residues for peptides IND, CP10A, and CP11, respectively. Choline groups are circled by a dotted line.

Table 2. Parameters extracted from radial distribution functions between the CM of Trp side chains and nitrogen atoms of choline groups.

Residue	N_{Chol}	D_{max} (nm)
	IND	
Trp4	2.0	0.46
Trp8	1.0	0.44
	CP10A	
Trp8	1.0	0.48
Trp9	1.0	0.46
Trp11	1.1	0.48
	CP11	
Trp5	1.0	0.46
Trp7	2.0	0.48
Trp10	2.1	0.45

CURRENT DEVELOPMENTS IN STRUCTURAL BIOINFORMATICS

One of the most important tools in the theoretical studies of bio(macro) molecule is the MD simulation. MD calculates the time dependent properties of a (bio)molecular system. Current MD simulations of biomolecules are directed to investigate structural and dynamics properties of systems composed of very large number of atoms over timescalesof many nanoseconds to microseconds (case of peptides folding). The continued growth of computer power and the advent of new algorithms like GPU-accelerated calculations are closely related to the more realistic approach of the molecular systems that allows chemical processes in condensed phases to be studied in an accurate and unbiased way. For example, MD trajectories can be generated with forces obtained from accurate electronic structure details (*ab initio* MD), as well as the improvement of the calculation efficiency when dealing with the complexity of biological problems. In this context we can cite a pharmacological interesting system, the case of quaternary structures of virus proteins and their interactions with active biological molecules, and the protein-carbohydrate interactions in glycoproteins. Future bioinformatics oriented studies will likely provide us with valuable tools for validating, improvement, prediction and development of a variety of biotechnological applications.

REFERENCES

1. Aehle, W. (2004). Enzymes in Industry – Production and Applications, John Wiley, ISBN 352-729-592-5, New York.
2. Alberts, B.; Bray, D.; Lewis, J.; Raff, M.; Roberts, K. & Watson, J.D. (1989). *In: The Immune System.* (Robertson, M., Ed.), Garland, New York.
3. Allen, M. P. & Tildesley, D. J. (1992). *Computer Simulation of Liquids*, Claredon Press, ISBN 978-019-855645-9, Oxford.
4. Åqvist, J. (April 13 2007). The Åqvist Group, May 28 2011, Available from: <http://xray.bmc.uu.se/~aqwww/> .
5. Beg, Q. K.; Kapoor, M.; Mahajan, L.; Hoondal, G. S. (2001). Microbial xylanases and their industrial applications: a review. *Appl Microbiol Biotechnol.*, 56, April 27, pp. 326-338 ISSN 0175-7598.
6. Cardoso, D. F.; Nato, F.; England, P.; Ferreira, M. L.; Vaughan, T. J.; Mota, I.; Maize, J. C.; Choumet, V. & Lafaye, P. (2000). Neutralizing human anti crotoxin scFv isolated from a nonimmunized phage library. *Scandinavian Journal of Immunology,*, 51, pp.337-344, ISSN 1365-3083.
7. Choumet, V.; Faure, G.; Robbe-Vincent, A.; Saliou, B.; Mazié, J. C. & Bon, C. (2003). Immunochemical analysis of a snake venom phospholipase A2 neurotoxin, crotoxin, with monoclonal antibodies. *Molecular Immunology*, 29, pp. 871-882, ISSN 0161-5890.

8. Cientistas Associados, (2008) *Intelligent System for developing antibodies based on the structure of the antigen (AbEvo)*, last accessed 28/07/2011, Avaiable from:

9. <http://www.cientistas.com.br/projetos/projects.htm>.

10. Degrève, L.; Brancaleoni, G. H.; Fuzo, C. A.; Lourenzoni, M. R.; Mazzé, F. M.; Namba, A. M. & Vieira, D. S. (2004). On the role of water in the protein activity. *Brazilian Journal of Physics*, 34, March 2004, pp. 102-115, ISSN 0103-9733.

11. Dowhan, W. (1997). Molecular basis for membrane phospholipid diversity: why are there so many lipids? *Annual Review of Biochemistry*, 66, July 1997, pp. 199-232, ISSN 0066-4154.

12. Faure, G.; Guillaume, J. L.; Camoin, L.; Saliou, B. & Bon, C. (1991). Multiplicity of acidic subunit isoforms of crotoxin, the phospholipase A2 neurotoxin from Crotalus durissus terrificus venom, results from posttranslational modifications. *Biochem*istry, 32, pp. 8074-8083, ISSN 0006-2960.

13. Fernández, C. & Wider, G. (2003). TROSY in NMR studies of the Structure and Function of Large Biological Macromolecules. *Current Opinion in Structural Biology*, 13, October 2003, pp. 570-580 ISSN 0959-440x.

14. Fuzo, C.A. & Degrève, L. (2011). Study of the antimicrobial peptide indolicidin and mutants in eukaryotic modelled membrane by molecular dynamics simulations. *Molecular Physics*, 109, January 2011, pp. 289-300, ISSN 1362–3028.

15. Fuzo, C.A. (2009). *Molecular simulation studies of peptide/bilayer systems: application to structure/activity relationship of the indolicidin and mutants.* Doctoral Thesis, University of São Paulo, Brazil.

16. Fuzo, C.A.; Castro, J.R.M & Degrève, L. (2008). Searching the global minimum of a peptide/bilayer potential energy surface by fast heating and cooling cycles of simulated annealing. *International Journal of Quantum Chemistry*, 108, June 2008, pp. 2403-2407, ISSN 1097-461x.

17. Georis, J.; Esteves, F. L.; Lamotte-Brasseur, J.; Bougnet,V.; Devreese, B.; Giannotta, F.; Granier, B. (2000). An additional aromatic interaction improves the thermostability and thermophilicity of a mesophilic family 11 xylanase: structural basis and molecular study. *Protein Sci.*, 9, September 8, pp. 466-475 ISSN 0961-8368.

18. Halevy, R.; Rozek, A.; Kolushevav, S.; Hancock, R.E.W. & Jelinek, R. (2003). Membrane binding and permeation by indolicidin analogs studied by a biomimetic lipid/polydiacetylene vesicle assay. *Peptides*, 24, November 2003, pp. 1753-1761, ISSN 0196-9781.

19. Jorgensen, W. L.; Chandrasekhar, J.; Madura, J. D.; Impey, W. R.; Klein, M. L. (1983). Comparison of simple potential functions for simulating liquid water. *J. Chem. Phys.*, 79, July 15, pp. 926-938 ISSN 0021-9606.

20. Laskowski, R. A.; Macarthur, M. W.; Moss, D. S. & Thornton, J. M. (1993). PROCHECK: a program to check the stereochemical quality of protein structures. *Journal of Applied Crystallography*, 26, pp. 283-291, ISSN 1600-5767.

21. Leach, A. R. (2001). Molecular Modelling: Principles and Applications, Prentice Hall, ISBN 058-238-210-6, Harlow, England.

22. Luthy, R.; Bowie, J. U. & Eisenberg, D. (1992). Assessment of protein models with threedimensional profiles. *Nature*, 356, p.83-85, ISSN 0028-0836.

23. Maitland, G. C. & Smith, G. C. (1971). The Intermolecular Pair Potential of Argon. *Molecular Physics*, 22, January 1971, pp. 861-868 ISSN 1362-3028.

24. Maitland, G. C.; Rigby, M., Smith, E., B., & Wakeham, W. A. (1981). *Intermolecular Forces : Their Origin and Determination,* Claredon Press, ISBN 019-855-611-x, Oxford.

25. Mancini, E. J.; Clarke, M.; Gowen, B. E.; Rutten, T. & Fuller, S. D. (2000). Cryo-Electron Microscopy Reveals the Functional Organization of an Enveloped Virus, Semliki Forest Virus. *Molecular Cell*, 5, February 2000, pp. 255-266 ISSN 1097-2765

26. Mateo, C.R.; Gómez, J.; Villalaín, J. & Ros, J.M.G. (2006). *Protein-Lipid Interactions: New Approaches and merging Concepts.* Springer-Verlag, ISBN 3-540-28400-1, Germany.

27. McQuarrie, D. A. (1976). *Statistical Mechanics*, Harper and Row, New York.

28. Morokuma, K. (1977). Why do Molecules Interact? The Origin of Electron Donor-Acceptor Complexes, Hydrogen Bonding and Proton Affinity. *Accounts of Chemical Research*, 10, August 77, pp. 294-300 ISSN 0001-4842..

29. Rapaport, D. C. (2005). *The Art of molecular Dynamics Simulation,* Cambridge University Press, ISBN 052-182-568-7, Cambridge, England.

30. Ritchie, D.W. (2003). Evaluation of protein docking predictions using *Hex* 3.1 in CAPRI rounds 1 and 2, *Proteins: Structure, Function, and Bioinformatics,* 52, pp. 98-106, ISSN 1097-0134.

31. Rocco, A. G.; Mollica, L.;Ricchiuto, P.; Baptista, A. M.; Gianazza, E.; Eberini, I. (2008). Characterization of the Protein Unfolding Processes Induced by Urea and Temperature. *Biophysical Journal*, 94, November 2010, pp. 2241-2251 ISSN 0006-3495.

32. Roitt, I. M. (1997). *Essential immunology*, Oxford: Blackwell Science, Malden, MA.

33. Šali A. & Blundell, T. L. (1993). Comparative protein modelling by satisfaction of spatial restraints. *Journal of Molecular Biology*, 234, pp.779-815, ISSN 0022-2836.

34. Sánchez, R. & Šali A. (1997). A. Evaluation of comparative protein structure modeling by MODELLER-3. *Proteins*, 1, pp. 50-58.

35. Selsted, M. E.; Novotny, M. J.; Morris, W. L.; Tang, Y. Q.; Smith, W. & Cullor, J. S. (1992). Indolicidin, a novel bactericidal tridecapeptide amide from neutrophils. *Journal of Biological Chemistry*, 267, March 1992, pp. 4292-4295, ISSN 1083-351x.

36. Shaw, D. E.; Maragakis, P.; Lindorff-Larsen, K.; Piana, S.; Dror, R. O.; Eastwood, M. P. ; Bank, J. A.; Jumper, J. M.; Salmon, J. K.; Shan, Y. & Wriggers, W. (2010) Atomic- Level Characterization of the Structural Dynamics of Proteins. *Science*, 330, October 2010, pp. 341-346 ISSN 0036-8075.

37. Torrez, M.; Schultehenrich, M.; Livesay, D. R. (2003) Conferring Thermostability to Mesophilic Proteins through Optimized Electrostatic Surfaces. *Biophys J.*, 85, November 1, pp. 2845-2853 ISSN 0006-3495.

38. van Gunsteren, W. F.; Dolenc, J. & Mark, A. E. (2008). Molecular Simulation as an Aid to Experimentalist. *Current Opinion in Structural Biology*, 18, February 2008, pp. 149- 153 ISSN 0959-440x.

39. Vieira, D. S., Degrève, L. (2009). An Insight into the Thermostability of a Pair of Xylanases : The Role of Hydrogen Bonds. *Molecular Physics*, 107, February 26, pp. 59-69 ISSN 0026-8976.

40. Wakarchuk, W. W.; Sung, W. L.; Campbell, R. L.; Cunningham, A.; Watson, D. C.; Yaguchi M. (1994). Thermostabilization of the *Bacillus circulans* xylanase by the introduction of disulfide bonds. *Protein Eng.*, 7, July 17, pp. 1379-1386 ISSN 0269-2139.

41. Wang, Z. & Wang, G. (2004). APD: the Antimicrobial Peptide Database. *Nucleic Acids Research*, 32, January 2004, pp. D590-D592, ISSN 1362-4962.

42. Zhang, L.; Rozek, A. & Hancock, R.E.W. (2001). Interaction of cationic antimicrobial peptides with model membranes. Journal of Biological Chemistry, 276, July 2001, pp. 35714-35722, ISSN 1083-351x.

CHAPTER-19

UTILIZING HIGH PERFORMANCE ION EXCHANGE CHROMATOGRAPHY FOR THE DEVELOPMENT AND DETAILED ANALYSIS OF MONOCLONAL ANTIBODIES

INTRODUCTION: AN OVERVIEW

The process of monoclonal antibody development and its subsequent physicochemical characterization is significantly enhanced through the application of high-performance ion exchange chromatography. This powerful analytical technique plays a crucial role in the isolation, purification, and detailed analysis of monoclonal antibodies. It allows researchers to separate these antibodies based on their charge properties, offering precise insights into their structure and purity. The physicochemical characterization, facilitated by high-performance ion exchange chromatography, encompasses a wide range of factors such as charge heterogeneity, post-translational modifications, and impurity profiles. This in-depth analysis not only ensures the production of highly pure and functionally active monoclonal antibodies but also contributes to their successful application in various therapeutic, diagnostic, and research fields.

Monoclonal antibodies (mAbs) represent a significant portion of products in the biopharmaceuticals market (Reichert, 2011; Scolnik, 2009). mAbs have been developed to treat a variety of indications to address significant unmet medical needs (Waldmann, 2003; Reichert & Valge-Archer, 2007; Ziegelbauer et al., 2008), and are generally target-specific and well tolerated with a relatively long half-life, contributing to the success of the molecule class for drug development (Scolnik, 2009). Of the classes of antibodies, or immunoglobulins, IgG1 is the most common immunoglobulin used for pharmaceutical and biomedical purposes (Reichert et al., 2005); however, other immunoglobulin types (e.g., IgG2, IgG4) and mAb-related products (e.g., Fc-fusion proteins, Fabs, etc.) are also being used for therapeutic purposes (Hudson & Souriau, 2003).

While a successful mAb product can generate upwards of a billion dollars or more in sales annually (Reichert, 2009), it takes a significant amount of time and resources to develop a new therapeutic mAb; current estimates indicate that it can take about 10-15 years (Dickson & Gagnon, 2004; DiMasi et al., 2003) and over $1 billion to bring a new biologic drug to market (DiMasi & Grabowski, 2007). Addressing bottlenecks and making improvements in the development process is essential to save time and money, expediting the delivery of new drugs to the clinic.

This chapter will briefly cover monoclonal antibody development, production and purification, and then focus on antibody characterization, particularly charge-species analysis using high performance ion exchange chromatography (IEC). Method development strategies, method robustness, validation and automation will be discussed. This chapter aims to be a reference text demonstrating the utility of IEC as well as providing strategies for developing rugged IEC methods for the characterization of therapeutic mAbs.

MONOCLONAL ANTIBODY DEVELOPMENT

Antibodies are physiological blood components that are produced by B lymphocytes, intended to bind to and neutralize foreign antigens and pathogens. Antibodies bind to a corresponding antigen in a highly specific manner. Although not covered in this chapter, potential mechanisms of action for mAbs have been described (Green et al., 2000). Polyclonal antibodies are a combination of immunoglobulin molecules secreted against a specific antigen, each identifying a different epitope. In contrast, monoclonal antibodies are derived from a single cell line that are all clones of a unique parent cell.

The first monoclonal antibody product was approved in 1986 and was a murine antibody for the prevention of kidney transplant rejection; however, patients frequently developed antibodies against the mouse-derived mAbs, which limited their effectiveness (Jones et al., 2007; Kuus-Reichel et al., 1994; Shawler et al., 1985). Advances in antibody engineering yielded techniques for generating chimeric mAbs that contain sequences from both human and murine sources (Morrison et al., 1984; Reichmann et al., 1988). Many of the mAbs approved for commercialization in the 1990s and early 2000s were chimeric antibodies. Chimeric antibody products are superior to murine antibodies, but they still pose a risk of immunogenicity to patients from their residual murine components (Carter, 2001). New technologies were developed to produce humanized mAbs, which contain approximately 95%

human components and 5% murine components (Carter & Presta, 2000, 2002). In these humanized mAbs, the CDRs of a human antibody gene were replaced by those from the CDR of a murine mAb gene (Figure 1). The resulting humanized antibody has the same antigen binding properties as the original murine antibody but contains minimal murine sequences (Co & Queen, 1991).

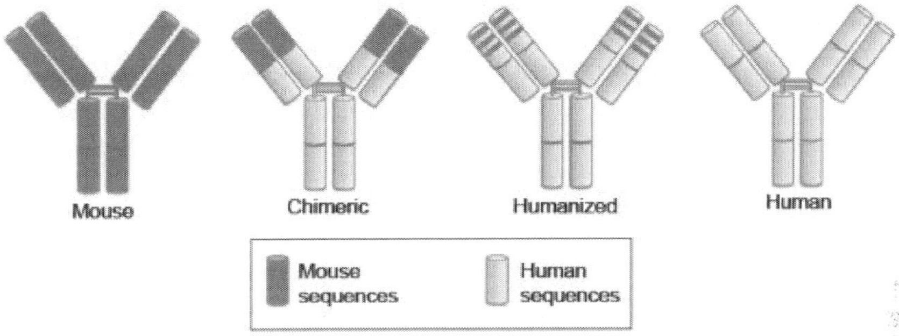

Fig. 1. Schematic representation of the sequence composition of murine, chimeric, humanized, and human antibodies.

More recently, less immunogenic therapeutic antibody products were developed by creating fully human mAbs. Several technologies exist to develop fully human antibodies, each falling into one of the two general classes—*in vivo* approaches using a murine system in which the immunoglobulin genes have been replaced by their human counterparts, or *in vitro* approaches such as phage display libraries containing millions of variations of antibody sequences coupled with a mechanism to express and screen these antibodies *in vitro* (Lonberg, 2005; McCafferty et al., 1990). Combining *in vivo* and *in vitro* discovery and molecular engineering technologies allows exquisite control of the antibody sequences and properties that was not possible when mAbs were first developed. Because of these advanced antibody engineering technologies, almost all antibody products currently in development are humanized or fully human mAbs and their derivatives.

MONOCLONAL ANTIBODY PRODUCTION

Recombinant monoclonal antibodies are typically produced in mammalian cell lines under defined cell culture conditions. Commercial scale production processes vary depending on the mAb, but generally, cells are taken from a master cell bank and inoculated into small- scale bioreactors. The cell culture is transferred to increasingly larger bioreactors until it reaches the final commercial scale bioreactor. Currently, final scale reactors have volumes ranging from 12,000 L to 24,000 L (Gottschalk, 2009). The cells are cultured in a controlled environment for days to weeks, and then the cell culture fluid is harvested by centrifugation (Shukla & Kandula, 2008). In mammalian cells, the product monoclonal antibodies are secreted from the cells into the supporting fluid medium. Centrifugation separates the cells from the fluids and facilitates simpler recovery procedures downstream.

Commercial mAb production requires considerable preproduction effort to ensure that the cell line is stable and can produce commercially appropriate quantities of antibody. In addition, the commercial production process must produce a product that meets the quality expectations of regulatory authorities. In the past few years, improvements have been made in critical areas, such as cell line generation and large-scale cell culture production, to maximize specific antibody productivity from a given cell line and improve overall productivity in bioreactors. These advances include the use of new expression vectors and transfection technology, high-throughput, robust screening technologies to select the highest producing clones rapidly and more effectively, improvements in cell culture and optimized bioreactor processes (Li et al., 2010; Schlatter et al., 2005). As a result, the production of cell lines expressing multigram quantities of antibody per liter of culture medium is now routine.

The product quality and product heterogeneity of every mAb is highly dependent on its manufacturing process (Abu-Absi et al., 2010; Horvath et al., 2010). The ideal manufacturing conditions would have optimal production levels of product in conjunction with the desired product quality profile. Attributes that are typically deemed critical in selecting stable clones and cell culture conditions are the product titer and product heterogeneity, including charged species and aggregates. Production titers directly correlate to the costs of the process and are desired to be as high as possible with minimal impact to other quality attributes of the product (Kelley, 2009). Critical quality attributes of the product, such as the level of aggregation, are carefully monitored, as failure to control critical quality attributes may pose a safety risk to the patient (Rosenberg, 2006).

MONOCLONAL ANTIBODY PURIFICATION AND FORMULATION

Once monoclonal antibodies are produced in cells, the mAbs must be recovered and purified. Recovery and purification processes vary widely depending on the manufacturing process and specific mAb characteristics, but generally, the isolation and purification of mAbs involve a centrifugation step to separate the cells from the cell culture fluid containing the mAb product, one or more chromatography steps, which can include affinity chromatography, cation or anion exchange chromatography, hydrophobic interaction chromatography (HIC) and displacement chromatography (Shukla et al., 2007), and filtration or precipitation steps (Gottschalk, 2009). Many of the purification steps are designed to remove contaminants and adventitious agents (e.g., bacteria, fungi, viruses, and mycoplasma).

After elution from the final chromatographic purification step, a unit operation is required to exchange the components of the chromatography elution buffer with the chosen formulation components. The predominant technology that has been used in the industry for buffer exchange and concentration is ultrafiltration/diafiltration using tangential-flow filtration (Genovesi, 1983; Shiloach, 1988; van Reis, 2001). After this step, the drug substance is filtered and typically frozen as bulk for storage until filling occurs to produce the final drug product.

The formulation of the mAb therapeutic is chosen in part to ensure product quality during shelf life. Formulations are designed to minimize protein aggregation, decrease viscosity, and increase shelf life through preventing degradation (Shire, 2009). High protein concentration formulations are being developed to allow for subcutaneous or intramuscular delivery of mAb products (Shire et al., 2004). Historically, the most conventional route of delivery for protein drugs has been intravenous administration because of poor bioavailability by most other routes, greater control during clinical administration, and faster pharmaceutical development. Subcutaneous delivery allows for home administration and improved patient compliance. However, development of high protein concentration formulations involves unique manufacturing challenges compared to low concentration formulations, such as higher viscosities and necessary changes to unit operation steps.

MONOCLONAL ANTIBODY CHARACTERIZATION AND RELEASE TESTING

Biopharmaceutical manufacturing of monoclonal antibodies produces a heterogeneous product of structurally related species. Antibody speciation can occur throughout the manufacturing process at various steps, including cell culture, harvest, purification, formulation, filling and during shelf life. Full-length monoclonal antibodies are high molecular weight proteins (around 150,000 Da), and have highly complex secondary and tertiary structures, subject to post-translational modifications. Therefore, product characterization and quality control testing are required at critical points throughout clinical development and manufacturing to control for these species (Harris et al., 2004). Figure 2 depicts the structure of a monoclonal antibody compared to a small molecule drug, illustrating the increased complexity of a biologic compared to a small molecule therapeutic.

Antibodies can be characterized by many physicochemical properties including hydrated size (Stokes radius), molecular weight, charge, hydrophobicity, electrophoretic mobility, isoelectric point (pI), sedimentation velocity, glycosylation, and spectral properties. The nature of each species can be related to differences in their primary, secondary, tertiary, or quaternary protein structures. In addition, monoclonal antibodies are susceptible to chemical or enzymatic modification, particularly at sites that are exposed to the protein- liquid interface. Product heterogeneity can be caused by a number of modifications, such as C-terminal processing of lysine residues (Harris, 1995; Santora et al., 1999; Weitzhandler et al., 1998), deamidation (Di Donato et al., 1993; Hsu et al., 1998), glycation (nonenzymatic glucose addition) (Quan et al., 2008), amino acid sequence variations (Yang et al., 2010), and noncovalent complexes (Santora et al., 2001).

Antibody products are characterized by physicochemical, immunochemical, and biological methods. Guidance documents have been issued by regulatory agencies and industry representatives recommending approaches for protein characterization (International Conference on Harmonisation of Technical Requirements for the Registration of Pharmaceuticals for Human Use (ICH), 1999; Schnerman et al., 2004). These orthogonal assays include potency, identity and purity assays, which evaluate "critical quality attributes" such as size and charge heterogeneity. These critical quality attributes are part of the overall target product profile, which is based on the desired clinical performance. The extent of characterization is linked to the level of risk associated with each phase of drug development. For example, while there may not be sufficient time or resources for extensive characterization of an antibody during early stage development, it is expected that the molecule will be well-characterized before the Biologic License Application (BLA) is submitted to the regulatory agencies.

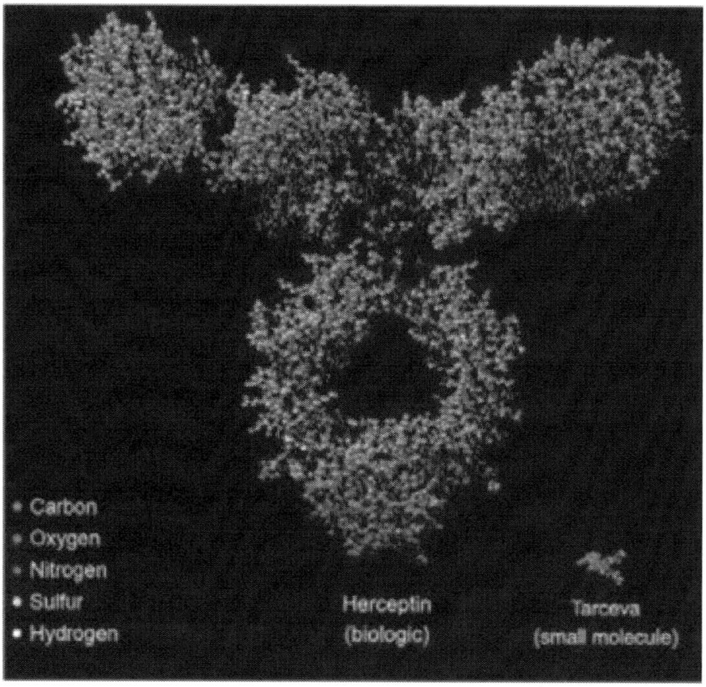

Fig. 2. Comparison of the structures of a mAb (Herceptin) and a small molecule therapeutic (Tarceva).

Many of the recommended protein characterization assays are based on liquid chromatography methods, such as ion exchange chromatography (IEC) for charge heterogeneity analysis, size exclusion chromatography (SEC) for size heterogeneity, and reversed-phase high performance liquid chromatography (RP-HPLC) for peptide mapping (Chirino & Mire-Sluis, 2004). The remainder of this chapter will primarily focus on ion exchange chromatography methods for analyzing charge heterogeneity for characterization and support of formulation and process development, as well as for lot release testing of drug substance and drug product (Schnerman et al., 2004).

ANALYZING MAB CHARGE HETEROGENEITY USING IEC

As mentioned previously, monoclonal antibodies are large proteins that are quite complex. While the light chain and heavy chain sequences of a particular mAb may be known, a number of modifications can introduce heterogeneity in the product. Thus, it is important to develop appropriate analytical methods to resolve the minor forms of the product. Analytical biochemists routinely use IEC for resolving charge variants of the protein. The scientist must then utilize orthogonal analytical methods to characterize the separated peaks of the ion exchange chromatogram. The characterization of a mAb is particularly important if the modifications occur in the complementarity-determining regions (CDR), as modifications in the CDR can affect the binding activity and potency of the mAb.

A strategy for the assignment of peaks from a weak cation exchange (WCX) mAb separation using a salt gradient has been published (Harris et al., 2001). Seven forms of a therapeutic recombinant antibody were resolved by cation-exchange chromatography. The peak fractions were collected, and structural differences were assigned by peptide mapping, which involves digesting the mAb with an enzyme and injecting the digest onto a reverse-phase column coupled to a mass spectrometer, and by hydrophobic interaction chromatography (HIC) after papain digestion. The peaks in this particular case were attributed to deamidation, isomerization, and succinimide intermediates. Other orthogonal analytical methods were used to characterize the IEC peaks; one of these methods—potency testing—determined that one minor peak demonstrated much lower potency than the main peak.

In another study, a recombinant humanized monoclonal IgG1 antibody with different states of glycosylation on the conserved asparagine residue in the CH2 domain was analyzed by cation exchange chromatography (Gaza-Bulseco et al., 2008). Two major peaks were observed and were further characterized by enzymatic digestion and mass spectrometry. It was found that this recombinant monoclonal antibody contained three glycosylation states—zero, one or two glycosylated heavy chains. The peak that eluted earlier on the cation exchange column contained antibodies with two glycosylated heavy chains containing fucosylated biantennary complex oligosaccharides with zero, one or two terminal galactose residues. The peak that eluted later from the column contained antibodies with zero, one or two glycosylated heavy chains. The oligosaccharide on the antibodies that eluted in the later peak was composed of only two GlcNAc residues. These results indicate that conformational changes, caused by different types of neutral oligosaccharides as well as the absence of certain oligosaccharides, can be differentiated by cation exchange column chromatography.

Lot release testing of mAbs

Once the mAb is purified and formulated, the resulting drug substance must be tested prior to lot release. A set of tests and acceptance criteria are established based on mAb characterization and regulatory requirements in order to ensure product quality (Food & Drug Administration (FDA), 1999). These tests typically include appearance, identity, purity, protein concentration, potency of the molecule, microbial limits or bioburden, and bacterial endotoxins (Table 1). IEC is one of the most frequently used lot release methods for purity for mAbs (Schnerman et al., 2004). Once these tests are performed and the results meet the established acceptance criteria, a Certificate of Analysis (COA) is generated and the lot is released for use. Finally, adequate stability studies should be performed on the mAb drug substance (e.g. frozen bulk for storage) and drug product (e.g. final vial) according to regulatory guidelines (Food & Drug Administration (FDA), 2003).

Ion exchange chromatography (IEC) has been a platform for monoclonal antibody purification and characterization for many years. For the analysis of charged species of proteins, IEC is a popular method due to the fact that it preserves the native conformation and maintains bioactivity of the protein, is relatively easy of use, is supported by the maturity of the equipment and consumables market, and has widespread use in the biopharmaceutical industry (Rea et al., 2010).

Charge-based methods are an integral component of characterization studies and quality control strategies because they are sensitive to many types of modifications. Charge profiling of intact antibodies can resolve species related to protein conformation, size, sequence species, glycosylation and post-translational modifications (Gaza-Bulseco et al., 2008; Harris et al., 2001; He et al., 2010). Although IEC can be used to track specific species, it is common to group all species not associated with the main peak and report them as either acidic or basic species (Figure 3). In addition, fractions collected from an IEC run can often be directly injected onto orthogonal columns for further analysis, such as reverse-phase and size exclusion chromatography columns, or submitted for potency testing.

Table 1. Commonly used tests found on a Certificate of Analysis for lot release; a selected subset is used for stability testing of mAbs.

Attribute	Test Name
Appearance	Color, Opalescence and Clarity
Identity	Peptide Mapping by RP-HPLC (Reverse-Phase HPLC), or
	MALDI (Matrix-Assisted Laser Deionization) Mass Spectrometry, or
	UV Spectroscopy (2nd Derivative)
Purity	Limulus Amebocyte Lysate (Endotoxin)
	Size Exclusion Chromatography (SEC)
	CE-SDS (Capillary Electrophoresis-Sodium Dodecyl Sulfate)
	IEC (Ion Exchange Chromatography) or icIEF (Imaged Capillary Isoeletric Focusing)
	Glycosylation Profile
	Peptide Mapping by RP-HPLC
Potency	Potency (ELISA/Cell-Based Assay)
Strength	UV Spectroscopy
General Tests	Osmolality
	pH
	Surfactant Concentration (e.g. Polysorbate 20)

ION EXCHANGE CHROMATOGRAPHY MECHANISM FOR MONOCLONAL ANTIBODIES (MABS)

IEC separates proteins based on differences in the surface charge of the molecules, with separation being dictated by the protein interaction with the stationary phase. The two main categories of ion exchange chromatography are cation exchange (CEX) and anion exchange (AEX). Cation exchange chromatography retains biomolecules by the interaction of the negatively-charged resin with histidine (pK ~ 6.5), lysine (pK ~ 10) and arginine (pK ~ 12) in the protein. Anion exchange chromatography primarily retains biomolecules by the interaction of the positively-charged resin with aspartic or glutamic acid side chains, which have pKa of ~4.4. In addition to the amino acid residues, cation exchange columns can also separate deamidated, glycated and other charged variants. Anion exchange columns have also been useful for separating phosphorylated and hydroxyl modified amino acids. When the pH equals the pI value of the protein, the net charge on the molecule is zero. However, significant retention can occur for proteins even when the pH of the mobile phase is equal to the pI of the molecule; despite an overall net charge of zero, only a portion of the mAb molecule will interact with the stationary phase, and there will be a net charge on that portion of the molecule because of an uneven distribution of charged groups throughout the molecule (Vlasak & Ionescu, 2008). Thus, it is possible to separate proteins having very similar charge (Figure 4), or even structural isomers with identical pI values, by ion exchange chromatography.

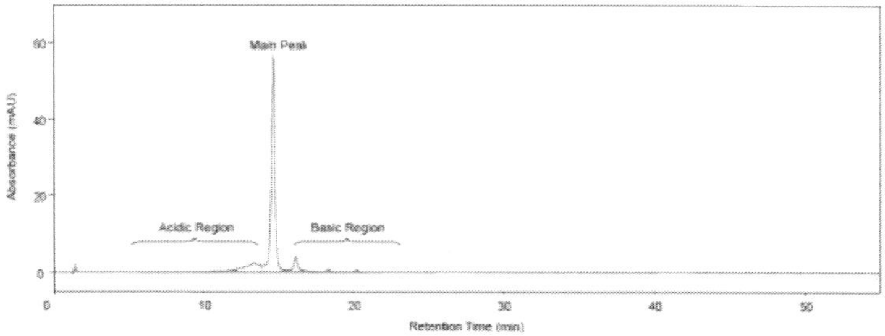

Fig. 3. Typical cation exchange chromatogram for analytical characterization of a mAb. Integration is shown, and main peak, acidic and basic regions are denoted.

There are two ways to elute the protein from the IEC column: 1) increasing salt concentration with time or 2) by varying the mobile phase pH value as a function of time. Increasing the salt concentration elutes the protein by increasing the ionic strength of the mobile phase, thus affecting the charge interaction of the mAb and the stationary phase. A pH gradient elutes the protein by changing the charge on the molecule, thus affecting the binding of the molecule to the stationary phase. While conventional salt gradient cation exchange chromatography is regarded as the gold standard for charge sensitive antibody analysis (Vlasak & Ionescu, 2008), method parameters such as column type, mobile phase pH, and salt concentration gradient often need to be optimized for each individual antibody.

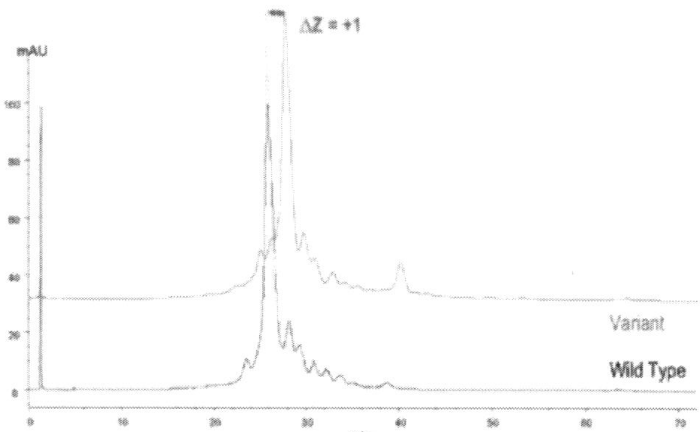

Fig. 4. Separation of mAbs differing by only one charge, a single amino acid change to primary structure. The elution buffer (0.5 M NaCl in 20 mM Tris, pH 7.3) was increased linearly on a ProPac WCX-10 column (4 x 250 mm), which was held at 50 °C and had a flow rate of 1 mL min^{-1}.

A recent publication described a multi-product pH gradient IEC method for the separation of mAb charge species for a variety of mAbs using a single method

(Farnan &Moreno, 2009). The following sections will discuss both salt-gradient and pH-gradient based elution methods, and the combination of the two modalities (hybrid methods).

DEVELOPING A SALT-BASED IEC METHOD

Salt-based IEC separations are developed by choosing a cation or anion exchange column and varying the buffer system, mobile phase pH value, and ionic strength gradient of the elution buffer. Figure 5 shows a typical development workflow for salt-based IEC and pH- based IEC development, and can serve as a guide for initial IEC method development. The following sections will cover in more detail the outputs to consider when screening various parameters during development. More general considerations regarding HPLC method development can be found in various texts (Kastner, 2000; Snyder et al., 1997).

Salt Gradient IEC pH Gradient IEC

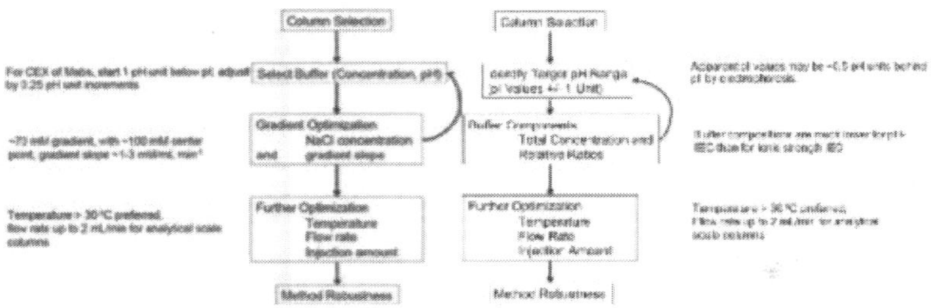

Fig. 5. Sequential salt-gradient IEC and pH-gradient IEC method development and optimization work flow.

Column selection, buffers and operating parameters for salt gradient IEC

Column selection is perhaps the most subjective part of the optimization process; picking between the different vendor offerings and functionalities can be difficult. Prior experience, data in the literature or unpublished results within the organization are often the best starting points.

Analytical ion exchange chromatography of proteins is typically carried out using mobile phases that are relatively neutral in pH values, 5.5 to 8.5 This general practice is recommended because at pH extremes, the protein is more likely to degrade. The selection of whether to use anion or cation exchange chromatography is also driven by the isoelectric point of the protein (pI) and the species to be resolved, e.g., phosphorylated species, C- terminal lysine variants, etc.

If the pI value of the mAb is greater than 8, a CEX column is evaluated at pH 6-7 initially. CEX primarily retains mAbs by the interaction of acid groups on the

CEX resin with lysine, arginine and histidine side chains on the mAb. Since mAbs are positively charged at a mobile phase pH below their pI, the mAb species would likely be retained and resolved on a CEX column under the recommended mobile phase pH range.

If the pI value of the mAb is less than 6, an AEX column is evaluated at a pH above 6 initially. AEX primarily retains biomolecules by the interaction of amine groups on the ion exchange resin with aspartic or glutamic acid side chains. Since mAbs are negatively charged at a mobile phase pH above their pI, the mAb species would likely be retained and resolved on an AEX column.

For intermediate pI values of 6-8, both CEX and AEX are evaluated because of the possibility that the portion of the mAb that interacts with the stationary phase, typically the side chains that are exposed to the mobile phase, has a different charge than the pI would suggest, e.g., the surface charge of the mAb is positive despite the entire mAb having an overall negative charge. Ultimately, the species of interest that are to be resolved determine whether CEX or AEX is chosen for molecules with intermediate pI's; the separation mode that better separates the species of interest is usually the one that is chosen for mAb analysis.

Figure 6 shows CEX and AEX chromatograms of a Fab (mAb fragment) reference sample and thermally stressed sample. In this case, the Fab molecule has a nominal pI value for the main species of 7.6. It should be noted that the separations on the AEX and CEX columns were each optimized independently for column type, pH value and salt gradient. It should also be noted that the terms "strong" and "weak" (in SAX, strong anion exchange, and WCX, weak cation exchange) refer to the extent of variation of ionization with pH due to the functional groups on the resin and not the strength of binding. Strong ion exchangers are completely ionized over a wide pH range whereas with weak ion exchangers, the degree of dissociation and thus exchange capacity varies much more markedly with pH. For this example, SAX results in significantly more peaks and much better resolution of the charge species in comparison to the WCX chromatogram. Particularly interesting is that the difference between the WCX and SAX elution profiles are much more vivid for the stressed samples than for the reference materials. We have seen examples where the converse is true and the CEX separation is better than that observed on the AEX. This contrast between the AEX and CEX profiles highlights an important feature of IEC that electrophoretic methods don't exhibit, which is the ability to magnify particular aspects of the protein structure and accentuate the separation of species relating to particular motifs (Vlasak & Ionescu, 2008).

In general, we have observed that for the separation of mAb variants using ion-exchange chromatography, the optimized chromatogram has a relatively shallow gradient over a narrow range of salt concentration. A typical method results in 100 mM NaCl as the center point of the gradient, with salt concentration increasing over 70 mM NaCl in a linear gradient. It is recommended to perform iterative gradient optimizations to narrow the NaCl gradient down to around 2 mM/mL min^{-1}. Iterative cycles are quicker and more predictive than performing a very long shallow gradient.

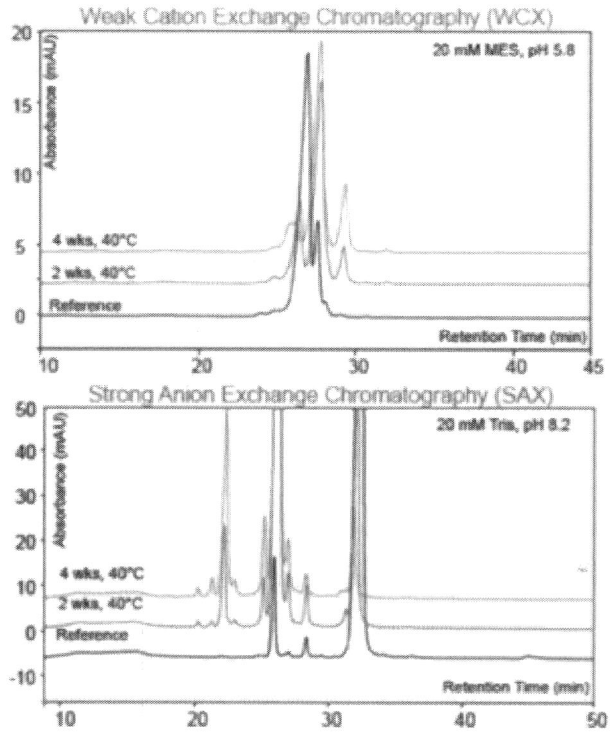

Fig. 6. Separation of Fab charge species using a weak cation exchange column (WCX) and a strong anion exchange column (SAX). Thermally stressed samples are labeled by incubation time and temperature of incubation.

Chromatograms obtained during the mobile phase pH value optimization for a mAb with a pI value around 9.5 are shown in Figure 7. Buffer species and buffer concentration for salt- gradient IEC are generally not significant factors, but should be chosen considering target pH and buffer pKa.

Although temperature does not significantly affect electrostatic interactions, it often affects the pH value of the mobile phase. This is particularly of concern for a Good's buffer system (group of buffers described in the research of Dr. Norman Good et al. in 1966, often used for IEC and other biochemistry applications) (Good et al., 1966), which can exhibit a change in pH value of around 0.02 per °C temperature change. This sensitivity creates a need to control the column temperature carefully.

A column compartment is always used, typically set at a value greater than 30°C to ensure good temperature stability in compartments that can only apply heat. Above 30°C, temperature control within +/- 1°C is readily achievable with commercially available equipment.

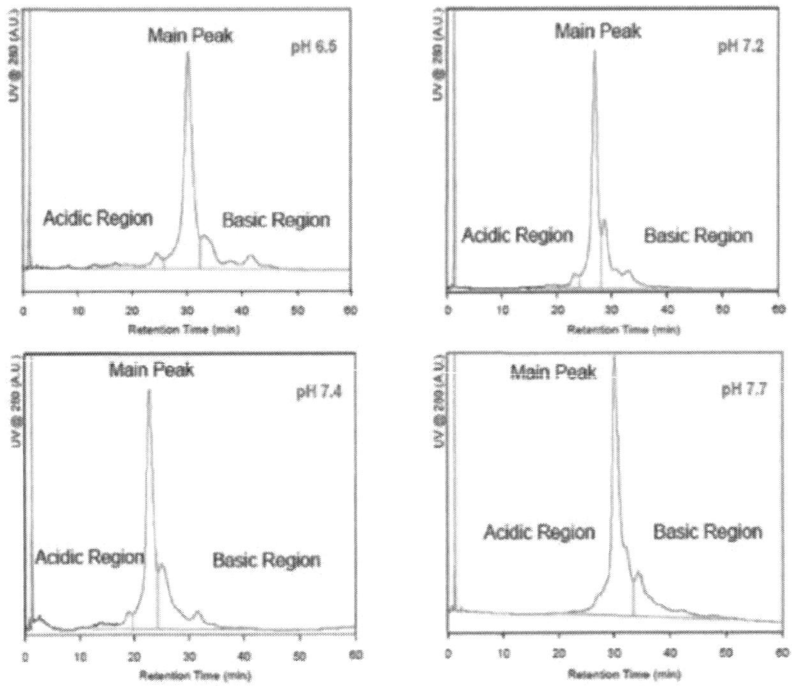

Fig. 7. Effect of mobile phase pH on mAb separation by WCX. The elution buffer (0.5 M NaCl) was increased linearly at 1 mM min^{-1} at a flow rate of 1 mL min^{-1} on a ProPac WCX-10 column (4 x 250 mm), which was held at 30 ºC. Different initial salt concentrations were optimized for each pH value. Integration is shown, and main peak, acidic and basic regions are denoted.

Subtle variations in selectivity with temperature may result from temperature-induced changes in mobile phase pH value (Figure 8). In Figure 8, the elution profile changes in two distinct regions as a function of temperature. Below 40°C, subtle changes in elution profile and retention times are observed consistent with minor changes to the mobile phase pH value as a function of temperature. However, above 40°C, the profiles exhibit much more radical changes with increasing temperature. This is interpreted to be related to the mAb having lost higher order structure at those elevated temperatures due to protein denaturing. For the mAb in Figure 8, it is clear that moderately elevated temperatures are not possible while maintaining the higher order structure; in general for IgG1 mAbs, chromatography at temperatures up to 55°C is readily possible. In summary, while mobile phase temperature does not affect protein charge directly, temperature can affect mobile phase pH and the structure of the protein, which can affect chromatographic separations. Thus, column temperature should be optimized considering these temperature effects.

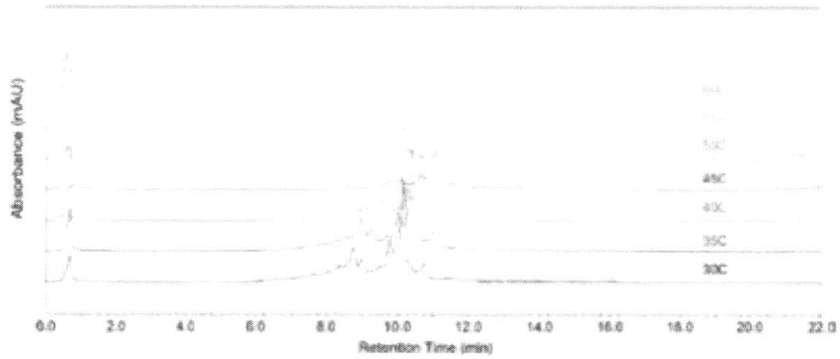

Fig. 8. Effect of temperature on mAb separation by CEX. The elution buffer (0.2 M sodium sulfate) was increased linearly on a ProPacWCX-10 column (4 x 250 mm).

DEVELOPING A PH GRADIENT-BASED IEC METHOD

Despite good resolving power and robustness, salt-based ion exchange separations are usually protein-specific and time-consuming to develop. A novel pH-based separation of proteins by cation exchange chromatography that was multi-product, high-resolution, and robust against variations in sample matrix salt concentration and pH was recently reported (Farnan & Moreno, 2009). A pH gradient-based separation method using cation exchange chromatography was also evaluated in a mock validation and deemed highly robust (Rea et al., 2011). Figure 9 depicts the separation of 16 mAbs by pH gradient IEC (pH-IEC). Each mAb was injected sequentially, demonstrating that in contrast to salt-based IEC, pH-IEC can be used to analyze multiple mAbs with a single method.

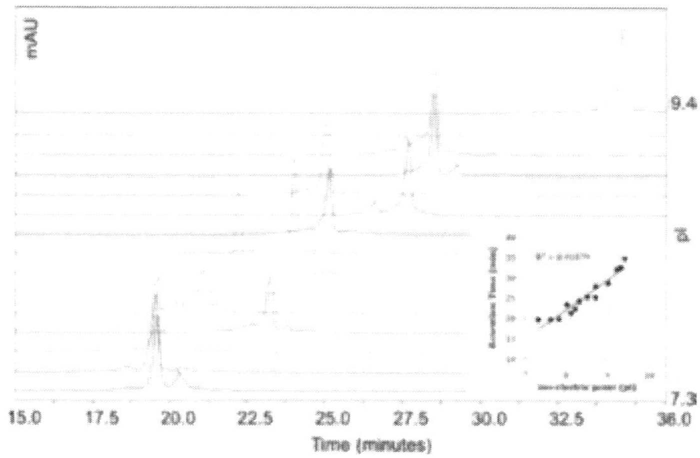

Fig. 9. Separation of 16 mAbs using a ProPac WCX-10 column by pH gradient IEC. Each mAb was analyzed using the same pH-IEC method, and each mAb was injected sequentially. mAb pI values ranged from pI 7.3 to pI 9.4.

Similar to salt-gradient IEC methods, pH-IEC separations are developed by choosing a cation or anion exchanging column and varying the buffer system, pH of the mobile phases, and other operating parameters, such as temperature and flow rate. Figure 5 shows a typical development workflow for pH-IEC, and can serve as a guide for initial pH-IEC method development.

Column selection, buffers and operating parameters for pH gradient IEC

Like conventional IEC, the conditions chosen for pH-IEC separations, such as buffer, pH, column temperature, and sample load, are dependent on the type of column selected. To choose a column, the pI of the mAb and the expected charge species should be considered. Considerations for column selection may differ slightly for pH-IEC compared to conventional IEC. For example, because the column will be exposed to a pH gradient, the column must be able to perform adequately over a large pH range, i.e., the charged groups on the chromatography resin must maintain their charge over the operating pH range. Also, buffer strength can affect resolution, and pH-IEC mobile phases typically have lower buffer strengths than conventional salt-gradient IEC. Several pH-IEC buffer systems have been published for mAb separations; these buffer systems can be used as starting points for formulating buffers for pH-IEC methods (Farnan and Moreno, 2009; Rea et al., 2011; Rozhkova, 2009).

High-throughput multi-product separations using pH-IEC

To increase the throughput of the analytical methods, smaller particle sizes and shorter column lengths are being utilized to reduce run time. In Figure 10, the utilization of a 3 μm particle size column reduced analysis time 16-fold compared to a 10 μm particle size column. Analysis times are greatly reduced using smaller particle sizes because as the particle size decreases, there a significant gain in column efficiency, and the efficiency does not decrease at increased flow rates or linear velocities (Swartz, 2005). In addition, because different mAbs can be analyzed using the same pH-IEC method in the same sequence, these high-throughput methods are capable of analyzing hundreds of mAbs per day, which is not possible with conventional, product-specific salt-based IEC.

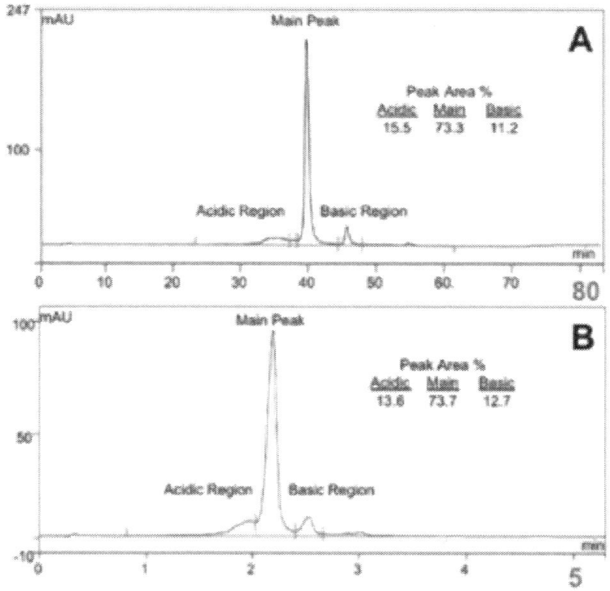

Fig. 10. Separation of a mAb using (A) a WCX column, 10 µm and (B) a SCX column, 3 µm, by pH gradient IEC. Each mAb was analyzed using the same buffers and gradient volume.

HYBRID/COMBINATION MODES OF IEC

Salt and pH may be combined to elute proteins from IEC columns. Combination or hybrid methods can be employed if either salt-based or pH-based methods prove inadequate for resolving species of interest, especially at extreme pHs. When pH increases above their pKa, amines, as used exclusively in the pH-IEC piperazine/imidazole/tris buffer system, become deprotonated and uncharged, resulting in decreased ionic strength. The bound proteins will also deprotonate and carry less charge. However, adequate amounts of positively charged ions are required to displace the bound proteins and to elute them off the cation exchange resin. Since the buffer salts alone can not provide enough positively charged ions at higher pH, additional salt is added to the pH-IEC elution buffers to maintain ionic strength. Figure 11 depicts measured conductivity as a function of elution time in pH-IEC with and without salt. In this case, adding salt to the elution buffer will compensate for the loss of ionic strength (represented by conductivity) due to deprotonation of buffer ions.

In Figure 12, separation of the charge species of three mAbs using pH-IEC with and without salt is compared. Without salt, mAb-1 with a pI of 9.4 did not elute, and mAb-3, with a low pI of 6.2, showed a very broad peak with significant tailing and no resolution of charge species. With the addition of salt, adequate separation of charge species is obtained for both high pI and low pI mAbs.

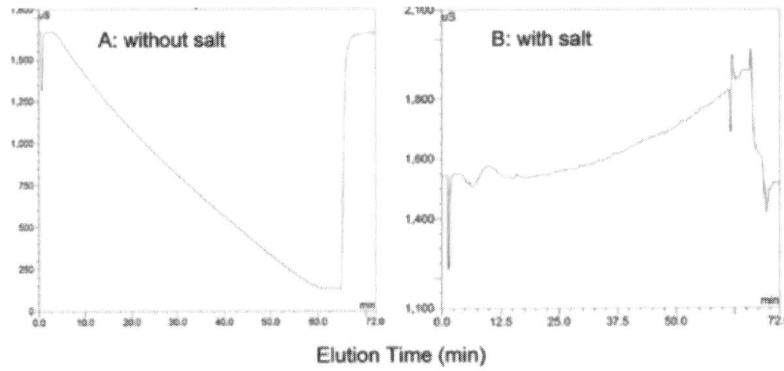

Fig. 11. Measured conductivity as a function of elution time in pH-IEC A) without, and B) with salt. Buffers are A) 11.6 mM piperazine, 1.5 mM imidazole, 2.4 mM Tris, B) 4 mM piperazine, 4 mM imidazole, 4 mM Tris, 16 mM NaCl.

EQUIPMENT CONFIGURATIONS TO ACCELERATE DEVELOPMENT

Ionic strength gradient ion-exchange methods are typically product-specific, with each method requiring a unique pair of mobile phases and experimental conditions. As discussed above, a significant number of mobile phase pH values and gradient profiles need to be evaluated. Changing mobile phase pH values normally requires user intervention to supply new mobile phase pairs; the time needed to manually change the system can slow downdevelopment. In order to more efficiently develop analytical IEC methods, alternative equipment configurations have been utilized to accelerate the selection of operational parameters, including quaternary buffer systems and customized solvent selection valves on the pump inlets to allow selection from an array of available solvents. Such modifications to the equipment and workflows can allow consecutive performance of significantly more experiments without requiring user intervention.

A quaternary buffer system can be utilized to develop a method and reduce the amount of user intervention by using a pair of buffer solvents (solvent lines A and B) to allow the pump to admix to achieve the desired mobile phase pH, and using two other solvents (solvent lines C and D) to generate the ionic strength gradient for elution. The quaternary system can apply different combinations of salt and pH by automatically programming the percentages of the four solvents to be mixed and applied to the column. Thus, programs can be generated to screen a variety of salt and pH conditions in a single sequence using only four buffers.

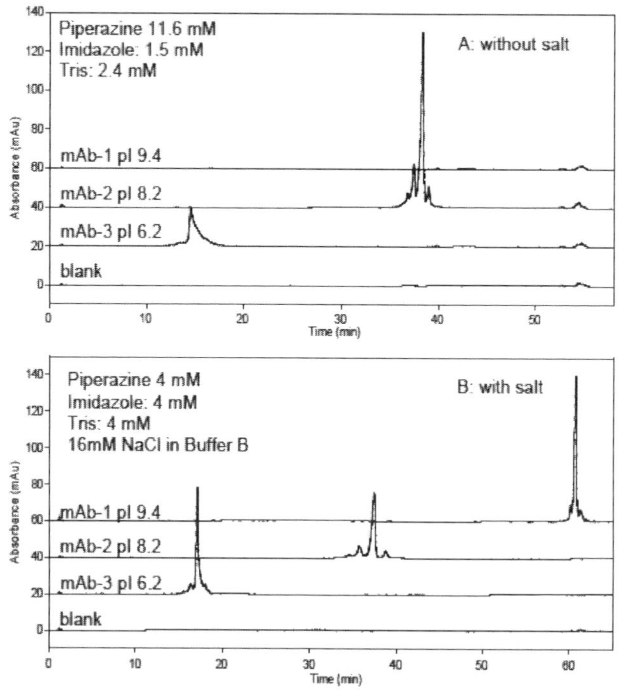

Fig. 12. Separation of the charge species of three mAbs using pH gradient with and without salt in a ProPac WCX-10 column. A) pH 5 to 9.5 in 45 minutes, gradient 0.1 pH unit/min; B) pH 5 - 10.8 in 58 minutes, gradient 0.1 pH unit/min.

Another approach, which is particularly important for binary pump systems, is to add a multi-port solvent selection valve to the system prior to the pump. Although the customized valve system requires the production of many buffers, the multiple valve configurations can allow users to further customize buffer components and concentrations (as opposed to only salt gradient and pH) compared to the quaternary system. Such a system allows up to a dozen more buffer combinations to be evaluated without intervention.

Because pH-IEC is performed by using a pH gradient and not a salt gradient, simply reversing the pH gradient allows for the chromatography mode to be switched between CEX and AEX. This reversal of gradient can be automatically performed through the chromatography software. Thus, multiple CEX and AEX columns can be screened by using only two buffers at different pH's and using a column switching valve to screen different column types (Figure 13). During development, it is helpful to have online pH and conductivity meters to ensure that the pH gradient is roughly linear and that the conductivity does not interfere with the separation efficiency.

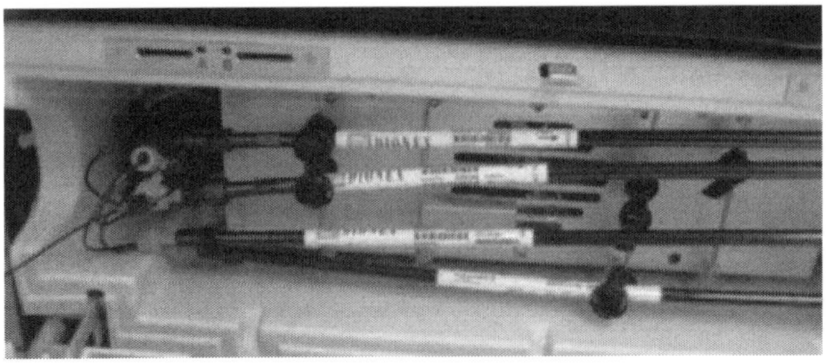

Fig. 13. HPLC column compartment equipped with a 6-port column switching valve, for screening of up to six different columns for pH-IEC without the need to change buffers or columns.

METHOD ROBUSTNESS AND VALIDATION

The robustness of an analytical procedure is a measure of its capability to remain unaffected by small, but deliberate variations in method parameters and provides an indication of its reliability during normal usage. For IEC, robustness can be evaluated by varying parameters such as injection volume, buffer pH, flow rate, and column temperature. In addition to robustness, intermediate precision can be demonstrated by evaluating inter-laboratory variations, such as different days of analysis and different analysts. Furthermore, the ability to use different instrument and column manufacturers for a particular method greatly reduces the business risk of the method; if a column supplier cannot meet demand or if an instrument manufacturer ceases production of a particular instrument model, method transfer to other instruments and columns can occur without loss of performance.

Obtaining robust performance

Obtaining robust performance of an IEC method often goes beyond the design of the method itself, and involves good equipment hygiene, elimination of metal corrosion (e.g. formation of iron oxide) and contamination (e.g. presence of metal ions such as Fe^{3+} ions), and mitigates the differences between instrument types. Problematic metal contamination typically results from corrosion of the fluid-contacting metal parts and can be avoided by using PEEK or titanium materials in the fluid paths. Good practices on obtaining robust method performance are discussed in the following sections.

Equipment hygiene

Maintaining good equipment hygiene is important in order to achieve robust performance. The following are good practices to ensure instrument hygiene:

- Filter mobile phases that are amenable to microbial growth with 0.2 mm filters prior to use; replace solvent reservoir filters (sinkers) each time mobile phase bottles are replenished;

- Flush and store HPLC system in 10% isopropanol in water when not in use, to prevent growth of microbes;
- Leave the system running at low flow rates to prevent salt build up and clogging;
- Keep all lines flushing as opposed to just a single channel;
- Flush auto-sampler components as needed with 10% isopropanol in water;
- Follow manufacturer's instructions regarding proper maintenance of HPLC instrumentation.

Metal contamination

Metals can negatively affect the ion-exchange chromatography of proteins. Protein chelation with metals are a secondary retention mechanism to the primary electrostatic interaction of ion-exchange chromatography. This secondary interaction results in peak tailing. These interactions can either occur with metal contaminating the column or with corroded surfaces within the HPLC. In addition to affecting separation, corrosion can result in physical damage to system, such pump seal failure and compromised performance of the detector cells. Halide containing eluents readily corrode HPLC systems manufactured from stainless steel, as stainless steel has the propensity to form rust (Collins et al., 2000a). Sodium acetate or sodium sulfate can be used as an eluting salt instead of halides; however, sulfate is divalent, thus concentrations in the eluting mobile phase would be different compared to using a halide, as halides are monovalent.

Metal contamination may be reversed by flushing with chelating agents such as oxalic acid dihydrate (Rao & Pohl, 2011). Also, stainless steel systems may require periodic passivation for reliable usage (Collins et al., 2000b). In light of the drawbacks of using a stainless steel HPLC system, more manufacturers are including biocompatible equipment (e.g. Titanium or PEEK) for analyzing mAbs and other protein products.

Transferring methods between instrument types

Transferring methods between instruments from different manufacturers can pose challenges due to the differences between instruments. As mentioned previously, equipment composition (e.g. stainless steel vs. titanium) is one of the factors to be considered when transferring a method between instrument types, in addition to gradient delay, mixing volumes, pump capabilities, and column compartment temperature ranges. Gradient delay and mixing volumes can differ between instruments, but they are generally only a significant concern for very fast gradient separations. In addition, shallow IEC gradients can challenge the performance of an HPLC; however, most gradients are >70 mM salt and can be proportioned over

30-40% of the pump range, well within the capabilities of modern pumps. Often a gradient hold for 5 minutes at the initial salt concentration is included, just in case a method is particularly sensitive when being transferred from one equipment type to another. In such cases the hold time can be adjusted to compensate for differences in the gradient delay volume between the instruments.

Temperatures inside the column are dependent on oven design and plumbing configuration. Having a pre-column heat exchanger in line or out of line could make a several degree difference in the temperature at which the column chemistry occurs. This is particularly concerning for buffers with which the pH can change rapidly with temperature. Figure 14 shows a comparison of column compartment temperature settings for two different instruments from different manufacturers. To make the correlation, thermocouples were fitted into T-pieces in the fluid path inside the column oven, but just prior to the column, and temperatures were measured for a range of column compartment set points and mobile phase flow rates. These measurements were used to estimate the temperature of mobile phase going through the column for each set point. By equating the measured fluid temperatures for each flow rate, the correlation of column compartment temperatures were plotted. It is noted in this correlation that there was also a significant effect of the mobile phase flow rate on the correlation.

Different detectors can sometimes yield differences in baseline slope. This can occur when moving from a single/double wavelength detector with a reference beam to a photodiode array (PDA) detector. The selection of an appropriate reference wavelength and bandwidth on the PDA can overcome detector variance.

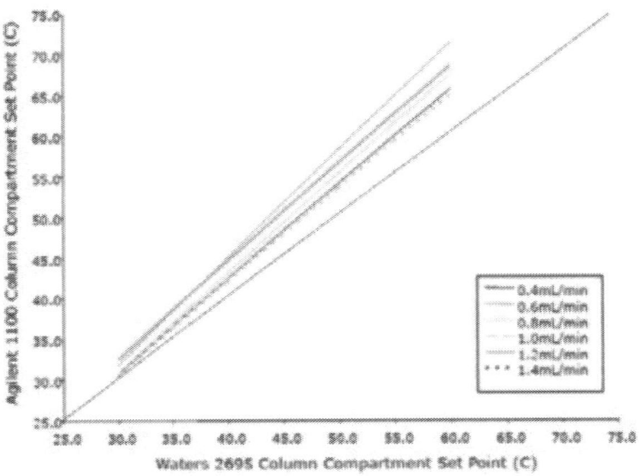

Fig. 14. Comparison of column compartment temperature settings required to achieve the same columns compartment temperature for two different HPLC models at different flow rates. Results are shown for a 4 x 250 mm Dionex ProPac column. The Agilent 1100 HPLC was configured using only the left hand side heat exchanger. The Waters 2695 HPLC was configured with the solvent pre-heater in-line.

Method validation

Before an analytical method can be incorporated into a characterization platform or a quality control system, it must first be demonstrated that the method is suitable for its intended purpose. Guidelines for validation of analytical methods have been published in the United States Pharmacopeia, by the International Conference on Harmonization (ICH), US Food and Drug Administration (FDA), and in published reviews (Bakshi & Singh, 2002). Methods must be evaluated considering regulatory requirements and validation procedures. In other words, the "validatability" of these methods must be assessed before implementation. Validation tests include precision, accuracy, and linearity. Intermediate precision is tested by using multiple instruments, multiple analysts, and multiple column lots. Methods must be validated and documented according to regulatory requirements prior to implementation into a control system for lot release of drug substance and drug product. Robustness studies can also be performed in conjunction with method validation. It has been our experience that the most significant effects on method robustness are: mobile phase pH value, column temperature, metal contamination and column age.

A system suitability range can be obtained from robustness studies. This range is often based on the standard deviation of the mean for a particular measured component, such as main peak relative area. The system suitability range indicates the precision of the method. pH-IEC may demonstrate an improvement in precision over conventional salt-based IEC (Rea et al., 2011). The 6σ ranges in Figure 15, which predicts a 99% method success rate, demonstrate the improved precision of the pH-gradient IEC method over conventional IEC, which can have a 6σ range of up to 8% main peak relative area (Figure 15).

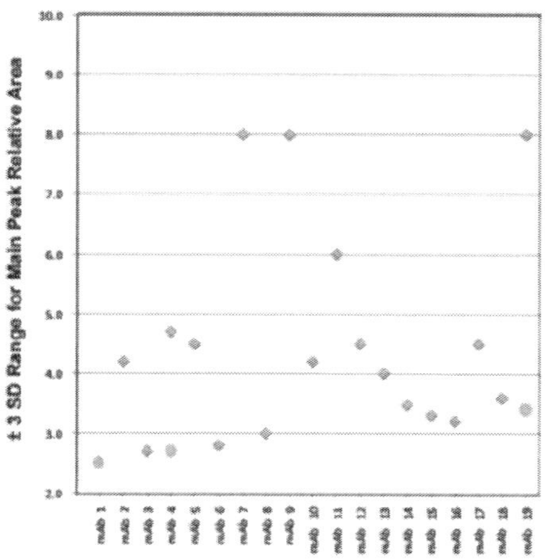

Fig. 15. Six sigma range (±3SD) for main peak relative area for salt gradient IEC (diamonds) and pH gradient IEC (circles) for a variety of mAbs.

AUTOMATION IN SAMPLE PREPARATION AND DATA HANDLING

In addition to high-throughput and multi-product analytical methods, the use of robotics for sample preparation automation may further reduce sample analysis time and cost. There are several companies that provide liquid handling automation instruments, including LEAP Technologies and TECAN. The LEAP Technologies CTC PAL liquid handling system is capable of on-the-fly sample preparation, such as protein dilution and digestion. On-the-fly sample preparations are viable if the sample preparation takes less time than the analytical method. For sample preparations that take longer than the analytical run time, batch sample preparation can be performed using robotic liquid handling systems such as the TECAN Freedom EVO, which can handle multi-well plates for increased sample throughput. Robotic liquid samplers can increase reproducibility, efficiency and safety compared to manual handling of samples.

The final steps to most characterization workflows include data analysis and report generation. Several software packages are available that are designed to reduce the time necessary to complete post-data acquisition tasks. For liquid chromatography applications, commercially available chromatography data software, such as Dionex's Chromeleon Chromatography Management Software and Waters Corporation's Empower Chromatography Data Software, include features such as automated peak integration and one-click report generation. In addition, laboratories are increasingly implementing electronic laboratory notebooks, which has advantages over traditional laboratory notebooks, including ease of data sharing and collaboration, streamlined review and witnessing processes, standardized documentation, and long-term data preservation.

SUMMARY

Monoclonal antibodies are valuable therapeutic products that are approved for a variety of indications. In this chapter, mAb development, production, purification, formulation, characterization and regulatory requirements were discussed, followed by a more detailed discussion on charge species analysis using IEC. Method development strategies, method robustness, validation and automation, as well as applications of salt-gradient and pH- gradient IEC methodologies for the analysis of mAbs were also covered. This chapter is intended to be a reference text for scientists such that a concise strategy can be implemented for developing robust IEC methods for the characterization of therapeutic mAbs, resulting in shorter method development times and enabling faster analysis of mAb products to support biopharmaceutical pipelines.

REFERENCES

1. Abu-Absi, S.F., Yang, L., Thompson, P., Jiang, C., Kandula, S., Schilling, B., & Shukla, A.A. (2010). Defining Process Design Space for Monoclonal Antibody Cell Culture. *Biotechnology and Bioengineering*, Vol. 106, No. 6, (August 2010), pp. 894-905, ISSN 1097-0290
2. Bakshi, M., & Singh, S. (2002). Development of Validated Stability-Indicating Assay Methods—Critical Review. *Journal of Pharmaceutical and Biomedical Analysis*, Vol. 28, No. 6, (June 2002), pp. 1011-1040, ISSN 0731-7085
3. Carter, P.J., & Presta, L.G. (2000). *Humanized antibodies and methods for making them.* U.S. Patent No. 6,054,297, Washington, DC, USA
4. Carter, P.J. (2001). Improving the Efficacy of Antibody-Based Cancer Therapies. *Nature Reviews Cancer*, Vol. 1, No. 2, pp. 118-129, (November 2001), ISSN 1474-175X
5. Carter, P.J., & Presta, L.G. (2002). *Method for making humanized antibodies.* U.S. Patent No. 6,407,213, Washington, DC, USA
6. Chirino, A.J., & Mire-Sluis, A. (2004). Characterizing Biological Products and Assessing Comparability Following Manufacturing Changes. *Nature Biotechnology*, Vol. 22, No. 11, (November 2004), pp. 1383-1391, ISSN 1087-0156
7. Co, M.S., & Queen, C. (1991). Humanized Antibodies for Therapy. *Nature*, Vol. 351, No. 6326, (June 1991), pp. 501–502, ISSN 0028-0836
8. Collins, K.E., Collins, C.H., & Bertran, C.A. (2000). Stainless Steel Surfaces in LC systems: I. Corrossion and Erosion. *LC–GC*, Vol. 18, No. 6, (June 2000), pp. 600–608, ISSN 0888-9090
9. Collins, K.E., Collins, C.H., & Bertran, C.A. (2000). Stainless Steel Surfaces in LC systems: II. Passivation and Practical Recommendations. *LC–GC*, Vol. 18, No. 6, (June 2000), pp. 688–692, ISSN 0888-9090
10. Dickson, M., & Gagnon, J.P. (2004). Key Factors in the Rising Cost of New Drug Discovery and Development. *Nature Reviews Drug Discovery*, Vol. 3, No. 5, (May 2004), pp. 417–429, ISSN 1474-1776
11. Di Donato, A., Ciardiello, M.A., de Nigris, M., Piccoli, R., Mazzarella, L., & D'Alessio, G. (1993). Selective Deamidation of Ribonuclease A. Isolation and Characterization of the Resulting Isoaspartyl and Aspartyl Derivatives. *Journal of Biological Chemistry*, Vol. 268, No. 7, (March 1993), pp. 4745-4751, ISSN 0021-9258
12. DiMasi, J.A., Hansen, R.W., & Grabowski, H.G. (2003). The Price of Innovation: New Estimates of Drug Development Costs. *Journal of Health Economics*, Vol. 22, No. 2, (March 2003), pp. 151–185, ISSN 0167-6296

13. DiMasi, J.A., & Grabowski, H.G. (2007). The Cost of Biopharmaceutical R&D: Is Biotech Different? *Managerial and Decision Economics,* Vol. 28, No. 4-5, (August 2007), pp. 469–479, ISSN 0143-6570

14. Farnan, D., & Moreno, G.T. (2009). Multiproduct High-Resolution Monoclonal Antibody Charge Variant Separations by pH Gradient Ion-Exchange Chromatography. *Analytical Chemistry*, Vol. 81, No. 21, (November 2001), pp. 8846-8857, ISSN 0003-2700

15. Food & Drug Administration (FDA). (1999). *Guidance for Industry: Q6B Test Procedures and Acceptance Criteria for Biotechnological/Biological Products.* FDA, Silver Spring, MD, USA

16. Food & Drug Administration (FDA). (2003). *Guidance for Industry: Q1A(R2) Stability Testing of New Drug Substances and Products.* FDA, Silver Spring, MD, USA

17. Gaza-Bulseco, G., Bulseco, A., Chumsae, C., & Liu, H. (2008). Characterization of the Glycosylation State of a Recombinant Monoclonal Antibody Using Weak Cation Exchange Chromatography and Mass Spectrometry. *Journal of Chromatography B*, Vol. 862, No. 2, (February 2008), pp. 155-160, ISSN 1570-0232

18. Genovesi, C.S. (1983). Several Uses for Tangential-Flow Filtration in The Pharmaceutical Industry. *Journal Parenteral Science and Technology*, Vol. 37, No. 3, (May-June 1983), pp. 81-86, ISSN 0279-7976

19. Good, N.E., Winget, G.D., Winter, W., Connolly, T.N., Izawa, S. & Singh, R.M.M. (1966). Hydrogen Ion Buffers for Biological Research. *Biochemistry,* Vol. 5, No. 2, (February 1966), pp. 467–477, ISSN 0001-527X

20. Gottschalk, U. (2009). *Process Scale Purification of Antibodies.* John Wiley & Sons, ISBN 978-0- 470-20962-2, Hoboken, New Jersey, U.S.A.

21. Green, M.C., Murray, J.L., & Hortobagyi, G.N. (2000). Monoclonal Antibody Therapy for Solid Tumors. *Cancer Treatment Reviews*, Vol. 26, No. 4, (August 2000), pp. 269-286, ISSN 0305-7372

22. Harris, R.J. (1995). Processing of C-terminal Lysine and Arginine Residues of Proteins Isolated from Mammalian Cell Culture, *Journal of Chromatography A*, Vol. 705, No. 1, (June 1995), pp. 129-134, ISSN 0021-9673

23. Harris, R.J., Kabakoff, B., Macchi, F.D., Shen, F.J., Kwong, M., Andya, J.D., Shire, S.J., Bjork, N., Totpai, K., & Chen, A.B. (2001). Identification of Multiple Sources of Charge Heterogeneity in a Recombinant Antibody. *Journal of Chromatography B*, Vol. 752, No. 2, (March 2001), pp. 233-245, ISSN 1570-0232

24. Harris, R.J., Shire, S.J., &Winter, C.W. (2004). Commercial Manufacturing Scale Formulation and Analytical Characterization of Therapeutic Recombinant

Antibodies. *Drug Development Research*, Vol. 61, No. 3, (March 2004), pp. 137-154, ISSN 1098- 2299

25. He, Y., Lacher, N.A., Hu, W., Wang, Q., Isele, C., Starkey, J., & Ruesch, M. (2010). Analysis of Identity, Charge Variants, and Disulfide Isomers of Monoclonal Antibodies with Capillary Zone Electrophoresis in an Uncoated Capillary Column. *Analytical Chemistry*, Vol. 82, No. 8, (April 2010), pp. 3222-3230, ISSN 0003-2700
26. Horvath, B., Mun, M., & Laird, M.W. (2010). Characterization of a Monoclonal Antibody Cell Culture Production Process Using a Quality by Design Approach. *Molecular Biotechnology*, Vol. 45, No. 3, (July 2010), pp. 203-206, ISSN 1559-0305
27. Hsu, Y.R., Chang, W.C., Mendiaz, E.A., Hara, S., Chow, D.T., Mann, M.B., Langley, K.E., & Lu, H.S. (1998). Selective Deamidation of Recombinant Human Stem Cell Factor During In Vitro Aging: Isolation and Characterization of the Aspartyl and Isoaspartyl Homodimers and Heterodimers. *Biochemistry*. Vol. 37, No. 8, (February 1998), pp. 2251-2262, ISSN 0001-527X
28. Hudson, P.J. & Souriau, C. (2003). Engineered Antibodies. *Nature Medicine,* Vol. 9, No. 1, (January 2003), pp. 129-134, ISSN 1078-8956
29. International Conference on Harmonisation of Technical Requirements for the Registration of Pharmaceuticals for Human Use (ICH). (1999). *ICH Topic Q6B: Specifications: Test Procedures and Acceptance Criteria for Biotechnological/ Biological Products.* ICH, Geneva, Switzerland, 1999.
30. Jones, S.D., Castillo, F.J., & Levine, H.L. (2007). Advances inthe Development of Therapeutic Monoclonal Antibodies. *BioPharm International,* Vol. 20, No. 10, (October 2007), pp. 96-114, ISSN 1542 -166X
31. Kastner, M. (2000). *Protein Liquid Chromatography (Journal of Chromatography Library).* Elsevier Science, ISBN 0-444-50210-6, Amsterdam, The Netherlands
32. Kelley, B. (2009). Industrialization of mAb Production Technology: The Bioprocessing Industry at a Crossroads. *Mabs,* Vol. 1, No. 5, (September 2009), pp. 443-452, ISSN 1942-0870
33. Kuus-Reichel, K., Grauer, L.S., Karavodin, L.M., Knott, C., Krusemeier, M., & Kay, N.E. (1994). Will Immunogenicity Limit the Use, Efficacy, and Future Development of Therapeutic Monoclonal Antibodies? *Clinical and Diagnostic Laboratory Immunology*, Vol. 1, No. 4, (July 1994), pp. 365–372, ISSN 1071-412X
34. Li, F., Vijayasankaran, N., Shen, A.Y., Kiss, R., & Amanullah A. (2010). Cell Culture Processes for Monoclonal Antibody Production. *Mabs*, Vol. 2, No. 5, (November 2010), pp. 466-479, ISSN 1942-0870

35. Lonberg, N. (2005) Human Antibodies from Transgenic Animals. *Nature Biotechnology,* Vol. 23, No. 9, (September 2005), pp. 1117–1125, ISSN 1087-0156

36. McCafferty, J., Griffiths, A.D., Winter, G. & Chiswell, D.J. (1990). Phage Antibodies: Filamentous Phage Displaying Antibody Variable Domains. *Nature,* Vol. 348, No. 6301, (December 1990), pp. 552–554, ISSN 0028-0836

37. Morrison, S.L., Johnson, M.J., Herzenberg, L.A. & Oi, V.T. (1984). Chimeric Human Antibody Molecules: Mouse Antigen-Binding Domains with Human Constant Domains. *Proceedings of the National Academy of Sciences USA,* Vol. 81, No. 21, (November 1984), pp. 6851–6855, ISSN 0027-8424

38. Quan, C., Alcala, E., Petkovska, I., Matthews, D., Canova-Davis, E., Taticek, R., & Ma, S. (2008). A Study in Glycation of a Therapeutic Recombinant Humanized Monoclonal Antibody: Where It Is, How It Got There, and How It Affects Charge- Based Behavior. *Analytical Biochemistry*, Vol. 373, No. 2, (February 2008), pp. 179-91, ISSN 0003-2697

39. Rao, S., & Pohl, C. (2011). Reversible Interference of Fe3+ with Monoclonal Antibody Analysis in Cation Exchange Columns. *Analytical Biochemistry*, Vol. 409, No. 2, (February 2011), pp. 293-295, ISSN 0003-2697

40. Rea, J.C., Moreno, G.T., Lou, Y., Parikh, R., & Farnan, D. (2010). High-Throughput Multi- Product Liquid Chromatography for Characterization of Monoclonal Antibodies. *BioPharm International*, Vol. 23, No. 11, (November 2010), pp. 44-51, ISSN 1542 -166X

41. Rea, J.C., Moreno, G.T., Lou, Y., & Farnan, D. (2011). Validation of a pH Gradient-Based Ion- Exchange Chromatography Method for High-Resolution Monoclonal Antibody Charge Variant Separations. *Journal of Pharmaceutical and Biomedical Analysis*, Vol. 54, No. 2, (January 2011), pp. 317–323, ISSN 0731-7085

42. Reichert, J.M., Rosensweig, C.J., Faden, L.B., & Dewitz, M.C. (2005). Monoclonal Antibody Successes in the Clinic. *Nature Biotechnology*, Vol. 23, No. 9, (September 2005), pp. 1073-1078, ISSN 1087-0156

43. Reichert, J.M. & Valge-Archer, V.E. (2007). Development Trends for Monoclonal Antibody Cancer Therapeutics. *Nature Reviews Drug Discovery,* Vol. 6, No. 5, (May 2007), pp. 349-356, ISSN 1474-1776

44. Reichert, J.M. (2009). Global Antibody Development Trends. *Mabs*. Vol. 1, No. 1, (January/February 2009), pp. 86-87, ISSN 1942-0870

45. Reichert, J.M. (2011). Antibody-Based Therapeutics to Watch in 2011. *MAbs*. Vol. 3, No. 1, (January-February 2011), pp. 76–99, ISSN 1942-0870

46. Reichmann, L., Clark, M., Waldmann, H. & Winter, G. (1988). Reshaping Human Antibodies for Therapy. *Nature,* Vol. 332, No. 6162 (March 1988), pp. 323–327, ISSN 0028-0836

47. Rosenberg, A.S. (2006). Effects of Protein Aggregates: An Immunologic Perspective. *AAPS Journal,* Vol. 8, No. 3, (August 2006), pp. E501-E507, ISSN 1550-7416

48. Rohzkova, A. (2009). Quantitative Analysis of Monoclonal Antibodies by Cation-Exchange Chromatofocusing. *Journal of Chromatography A*, Vol. 1216, No. 32, (August 2009), pp. 5989-5994, ISSN 0021-9673

49. Santora, L.C., Krull, I.S., & Grant, K. (1999). Characterization of Recombinant Human Monoclonal Tissue Necrosis Factor-α Antibody Using Cation-Exchange HPLC and Capillary Isoelectric Focusing, *Analytical Biochemistry*, Vol. 275, No. 1, (November 1999), pp. 98-108, ISSN 0003-2697

50. Santora, L.C., Kaymakcalan, Z., Sakorafas, P., Krull, I.S., & Grant, K. (2001). Characterization of Noncovalent Complexes of Recombinant Human Monoclonal Antibody and Antigen Using Cation Exchange, Size Exclusion Chromatography, and BIAcore. *Analytical Biochemisty*, Vol. 299, No. 2, (December 2001), pp. 119-129, ISSN 0003-2697

51. Schlatter, S., Stansfield, S.H., Dinnis, D.M., Racher, A.J., Birch, J.R., & James, D.C. (2005). On the Optimal Ratio of Heavy to Light Chain Genes for Efficient Recombinant Antibody Production by CHO cells. *Biotechnology Progress,* Vol. 21, No. 1, (January 2005), pp. 122-133, ISSN 8756-7938

52. Schneider, C.K. (2008). Monoclonal Antibodies - Regulatory Challenges. *Current Pharmaceutical Biotechnology*, Vol. 9, No. 6, (December 2008), pp. 431-438, ISSN 1389-2010

53. Schnerman, M.A., Sunday, B.R., Kozlowski, S., Webber, K., Gazzano-Santoro, H., & Mire- Sluis, A. (2004). CMC Strategy Forum Report: Analysis and Structure Characterization of Monoclonal Antibodies. *BioProcess International*, Vol. 2, No. 2, (February 2004), pp. 42–52, ISSN 1542-6319

54. Scolnik, P.A. (2009). MAbs: A Business Perspective. *Mabs,* Vol. 1, No. 2, (March 2009), pp. 179-184, ISSN 1942-0870

55. Shawler, D.L., Bartholomew, R.M., Smith, L.M. & Dillman, R.O. (1985). Human Immune Response to Multiple Injections of Murine Monoclonal IgG. *Journal of Immunology*, Vol. 135, No. 2, (August 1985), pp. 1530–1535, ISSN 0022-1767

56. Shiloach, J., Martin, N., & Moes, H. (1988). Tangential Flow Filtration. *Advances in Biotechnological Processes*, Vol. 8, pp. 97-125, ISSN 0736-2293

57. Shire, S.J., Shahrokh, Z., & Liu, J. (2004). Challenges in the Development of High Protein Concentration Formulations. *Journal of Pharmaceutical Sciences*, Vol. 93, No. 6, (June 2004), pp. 1390-1402, ISSN 0022-3549
58. Shire, S.J. (2009). Formulation and Manufacturability of Biologics. *Current Opinion in Biotechnology*, Vol. 20, No. 6, (December 2009), pp. 708-714, ISSN 0958-1669
59. Shukla, A.A., Hubbard, B., Tressel, T., Guhan, S., & Low, D. (2007). Downstream Processing of Monoclonal Antibodies—Application of Platform Approaches. *Journal of Chromatography B*, Vol. 848, No. 1, (March 2007), pp. 28-39, ISSN 1570-0232
60. Shukla, A.A. & Kandula, J.R. (2008). Harvest and Recovery of Monoclonal Antibodies from Large-Scale Mammalian Cell Culture. *BioPharm International*, Vol. 21, No. 5, (May 2008), pp. 18-25, ISSN 1542 -166X
61. Snyder, L.R., Kirkland, J.J., & Glajch, J.L. (1997). *Practical HPLC Method Development*. John Wiley & Sons, ISBN 0-471-00703-X, Hoboken, New Jersey, U.S.A.
62. Swartz, M.E. (2005). Ultra Performance Liquid Chromatography (UPLC): An Introduction. *LC-GC North America*, Vol. 23, No. 5, pp. 8-14, ISSN 1527-5949
63. vanReis,R.,&Zydney,A.(2001).MembraneSeparationsinBiotechnology. *Current Opinion in Biotechnology*, Vol. 12, No. 2, (April 2001), pp. 208-211, ISSN 0958-1669
64. Vlasak, J., & Ionescu, R. (2008). Heterogeneity of Monoclonal Antibodies Revealed by Charge-Sensitive Methods. *Current Pharmaceutical Biotechnology*, Vol. 9, No. 6, (December 2008), pp. 468–481, ISSN 1389-2010
65. Waldmann, T.A. (2003). Immunotherapy: Past, Present and Future. *Nature Medicine*, Vol. 9, No. 1, (January 2003), pp. 269-277, ISSN 1078-8956
66. Weitzhandler, M., Farnan, D., Horvath, J., Rohrer, J.S., Slingsby, R.W., Avdalovic, N., & Pohl, C. (1998). Protein Variant Separations Using Cation Exchange Chromatography on Grafted, Polymeric Stationary Phases, *Journal of Chromatography A*, Vol. 828, No. 1-2, (December 1998), pp. 365-372, ISSN 0021-9673
67. Yang, Y., Strahan, A., Li, C., Shen, A., Liu, H., Ouyang, J., Katta, V., Francissen, K. & Zhang, B. (2010). Detecting Low Level Sequence Variants in Recombinant Monoclonal Antibodies. *MAbs*, Vol. 2, No. 3, (May/June 2010), pp. 285–298, ISSN 1942-0870
68. Ziegelbauer, K., & Light, D.R. (2008). Monoclonal Antibody Therapeutics: Leading Companies to Maximise Sales and Market Share. *Journal of Commercial Biotechnology*, Vol. 14, No. 1, (January 2008), pp. 65-72, ISSN 1462-8732

INDEX

D

E

R

S